D1561347

Size and Scaling in Primate Biology

ADVANCES IN PRIMATOLOGY

Series Editors:

W. PATRICK LUCKETT
University of Puerto Rico
San Juan, Puerto Rico

CHARLES R. NOBACK
Columbia University
New York, New York

Editorial Board:

RUSSELL L. CIOCHON, University of California, Berkeley, California

JOHN F. EISENBERG, Florida State Museum, Gainesville, Florida

MORRIS GOODMAN, Wayne State University School of Medicine, Detroit, Michigan

F. A. JENKINS, Jr., Harvard University, Cambridge, Massachusetts

FREDERICK S. SZALAY, Hunter College, New York, New York

THE PRIMATE BRAIN
Edited by Charles R. Noback and William Montagna

MOLECULAR ANTHROPOLOGY: Genes and Proteins in the Evolutionary Ascent of the Primates
Edited by Morris Goodman and Richard E. Tashian

SENSORY SYSTEMS OF PRIMATES
Edited by Charles R. Noback

NURSERY CARE OF NONHUMAN PRIMATES
Edited by Gerald C. Ruppenthal

COMPARATIVE BIOLOGY AND EVOLUTIONARY RELATIONSHIPS OF TREE SHREWS
Edited by W. Patrick Luckett

EVOLUTIONARY BIOLOGY OF THE NEW WORLD MONKEYS AND CONTINENTAL DRIFT
Edited by Russell L. Ciochon and A. Brunetto Chiarelli

NEW INTERPRETATIONS OF APE AND HUMAN ANCESTRY
Edited by Russell L. Ciochon and Robert S. Corruccini

SIZE AND SCALING IN PRIMATE BIOLOGY
Edited by William L. Jungers

A Continuation Order Plan is available for this series. A continuation order will bring delivery of each new volume immediately upon publication. Volumes are billed only upon actual shipment. For further information please contact the publisher.

Size and Scaling in Primate Biology

Edited by
WILLIAM L. JUNGERS
State University of New York at Stony Brook
Stony Brook, New York

PLENUM PRESS • NEW YORK AND LONDON

Library of Congress Cataloging in Publication Data

Main entry under title:

Size and scaling in primate biology.

(Advances in primatology)
Includes bibliographical references and index.
1. Primates—Size. 2. Allometry. 3. Mammals—Size. I. Jungers, William L., 1948- . II. Series.
QL737.P9S59 1984 599.8 84-16099
ISBN 0-306-41560-7

QL
737
.P9
S59
1985

© 1985 Plenum Press, New York
A Division of Plenum Publishing Corporation
233 Spring Street, New York, N.Y. 10013

All rights reserved

No part of this book may be reproduced, stored in a retrieval system, or transmitted in any form or by any means, electronic, mechanical, photocopying, microfilming, recording, or otherwise, without written permission from the Publisher

Printed in the United States of America

To my mother
and
To the memory of my father

Contributors

R. McNeill Alexander
Department of Pure and Applied Zoology
University of Leeds
Leeds LS2 9JT, England

Este Armstrong
Department of Anatomy
Louisiana State University Medical Center
New Orleans, Louisiana 70112

Fred L. Bookstein
Center for Human Growth and Development
and Department of Radiology
University of Michigan
Ann Arbor, Michigan 48109

James M. Cheverud
Departments of Anthropology,
Cell Biology and Anatomy,
and Ecology and Evolutionary Biology
Northwestern University
Evanston, Illinois 60201

D. J. Chivers
Sub-department of Veterinary Anatomy
University of Cambridge
Cambridge CB2 1QS, England

T. H. Clutton-Brock
Large Animal Research Group
Department of Zoology
University of Cambridge
Cambridge CB2 3EJ, England

Larry R. Cochard
Department of Cell Biology and Anatomy
Northwestern University Medical School
Chicago, Illinois 60611

Robert S. Corruccini
Department of Anthropology
Southern Illinois University
Carbondale, Illinois 62901

Brigitte Demes
Arbeitsgruppe Funktionelle Morphologie
Ruhr-Universität Bochum
4630 Bochum 1, Federal Republic of Germany

John G. Fleagle
Department of Anatomical Sciences
School of Medicine
State University of New York
Stony Brook, New York 11794

Susan M. Ford
Department of Anthropology
Southern Illinois University
Carbondale, Illinois 62901

Philip D. Gingerich
Museum of Paleontology
University of Michigan
Ann Arbor, Michigan 48109

Paul H. Harvey
School of Biological Sciences
University of Sussex
Falmer, Brighton BNI 9QG
Sussex, England

Norman C. Heglund
Concord Field Station
Museum of Comparative Zoology

Harvard University
Cambridge, Massachusetts 02138

C. M. Hladik
C.N.R.S.
Muséum National d'Histoire Naturelle
Laboratoire d'Ecologie Générale
91800 Brunoy, France

William L. Jungers
Department of Anatomical Sciences
School of Medicine
State University of New York
Stony Brook, New York 11794

Russell Lande
Department of Biophysics and Theoretical Biology
University of Chicago
Chicago, Illinois 60637

Susan G. Larson
Department of Anatomical Sciences
School of Medicine
State University of New York
Stony Brook, New York 11794

Walter Leutenegger
Department of Anthropology
University of Wisconsin
Madison, Wisconsin 53706

A. M. MacLarnon
Department of Anthropology
University College
London WC1E 6BT, England

R. D. Martin
Department of Anthropology
University College
London WC1E 6BT, England

Holger Preuschoft
Arbeitsgruppe Funktionelle Morphologie
Ruhr-Universität Bochum
4630 Bochum 1, Federal Republic of Germany

Brian T. Shea
Departments of Anthropology and Cell Biology and Anatomy
Northwestern University
Evanston, Illinois 60201

B. Holly Smith
Center for Human Growth and Development
University of Michigan
Ann Arbor, Michigan 48109

Richard J. Smith
Department of Orthodontics
School of Dental Medicine
Washington University
St. Louis, Missouri 63110

Karen Steudel
Department of Zoology
University of Wisconsin
Madison, Wisconsin 53706

Milford H. Wolpoff
Department of Anthropology
University of Michigan
Ann Arbor, Michigan 48109

Preface

In very general terms, "scaling" can be defined as the structural and functional consequences of differences in size (or scale) among organisms of more or less similar design. Interest in certain aspects of body size and scaling in primate biology (e.g., relative brain size) dates to the turn of the century, and scientific debate and dialogue on numerous aspects of this general subject have continued to be a primary concern of primatologists, physical anthropologists, and other vertebrate biologists up to the present. Indeed, the intensity and scope of such research on primates have grown enormously in the past decade or so. Information continues to accumulate rapidly from many different sources, and the task of synthesizing the available data and theories on any given topic is becoming increasingly formidable. In addition to the formal exchange of new ideas and information among scientific experts in specific areas of scaling research, two of the major goals of this volume are an assessment of our progress toward understanding various size-related phenomena in primates and the identification of future prospects for continuing advances in this realm.

Although the subject matter and specific details of the issues considered in the 20 chapters that follow are very diversified, all topics share the same fundamental and unifying biological theme: body size variation in primates and its implications for behavior and ecology, anatomy and physiology, and evolution. The contributors have been careful to distinguish precisely among the various levels of scaling or allometry (ontogenetic, adult intraspecific, and interspecific including phylogenetic), and several chapters have explored the biological and statistical relationships among these different levels. In order to place primates within their proper mammalian context and thereby provide a broader zoological perspective to the intentionally "primatocentric" focus of this volume, several authors have examined primate scaling trends in explicit comparisons to allometric patterns characteristic of other mammalian groups. Such contrasts allow us to ask if primates follow general mammalian scaling trends or if there is something unique to particular aspects of primate allometry. Both the promise and limitations of the allometric approach in the study of size and adaptation in fossil primates are also given careful consideration.

Size itself and the effects of differences in size or scale appear to be inextricably linked to almost every aspect of primate biology that we may wish to investigate. Owing to the research efforts of the numerous contributors, I believe that the contents of this volume add greatly to our current understanding of this complex linkage in primates. At the same time, I hope that future inquiries and insights into the relationships between size and scaling will be stimulated by reading and reflecting on these contributions.

William L. Jungers

Stony Brook, New York

Contents

Size and Scaling in Primate Biology

Size and Adaptation in Primates

1

JOHN G. FLEAGLE

When we try to pick out anything by itself, we find it hitched to everything else in the universe.

John Muir

Introduction

Living primates vary in size from species averaging less than 100 g to some averaging well over 100 kg (Fig 1). The fossil record provides evidence of both much larger and much smaller species (e.g., Fleagle, 1978; Gingerich *et al.,* 1982; Gunnell, 1983; Simons and Ettel, 1970). There is no doubt that much of the diversity that these species show in structure, physiology, behavior, and ecology is intimately related to differences in body size (e.g., Schmidt-Nielson, 1975; Clutton-Brock and Harvey, 1983). The details of this relationship are the domain of allometry. Thus, in its broadest sense, allometry is the study of the relationship between size and adaptation.

As the name implies, allometric studies are necessarily quantitative; they are also frequently theoretical and statistically sophisticated—all the features that we would hope for in "hard science." Indeed, allometric analysis is probably as close to the physical sciences as organismic biology can ever come.

JOHN G. FLEAGLE • Department of Anatomical Sciences, School of Medicine, State University of New York, Stony Brook, New York 11794.

Fig. 1. A mouse lemur (*Microcebus murinus*), the smallest living primate species, and a gorilla (*Gorilla gorilla*), the largest living primate species.

However, in our pursuit of design criteria, optimization, large sample sizes, and statistical sophistication, it is important that we maintain a biological or even naturalistic perspective on both the questions we are asking and the individual animals that form our data points. As many authors (e.g., Mayr, 1976; Gould, 1980) have so eloquently reminded us, evolution by natural selection has ensured that there is an element of serendipity in nature that behooves caution in our attempts to treat the biological world as a physics experiment.

The purpose of this chapter is to look closely at this very complex relationship between size and adaptation. In particular I have three main objectives: (1) to emphasize the need for a naturalistic perspective in allometric studies, (2) to discuss the differing approaches to the interpretation of size-related differences usually taken in morphological studies on the one hand and behavioral or ecological studies on the other, and finally, (3) to point out the potential difficulty of studying size-related differences in morphology without careful consideration of any differences in ecological adaptation.

Types of Allometric Studies

In their simplest and most common form, allometric analyses involve a bivariate regression of some biological variable on the ordinate and some estimate of size on the abscissa (Fig. 2). As the chapters of this volume attest, there is considerable diversity of opinion about what constitutes an appropriate measure of size and what is the proper regression technique. The variables that can be represented on the ordinate seem limited only by the imagination of the analyst. In this chapter, however, I want to concentrate less on the axes and the regression techniques and more on the actual points themselves—the individuals, populations, or species that are being compared. From this perspective, there are three major types of allometric analyses: (1) *ontogenetic* studies, in which the individuals being compared represent a growth series (size differences are age differences); (2) *intraspecific* studies, which compare size-related changes in different-sized adult individuals of the same species; and (3) *interspecific* studies, in which the units of comparison are usually adults of different species. Each of these types of allometric analysis has its own peculiarities dictated by the nature and biological interrelationships of the units being studied.

In *ontogenetic* or *growth allometry,* the individuals (data points) being compared are growth stages in a single species (Fig. 3). On the abscissa, absolute size, absolute age, or developmental age are often used interchangeably [and incorrectly; see Shea (1983)]. The individual data points may be drawn from the growth record of a single individual (longitudinal studies), different individuals of the same species (cross-sectional studies), or some combination (mixed longitudinal). Regardless of the details of the data base, it is reasonable to assume in studies of growth allometry that a line fit to these points approximates an actual ontogenetic pathway traversed by all normal individuals of that species during their lifetime [but see Tanner (1951)]. In this regard, a regression line in growth allometry describes a readily interpretable biological phenomenon—ontogeny. It is important to keep in mind, however, that individuals of different ages may differ considerably in many aspects of their behavior as well as their morphology (e.g., Rawlins, 1976).

Intraspecific allometry also compares members of a single species, but in this case the comparison is among different-sized adults (Fig. 4). Although all of the individuals being compared are presumably the endpoints of roughly similar (species-specific) ontogeny, the regression line fit to these points is not so easily interpretable as that in ontogenetic allometry. Some have argued that

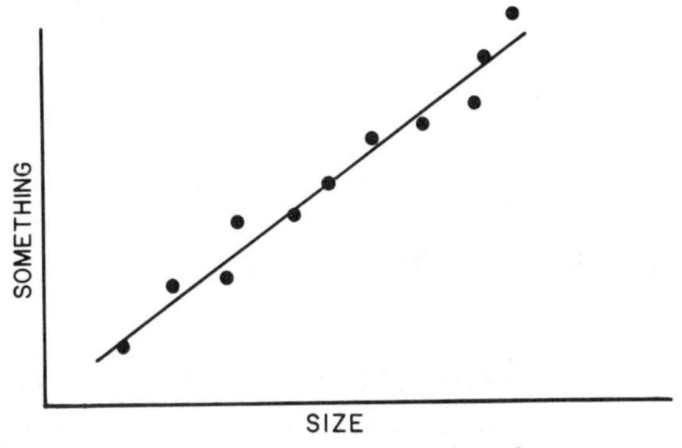

Fig. 2. A generalized bivariate plot.

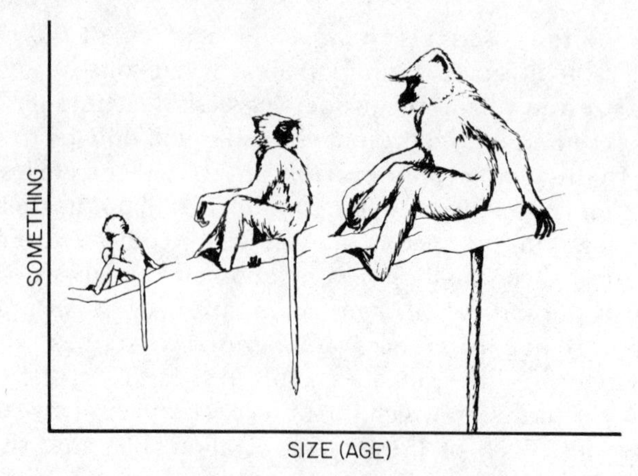

Fig. 3. A representation of growth allometry or ontogenetic allometry, in which the size axis (the abscissa) also represents age and development.

adult (or static) intraspecific allometry represents a true measure of size-related morphological changes that are independent of adaptation, assuming that all members of the same species are close to identical in their adaptive needs (e.g., Gould, 1978). Others have argued that an intraspecific comparison just describes sexual dimorphism (and any associated behavioral or ecological differences), different heritabilities of the features being examined (e.g., Lande, 1979), or just an artifact of developmental "noise" (e.g., Smith, 1980).

Interspecific allometry examines size-related changes among species that differ in body size—the common mouse-to-elephant, or galago-to-gorilla curve (Fig. 5). Obviously, the units of analysis (species) used for interspecific comparisons are in most regards (e.g., genetics, behavior) less homogeneous than those of either ontogenetic or intraspecific comparisons. This lack of genetic and behavioral homogeneity could theoretically have one of two very different effects on the way one interprets interspecific comparisons. On the one hand, one might argue that anything that such a diverse group of animals has in common must be some basic biological relationship, independent of their particular adaptive differences. Alternately, a regression line derived

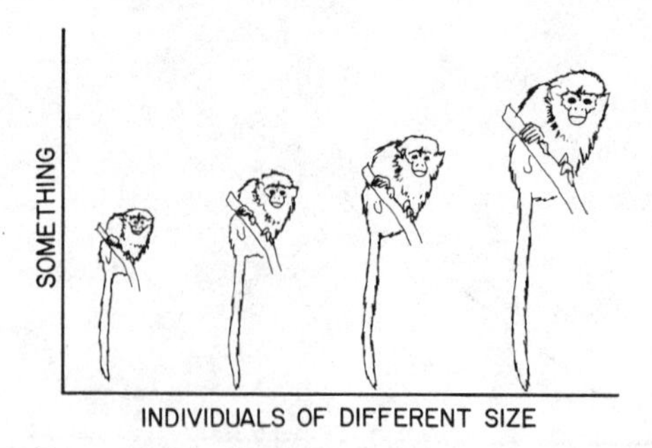

Fig. 4. A representation of intraspecific allometry or static adult allometry, in which comparisons are made among different-sized adult individuals of the same species.

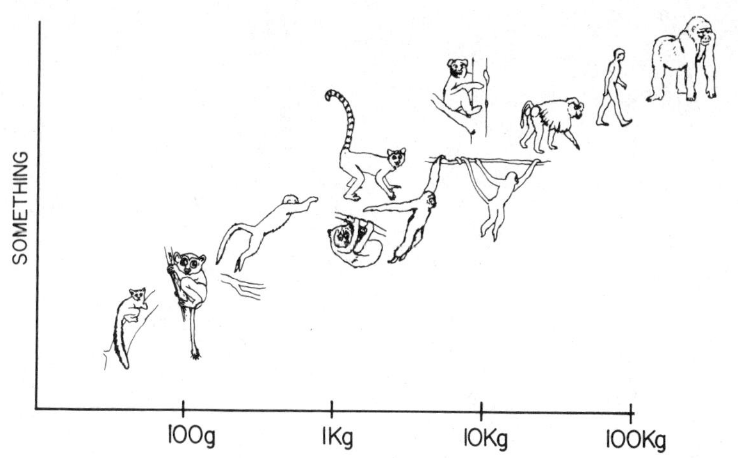

Fig. 5. A representation of interspecific allometry, in which comparisons are made between different-sized species.

from a hodge-podge collection of very diverse animals may in fact be nothing more than a statistical fit that describes the pattern of size-related adaptive differences shown by one particular sample (e.g., Smith, 1980). The following sections of this chapter are largely concerned with distinguishing and reconciling these alternate ways of interpreting the role of adaptive differences in interspecific allometry.

Two Approaches to Allometry

Traditionally, allometric studies have followed one of two distinct approaches, usually considered mutually exclusive (e.g., Clutton-Brock and Harvey, 1979). Anatomists and physiologists (but also some ecologists) have usually taken what I would call an "engineering approach," and have used allometric analyses to identify aspects of size-related change (usually in structure or physiology) that are *independent of specific ecological or behavioral differences,* but rather represent underlying "design features" inherent in the way primates (or other organisms) are built. In other words, they are looking for changes in shape or function that enable animals that differ in size to accomplish the same behavior. In contrast, many ecologists and behaviorists have used allometric analyses to identify consistent patterns in the way in which aspects of ecology and behavior (such as diet, locomotion, or social organization) *vary in conjunction with size.* However, not only are these two approaches largely complementary, they are also very much intertwined. Just as many of the size-related adaptive changes in ecology and behavior seem to be linked to more general structural or physiological limitations imposed by geometry, many size-related structural changes commonly seen among primates may well just reflect size-related behavioral differences imposed by the dimensions of the environment. As McMahon (1975) has emphasized, the key

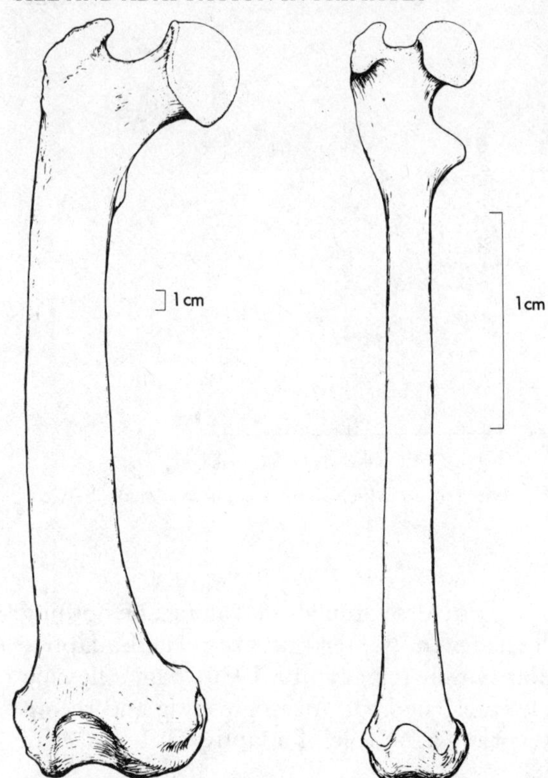

Fig. 6. A femur of a gorilla (*Gorilla gorilla*) and a pygmy marmoset (*Cebuella pygmaea*), the largest and smallest living anthropoids, drawn to the same size. Note that the gorilla femur is relatively much thicker than that of the marmoset.

to understanding the true nature of size-related differences in structure or behavior lies in a careful consideration of the concept of *functional equivalence.*

The Engineering Approach

In examining size-related differences among primates or other organisms, anatomists, physiologists, and engineers like McMahon have generally asked, "What changes are required for animals of different sizes to function in the same way?" or "How do animals of different size maintain functional equivalence?" At a gross level, many of the commonly observed structural differences between small and large animals surely reflect simple geometric considerations. Thus the relative thickness of the bones of gorillas compared with those of pygmy marmosets (Fig. 6) is roughly related to the fact that cross-sectional area must scale at greater than the length squared to support the mass of the larger animal, which theoretically scales as length cubed (e.g., Galilei, 1914). In reality, the scaling of bone shape is not so simply a matter of geometry (e.g., McMahon, 1973; Alexander *et al.,* 1979; Aiello, 1981), and sorting out the actual scaling parameters underlying size and shape changes

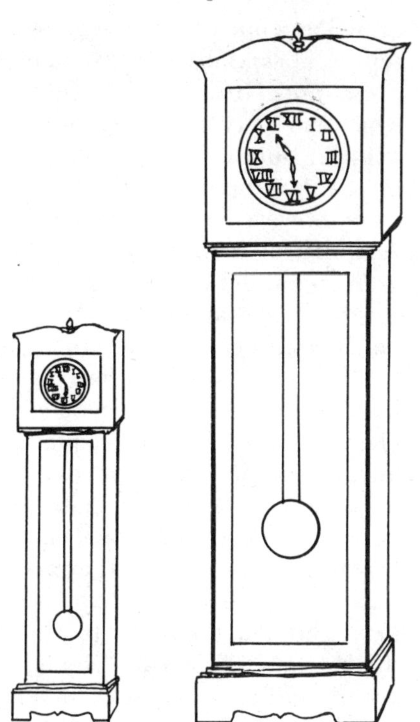

Fig. 7. A large and a small pendulum clock. The period of the pendulum varies in proportion to the square root of its length. In order for the two clocks to keep the same time, the gears cannot be simple geometric size variates, but must be designed with different ratios to take account of the differing periods of the pendula.

in animal skeletons involves more rigorous theoretical (e.g., McMahon, 1973, 1975), descriptive, and experimental (e.g., Lanyon, 1981; Biewener, 1982) investigations of the physical parameters or conditions that are being optimized or kept constant at different sizes. Before we can understand how animals of different size are able to maintain functional equivalence, the *criteria* of functional equivalence must be ascertained.

To take an example from everyday life, the criterion of functional equivalence for small and large clocks is that they keep the same time. If we want a small pendulum clock to keep the same time as a large grandfather clock (Fig. 7), the ratios of the gears driving the hands must be modified to take account of the fact that the period of the pendulum changes with its length (actually, in proportion to the square root of the length). Similarly, McMahon (1973) has argued that the functional equivalence for trees or animal bones is the ability to keep from buckling under imposed weight. Thus, a tree trunk or animal bone must increase in diameter according to the 0.67 power of its length.

For either trees or clocks, this concept of functional equivalence is an underlying assumption in any attempt to identify the structural or physiological design criteria associated with size changes. However, often in empirical, rather than theoretical, studies the "allometric baseline" or "criterion of subtraction" (Gould, 1975, 1978) used in both morphological and ecological studies is the regression line itself. The regression line is usually interpreted as a line of functional equivalence, and deviations from the regression represent specialized, presumably adaptive differences (e.g., Kay, 1975*b;* Andrews and Groves, 1975; Delson and Andrews, 1975). For those cases in which the regression lines for various groups or subgroups of mammals show a similar slope or one that meets theoretical expectations predicted from some mechanical mode, such an interpretation seems quite reasonable. In many cases, however, the functional equivalence represented by such a baseline is very difficult to identify, and we cannot ignore the possibility that the baseline itself might reflect adaptive differences in either behavior or ecology. As Smith (1980) has emphasized, there is no *a priori* reason for attributing functional equivalence to an empirical (statistical) regression line.

For example, by what type of criterion can we consider that a large number of primate species ranging in size from galagos to gorillas (Fig. 5) are likely to be functionally equivalent? They certainly all maintain structural integrity in the face of gravity, and if one is testing a theory that gravity is the major force influencing limb proportions (e.g., McMahon, 1973, 1975), then an assumption of functional equivalence for such a broad comparison is clearly reasonable. However, they are certainly not "equivalent" in their diet or the way they use their teeth (e.g., Kay, 1975*a,b*), in the way they use their limbs to move about, in the forces to which they subject them during locomotion, or in the habitats they occupy. To have confidence that the allometric baseline or criterion of subtraction is "independent of adaptive differences" we have to be very careful in both our choice of parameters and our choice of species (e.g., McMahon, 1975; Jungers, 1979; Kay, 1975*a,b*).

The Ecological Approach

In contrast with the "engineering" approach of looking for those size-related changes that are independent of adaptation, many ecologists and other naturalists have used allometric studies in a more descriptive fashion to examine and document the adaptive potential of size differences. Thus an ecologist is more likely to ask, "How are large animals behaviorally or ecologically different from small ones" (Bourlière, 1975; Western, 1979; Clutton-Brock and Harvey, 1983) or "What can (or must) large primates do better than small ones, and vice versa?"

For example, numerous studies have shown that the importance of invertebrates in the diet of primates is inversely proportional to body size, while the proportion of foliage in the diet increases with size. Likewise, style of arboreal locomotion is related to body size. Leaping decreases with size, while suspensory behavior increases (Fleagle and Mittermeier, 1980; Garber and Easley, 1984). Thus, for an ecologist, size is a guide to adaptive differences and can also serve as a rough predictor of behavior in paleontological reconstructions (e.g., Fleagle, 1978; Gingerich *et al.*, 1982; Kay and Simons, 1980).

The difference between the ecological approach and the engineering approach becomes more obvious if we again focus on the concept of equivalence. In the ecological approach, the criterion of functional equivalence is usually nothing more or less than existence or survival. The diet and locomotion of a galago or mouse lemur and a gorilla (Fig. 1) are equivalent in the sense that they allow each species to live and reproduce. From an ecological point of view a galago-to-gorilla comparison is more descriptive than analytical and illuminates adaptive differences rather than "design criteria."

In the examples discussed so far, the contrast between the engineering approach and the ecological approach is relatively clear. While these different approaches are real, the distinction is largely artificial and provides a much simpler picture of the relationship between size and adaptation than actually exists. Just as anatomists and physiologists have probably assumed functional equivalence in many instances when it was unwarranted, ecologists have probably failed to look for criteria of functional equivalence when they were appropriate and useful for explaining the observed size-related differences. For example, consider two aspects of primate behavior for which size-related adaptive differences have been clearly documented—diet and locomotion (e.g., Fleagle, 1984; Kay, 1984).

Size and Diet

In general, primate diets are closely linked with body size (Fig. 8). Primates that specialize in eating insects tend to be relatively small, whereas primates that eat leaves tend to be relatively large. Likewise, fruit-eaters tend to supplement their diets with either insects or leaves, depending on their size. These patterns are the result of the interaction of several independent size-

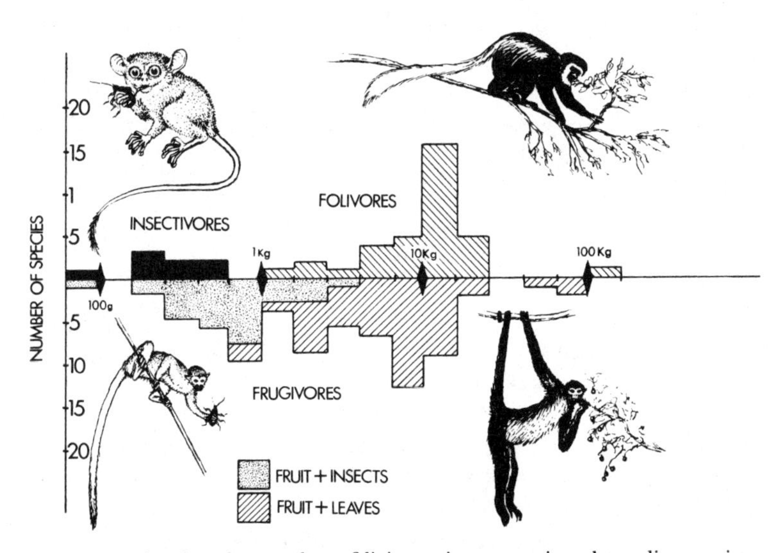

Fig. 8. A histogram showing the number of living primate species whose diet consists predominantly of insects, fruit, or leaves plotted against the average mass of the species. Note that insect-eaters tend to be small, leaf-eaters tend to be large, and fruit-eaters tend to supplement their diet with either insects or leaves, depending upon their body size. (Modified after Kay, 1984.)

related phenomena. First of all, primates need a balanced diet that not only meets their caloric (energy) needs, but also contains other nutritional requirements, such as protein and a variety of trace elements and vitamins. While fruits are high in calories, they are very low in protein content; thus, most primates must turn to other sources for their protein needs. The two most abundant sources of dietary protein for primates are other animals (such as invertebrates) or folivorous materials such as leaves, shoots, and buds. Why, then, do small primates tend to eat insects and large ones folivorous material? Although these two protein strategies are in a nutritional sense probably comparable, the scaling problems that they present are very different and must be considered separately.

First, consider insect-eating. Insects (and animal material in general) are an excellent source of nutrients. They contain virtually all the nutritional requirements a primate needs. After all, one is really just shuffling building blocks like amino acids from one organism to another. Furthermore, they are relatively high in calories per unit weight. This is particularly important for a small animal since small animals have relatively higher energy requirements than large ones because of the way metabolism scales in all mammals (recall the shrew, which must eat several times its body weight in food every day). Insects are so good as a food source that the real question is not why small primates eat them, but why large ones do not.

The answer seems to lie in their availability. No primates have evolved the specialized abilities of creatures such as anteaters to prey on large colonies of social insects; rather, they must depend upon locating and catching isolated individuals. It has been suggested, and seems quite reasonable, that the number of insects that a primate can find and catch in a given day (or night) probably does not vary much from species to species (Hladik and Hladik, 1969), provided of course that they each look in the appropriate places and have the right attributes, such as grasping abilities and eyesight. Thus, in an 8-hr active period, any two primates might be able to ingest 20 insects of one type or another. For a small prosimian this could supply all of its energy requirements for the day, but for a larger monkey it might meet its protein needs, but would leave it short of energy. Thus while larger primates might supplement their fruity (high-energy) diet with insects, they could not rely solely on insects in the way a small primate could.

What about leaves? Unlike insects, they are neither cryptic nor hard to catch. However, leaves pose their own problems (e.g., Montgomery, 1978; Milton, 1979). Although relatively high in protein (particularly young leaves, buds, and shoots), they also contain large amounts of less palatable components, such as cellulose or toxins. Compared with either insects or fruits, leaves are generally low in energy yield in proportion to their weight. Large body size to some degree enables a primate to overcome some of these problems inherent in a leafy diet. First of all, large animals need less energy per kilogram of mass than do small animals, so they can more easily afford to have a diet that is relatively low in energy (e.g., Kleiber, 1961). Second, although primates do not have the enzymes needed to break down the cellulose in leaves, like many other animals, they are sometimes able to maintain colonies of bacteria in their guts to perform the task for them (Bauchop, 1978). However, such digestion takes time, and the time that food spends traveling through an animal's gut is roughly proportional to the length of its gut and thus to its size (Parra, 1978). Thus a small primate with a short gut hardly has an opportunity to digest plant fibers, while a larger animal with a longer gut does. Furthermore, these longer, slower guts with special chambers for fermenting cellulose also seem to help detoxify some of the poisons that plants have put into their leaves to discourage leaf-eaters.

Thus, while the upper size limit of insect-eaters seems to be imposed by the apparently random nature of insect distribution, the lower size limit of folivores is imposed by metabolic and digestive parameters. The size-related differences in primate diets can on the one hand be viewed as behavioral adaptations to circumvent "design criteria" that are inherent in mammalian geometry and physiology (gut proportions and metabolism) as well as those of the environment (insect abundance). Alternately, of course, one could view size as a gross morphological adaptation (with many physiological ramifications) for exploiting different ecological niches. Recognizing the role of size in dietary habits does not eliminate the adaptive differences inherent in the size differences, but it clouds much of the distinction between the two traditional approaches to allometry.

If this analysis of the relationship between size and diet in primates [and other mammals as well (e.g., Montgomery, 1978)] is correct, what relevance does it have for interpreting allometric scaling in morphological features of the digestive system? For example, it hardly seems appropriate to plot dental dimensions as a function of body size for a large group of living primates and assume that the regression line represents a line of functional equivalence (e.g., Pilbeam and Gould, 1974; Gould, 1975) when we suspect in advance that diet (and tooth function) will change with size. Rather, a more appropriate strategy for investigating the functional allometry of tooth scaling would be to look at dental scaling in subgroups with dietary homogeneity (e.g., Kay, 1975*a,b,* 1978). The same is true for investigating the scaling of the digestive tract (e.g., Chivers and Hladik, 1981). In the face of documented size-related dietary differences, partitioning the functional correlates of size from those of adaptation requires a carefully designed analytic approach, which incorporates ecological differences.

Recognizing that size differences often involve both morphological and behavioral considerations, we can also consider the concept of functional equivalence at another level. I have suggested that in an "ecological approach" concerned with the differences between large and small animals, the criterion of equivalence has often meant little more than survival. In that sense the differences between large and small animals must enable them to be functionally equivalent at different sizes. However, in analysis of specific systems we can consider ecological or behavioral equivalence at a finer level. For example, in the discussion of dietary scaling above, the differences between large and small primates are more specifically adaptations that enable them to be *nutritionally equivalent* at different sizes, that is, to meet their caloric and protein requirements.

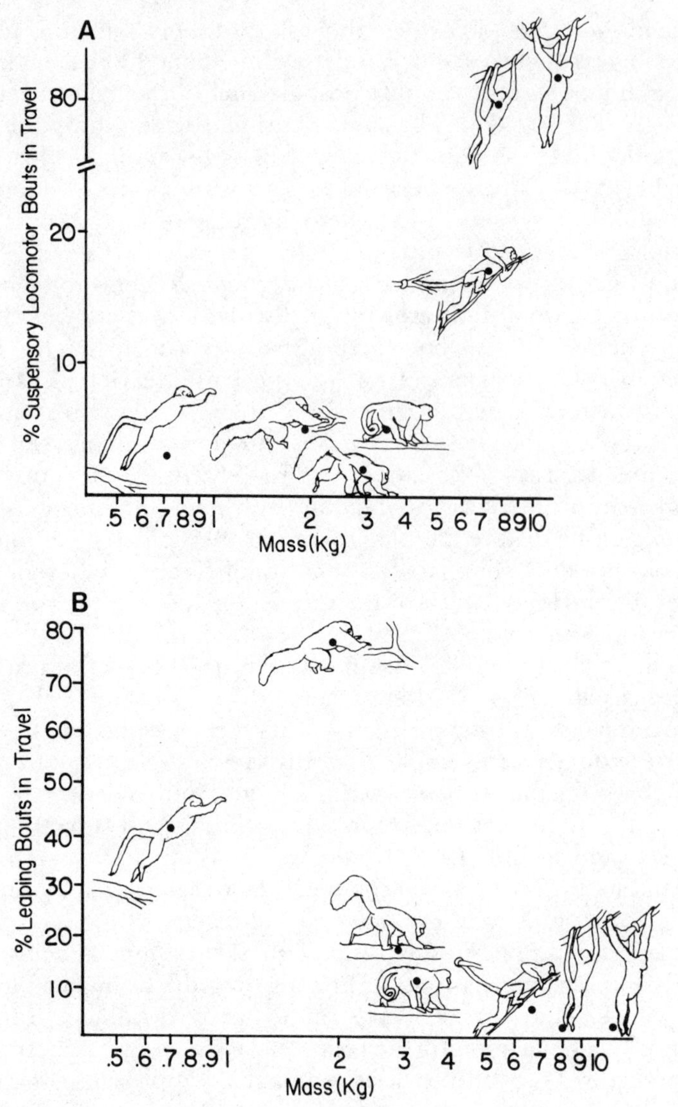

Fig. 9. Bivariate plots of the relative proportion of (A) suspensory and (B) leaping behavior in the locomotor activity of seven species of New World monkeys. Note that with increasing size, the frequency of leaping decreases and the frequency of suspensory behavior increases. (Modified from Fleagle and Mittermeier, 1980.)

Size and Arboreal Locomotion

Among arboreal primates there are size-related trends in the use of different types of locomotion. Although we lack the extensive quantitative data on primate locomotion that we have for diet, the allometry of locomotor behavior has been quantitatively assessed for a good sample of South American monkeys (Fleagle and Mittermeier, 1980). Broadly similar patterns seem to hold for the order as a whole (e.g., Garber and Easley, 1984) and to a lesser extent within various other major radiations. In general, leaping is more

Fig. 10. The same forest would appear very different to a small monkey and a large one. The smaller monkey would encounter more gaps that it would have to cross by leaping, while the larger monkey would encounter relatively more gaps that could be crossed by bridging.

common among small primates (Fig. 9A), while suspensory behavior is more common in larger species (Fig. 9B). Like fruit-eating, quadrupedal walking and running does not seem to show any pattern with respect to body size, and in some sense can be regarded as an alternate strategy of locomotion.

These trends in leaping and suspensory behavior seem to be primarily the results of simple mechanical phenomena (Fleagle and Mittermeier, 1980; Cartmill and Milton, 1977). If we imagine two primates, a small one and a large one, traveling through the forest canopy, they will both encounter a number of gaps between trees that they somehow must cross to continue their journey (Fig. 10). In the same forest, the smaller one will more frequently encounter gaps that it can only cross by leaping, while the larger one will more frequently encounter gaps that could be crossed by bridging or suspending itself between the terminal supports (Fig. 10). Similarly, leaping involves the generation of high propulsive forces from the hindlimbs alone; smaller animals would find more supports that could sustain their leaps than would larger animals. Likewise, during both locomotion and feeding, larger animals will more frequently encounter supports too narrow and too weak to support their larger bodies and would more readily need to suspend themselves below multiple branches both for support and balance (Fig. 11). Finally, it has been argued that should a primate fall from a tree, the energy its body must dissipate upon hitting the ground is proportional to its mass (Cartmill and Milton, 1977). Therefore, large animals need to be much more cautious in their locomotion.

Fig. 11. During feeding, a small monkey would more frequently encounter single branches that could support its weight, whereas the larger monkey would more frequently need to distribute its weight among several branches both for support and balance.

Like the size-related dietary differences, size-related locomotor differences can be viewed as adaptive responses to overcome design criteria that are built into either their own geometry or the environment (e.g., distribution of supports, gravity). Only through locomotor differences can large and small arboreal primates maintain locomotor (functional) equivalence in being able to inhabit the same environment. Alternately, again, size might be viewed as a gross morphological adaptation that facilitates different locomotor abilities and the utilization of different substrate environments. As in the previous example, consideration of the scaling of behavioral features complicates the distinction between the functional correlates of size and those of adaptive differences.

For example, consider the limb proportions, locomotor behavior, and ecology of a diverse sample of clawless neotropical monkeys, six species from Surinam, plus the muriqui or wooly spider monkey from southeastern Brazil. If we plot limb proportions as a function of body mass, intermembral index (forelimb length × 100/hindlimb length) clearly increases proportionally with increasing size (Fig. 12). From an engineering point of view, one might in-

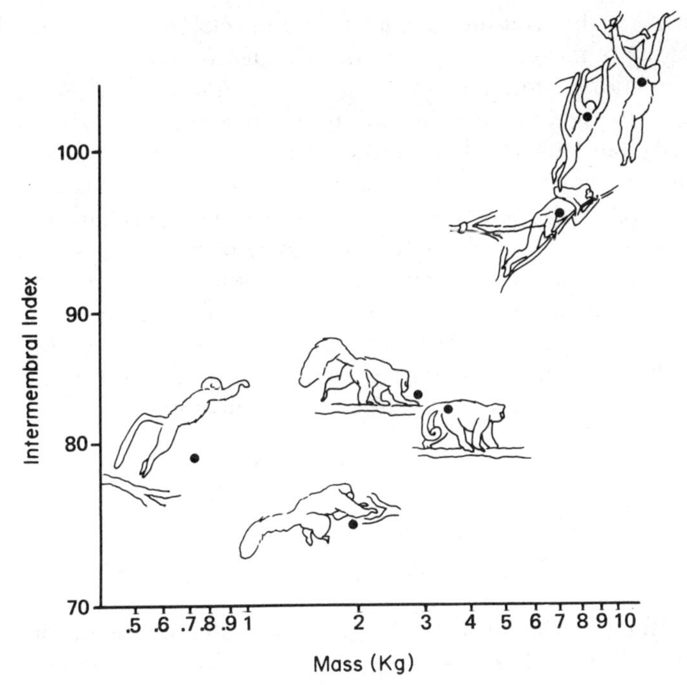

Fig. 12. Intermembral index (forelimb length/hindlimb length) plotted against mass for seven species of New World monkeys.

terpret this plot as demonstrating that larger monkeys require relatively longer forelimbs than smaller ones for structural reasons in order to maintain functional equivalence (e.g., Biegert and Mauer, 1972; Delson and Andrews, 1975). However, if we look closely at the locomotor behavior of these particular species, the situation becomes more complicated. Just as these eight species show differences in limb proportions that are correlated with body size, they also show detailed differences in locomotor behavior as described above (Figs. 9A, 9B, and 12). Their use of suspensory behavior (brachiation and climbing) increases with body size, while leaping decreases with an increase in body size. Either or both of these behavioral trends could be offered as causal explanations for the observed morphological trends and supported by well-established biomechanical arguments that explain why leapers should have relatively long legs and short arms, while suspensory species should have relatively longer arms (e.g., Fleagle, 1977; Preuschoft and Demes, 1984), as well as many detailed differences in finer aspects of their skeletal anatomy.

It thus seems likely that many size-related proportional changes do not reflect functionally equivalent structural conditions, but functionally different structural conditions related to different behavioral abilities and habits. The small leapers have relatively longer hindlimbs for leaping, while the larger suspensory forms have relatively longer forelimbs for suspensory behavior. In this case, as in the previous discussion of diet, the proportional changes enable the different species to maintain some type of equivalence, but in an ecological rather than a structural context.

However, there are even further possible interpretations for these same results. Jungers (this volume) argues that size-related changes in the intermembral index of most major groups of primates are the result of selection to maintain functional competence in vertical climbing [following Cartmill (1974)]. Like the behavioral explanation, this one accords with the observed proportional changes with size and has the appealing advantage of a mechanical basis and potentially broad applicability. Since the two explanations are broadly concordant in their predictions, the correct one is hard to identify at present. The main justifications for my choice of an ecological explanation over the "climbing" argument are that (1) climbing is a relatively rare activity for many of the smaller species, and (2) climbing could never account for the absolutely long legs of the leapers or any of the more detailed differences in muscle and bone architecture, only the relative increase in forelimb to hindlimb length with increasing size. They are not incompatible.

Discussion

These examples demonstrate the great difficulty of interpreting the actual significance of any allometric regression beyond its basic descriptive value. An "engineering" approach requires some consideration of the behavior and ecology of the species being compared, because without those data it is very difficult to generate a meaningful concept of the way in which different species are functionally equivalent. A vague concept of structural allometry should not be the null hypothesis for a regression line.

Usually an appreciation of the behavioral and ecological diversity exhibited by the sample being compared will generate a whole new set of competing explanations for the observed trends. Such complications are, in my opinion, unavoidable because of the way evolution by natural selection has operated to produce our world. In an evolutionary framework, there are two ways in which animals can adapt to the scaling problems imposed by simple geometry and by features of the environment beyond their control. They can change their proportions in ways that depart from geometric scaling ("engineering allometry") or they can adopt different lifestyles at different sizes that capitalize to some extent on these geometric factors ("behavioral allometry"). Not surprisingly, primates and most other organisms appear to have followed both of these options, and multiple explanations of size-related trends should be an incentive toward more rigorous evaluations of criteria of functional equivalence and ways of falsifying alternative explanations.

Such a balanced consideration of both engineering and ecological approaches is critical if we are ever going to make any real progress in the understanding of biological scaling. Just as extreme adaptationists have perhaps been overly eager to find an adaptive story for any and all aspects of an animal's biology (Gould and Lewontin, 1979; but see Mayr, 1981), students of allometry have been far too ready to attribute explanatory powers to a bivariate regression (Smith, 1980). Gould (1978) has argued that "allometric studies often leave nothing after the subtraction of size." I would suggest that the converse is just as true—allometric studies often leave nothing after the subtraction of adaptive differences. Oxnard (1978) has expressed similar doubts about separating the size-related aspects of shape from those reflecting adaptation.

Attributing the differences between galagos and gorillas to allometry just because they lie at the endpoints of a regression line is a common phenomenon, but one that has little power as an explanation. As many of the contributors to this volume emphasize, an allometric comparison is a *description* of size-related differences, not an explanation of them. For many size-related structural changes, the explanation may be that the animals behave differently; and the explanation for many size-related behavioral changes may well be that they enable the animals to function in an equivalent manner in an environment with its own dimensionality.

How then can we investigate biological scaling with an eye toward realistically interpreting the functional relationships between size and adaptation in both morphology and ecology? First, it is important to realize the critical role that theory must play in allometric studies. An allometric regression is only *potentially* a statement of functional equivalence, in some form or other. It is an investigator's obligation to demonstrate that he or she is indeed testing some aspect of functional equivalence and to explain how an empirical regression line accords with or departs from the predicted slope. Without theory, an allometric comparison is only a graphic or statistical description of size-related differences and has no real interpretive value for questions of functional equivalence or adaptation.

However, even in a descriptive approach, careful choice of subjects can certainly enhance the likelihood that an allometric regression may reflect a condition of functional equivalence. Thus, Jungers' (1979, 1984, and this volume) studies of limb proportions *within* taxonomic families of primates or McMahon's (1975) study of limb dimensions within families and orders of ungulates provide a greater "control" over adaptive and genetic diversity than a broader comparison. We are more likely to gain insight into the effects of size changes within a relatively uniform adaptive radiation (and a restricted genetic base) than through a more random choice of subjects.

Acknowledgments

This chapter has profitted greatly from seemingly endless discussions and arguments with Jack T. Stern, Jr., Norman Creel, Susan Larson, David Krause, Richard Kay, and especially Bill Jungers. I am additionally indebted to Bill Jungers for inviting me to participate in the symposium at the Congress of the International Primatological Society in Atlanta upon which this book is based. Lucille Betti and Stephen Nash drafted the figures. Much of the work described in this paper was supported by research grants BNS 7724921, BNS 7924149, and BNS 8210949 from the National Science Foundation and a fellowship from the John Simon Guggenheim Memorial Foundation.

References

Aiello, L. C. 1981. The allometry of primate body proportions. *Symp. Zool. Soc. Lond.* **48:**331–358.

Alexander, R. McN., Jayes, A. S., Maloiy, G. M. O., and Wathuta, E. M. 1979. Allometry of the limb bones of mammals from shrew (*Sorex*) to elephant (*Loxodonta*). *J. Zool. Lond.* **189:**305–314.

Andrews, P., and Groves, C. 1975. Gibbons and brachiation. *Gibbon and Siamang* **4:**167–218.

Bauchop, T. 1978. Digestion of leaves in vertebrate arboreal folivores, in: *The Ecology of Arboreal Folivores* (G. G. Montgomery, ed.), pp. 193–204, Smithsonian Institution Press, Washington.

Biegert, J., and Maurer, R. 1972. Rampfskelettlange, Allometrien und Korperproportionen bei catarrhinen Primaten. *Folia Primatol.* **17:**142–156.

Biewener, A. A. 1982. Bone strength in small mammals and bipedal birds: Do safety factors change with body size?. *J. Exp. Biol.* **98:**289–301.

Bourliere, F. 1975. Mammals, small and large: The ecological implications of size, in: *Small mammals: their productivity and population dynamics.* (F. B. Golley, K. Petrusewicz, and L. Ryszkowski, eds.), pp. 1–8, Cambridge University Press, Cambridge.

Cartmill, M. 1974. Pads and claws in arboreal locomotion, in: *Primate Locomotion* (F. A. Jenkins, Jr., ed.), pp. 45–83, Academic Press, New York.

Cartmill, M., and Milton, K. 1977. The lorisiform wrist joint and the evolution of "brachiating" adaptations in the Hominoidea. *Am. J. Phys. Anthropol.* **47:**249–272.

Chivers, D. J., and Hladik, C. M. 1981. Morphology of the gastrointestinal tract in primates: Comparisons with other mammals in relation to diet. *J. Morphol.* **166:**337–386.

Clutton-Brock, T. H., and Harvey, P. H. 1979. Comparison and adaptation. *Proc. R. Soc. Lond. B* **205:**547–565.

Clutton-Brock, T. H., and Harvey, P. H. 1983. The functional significance of variation in body size among mammals, in: *Advances in the Study of Mammalian Behavior,* (J. F. Eisenberg and D. G. Kleiman, eds.), pp. 632–665, American Society of Mammalogy.

Delson, E., and Andrews, P. 1975. Evolution and interrelationships of the catarrhine primates, in: *Phylogeny of the Primates* (W. P. Luckett and F. S. Szalay, eds.), pp. 405–446, Plenum Press, New York.

Fleagle, J. G. 1977. Locomotor behavior and skeletal anatomy of sympatric Malaysian leaf-monkeys (*Presbytis obscura* and *Presbytis melalophos*). *Yearb. Phys. Anthropol.* **20:**440–453.

Fleagle, J. G. 1978. Size distributions of living and fossil primate faunas. *Paleobiology* **4:**67–76.

Fleagle, J. G. 1984. Primate locomotion and diet, in: *Food Acquisition and Processing in Primates* (D. J. Chivers, B. A. Wood, and A. Bilsborough, eds.), pp. 105–117, Plenum Press, New York.

Fleagle, J. G., and Mittermeier, R. A. 1980. Locomotor behavior, body size, and comparative ecology of seven Surinam monkeys. *Am. J. Phys. Anthropol.* **52:**301–314.

Galilei, G. 1914. *Dialogues Concerning Two New Sciences* (H. Crew and A. DeSalvio, transl.), Macmillan, New York (original publication, 1638).

Garber, P. A., and Easley, S. P. 1984. Substrate preferences, body weight, and positional behavior in arboreal primates. *Am. J. Phys. Anthropol.* (in press).

Gingerich, P. D., Smith, B. H., and Rosenberg, K. 1982. Allometric scaling in the dentition of primates and the prediction of body weight from tooth size in fossils. *Am. J. Phys. Anthropol.* **58:**81–100.

Gould, S. J. 1975. Allometry in Primates with emphasis on scaling and the evolution of the brain, in: *Approaches to Primate Paleobiology* (Contrib. Primatol., Vol. 5, F. Szalay, ed.), pp. 244–292, S. Karger, Basel.

Gould, S. J. 1978. Generality and uniqueness in the history of life: An exploration with random models. *Bioscience* **28:**277–281.

Gould, S. J. 1980. *The Panda's Thumb,* W. W. Norton, New York.

Gould, S. J., and Lewontin, R. C. 1979. The spandrels of San Marco and the Panglossian paradigm: A critique of the adaptationist programme. *Proc. R. Soc. Lond. B* **205:**147–164.

Gunnell, G. 1983. Body size and peleobiology of early Tertiary North American primates. *Am. J. Phys. Anthropol.* **60:**202.

Hladik, A., and Hladik, C. M. 1969. Rapports trophique entre vegetation et primates dans la foret de Barro Colorado (Panama). *Terre Vie* **23:**25–117.

Jungers, W. L. 1979. Locomotion, limb proportions, and skeletal allometry in lemurs and lorises. *Folia Primatol.* **32:**8–28.

Jungers, W. L. 1984. Scaling of the hominoid locomotor skeleton with special reference to the lesser apes, in: *The Lesser Apes: Evolutionary and Behavioral Biology* (H. Preuschoft, D. Chivers, W. Brockelman, and N. Creel, eds.), pp. 146–169, Edinburgh University Press, Edinburgh.

Kay, R. F. 1975*a*. Allometry and early hominids. *Science* **189:**63.

Kay, R. F. 1975*b*. The functional adaptations of primate molar teeth. *Am. J. Phys. Anthropol.* **43:**195–216.

Kay, R. F. 1978. Molar structure and diet in extant cercopithecidae, in: *Development, Function, and Evolution of Teeth* (P. M. Butler and K. A. Joysey, eds.), pp. 309–339, Academic Press, New York.

Kay, R. F. 1984. On the use of anatomical features to infer foraging behavior in extinct primates, in: *Adaptations for Foraging in Nonhuman Primates* (J. Cant and P. Rodman, eds.), pp. 21–53, Columbia University Press, New York.

Kay, R. F., and Simons, E. L. 1980. The ecology of Oligocene African Anthropoidea. *Int. J. Primatol.* **1:**21–38.

Kleiber, M. 1961. *The Fire of Life. An Introduction to Animal Energetics,* Wiley, New York.

Lande, R. 1979. Quantitative genetic analysis of multivariate evolution, applied to brain: body size allometry. *Evolution* **33:**402–416.

Lanyon, L. E. 1981. Locomotor loading and functional adaptation in limb bones. *Symp. Zool. Soc. Lond.* **48:**305–330.

McMahon, T. A. 1973. Size and shape in biology. *Science* **179:**1201–1204.

McMahon, T. A. 1975. Allometry and biomechanics: limb bones in adult ungulates. *Am. Nat.* **109:**547–563.

Mayr, E. 1976. *Evolution and the Diversity of Life: Selected Essays,* Belknap Press, Cambridge, Massachusetts.

Mayr, E. 1981. How to carry out the adaptationist program?. *Am. Nat.* **121:**324–334.

Milton, K. 1979. Factors influencing leaf choice by howling monkeys: A test of some hypotheses of food selection by generalist herbivores. *Am. Nat.* **114:**362–378.

Montgomery, G. G. (ed.) 1978. *The Ecology of Arboreal Folivores,* Smithsonian Institution Press. Washington, D.C.

Oxnard, C. E. 1978. One biologist's view of morphometrics. *Annu. Rev. Ecd. Syst.* **9:**219–242.

Parra, R. 1978. Comparison of foregut and hindgut fermentation in herbivores, in: *The Ecology of Arboreal Folivores* (G. G. Montgomery, ed.), pp. 205–230, Smithsonian Institution Press, Washington, D.C.

Pilbeam, D. R., and Gould, S. J. 1974. Size and scaling in human evolution. *Science* **186:**892–900.

Preuschoft, H., and Demes, B. 1984. The biomechanics of brachiation, in: *The Lesser Apes: Evolutionary and Behavioral Biology* (H. Preuschoft, D. Chivers, W. Brockelman, and N. Creel, eds.), pp. 96–118, University of Edinburgh Press, Edinburgh.

Rawlins, R. 1976. Locomotor ontogeny in *Macaca mulatta:* I. Behavioral strategies and tactics. *Am. J. Phys. Anthropol.* **44:**201.

Schmidt-Nielsen, K. 1975. Scaling in biology: The consequences of size. *J. Exp. Zool.* **194:**287–307.

Shea, B. T. 1983. Allometry and heterochrony in the African apes. *Am. J. Phys. Anthropol.* **62:**275–290.

Simons, E. L., and Ettel, P. C. 1970. *Gigantopithecus. Sci. Am.* **222:**76–85.

Smith, R. J. 1980. Rethinking allometry. *J. Theor. Biol.* **87:**97–111.

Tanner, J. M. 1951. Some notes on the reporting of growth data. *Hum. Biol.* **23:**93–159.

Western, D. 1979. Size, life history, and ecology in mammals. *Afr. J. Ecol.* **17:**185–204.

Genetic and Evolutionary Aspects of Allometry

2

RUSSELL LANDE

Introduction

Huxley (1932) consolidated the study of allometry by providing a theoretical formulation of allometric growth and by analyzing many examples of allometric relationships, including the static allometry of adults within a population and the evolutionary allometry between populations and species. Unaided by an explicit theory connecting the different levels of allometry, Huxley (1932, pp. 212–224), Gould (1975, pp. 277–280), and others have attempted to suggest possible relationships between levels of allometry, usually by a simple extrapolation from one level to another.

The present chapter considers the theoretical foundations of ontogenetic, static, and evolutionary allometry and the relationships among them. Emphasis is placed on the potential contribution of quantitative genetic methods to an integrated understanding of levels of allometry. To illustrate these methods, data on phenotypic and genetic variation in brain and body weights in mammals and primates are analyzed, and an explanation of von Baer's law of comparative embryology is given in terms of quantitative developmental genetics.

Ontogenetic Allometry

The basic formula of allometric growth is derived by postulating that during individual development the growth rate of any part of an organism is

---

RUSSELL LANDE • Department of Biophysics and Theoretical Biology, University of Chicago, Chicago, Illinois 60637.

proportional to a specific rate constant times the present size of the part and to a general growth function F that may depend on various characters and total body size as well as age or time. For two characters X and Y the growth equations

$$\frac{dX}{dt} = \alpha X F(X,Y,\ldots;t), \quad \frac{dY}{dt} = \beta Y F(X,Y,\ldots;t) \tag{1}$$

imply

$$\frac{dY}{Y} = \frac{\beta\, dX}{\alpha X}$$

which upon integration yields

$$Y/Y_0 = (X/X_0)^{\beta/\alpha}$$

where X_0 and Y_0 are the initial values of the characters. Defining the constants $k = \beta/\alpha$ and $b = Y_0/X_0^{\beta/\alpha}$, one customarily writes this equation in the form

$$Y = bX^k \tag{2a}$$

or

$$\log Y = k \log X + \log b \tag{2b}$$

which describes a straight line on logarithmic coordinates (Huxley, 1932). In most allometric studies the independent character X is a measure of overall body size, such as total weight or length, although Huxley (1924, 1932) used it to represent total body weight minus the organ weight Y.

It is not widely appreciated that the basic allometric formulas (2a) and (2b) are only approximate. Haldane (in Huxley, 1932, p. 81) pointed out that a sum of parts growing at different rates with respect to body size cannot itself exactly follow an allometric relationship with body size (cf. Reeve and Huxley, 1945). Huxley (1932, pp. 81–82) mentions, however, that if the relative growth rates of the parts are not very different, the allometric formula should be sufficiently accurate for most purposes.

To see this problem more clearly, suppose that during some stage of its development an individual organism is composed of a number of primary parts $Z_1, \ldots, Z_n$ growing with different specific rates β_i and influenced by the same general growth function F,

$$\frac{dZ_i}{dt} = \beta_i Z_i F(Z_1, \ldots, Z_n; t) \tag{3a}$$

The primary parts then grow allometrically with respect to each other. Now consider two characters composed of (weighted) sums of the primary parts,

$$X = \sum_{i=1}^{n} a_i Z_i, \qquad Y = \sum_{i=1}^{n} c_i Z_i \tag{3b}$$

in which a_i and c_i are constants. Letting the logarithms of these two characters be designated as $x = \log X$ and $y = \log Y$, at any point in development the slope of the allometric relation between the characters is

$$k = \frac{dy}{dx} = \frac{dY}{Y\,dt} \bigg/ \frac{dX}{X\,dt} = \bar{\beta}_y / \bar{\beta}_x \tag{4a}$$

where

$$\bar{\beta}_x = \sum_{i=1}^{n} \beta_i a_i Z_i \bigg/ \sum_{i=1}^{n} a_i Z_i \tag{4b}$$

and a similar expression defines $\bar{\beta}_y$. Thus at any point in development the slope of the allometric relation between two characters is the ratio of weighted averages of the specific growth rates of their parts, with the weighting factor for each part being its proportional contribution to the character. The allometric slope (4a) generally will change during individual development because in time the more rapidly growing parts will contribute proportionally more to the weighted average growth rate of each composite character.

The magnitude of departure from simple allometry can be assessed by the amount of change in the allometric slope k with a change in x. This is the curvature of the allometric relation, which can be derived from (3) and (4) as

$$\frac{d^2y}{dx^2} = \frac{dk}{dx} = \frac{dk}{dt} \bigg/ \frac{dX}{X\,dt} = (\sigma^2_{\beta_y} - k\sigma^2_{\beta_x}) / \bar{\beta}_x^{\,2} \tag{5a}$$

where

$$\sigma^2_{\beta_x} = \left(\sum_{i=1}^{n} \beta_i^2 a_i Z_i \bigg/ \sum_{i=1}^{n} a_i Z_i \right) - \bar{\beta}_x^{\,2} \tag{5b}$$

is the weighted variance of the specific growth rates of the primary parts of character X, and an analogous formula defines $\sigma^2_{\beta_y}$. The allometric relation between two composite characters, such as a major organ and body size, may thus show positive or negative curvature, or more complex undulations during ontogeny.

Similar considerations apply to the analysis of static and evolutionary allometry. The ultimate justification for the use of the basic allometric equations (2a) and (2b) in any study must therefore be the empirical fit of the data to a straight line on logarithmic coordinates.

Static Allometry

Individual differences in development produce phenotypic variation within populations, which is the raw material for natural selection and evolution. Both genetic and environmental factors and their interactions contribute to individual variation within populations. At any given age or stage of development a static allometry coefficient can be estimated from the joint distribution of a character with body size.

A commonly used measure of allometry is the regression coefficient of logarithmically transformed data. Denoting the logarithms of body size and another character as $x = \log X$ and $y = \log Y$, with standard deviations σ_x and σ_y and correlation ρ_{xy}, we have that the phenotypic regression coefficient (which minimizes squared vertical deviations from the allometry line) is

$$P_{xy}/P_{xx} = \rho_{xy}\sigma_y/\sigma_x \tag{6}$$

in which P_{xy} is the phenotypic covariance of x and y.

Occasionally, other measures of relationship between characters are used, such as the major axis, or the reduced major axis with slope σ_y/σ_x, which minimizes squared vertical and horizontal deviations from the allometry line (Kermack and Haldane, 1950). The different forms of regression are nearly equivalent when the characters are very highly correlated ($\rho_{xy} \simeq 1$) as in most ontogenetic data, but for the static allometry of adults within a population the correlation between the characters is often low or negative, and the standard regression formula (6) seems most informative. A multivariate version of allometry proposed by Jolicoeur (1963) employs the first principal component of variation as a generalized size factor against which each trait is compared. A serious limitation of this method is that the allometric coefficient for a given trait depends on which other characters have been included in the analysis (Jungers and German, 1981). It therefore seems best to use an unambiguous measure of body size, like total weight or length, as a standard of comparison for other characters.

The static allometry of adults within a population depends on the variances and covariances of the characters (6), which can be expressed in terms of their initial values and growth increments. Variances and covariances of the latter bear no necessary relationship to the average ontogenetic trajectory in a population, except in the trivial case when all individual variation corresponds to different degrees of progression along a single developmental pathway. Despite the frequent similarity between levels of allometry, Cock (1966) and Cheverud (1982) have given theoretical and empirical examples of major differences between ontogenetic and static allometry.

Evolutionary Allometry

Allometric relationships among characters in closely related populations or species often result from secondary adaptations to evolutionary changes in body size (Rensch, 1959, Chapter 6; Gould, 1966). A common null hypothesis in the comparison of closely related species is that they diverged morphologically due to natural selection acting only on overall body size, and that any shape changes were a passive, correlated response to selection on size. This null hypothesis actually implies a precise connection between the coefficient of evolutionary allometry and the *genetic* variances and covariances of the characters within populations. To derive this relationship, assume that the characters are measured on logarithmic scales, and that on these scales the phenotypic variances and covariance remain nearly constant during the evolution of the mean values, as is often empirically true (Wright, 1968, Chapter 15; Falconer, 1981, Chapter 17). The strength of selection on body size is described by the selection differential S_x, the difference in mean body size between selected and unselected adults. The evolutionary response in average body size during one generation is

$$\Delta\bar{x} = (G_{xx}/P_{xx})S_x = h_x^2 S_x \tag{7a}$$

in which h_x^2 is the heritability of body size, the ratio of its additive genetic variance G_{xx} to its total phenotypic variance P_{xx}. The correlated response in the mean of another character during one generation of selection on body size alone is

$$\Delta\bar{y} = (G_{xy}/P_{xx})S_x = h_x h_x \gamma_{xy}(\sigma_y/\sigma_x)S_x \tag{7b}$$

where G_{xy} and γ_{xy} are the additive genetic covariance and correlation between x and y. The genetic parameters in (7) can be estimated from data on phenotypic correlations between relatives, i.e., parents and offspring or full and half-siblings (Falconer, 1981). These equations are based on the assumptions of no genotype–environment correlations and linearity of the regressions of additive genetic values of the characters on body size, which are likely to be satisfied for polygenic characters that follow a bivariate normal distribution.

The slope of the evolutionary trajectory when selection acts solely on body size is

$$\Delta\bar{y}/\Delta\bar{x} = G_{xy}/G_{xx} = \gamma_{xy}h_y\sigma_y/h_x\sigma_x \tag{8}$$

The selection differential S_x has canceled from the equation, hence any fluctuations in the rate and direction of evolution of body size have no effect on the slope of the line of evolutionary allometry. If the ratio of the genetic covariance to the genetic variance in body size remains nearly constant, the evolutionary trajectory will tend to be a straight line on logarithmic coordinates, that is, "allometric" in the sense of Huxley (1932). Clearly, under the assumptions of this model, when selection acts only on body size the coefficient of evolutionary allometry is given by the static genetic regression coefficient (8) rather than the static phenotypic regression coefficient (6), as first noted by Reeve (1950). Static genetic and phenotypic regressions do not necessarily coincide, and in some cases they may differ greatly (Falconer, 1981, Chapter 19; Atchley and Rutledge, 1980; Cheverud, 1982; Lande, 1982).

Thus the suggestion by Gould (1975, pp. 277–280) and others that selection on body size alone would extrapolate static phenotypic (or ontogenetic) lines of allometry is incorrect in theory and may be seriously misleading in practice. However, there are some organ systems, such as linear skeletal measurements, for which genetic and phenotypic parameters tend to be similar (Bailey, 1956; Leamy, 1977; Atchley *et al.,* 1981).

Patterns of interspecific variation are influenced both by phenotypic change within lineages and by differential extinction and speciation. But branching and extinction of lineages do not themselves create character divergence, which generally arises only from the accumulation of genetic variation within lineages (excluding direct effects of the environment, and some non-Mendelian forms of inheritance). On the assumption that phyletic changes were a quantitative genetic response to natural selection, the net selective forces can be analyzed for any part of a phylogeny, and in particular those lineages leading to modern species.

Another null hypothesis that may be considered in the comparison of closely related populations or species is that they diverged without any selection at all, by random genetic drift in small populations. The expected coefficient of evolutionary allometry in this case is virtually identical to that for divergence by selection on body size alone, Eq. (8), and discriminating between these two null hypotheses requires more sophisticated tests. The theory of selection and random genetic drift has been generalized to multiple correlated characters, and can be employed to analyze the divergence between contemporary populations in addition to strictly phylogenetic sequences (Lande, 1979, 1982).

The Evolution of Ontogeny: von Baer's Law

An important generalization in comparative embryology is von Baer's law that the early developmental stages of closely related species tend to resemble each other more than the later stages (de Beer, 1958, p. 3). Several exceptions to this law are known, where related species diverge most in their early development (de Beer, 1958; Cock, 1966). However, the pattern is sufficiently general that it, as well as exceptions to it, deserve an explanation in modern evolutionary and genetic terms. Darwin (1859, Chapter 8) noted that variations appearing in individuals at a certain age tend to be manifested at a corresponding age in their offspring, and that most species-specific characters appear late in ontogeny. Genes that act late in development to change the relative growth rates of different organs are likely to produce small, quantitative variations that are the basis for most morphological evolution in higher animals (Wright, 1968, Chapter 15; Falconer, 1981, Chapter 12; Lande, 1981).

A more detailed view of the evolution of ontogeny can be achieved by considering the pattern of phenotypic and genetic correlations between characters measured at a series of ages. For a particular trait, it is necessarily true that as the time between measurements decreases toward zero the phenotypic and genetic correlations must both approach unity, and it should be anticipated that with increasing time between measurements the correlations may decrease substantially. If the relative variation and heritability of the trait are roughly uniform at all ages, selection restricted to a single point in the life cycle would tend to produce the largest response in the age class under selection, and the magnitude of correlated responses would diminish at ages further removed from the one under selection.

In species with extended parental care of offspring, individuals are shielded from many sources of selective mortality during embryonic and early postnatal stages, and most selection on morphological traits probably operates during juvenile and adult stages. Such species, including most mammals and all primates, are expected to obey von Baer's law because selection is concentrated mainly on the later developmental stages, and the correlated responses in early developmental stages are likely to be comparatively small. For example, the prenatal growth of brain and body weights in various primate species, including *Homo sapiens,* occurs along virtually the same allometric line, with interspecific differences developing mainly in the postnatal stages (Count, 1947; Holt *et al.,* 1975). Some of the most notable exceptions to von Baer's law occur in taxa with minimal parental investment, where early developmental stages are free-living and fully exposed to selection, and hence the young are prone to evolve special adaptations, as in many insects (Darwin, 1859, p. 440; de Beer, 1958, Chapter 6; Cock, 1966, pp. 132–135).

Heritable quantitative variation in ontogeny has been measured within populations in a number of different species (e.g., Kidwell *et al.,* 1952, 1979; Cock, 1966; Atchley and Rutledge, 1980). Cheverud *et al.* (1982) estimated correlations between age-specific characters in a detailed study of variation in postnatal ontogeny of random-bred populations of rats. They found that phenotypic and genetic correlations between the same trait measured at different ages were quite high for ages close together, but decreased markedly with increasing time between measurements. This strongly supports the above explanation of von Baer's law.

Brain–Body Allometry in Mammals and Primates

At various taxonomic levels, mammalian brain and body weights tend to follow allometric curves. Individual ontogenetic trajectories for log brain weight against log body weight typically have a high prenatal slope, exceeding 1.0, which changes to a low postnatal slope, around 0.1 or 0.2 (Count, 1947; Sacher, 1982). The change of slope in the ontogenetic curve corresponds to the cessation of neuronal cell division around the time of birth (Epstein,

1979). Among adults within a mammalian species, the static allometric slope is usually in the range of 0.2–0.4. At progressively higher taxonomic levels the evolutionary allometric slope increases, from about 0.2–0.4 between subspecies or very closely related species to roughly 0.67 for mammalian orders (Bauchot and Stephan, 1964; Jerison, 1973; Gould, 1975; Martin, 1982). This pattern of increasing allometric slope among adult mammals at progressively higher taxonomic levels demands an evolutionary explanation.

To measure the relative variability of phenotypic traits, it is convenient to employ the natural log scale (base *e*) because on this scale of measurement the standard deviation of a character is approximately equal to the coefficient of variation of the untransformed data, $\sigma_{\ln X} \simeq \sigma_X/\bar{X}$, if the latter is less than about 30% (Wright, 1968, p. 229). Heritabilities and genetic and phenotypic correlations, being dimensionless quantities, are nearly invariant under logarithmic transformation.

Coefficients of variation for adult brain weights within populations of small mammals are in the range of about 6–7%, and those for adult body weights range from about 12 to 15% (Yablokov, 1974, pp. 49, 66, 72). In diverse mammalian species, including primates (for which the coefficients of variation are somewhat higher), relative variation in adult brain weight (within sexes for sexually dimorphic species) is roughly half as large as the relative variation in adult body weight, $\sigma_{\ln \text{ brain}}/\ \sigma_{\ln \text{ body}} \simeq 0.4\text{–}0.5$ (Latimer and Sawin, 1955, p. 532, Yablokov, 1974; Holloway, 1980). Breeding experiments with domesticated mammals show estimated heritabilities of body weight ranging from 0.3 in pigs to about 0.35 in mice and sheep, 0.5 in rats, and 0.65 in beef cattle (Falconer, 1960, 1981, Chapter 10; Atchley and Rutledge, 1980). The lower values are probably more representative of natural populations, which, presumably, are exposed to greater environmental variance. Selection on brain weight in a random-bred population of mice by Roderick *et al.* (1976) produced a realized heritability of 0.64 for adults aged 100–150 days. The higher heritability and lower coefficient of variation of body weight in comparison to those for brain weight reflect a component of environmental variation (nutritional level) with a greater effect on body weight than on brain weight.

Roderick *et al.* (1976) also recorded the correlated response in body weight during selection on brain weight, giving an evolutionary allometric slope of 0.77 for selection on brain size alone (Lande, 1979, Fig. 2). From their data and Falconer's (1973) selection experiments on body weight in mice, the genetic correlation between brain and body weights can be estimated as approximately $\gamma = 0.68$ (Lande, 1979). Atchley *et al.* (1983) demonstrated that in rats the average genetic correlation between adult cranial capacity and juvenile or subadult body weight within the sexes is about 0.5, but that involving adult body weight declines, decreasing more in females than in males. A rough conservatism of the genetic variation parameters of brain and body weights in natural populations of nonprimate mammals can be inferred from the approximate constancy of phenotypic variation patterns, since these characters are moderately to highly heritable.

The preceding information on the variation and inheritance of brain and body weights indicates that in most mammalian species (excluding primates) the static genetic regression of log brain weight against log body weight lies in the range of about 0.2–0.4, which is similar to the static phenotypic regression of adult allometry within populations. Because the evolutionary allometric slope between subspecies and closely related species usually falls in this same range, these data support the hypothesis that most closely related mammalian taxa have diverged by selection mainly on body size (and/or by random genetic drift). The progressively higher allometric slopes at higher taxonomic levels implies that the long-term diversification in brain and body weights involved a great deal of selection on brain sizes.

It is plausible that in an environment where temperature and food resources are fluctuating in time and space, short-term selective forces act much more strongly on body size than on brain size. Kurtén (1959, 1960) documented body size changes in several mammalian species that paralleled temperature oscillations during the Pleistocene, and Jerison (1973, pp. 358–360) has shown for La Brea Pleistocene mammals and their closest living relatives, which differ substantially in body size, that relative brain sizes scale with an allometric slope of about 0.2–0.4. The long-term trend toward increasing encephalization in mammals, discussed by Jerison (1973), implies that there is a weaker, but more sustained, selection pressure for increasing brain size. Thus it appears that on longer time scales, approaching tens of millions of years, the fluctuating selection on body size within lineages largely cancels itself out, and the cumulative forces of selection on brain size become increasingly important relative to the net selection on body size.

Within species of insectivores the phenotypic correlation between adult brain and body weight is moderately high, $\rho \simeq 0.7$–0.8 (Bauchot and Stephan, 1964). Similarly high values of the phenotypic correlation apparently apply to most mammalian species, other than primates. This can be deduced from formula (6) for the static allometry coefficient, knowing that the latter is about 0.2–0.4 and that $\sigma_{\ln \text{brain}}/\sigma_{\ln \text{body}} \simeq 0.4$–$0.5$, which puts ρ in the range of 0.4–1.0. The substantial phenotypic correlation between adult brain and body weight within most mammalian species is commensurate with the substantial genetic correlation between these characters, as estimated in mice and rats.

Within species of primates the average phenotypic correlation of adult brain and body weight for the separate sexes is considerably lower than in most mammals. In various macaques, baboons, pongids, and humans ρ ranges from about 0.1 to 0.3 (Sholl, 1948; Jerison, 1973, p. 398; Holloway, 1980). This indicates that during the evolutionary origin of primates much of the increase in brain size occurred in structures, such as the neocortex, that perform higher cognitive functions and integrate parts of the brain but have relatively little direct functional integration with other body parts. Assuming a similarity of genetic and phenotypic correlations, the phenotypic data further suggest that the genetic correlation between adult brain and body weight within primate species is also rather low. In evolutionary terms, this would mean that changes in brain size are only weakly coupled to changes in body size in primates.

Genetic uncoupling of brain and body sizes would have facilitated further encephalization in primates. In the human lineage, Pilbeam and Gould (1974) estimated that the increase of brain and body size was allometric with a slope of about 1.7. Such a steep evolutionary trajectory of increasing relative brain size in a nonprimate mammal population with a strong genetic correlation between brain and body sizes would require not only selection for larger brains, but also comparatively strong selection for smaller bodies to prevent an excessive correlated response from selection on the brain. Selection against increasing body size would then produce a negative correlated response in brain size, thereby hindering encephalization. But if the genetic correlation between brain and body size within populations in the human lineage was as low as suggested by the data on primates, hominids would have been enabled to rapidly increase brain size in response to selection for more complex behavior without the cost of antagonistic selection to prevent the evolution of gigantism.

Acknowledgments

I thank Dr. James M. Cheverud for comments on the manuscript. This work was supported by U.S. Public Health Service grant GM-27120.

References

Atchley, W. R., and Rutledge, J. J. 1980. Genetic components of size and shape. I. Dynamics of components of phenotypic variability and covariability during ontogeny in the laboratory rat. *Evolution* **34:**1161–1173.

Atchley, W. R., Rutledge, J. J., and Cowley, D. E. 1981. Genetic components of size and shape. II. Multivariate covariance patterns in the rat and mouse skull. *Evolution* **35:**1037–1055.

Atchley, W. R., Leamy, L., and Riska, B. 1983. Genetics of brain and body size associations: Data from the rat (unpublished manuscript).

Bailey, D. W. 1956. A comparison of genetic and environmental principal components of morphogenesis in mice. *Growth* **20:**63–74.

Bauchot, R., and Stephan, H. 1964. Le poids encephalique chez les insectivores Malagaches. *Acta Zool.* **45:**63–75.

Cheverud, J. M. 1982. Relationships among ontogenetic, static and evolutionary allometry. *Am. J. Phys. Anthropol.* **59:**139–149.

Cheverud, J. M., Rutledge, J. J., and Atchley, W. R. 1982. Quantitative genetics of development: Genetic correlations among age-specific trait values and the evolution of ontogeny. *Evolution* **37:**895–905.

Cock, A. G. 1966. Genetical aspects of metrical growth and form in animals. *Q. Rev. Biol.* **41:**131–190.

Count, E. W. 1947. Brain and body weight in man: Their antecedents in growth and evolution. *Ann. N.Y. Acad. Sci.* **46:**993–1122.

Darwin, C. 1859. *On the Origin of Species,* Murray, London.

de Beer, G. 1958. *Embryos and Ancestors,* 3rd ed., Oxford University Press, Oxford.

Epstein, H. T. 1979. Correlated brain and intelligence development in humans, in: *Development and Evolution of Brain Size* (M. E. Hahn, C. Jensen, and B. C. Dudek, eds.), pp. 111–131, Academic Press, New York.

Falconer, D. S. 1960. *Introduction to Quantitative Genetics,* Ronald Press, New York.

Falconer, D. S. 1973. Replicated selection for body weight in mice. *Genet. Res.* **22:**291–321.

Falconer, D. S. 1981. *Introduction to Quantitative Genetics,* 2nd ed., Longman, London.

Gould, S. J. 1966. Allometry and size in ontogeny and phylogeny. *Biol. Rev.* **41:**587–640.

Gould, S. J. 1975. Allometry in primates, with emphasis on scaling and the evolution of the brain, in: *Approaches to Primate Paleobiology* (Contrib. Primat., Vol. 5, F. Szalay, ed.), pp. 244–292, S. Karger, Basel.

Holloway, R. L. 1980. Within-species brain–body weight variability: A reexamination of the Danish data and other primate species. *Am. J. Phys. Anthropol.* **53:**109–121.

Holt, A. B., Cheek, D. B., Mellits, E. D., and Hill, D. E. 1975. Brain size and the relation of the primate to the nonprimate, in: *Fetal and Postnatal Cellular Growth* (D. B. Cheek, ed.), pp. 23–44, Wiley, New York.

Huxley, J. S. 1924. Constant differential growth ratios and their significance. *Nature* **114:**895–896.

Huxley, J. S. 1932. *Problems of Relative Growth,* Methuen, London (reprinted Dover, New York, 1972).

Jerison, H. J. 1973. *Evolution of the Brain and Intelligence.* Academic Press, New York.

Jolicoeur, P. 1963. The multivariate generalization of the allometry equation. *Biometrics* **19:**497–499.

Jungers, W. L., and German, R. Z. 1981. Ontogenetic and interspecific skeletal allometry in nonhuman primates: Bivariate versus multivariate analysis. *Am. J. Phys. Anthropol.* **55:**195–202.

Kermack, K. A., and Haldane, J. B. S. 1950. Organic correlation and allometry. *Biometrika* **37:**30–41.

Kidwell, J. F., Gregory, P. W., and Guilbert, H. R. 1952. A genetic investigation of allometric growth in Hereford cattle. *Genetics* **37:**158–174.

Kidwell, J. F., Herbert, J. G., and Chase, H. B. 1979. The inheritance of growth and form in the mouse. V. Allometric growth. *Growth* **43:**47–57.

Kurtén, B. 1959. Rates of evolution in fossil mammals. *Cold Spring Symp. Quant. Biol.* **24:**205–215.

Kurtén, B. 1960. Chronology and faunal evolution of the earlier European glaciations. *Comment. Biol. Soc. Sci. Fenn.* **21:**1–62.

Lande, R. 1979. Quantitative genetic analysis of multivariate evolution, applied to brain:body size allometry. *Evolution* **33:**402–416.

Lande, R. 1981. The minimum number of genes contributing to quantitative variation between and within populations. *Genetics* 99:541–553.

Lande, R. 1982. A quantitative genetic theory of life history evolution. *Ecology* **63:**607–615.

Latimer, H. B., and Sawin, P. B. 1955. The weight of the brain, of its parts and the weight and length of the spinal cord in the rabbit (Race X). *J. Comp. Neurol.* **103:**513–539.

Leamy, L. 1977. Genetic and environmental correlations of morphometric traits in random-bred mice. *Evolution* **31:**357–369.

Martin, R. D. 1982. Allometric approaches to the evolution of the primate nervous system, in: *Primate Brain Evolution: Methods and Concepts* (E. Armstrong and D. Falk, eds.), pp. 39–56, Plenum Press, New York.

Pilbeam, D., and Gould, S. J. 1975. Size and scaling in human evolution. *Science* **186:**892–901.

Reeve, E. C. R. 1950. Genetical aspects of size allometry. *Proc. R. Soc. Lond. B* **137:**515–518.

Reeve, E. C. R., and Huxley, J. S. 1945. Some problems in the study of allometric growth, in: *Essays on "Growth and Form" Presented to D'Arcy Wentworth Thompson* (W. E. le Gros Clark and P. B. Medawar, eds.), pp. 121–156, Clarendon Press, Oxford.

Rensch, B. 1959. *Evolution above the Species Level,* Columbia University Press, New York.

Roderick, T. H., Wimer, R. E., and Wimer, C. C. 1976. Genetic manipulation of neuroanatomical traits, in: *Knowing, Thinking and Believing* (L. Petrinovich and J. McGaugh, eds.), pp. 143–178, Plenum Press, New York.

Sacher, G. A. 1982. The role of brain maturation in the evolution of the primates, in: *Primate Brain Evolution: Methods and Concepts* (E. Armstrong and D. Falk, eds.), pp. 97–112, Plenum Press, New York.

Sholl, D. 1948. The quantitative investigation of the vertebrate brain and the applicability of allometric formulae to its study. *Proc. R. Soc. Lond. B* **135**:243–258.

Wright, S. 1968. *Evolution and the Genetics of Populations*, Vol. 1, *Genetic and Biometric Foundations*, University of Chicago Press, Chicago.

Yablokov, A. V. 1974. *Variability of Mammals*, Amerind, New York.

Sexual Dimorphism in Primates

3

The Effects of Size

WALTER LEUTENEGGER AND
JAMES M. CHEVERUD

Introduction

Secondary sexual differentiation in sexually reproducing organisms leads to more or less pronounced sexual dimorphism in an array of interrelated morphological, physiological, and behavioral features. Among primates morphological sexual differences occur in a wide range of characteristics, including (1) size (body weight and linear body dimensions such as trunk, head, and tail length), (2) dentition (such as canine size), (3) cranial features (such as prognathism, or sagittal and nuchal cresting), (4) locomotor apparatus (dimensions of axial and appendicular skeleton, muscular development), (5) internal organs (such as brain and heart size), (6) external features (such as pelage color and markings, shoulder capes and manes, permanent skin ridges, and coloration, particularly on the face), and (7) maturational, seasonal, or periodic morphological changes associated with reproductive cycles (Leutenegger, 1982*a*).

Sexual dimorphism not only occurs in a wide range of features but also shows considerable variation between species. The extent of body weight

---

WALTER LEUTENEGGER • Department of Anthropology, University of Wisconsin, Madison, Wisconsin 53706. JAMES M. CHEVERUD • Departments of Anthropology, Cell Biology and Anatomy, and Ecology and Evolutionary Biology, Northwestern University, Evanston, Illinois 60201.

dimorphism, for example, ranges from species in which males on the average are slightly more than twice as heavy as females to those in which females are slightly heavier than males (Clutton-Brock and Harvey, 1977; Leutenegger and Kelly, 1977). Similarly, canine size dimorphism ranges from species in which males have markedly larger canines than females to those in which canine size is virtually the same in the two sexes (Swindler, 1976; Leutenegger and Kelly, 1977; Harvey *et al.,* 1978*a,b*).

Two general theories offer causal explanations of such variations in sexual dimorphism. The more traditional one explains sexual dimorphism in terms of sexual selection (Darwin, 1871). The argument is based on the competition among members of one sex for those of the other and the different selective pressures affecting the competitor and the object of competition. The second principal explanation of sexual dimorphism is that differences between males and females are a means of reducing intraspecific competition for resources, especially food (Selander, 1972). While these concepts, either alone or in combination, explain sexual dimorphism to varying extents, it is also evident that in primates (Clutton-Brock and Harvey, 1977), as in other mammalian taxa (Ralls, 1977), a considerable portion of the variance in sexual dimorphism is not accounted for. In order to account more fully for the variance, additional variables, primarily ecological factors, such as habitat, diet, distribution and abundance of food resources, predator defense or avoidance, and positional behavior, have been invoked (Crook and Gartlan, 1966; Crook, 1972; Eisenberg *et al.,* 1972; Clutton-Brock and Harvey, 1977, 1978; Clutton-Brock *et al.,* 1977; Leutenegger and Kelly, 1977; Ralls, 1977; Leutenegger, 1978, 1982*a,b;* Alexander *et al.,* 1979; Leutenegger and Cheverud, 1982). Factors of evolutionary heritage, in the sense of "phylogenetic legacy" (Raup, 1972) or "phylogenetic inertia" (Wilson, 1975), may also affect sexual dimorphism (Clutton-Brock and Harvey, 1977; Harvey *et al.,* 1978*a;* Leutenegger and Cheverud, 1982).

A factor that has received only sporadic attention, but clearly should be considered in any analysis of sexual dimorphism, is body size. Rensch (1950, 1954) showed, 30 years ago, that sexual dimorphism in appendicular structures tends to increase with increasing body size in various arthropod and avian taxa. More recently, Ralls (1977) observed that extreme body size dimorphism evolves much more frequently in large species of mammals than in small ones. In preliminary reports both Clutton-Brock *et al.* (1977) and Leutenegger (1978) demonstrated positive allometry for body weight dimorphism in primates. While evidence for a positive relationship between body size dimorphism and body size is accumulating for various invertebrate (insects and spiders), reptilian (lizards), avian (pheasants, birds of paradise, grouse, and quails), and mammalian taxa (primates, pinnipeds, ungulates, rodents, and carnivores), its adaptive significance is not well understood and different interpretations have been offered (Clutton-Brock *et al.,* 1977; Leutenegger, 1978; Alexander *et al.,* 1979; Maiorana, 1984). Similarly, a positive relationship between canine size dimorphism and body size in both extant and extinct primates has elicited different arguments and interpretations (Fleagle *et al.*, 1980; Leutenegger, 1981, 1982*b;* Gingerich, 1981). Most recently, Leutenegger and Cheverud (1982) have attempted a quantification of correlates of sexual dimorphism in primates by testing for degree and significance of association between a series of ecological and size variables and sexual dimorphism in body weight and canine size. We found that variation in body weight dimorphism could be almost entirely attributed to body weight, while about half of the variation in canine size dimorphism could be explained in terms of canine size. The finding of the unexpectedly high contribution of size to sexual dimorphism led us to cast doubts on prevailing theories on the evolution of sexual dimorphism, and we tentatively suggested an alternative hypothesis.

In this chapter we reanalyze the effects of size together with those of ecological variables on sexual dimorphism in body weight and canine size in primates. Based on the findings, we critically examine previous theories on the evolution of sexual dimorphism and offer a detailed new selection model formulated and tested using principles of quantitative genetic theory.

Materials and Methods

The independent variables chosen are those generally recognized as potential contributors to sexual dimorphism and include mating system (as an indicator of intensity of sexual selection), activity rhythm, habitat, diet, taxon (as a measure of phyletic inertia), and size (as reflected by body weight and canine size). Data on these variables were collected for 70 nonhuman primate species representing all extant families, except Daubentoniidae and Tarsiidae. Ecological data are from the literature. Body weight data are those of adult males and females. Each species is represented by male and female means of at least 16 individuals, with approximately equal numbers of each sex. Although the data are taken from various sources and therefore not perfectly comparable, a major source of error usually found in body weight studies has been eliminated by disregarding the data from captive animals and using data only from wild-trapped and wild-shot animals. All the body weight data were extracted from Leutenegger (1973, 1978), Rothenfluh (1976), and Leutenegger and Kelly (1977). Canine size is represented by male and female means of the mesiodistal diameter of the maxillary permanent canine. The maxillary canine rather than the mandibular canine was selected because of its larger size and greater importance in behavioral displays. All the canine dimensions were extracted from Leutenegger (1976), Swindler (1976), and Leutenegger and Kelly (1977).

The independent variables are defined as follows. Mating system is identified as either polygynous or monogamous. This dichotomous classification of mating system is used instead of an interval scale measure, such as socionomic sex ratio (Clutton-Brock and Harvey, 1977), because interval level measures are often difficult to measure accurately due to their wide variability within species, and because of their often uncertain relationship to the intensity of intermale competition. Activity rhythm refers to the time of day in which the animal is most active, scored as either nocturnal or diurnal. Habitat refers to the milieu in which activity typically takes place, either arboreal or terrestrial. Diet refers to the predominant source of nutrition, each species being scored as frugivorous or folivorous. Insectivorous species were ultimately excluded from our sample because adequate data were available for only two such species. Weight refers to the $\log_{10}$ of mean female body weight, and canine size to the $\log_{10}$ of mean female canine size. Taxon is defined as the general taxonomic category to which a species belongs. Due to large differences in the number of adequately studied species in different taxonomic groups, the taxonomic categories are not all at the same level of the Linnaean hierarchy. The categories are: (1) Strepsirhini, (2) Callitrichidae, (3) Cebidae, (4) Cercopithecidae, (5) Hylobatidae, (6) Pongidae.

Initially, correlations were calculated among the independent variables described above in order to elucidate their interrelationships. Phi (ϕ) values were calculated between dichotomous variables, Cramer's V between dichotomous and polychotomous variables, eta (η) values between dichotomous and continuous variables, and Pearson product moment correlations among continuous variables (Blalock, 1972).

Variation in sexual dimorphism in weight and canine size was analyzed using analysis of variance and regression and the SPSS subprograms ANOVA and SCATTERGRAM. Weight dimorphism is defined as the $\log_{10}$ of the difference between male and female weights. This is a measure of absolute sexual dimorphism, in contrast to the more commonly used ratio of male to female weights, which attempts to measure sexual dimorphism relative to weight. However, since the effects of weight are removed from dimorphism by taking a ratio only when the regression slope of male with female weight is one (Huxley, 1932), this measure is inappropriate for the study of weight's effects on weight dimorphism. Also, ratios of normally distributed variables are typically skewed to the right and leptokurtotic, making their use in regression and analysis of variance questionable (Atchley *et al.*, 1976; Anderson and Lydic, 1977). The use of the difference between male and female means as a measure of dimorphism also allows an easier integration of between-species analyses with evolutionary theory and laboratory experiments in the evolution of sex dimorphism (Bird and Schaffer, 1972; Eisen and Hanrahan, 1972). Since the effects of weight on weight dimorphism interpreted in the light of evolutionary theory are an object of study in this report, we will define weight dimorphism as the $\log_{10}$ of the difference between male and female weights rather than their ratio. Comparisons between this study and previous work in which dimorphism is defined as a ratio should concentrate on our analysis of dimorphism after the effects of weight are controlled for, keeping in mind that ratios imperfectly control for magnitude.

Only species in which males were larger than females were analyzed, because when male weight minus female weight is zero or negative, a logarithm cannot be taken. The number of species for which body weight data are available and in which males are heavier than females is 44. Canine dimorphism is the $\log_{10}$ of the difference between male and female canine sizes. Only species in which male canines are larger than those of females were retained in the analysis, leaving a total of 38 species. Finally, correlations were calculated between weight dimorphism and canine size dimorphism.

Correlates of Sexual Dimorphism: Ecological and Size Variables

Association among Independent Variables

Correlations among the dichotomous and polychotomous traits indicate that taxon is significantly associated with mating system ($V = 0.953$), activity rhythm ($V = 0.757$), and habitat ($V = 0.531$), although none of these last three traits are associated with each other. The size of our sample, and perhaps of the entire order Primates, will not allow discrimination between the effects of highly correlated causes on sexual dimorphism. The effects of taxonomic assignment, and thus of phylogenetic inertia, therefore cannot be analyzed. Effects that we attribute to mating system, activity rhythm, and habitat may be confounded with phylogenetic inertia. The possibility among primates of a strong phylogenetic effect on ecological variables and sexual dimorphism has also been suggested by Clutton-Brock and Harvey (1977). The present study analyzes variability at the ordinal level. Patterns of variability within lower-level taxa may be different and will be the subject of future studies. The only other significant correlation among noncontinuous variables is between diet and habitat ($\phi = 0.298$), terrestrial species tending to be frugivorous relatively more often than arboreal species. However, this correlation is not very strong.

Correlations (η values) between the dichotomous and continuous independent variables are presented in Table 1. Weight is significantly associated with activity rhythm (diurnal species being larger than nocturnal ones) and habitat (terrestrial species being larger than arboreal ones). Canine size is significantly associated with mating system (polygynous species having larger canines) and habitat (terrestrial species having larger canines). Weight and canine size are very highly correlated across species ($r = 0.954$, $N = 24$).

Variation in Weight Dimorphism

As a first step, the independent variables were separately tested for association with weight dimorphism (Table 2, row 1) using analysis of variance for the dichotomous variables and regression for weight. Weight is very highly correlated ($r = 0.916$) with weight dimorphism, with a regression slope of 1.395 and an intercept of −2.01. The slope is significantly different from 1.00

Table 1. Correlations between Ecological and Size Variables[a]

	Mating system	Activity rhythm	Habitat	Diet	Flogwt	Flogcan
Mating system	1.000	0.008	0.267	0.214	0.200	0.320
Activity rhythm		1.000	0.180	0.187	0.520	0.430
Habitat			1.000	0.298	0.473	0.463
Diet				1.000	0.200	0.234
Flogwt					1.000	0.954
Flogcan						1.000

[a] ϕ values are between dichotomous traits, η values between dichotomous and continuous traits, and Pearson product moment correlations between continuous traits. The underlined values are significantly different from zero at the 0.05 level. Flogwt = $\log_{10}$ female weight and Flogcan = $\log_{10}$ female canine size.

at the 5% level and the intercept is significantly different from zero at the 0.1% level. This indicates positive allometry for weight dimorphism with respect to weight. Weight dimorphism is also significantly associated with mating system (polygynous species being more dimorphic than monogamous ones), activity rhythm (diurnal species being more dimorphic than nocturnal ones), and habitat (terrestrial species being more dimorphic than arboreal ones).

In the second phase of analysis, weight was regressed out of weight dimorphism and the residuals were tested for correlation with the rest of the independent variables as measured by standardized partial regression coefficients (Table 2, row 2). With all variation in weight dimorphism related to weight removed, weight dimorphism is significantly associated with mating system, habitat, and diet, frugivores being more dimorphic than folivores.

Table 2. Analysis of Variation for Body Weight Dimorphism[a]

	Flogwt	Mating system	Activity rhythm	Habitat	Diet	R^2
WD (independently)	0.916	0.42	0.55	0.62	0.01	—
WD (controlled for weight)		0.26	0.09	0.22	0.18	—
WD (controlled for other factors, not weight)		0.20	0.44	0.53	0.14	0.653
WD (controlled for all factors and weight)		0.26	0.07	0.08	0.16	0.916

[a] All values for factors are standardized partial regression coefficients; underlined values are significantly different from zero at the 0.05 level. R^2 values are proportion of variation among primates in weight dimorphism accounted for by the factors included. None of the interaction effects was significant. WD, Weight dimorphism = $\log_{10}$ (male weight minus female weight). Only species with males heavier than females are included ($N = 44$).

When weight is not regressed out and a four-way analysis of variance is performed, weight dimorphism is significantly associated only with activity rhythm and habitat (Table 2, row 3). This four-way analysis accounts for about 65% of the variance in weight dimorphism among species. None of the interaction effects tested were significant. When weight is regressed out and the residuals of weight dimorphism are submitted to a four-way analysis of variance (Table 2, row 4), weight dimorphism is significantly associated only with mating system and diet—the opposite of the results of the previous analysis, which did not take body weight into account. About 92% of the variance in weight dimorphism is explained by this analysis, which includes body weight. None of the interaction effects tested are significant.

The analyses in which weight was regressed out assume a theoretical primacy for the effects of body weight on weight dimorphism. Since weight is correlated with activity rhythm and habitat, we cannot statistically separate the effects of weight and these two variables on sexual dimorphism. In the analysis, we have taken the theoretical position that body weight has a direct effect on sexual dimorphism, while habitat and activity rhythm affect weight dimorphism indirectly through their effects on weight. However, it should be pointed out that direct effects of habitat and activity rhythm, independent of body weight, are allowed for in the model.

Variation in weight dimorphism among primates can be almost entirely attributed to weight (83% variance of weight dimorphism). Some small amount of variance can also be attributed to mating system (6.8%) and diet (2.5%). Activity rhythm and habitat both showed significant bivariate associations with weight dimorphism, but a large part of these associations can be attributed to the observation that terrestrial and/or diurnal primates are larger than their arboreal and/or nocturnal relatives. Thus, habitat and activity rhythm have an indirect effect on weight dimorphism.

Variation in Canine Size Dimorphism

As in the analysis of weight dimorphism, the relationships between the independent variables and canine size dimorphism were separately analyzed (Table 3, row 1). Canine size was fairly highly correlated with canine size dimorphism ($r = 0.705$) with a regression slope of 1.821, which is significantly different from 1.0 at the 0.1% level. This indicates strong positive allometry for canine size dimorphism with respect to canine size. Canine size dimorphism is also significantly associated with mating system (polygynous species showing more dimorphism), activity rhythm (diurnal species being more dimorphic), and habitat (terrestrial species being more dimorphic).

Next, canine size was regressed out of canine size dimorphism and the residuals were tested for association with the other factors with standardized partial regression coefficients (Table 3, row 2). With all variation in canine size removed, canine size dimorphism is significantly associated only with mating

Table 3. Analysis of Variation for Canine Size Dimorphism[a]

	Flogcan	Mating system	Activity rhythm	Habitat	Diet	R^2
CD (independently)	0.705	0.51	0.74	0.50	0.08	—
CD (controlled for size)		0.42	0.52	0.22	0.06	—
CD (controlled for other factors, not size)		0.29	0.58	0.33	0.01	0.736
CD (controlled for all factors, and size)		0.31	0.44	0.17	0.08	0.800

[a] All values for factors are standardized partial regression coefficients; underlined values are significantly different from zero at the 0.05 level. R^2 values are proportion of variation among primates in canine size dimorphism accounted for by factors included. None of the interaction effects was significant. CD, Canine dimorphism = $\log_{10}$ (male canine size minus female canine size). Only species with males larger than females are included (N = 38).

system and activity rhythm. When canine size is not regressed out and a four-way analysis of variance is performed, canine size dimorphism is again significantly associated with mating system, activity rhythm, and habitat (Table 3, row 3). This four-way analysis of variance accounts for 74% of the variation in canine size dimorphism. None of the interaction effects tested are significant. When canine size is regressed out and a four-way analysis of variance is performed on the residuals, canine size dimorphism is significantly associated with mating system and activity rhythm (Table 3, row 4). About 80% of the variance in canine size dimorphism is explained by this analysis, which included canine size. None of the interaction effects tested are significant.

Variation in canine size dimorphism among primates can be largely explained in terms of canine size (49% of variance in canine size dimorphism), mating system (10%), and activity rhythm (20%). Terrestrial species are more dimorphic in canine size than arboreal ones, but this relationship can be explained by the observation that terrestrial animals have larger canines than arboreal ones (Table 2) and species with larger canine size are more dimorphic (Table 3).

Finally, canine size dimorphism is fairly highly correlated with weight dimorphism (r = 0.857, N = 24) (Table 4). The regression slope of 0.459 is significantly different from one-third, the isometric slope of a linear/cubic relationship, at the 5% level. The intersept of 0.776 is significantly different from zero at the 0.1% level. This indicates positive allometry for canine size dimorphism with respect to weight dimorphism. Our observations show that both canine size dimorphism and weight dimorphism are strongly affected by their associated measures of size, canine size and weight, which are in turn strongly associated with each other (r = 0.954). So canine size dimorphism

Table 4. Correlations between Body Weight Dimorphism, Canine Size Dimorphism, and Associated Size Measures[a]

	Flogwt	Flogcan	WD	CD
Flogwt	1.000	0.954	0.916	0.615
Flogcan		1.000	0.863	0.705
WD			1.000	0.857
CD				1.000

[a]Correlations represent Pearson product moments. The underlined values are significantly different from zero at the 0.05 level.

and weight dimorphism are closely related through causal factors related to overall size.

Partial Contribution of Individual Variables

A major portion of the variance in sexual dimorphism in both body weight and canine size can be accounted for by a series of ecological and size variables. The partial contribution of individual variables is estimated as follows.

Mating System. In accordance with Darwinian expectations, sexual dimorphism is found to be significantly associated with differences in mating competition between the sexes; i.e., polygynous species are more dimorphic than monogamous ones. Quite surprisingly, however, only minor amounts of the variance in weight dimorphism and canine size dimorphism can be attributed to differences in mating system. This indicates that the intensity of sexual selection can no longer be viewed as the major contributor to the degree of sexual dimorphism.

Diet. An even smaller, but still significant amount of variance in weight dimorphism can be attributed to dietary differences; i.e., frugivorous species are more dimorphic than folivorous ones. This is in accordance with the hypotheses that energetic constraints—which presumably limit the increase in male size (Clutton-Brock and Harvey, 1977)—may be more restricting in some dietary groups than in others (Leutenegger and Kelly, 1977). Among nonhuman primates, canines are basically freed from dietary functions and are used principally in a social context (e.g., displays and threats in intermale competition or predator defense). The prediction that dietary differences should have no effect on canine size dimorphism is borne out by our results.

Activity Rhythm. Significant bivariate associations exist between activity rhythm and sexual dimorphism (in both body weight and canine size), diurnal species being more dimorphic than nocturnal ones. However, when size is removed, activity rhythm contributes significantly to canine size dimorphism only, while it does not play a role in weight dimorphism. We interpret the effect of differences in activity rhythm on canine size dimorphism in terms of visibility. Among nocturnal species, canine displays and threats are not so readily perceived as they are in diurnal species. Furthermore, nocturnal species rely on concealment in avoiding predators rather than on defense aided by displays.

Habitat. While significant bivariate associations exist between habitat differences and sexual dimorphism in body weight and canine size, these associations disappear when the associated size measure is removed. This refutes the thesis that terrestrial species are more dimorphic than their arboreal counterparts because of habitat differences and related differences in predator pressures (Clutton-Brock and Harvey, 1977; Harvey *et al.*, 1978*a,b*). It simply is larger size *per se* that results in increased sexual dimorphism.

Size. While the influence of size on sexual dimorphism has been recognized in general terms (Ralls, 1977), the precise effect of this variable has not been well understood. Here we demonstrate that size is not just one of numerous factors that affect sexual dimorphism; it is the major one. This indicates that sexual selection no longer can be considered the major cause of variability in sexual dimorphism as generally believed (Ralls, 1977). We further demonstrate that both body weight dimorphism and canine size dimorphism are positively allometric with respect to their associated size measures, body weight and canine size, which are in turn strongly correlated, thus confirming results of earlier studies (Clutton-Brock *et al.*, 1977; Leutenegger, 1978, 1982*b;* Leutenegger and Cheverud, 1982).

Allometry and Evolution of Sexual Dimorphism

Evolution of Sexual Dimorphism by Selection Dimorphism

The recognition of size as producing a major effect on sexual dimorphism in addition to the phenomenon of positive allometry in sexual dimorphism requires a reassessment of theories for the evolution of sexual dimorphism. Sexual dimorphism can only evolve if there is dimorphism in selection and/or dimorphism in genetic variances. Most of the workers considering the evolution of sexual dimorphism have attempted to explain it by means of intrasexual competition for mates or sexual selection, intersexual choice, lopsided reproductive selection (Maiorana, 1984) and resource diversification (Selander, 1972). All of these models attempt to explain sexual dimorphism in selection intensities that will generate sexual dimorphism for the phenotype in question. In contrast to the extensive attention paid to the role of dimorphism in selection intensities, relatively little attention has been directed toward the causes and effects of variance dimorphism on the evolution of sexual dimorphism (Leutenegger and Cheverud, 1982).

In this section we will present a general, quantitative genetic model for the evolution of sexual dimorphism (Lande, 1980) and then compare and constrast the effects of selection and variance dimorphisms on the evolution of character dimorphism and on size and scaling. The response of males R_M and females R_F to selection on any single phenotype are given by

$$R_M = (1/2)(h_M^2 S_{PM} i_M + h_M h_F r_a i_F S_{PM}) \tag{1}$$

and

$$R_F = (1/2)(h_F^2 S_{PF} i_F + h_M h_F r_a i_M S_{PF}) \tag{2}$$

where h^2 is the heritability, h is the square root of heritability, S is the standard deviation, i is the selection intensity (Falconer, 1960), and r_a is the genetic correlation between the sexes (Lande, 1980). The subscript M indicates male, F indicates female, and P indicates phenotypic standard deviation.

The response of sexual dimorphism R_{SD}, defined as the male mean minus the female mean, to selection is given by $R_M - R_F$,

$$R_{SD} = (1/2)[h_M^2 S_{PM} i_M - h_F^2 S_{PF} i_F + h_M h_F r_a (i_F S_{PM} - i_M S_{PF})] \tag{3}$$

Note that when there is no dimorphism in the variance terms h^2, h, or S or in the selection intensities i, R_{SD} equals zero. The evolutionary allometry of dimorphism K_{SD} with female size is given by R_{SD}/R_F,

$$K_{SD} = \frac{S_{AM}\ (h_M i_M + h_F r_a i_F)}{S_{AF}\ (h_F i_F + h_M r_a i_M)} - 1 \tag{4}$$

Subscript A indicates additive genetic standard deviations. When K_{SD} is greater than one, sexual dimorphism is positively allometric relative to female size.

First we will consider the traditional case of dimorphism in selection intensities ($i_M \neq i_F$) as a source of selection for character dimorphism. Since no variance dimorphism is posited in these models, none will be assumed here. Therefore, we set $S_{PM} = S_{PF}$ and $h_M^2 = h_F^2$. Under these conditions

$$R_M = (1/2)h^2 S_P (i_M + r_a i_F) \tag{5}$$

and

$$R_F = (1/2)h^2 S_P (i_F + r_a i_M) \tag{6}$$

The response of dimorphism to this type of selection is

$$R_{SD} = (1/2)h^2 S_P (i_M - i_F)(1 - r_a) \tag{7}$$

So sexual dimorphism increases whenever selection is stronger in males than it is in females, the traditional, obvious result. However, it is also important to note that the genetic correlation also affects R_{SD} and that the higher the correlation, the smaller the increase in dimorphism. For this type of evolution to produce positive allometry, as found in the analysis in the section, Correlates of Sexual Dimorphism,

$$i_M + i_F r_a > 2(i_F + i_M r_a) \tag{8}$$

In other words, the response to selection in males must be twice the response in females. With low genetic correlation of only about 0.5 and a selection intensity of one in females, the selection intensity in males would have to be 10, an unreasonably high value for this kind of selection to produce positive evolutionary allometry.

As a special case of the selection dimorphism model, we present the situation where $i_M \neq 0$ and $i_F = 0$ for consideration. This scenario fits many evolutionary arguments where selection, such as sexual selection or predator defense selection (DeVore and Washburn, 1963), acts primarily on the male while the same trait is neutral in the female. For this special case,

$$R_{SD} = (½)h^2 S_P i_M (1 - r_a) \tag{9}$$

and the evolutionary allometry is

$$K_{SD} = (1 - r_a)/r_a \tag{10}$$

In this situation positive allometry will only be produced if the genetic correlation between sexes is below 0.5. However, genetic correlations between homologous male and female characters are often quite high ($r_a > 0.9$) (Lande, 1980). High genetic correlation between sexes severely limits the evolution of sexual dimorphism by sexual selection and makes it impossible for this type of selection to produce positive evolutionary allometry between dimorphism and female size. Therefore, it is not a likely cause of the pattern of scaling of sexual dimorphism with size in primates noted above.

A second special case of selection dimorphism is resource divergence selection, where selection acts in opposite directions in males and females, or $i_M = -i_F$. Under these conditions

$$R_{SD} = h^2 S_P i_M (1 - r_a) \tag{11}$$

This value is twice that given in Eq. (9) because selection is on both sexes instead of just one. In principle, if $i_M - i_F$ is the same for these two special cases, they will produce the same evolutionary change in the level of sexual dimorphism. However, the divergent selection case selects for smaller females as dimorphism increases and thus leads to a negative correlation between dimorphism and female size, in contradiction to the relationship between size and dimorphism presented in the section, *Correlates of Sexual Dimorphism*.

The results of the models presented above for the evolution of sexual dimorphism by dimorphism in selection intensities indicate that this mode of selection is not very potent in producing sexual dimorphism when the genetic correlation between sexes is high (Lande, 1980) as is commonly found in empirical studies (Harrison, 1953; Korkman, 1957; Eisen and Legates, 1966; Eisen and Hanrahan, 1972; Frankham, 1968). Despite these theoretical and empirical results, most evolutionary speculation concerning the evolution of dimorphism consists of hypothesizing differences in the intensity of selection between the two sexes as the sole evolutionary source of sexual dimorphism. Also, it appears that selection dimorphism cannot easily produce the positive allometry between sexual dimorphism and female size found in the analysis presented in the section, *Correlates of Sexual Dimorphism* (Leutenegger and Cheverud, 1982). Therefore, we feel that dimorphism in selection intensities is not sufficient to explain the evolution of character dimorphism in primates.

Evolution of Sexual Dimorphism by Variance Dimorphism

Sexual dimorphism can also evolve through variance dimorphism (Leutenegger and Cheverud, 1982). This model assumes no dimorphism in selection intensities ($i_{\mathrm{M}} = i_{\mathrm{F}}$) but allows dimorphism in heritability and standard deviations. The response of males to selection under this model is

$$R_{\mathrm{M}} = (\tfrac{1}{2})i(h^2_{\mathrm{M}}S_{\mathrm{PM}} + h_{\mathrm{M}}h_{\mathrm{F}}r_a\mathrm{S}_{\mathrm{PM}}) \tag{12}$$

while the response of females is

$$R_{\mathrm{F}} = (\tfrac{1}{2})i(h^2_{\mathrm{F}}S_{\mathrm{PF}} + h_{\mathrm{M}}h_{\mathrm{F}}r_aS_{\mathrm{PF}}) \tag{13}$$

The evolution of character dimorphism R_{SD} is given by

$$R_{\mathrm{SD}} = (\tfrac{1}{2})i[h^2_{\mathrm{M}}\, S_{\mathrm{PM}} - h^2_{\mathrm{F}}S_{\mathrm{PF}} + h_{\mathrm{M}}h_{\mathrm{F}}r_a(S_{\mathrm{PM}} - S_{\mathrm{PF}})] \tag{14}$$

as in Leutenegger and Cheverud (1982). The evolutionary allometry K_{SD} of dimorphism is given by

$$K_{\mathrm{SD}} = \frac{S_{AM}\ (h_M + h_F r_a)}{S_{AF}\ (h_F + h_M r_a)} \quad 1 \tag{15}$$

As the genetic correlation r_a approaches one, dimorphism of heritabilities becomes relatively unimportant, so assuming $h^2_{\mathrm{M}} = h^2_{\mathrm{F}}$,

$$R_{\mathrm{SD}} = (\tfrac{1}{2})ih^2(S_{\mathrm{PM}} - S_{\mathrm{PF}})(1 + r_a) \tag{16}$$

and

$$K_{\mathrm{SD}} = (S_{\mathrm{PM}}/S_{\mathrm{PF}}) - 1 = (S_{\mathrm{AM}}/S_{\mathrm{AF}}) - 1 \tag{17}$$

From these simplified equations it is easily seen that the rate of increase in sexual dimorphism is directly dependent on the magnitude of variance dimorphism expressed in standard deviations $S_{PM} - S_{PF}$. Also, the higher the genetic correlation, the greater the increase in dimorphism, in direct contrast to the results derived for selection dimorphism [see Eqs. (7), (9), and (11)]. So the empirical finding of high genetic correlation between the sexes (Lande, 1980) indicates that variance dimorphism may be a very potent source of character dimorphism. The condition for the production of positive allometry between dimorphism and size is that the ratio of the standard deviations S_{PM}/S_{PF} or S_{AM}/S_{AF} be greater than two. This ratio is high relative to the empirical findings for primates (Leutenegger and Cheverud, 1982), where an average ratio of phenotype standard deviation is 1.5, but not extreme. Therefore, it appears as if variance dimorphism can lead to positive allometry much more easily than can selection dimorphism.

Variance dimorphism may also be a more potent producer of character dimorphism than selection dimorphism under natural conditions. Comparing the sexual selection model [$i_M \neq 0$, $i_F = 0$; $S_P = (½)(S_{PM} + S_{PF})$] and the variance dimorphism model ($i_M = 1$, $i_F = 1$) using the same overall selection intensity, it is apparent that when the genetic correlation between sexes is high, variance dimorphism is more potent than selection dimorphism. The condition under which variance dimorphism is the more potent is

$$2h_M h_F r_a/(h_M^2 + h_F^2) > S_{PF}/S_{PM} \tag{18}$$

or when heritabilities are assumed equal,

$$r_a > S_{PF}/S_{PM} \tag{19}$$

With a genetic correlation of 0.9, variance dimorphism is more potent whenever the ratio of standard deviations (female/male) is less than 0.9. The average ratio in primates is about 0.67. The same general results hold when

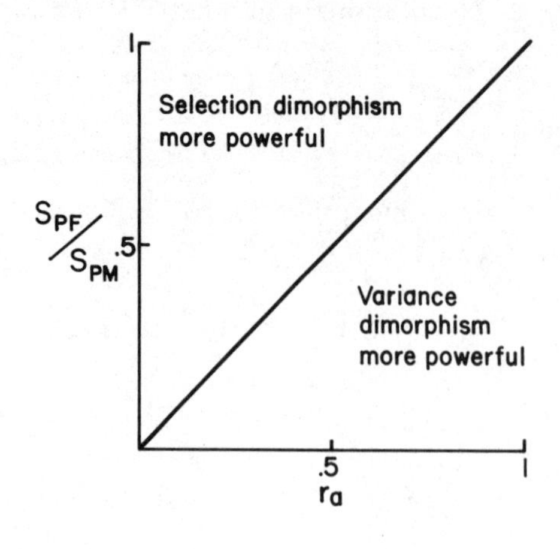

Fig. 1. Parameter values for which variance dimorphism is more powerful than selection dimorphism in the production of character dimorphism. S_{PF}, female phenotypic standard deviation; S_{PM}, male phenotypic standard deviation; r_a, genetic correlation between the sexes. For further explanation see text.

divergent selection is considered. Figure 1 displays the regions in which variance dimorphism and selection dimorphism are favored.

The models presented above indicate that variance dimorphism is a potent source for the evolution of character dimorphism that has been largely overlooked in evolutionary model-building. For example, it is well known that terrestrial primates have a larger body size and are more dimorphic than arboreal primates (section on *Correlation of Sexual Dimorphism*) (Clutton-Brock and Harvey, 1977; Leutenegger and Cheverud, 1982). If there was variance dimorphism in the ancestral species and selection occurred for larger body size after the arboreal–terrestrial transition (for such reasons as increased day range length or defense against predators), sexual dimorphism would also evolve, perhaps at a high rate. Thus selection for larger body size would, as an "unintended" result, also increase dimorphism. This may be the indirect mechanism through which terrestrial species came to have greater sexual dimorphism *and* larger body size than arboreal species.

Variance dimorphism also produces positive allometry between size and character dimorphism more easily than selection dimorphism. However, a combination of the sexual selection ($i_M = 2$, $i_F = 0$) and variance dimorphism ($S_{PM} = 2S_{PF}$) may be even more efficient in producing positive allometry. For example, in a total or combined selection model [see Eqs. (2)–(4)] with $i_M = 2$, $i_F = 0$, $S_{PM} = 2$, $S_{PF} = 1$, $h^2_M = h^2_F = 1$, and $r_a = 0.9$, the response of sexual dimorphism R_{SD} is 1.1 and the response of the female mean R_F is 0.9. So $K_{SD} = 1.222$. In a pure variance dimorphism model with the same overall selection intensity, $R_{SD} = 0.95$, $R_F = 0.95$, and $K_{SD} = 1.000$, while in a pure selection dimorphism model with the same overall level of variance $R_{SD} = 0.15$, $R_F = 1.35$, and $K_{SD} = 0.111$. Clearly neither the variance dimorphism model nor, especially, the selection dimorphism model by itself produces as strong a positive allometry as the combined model. One might say that their combined efforts are synergistic in this respect, because combined the response of the female mean is less than in either of the separate models.

Therefore, it seems most likely that the positive allometry of character dimorphism with female size is due to a combination of variance and selection dimorphism, with variance dimorphism accounting for most of the evolution of character dimorphism itself. This combination seems to be the most consistent with the empirical results described in the section on *Correlates of Sexual Dimorphism,* where size itself has a large effect on dimorphism, mating system has a relatively small effect, and dimorphism is positively allometric with size.

Summary

The effects of a series of ecological and size factors on the degree of sexual dimorphism in body weight and canine size were studied among subsets of 70 primate species. Variation in body weight dimorphism can be almost entirely attributed to body weight (83% of variance R^2 of weight dimorphism). Much smaller amounts of the variation can be attributed to mating system (R^2 = 6.8%, polygynous species being more dimorphic than monogamous ones) and diet (R^2 = 2.5%, frugivorous species being more dimorphic than folivorous ones). Habitat (arboreal versus terrestrial) and activity rhythm (nocturnal versus diurnal) have only an indirect effect on weight dimorphism. Variation in canine size dimorphism can largely be explained in terms of canine size (R^2 = 49%), activity rhythm (R^2 = 20%, diurnal species being more dimorphic than nocturnal ones), and mating system (R^2 = 10%). Habitat and diet do not play a significant role in canine size dimorphism. Sexual dimorphism in both body weight and canine size is positively allometric with respect to their associated size measures.

The recognition of size as producing a major effect on sexual dimorphism, in addition to the phenomenon of positive allometry, requires a reassessment of theories for the evolution of sexual dimorphism. Sexual dimorphism can only evolve if there is dimorphism in selection and/or dimorphism in genetic variances. Based on a general, quantitative genetic model for the evolution of sexual dimorphism, we demonstrate that, in contrast to traditional views, selection dimorphism is not very potent in producing character dimorphism or positive allometry between character dimorphism and size. Variance dimorphism, which has been largely overlooked in previous models, is demonstrated to be more powerful than selection dimorphism as a source of character dimorphism. Variance dimorphism also produces positive allometry between character dimorphism and size more easily than does selection dimorphism. However, a combination of sexual selection and variance dimorphism seems to be the most effective producer of positive allometry of sexual dimorphism, which is in accordance with the empirical results of this study.

ACKNOWLEDGMENTS

This study was supported in part by Faculty Research Grants from the Graduate School, University of Wisconsin at Madison, to W.L.

References

Alexander, R. D., Hoogland, J. L, Howard, R. D., Noonan, K. M., and Sherman, P. W. 1979. Sexual dimorphism and breeding systems in pinnipeds, ungulates, primates and humans, in: *Evolutionary Biology and Human Social Behavior: An Anthropological Perspective* (N. A. Chagnon and W. Irons, eds.), pp. 402–435, Duxbury Press, North Scituate, Massachusetts.

Anderson, D. and Lydic, R. 1977. On the effect of using ratios in the analysis of variance. *Biobehav. Rev.* **1**:225–229.

Atchley, W., Gaskins, C., and Anderson, D. 1976. Statistical properties of ratios. I. Empirical results. *Syst. Zool.* **25:**137–148.

Bird, M., and Schaffer, H. 1972. A study of the genetic basis of sexual dimorphism for wing length in *Drosophila melanogaster. Genetics* **72:**475–487.

Blalock, H. 1972. *Social Statistics,* McGraw-Hill, New York.

Clutton-Brock, T. H., and Harvey, P. H. 1977. Primate ecology and social organization. *J. Zool. Lond.* **183:**1–39.

Clutton-Brock, T. H., and Harvey, P. H. 1978. Mammals, resources and reproductive strategies. *Nature* **273:**191–195.

Clutton-Brock, T. H., Harvey, P. H., and Rudder, B. 1977. Sexual dimorphism, socionomic sex ratio and body weight in primates. *Nature* **269:**797–800.

Crook, J. H. 1972. Sexual selection, dimorphism, and social organization in the primates, in: *Sexual Selection and the Descent of Man, 1871–1971* (B. G. Campbell, ed.), pp. 231–281, Aldine, Chicago.

Crook, J. H., and Gartlan, J. S. 1966. On the evolution of primate societies. *Nature* **210:**1200–1203.

Darwin, C. 1871. *The Descent of Man and Selection in Relation to Sex,* Murray, London.

DeVore, I., and Washburn, S. L. 1963. Baboon ecology and human evolution, in: *African Ecology and Human Evolution* (F. C. Howell and F. Bourlière, eds.), pp. 335–367, Aldine, Chicago.

Eisen, E. J., and Hanrahan, J. P. 1972. Selection for sexual dimorphism in body weight in mice. *Aust. J. Biol. Sci.* **25:**1015–1024.

Eisen, E. J., and Legates, J. E. 1966. Genotype–sex interaction and the genetic correlation between the sexes for body weight in *Mus musculus. Genetics* **54:**611–623.

Eisenberg, J. F., Muckenhirn, N. A., and Rudran, R. 1972. The relation between ecology and social structure in primates. *Science* **176:**863–874.

Falconer, D. S. 1960. *An Introduction to Quantitative Genetics,* Roland Press, New York.

Fleagle, J. G., Kay, R. F., and Simons, E. L. 1980. Sexual dimorphism in early anthropoids. *Nature* **287:**328–330.

Frankham, R. 1968. Sex and selection for a quantitative character in *Drosophila.* I. Single-sex selection. *Aust. J. Biol. Sci.* **21:**1215–1223.

Gingerich, P. D. 1981. Cranial morphology and adaptations in Eocene Adapidae. I. Sexual dimorphism in *Adapis magnus* and *Adapis parisiensis. Am. J. Phys. Anthropol.* **56:**217–234.

Harrison, B. J. 1953. Reversal of a secondary sex character by selection. *Heredity* **7:**153–164.

Harvey, P. H., Kavanagh, M., and Clutton-Brock, T. H. 1978*a.* Sexual dimorphism in primate teeth. *J. Zool. Lond.* **186:**475–485.

Harvey, P. H., Kavanagh, M., and Clutton-Brock, T. H. 1978*b.* Canine tooth size in female primates. *Nature* **276:**817–818.

Huxley, J. S. 1932. *Problems of Relative Growth,* Methuen, London.

Korkman, N. 1957. Selection with regard to the sex difference of body weight in mice. *Hereditas* **43:**665–678.

Lande, R. 1980. Sexual dimorphism, sexual selection, and adaptation in polygenic characters. *Evolution* **34:**292–307.

Leutenegger, W. 1973. Maternal–fetal weight relationships in primates. *Folia primatol.* **20**: 280–293.

Leutenegger, W. 1976. Metric variability in the anterior dentition of African colobines. *Am. J. Phys. Anthropol.* **45:**45–52.

Leutenegger, W. 1978. Scaling of sexual dimorphism in body size and breeding system in primates. *Nature* **272:**610–611.

Leutenegger, W. 1981. Sexual dimorphism in early anthropoids. *Nature,* **290:**609.

Leutenegger, W. 1982*a.* Sexual dimorphism in nonhuman primates, in: *Sexual Dimorphism in Homo sapiens* (R. L. Hall, ed.), pp. 11–36, Praeger, New York.

Leutenegger, W. 1982*b.* Scaling of sexual dimorphism in body weight and canine size in primates. *Folia primatol.* **37:**163–176.

Leutenegger, W., and Cheverud, J. 1982. Correlates of sexual dimorphism in primates: Ecological and size varables. *Int. J. Primatol.* **3:**387–402.

Leutenegger, W., and Kelly, J. T. 1977. Relationship of sexual dimorphism in canine size and body size to social, behavioral, and ecological correlates in anthropoid primates. *Primates* **18:**117–136.

Maiorana, V. 1984. Size and the degree of sexual dimorphism: Lopsided reproductive selection (in preparation).

Ralls, K. 1977. Sexual dimorphism in mammals: Avian models and unanswered questions. *Am. Nat.* **111:**917–938.

Raup, D. M. 1972. Approaches to morphologic analysis, in: *Models in Paleobiology* (T. J. M. Schopf, ed.), pp. 28–44, Freeman, San Francisco.

Rensch, B. 1950. Die Abhängigkeit der relativen Sexualdifferenz von der Körpergrösse. *Bonn Zool. Beitr.* **1:**58–69.

Rensch, B. 1954. *Neuere Probleme der Abstammungslehre,* Enke, Stuttgart.

Rothenfluh, E. 1976. Ueberprüfung der Gewichtsangaben adulter Primaten im Vergleich zwischen Gefangenschafts- und Wildfangtieren, Diplomarbeit, Universität Zürich.

Selander, R. K. 1972. Sexual section and dimorphism in birds, in: *Sexual Selection and the Descent of Man, 1871–1971* (B. G. Campbell, ed.), pp. 180–230, Aldine, Chicago.

Swindler, D. R. 1976. *Dentition of Living Primates,* Academic Press, London.

Wilson, E. O. 1975. *Sociobiology,* Harvard University Press, Cambridge, Massachusetts.

Size, Sexual Dimorphism, and Polygyny in Primates 4

T. H. CLUTTON-BROCK

> With many of the Quadrumana, we have . . . evidence of the action of sexual selection in the greater size and strength of the males, and in the greater development of their canine teeth, in comparison with the females.
>
> Charles Darwin (1871)

Sexual Dimorphism and Polygyny

Among primates, the extent of sexual dimorphism in body size ranges from species where mature females are slightly larger than mature males, as in some of the marmosets and tamarins (Ralls, 1976), through species where males are slightly larger than females, like many of the diurnal lemurs and the arboreal colobines, to those where males are nearly twice as heavy as females, as in the larger cercopithecines, the gorilla and the orang (Clutton-Brock and Harvey, 1978). The tendency for polygynous mammals to show greater size dimorphism than monogamous ones was originally noticed by Darwin (1871), and quantitative studies have subsequently confirmed that in primates (Gautier-Hion, 1975; Clutton-Brock *et al.*, 1977; Clutton-Brock and Harvey, 1978) (see Fig. 1), pinnipeds and ungulates (Alexander *et al.*, 1979), as well as birds (Lack, 1968) and amphibians (Shine, 1979) monogamous species consistently show less dimorphism than polygynous ones.

T. H. CLUTTON-BROCK • Large Animal Research Group, Department of Zoology, University of Cambridge, Cambridge CB2 3EJ, England.

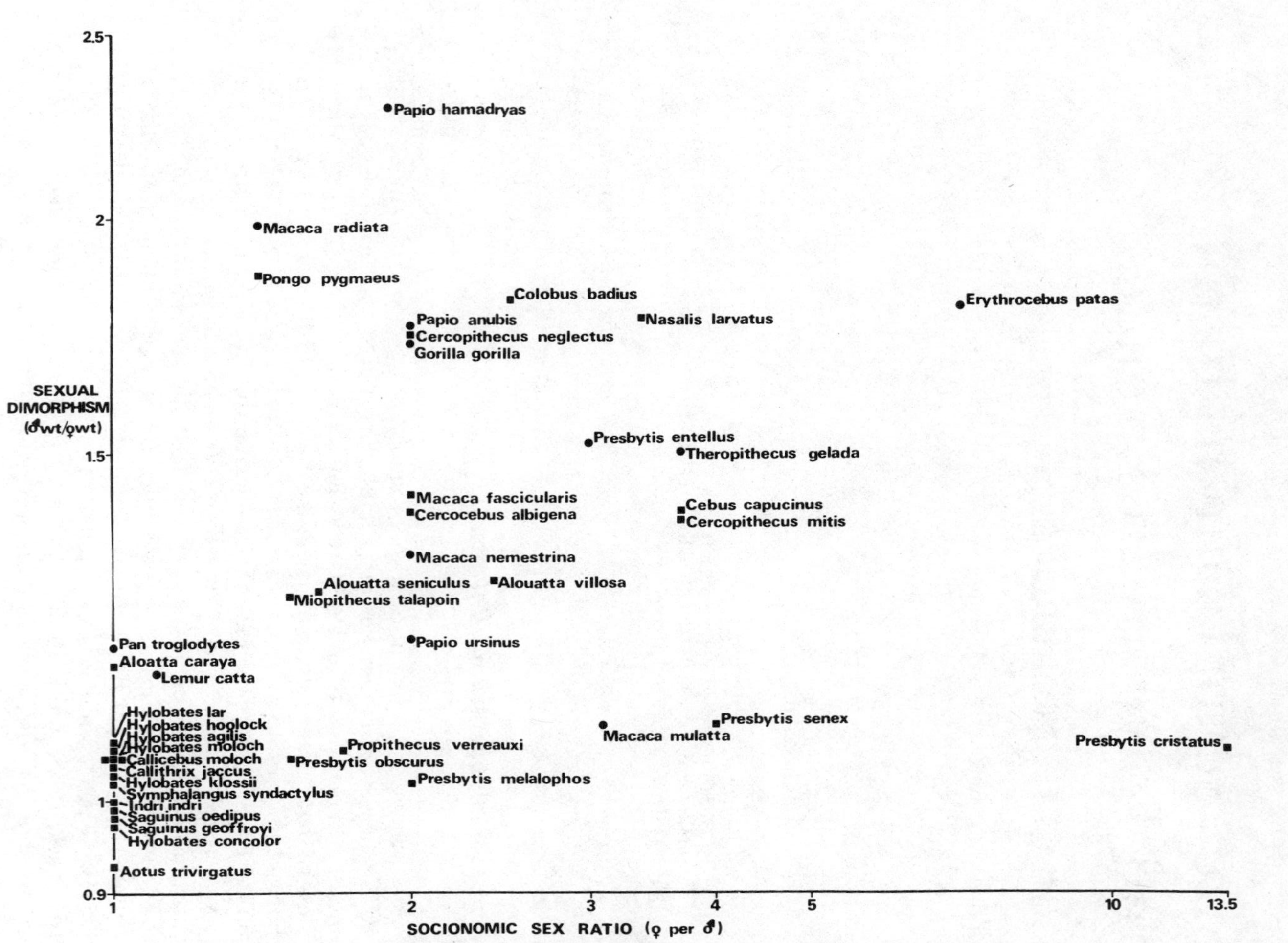

Fig. 1. Sexual dimorphism in body weight (weight of male/weight of female) plotted against socionomic sex ratio (females per adult

However, as Darwin (1871, p. 217) pointed out the association between polygyny and size dimorphism is not always a close one:

> That some relation between polygamy and the development of secondary sexual characters appears nearly certain. . . . Nevertheless many animals, which are strictly monogamous, especially birds, display strongly marked secondary sexual characters; whilst some few animals, which are polygamous, do not have such characters.

Pronounced sexual dimorphism in body size is found in at least one monogamous primate, the de Brazza monkey (*Cercopithecus neglectus*) (Moreno-Black and Maples, 1977; Gautier-Hion, 1980), while some polygynous primates (such as *Propithecus verreauxi* and *Chiropotes satanas*) show little size dimorphism (Sussman and Richard, 1974; M. Ayres, personal communication; Clutton-Brock and Harvey, 1978). Similar exceptions exist in other mammalian groups: among equids, Burchell's zebra shows little sexual dimorphism, though polygyny is well developed (Klingel, 1972), while in some polygynous carivores, including the spotted hyena, females are larger than males (Kruuk, 1972).

Thus any satisfactory functional explanation of the relationship between sexual dimorphism and polygyny needs to be able to accommodate both the general rule and a considerable number of exceptions to it. Most explanations of the rule rely on the argument that sexual selection is more intense in polygynous species than in monogamous ones without defining precisely what is meant by the intensity of sexual selection (Clutton-Brock *et al.*, 1977; Ralls, 1977; Clutton-Brock and Harvey, 1978; Alexander *et al.*, 1979; Leutenegger, 1978). The argument is usually based on three assumptions: that male breeding success varies more widely in polygynous species than in monogamous ones; that this is associated with an increase in the frequency and intensity of competition between males for mates; and that selection consequently favors the greater development of traits that enhance combative success in polygynous species. However, there are both theoretical and empirical difficulties with all three assumptions.

Polygyny and Variation in Male Success

First, it is not yet clear that polygyny is necessarily associated with a substantial increase in the extent to which breeding success varies among males (Clutton-Brock, 1983). Virtually all attempts to measure variation in reproductive success among male vertebrates have been based on measures collected over short periods of the lifespan (Clutton-Brock *et al.*, 1977; Alexander *et al.*, 1979) though it is through individual differences in *lifetime* breeding success that selection will operate (Maynard Smith, 1958; Grafen, 1982). In polygynous species, variation in male success measured over short periods of time is likely to overestimate variation in lifetime success, for the age of

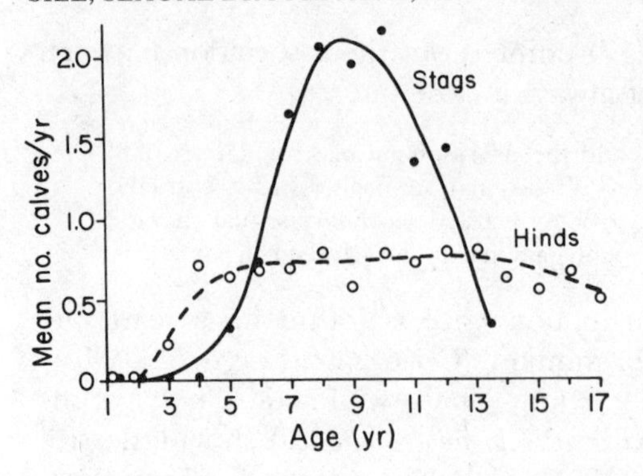

Fig. 2. Mean number of calves sired/borne per year by red deer hinds and stags of different ages. (From Clutton-Brock, 1983.)

males typically has a strong effect on their breeding success and young and old animals are substantially less successful than individuals in their prime.

An opposite bias is likely to occur for females. Where individual differences in breeding success are consistent across years and age exerts a weaker influence on breeding success than in males, short-term measures of variation will *underestimate* the extent to which lifetime success varies among females; consequently, they will tend to overestimate the extent to which variation in breeding success differs between the two sexes. For example, in the polygynous red deer (*Cervus elaphus*), male breeding success is strongly affected by age and peaks among 7- to 10-year-olds (Fig. 2). Among females, age has a less pronounced effect on breeding success and the same animals tend to be either consistently successful or consistently unsuccessful breeders (Clutton-Brock *et al.,* 1982). As a result, measures of variation in breeding success calculated across all mature individuals within seasons systematically overestimate variation in lifetime breeding success among males and underestimate it among females (see Tables 1 and 2).

Polygyny and Competition between Males

Second, direct competition between males need not necessarily be more intense in polygynous than monogamous species. Though differences in

Table 1. Variation in Number of Calves Sired, Calculated across all Male Red Deer ≥1 Year Old Present in Each of Seven Breeding Seasons[a]

	Breeding season						
	1974	1975	1976	1977	1978	1979	1980
$\overline{X}$	0.875	1.130	0.884	0.654	0.725	0.825	0.725
$\sigma^2/\overline{X}^2$	3.090	2.108	4.762	6.126	8.072	3.624	3.293
N	46	56	69	78	91	61	68

[a]The index of variation used $\sigma^2/\overline{X}^2$ is that recommended by Wade and Arnold (1980). These estimates do not take into account calves dying in their first year of life.

Table 2. Variation in Number of Calves Sired across Stags of the Same Age within Breeding Seasons[a]

	Age of male, years										
	1–4	5	6	7	8	9	10	11	12	13+	LRS
$\bar{X}$	0	0.32	0.72	1.65	2.07	1.95	2.17	1.35	1.43	0.33	10.15
$\sigma^2/\bar{X}^2$	—	7.21	2.17	1.61	1.02	1.22	1.35	0.71	2.26	2.25	1.22
N	—	28	29	41	28	19	12	10	7	8	28

[a]These estimates do not take into account calves dying in their first year of life. Variation in lifetime success calculated across all individuals reaching breeding age (5 years) is also shown (LRS). The index of variation used $\sigma^2/\bar{X}^2$ is that recommended by Wade and Arnold (1980).

breeding success among males in monogamous species may be caused principally by variation in mate or territory quality, while in polygynous systems they are caused mainly by variation in mate number (Bateman, 1948; Wade, 1979; Clutton-Brock *et al.*, 1982; Clutton-Brock, 1983), there is no *a priori* reason why, if their lifetime reproductive success is at stake, monogamous males should not compete as intensely for good mates or territories as do polygynous ones for large harems. Instead, the frequency of intermale fighting is likely to be determined by the conditions determining mate access and it is possible to imagine circumstances under which the frequency of fighting is likely to be common in monogamous species and rare in polygynous ones; one example would be cases where competition between polygynous males for mates is preempted by female choice.

Empirical data that would allow us to compare the frequency and intensity of fighting between monogamous and polygynous species are not yet available. In polygynous mammals, fights between males can be common and dangerous (Zuckerman, 1932; Geist, 1971; Nagel and Kummer, 1974; Hrdy, 1977; Clutton-Brock *et al.*, 1979), but fighting between males can also be intense in species where polygyny is not well developed (Sorenson, 1974; Charles-Dominique, 1974; Kleiman, 1977).

Polygyny and Sexual Selection

This raises the third and most important point. It is neither the extent of variation in male breeding success, the difference in variation between the sexes, nor sex differences in the frequency of fighting that will determine the degree of sexual dimorphism in body size that is likely to evolve. Instead, it is the comparative *effects* of body size on the lifetime breeding success of males and females (Clutton-Brock, 1983). For example, while sexual dimorphism in size will evolve where variation in breeding success is greater among males than females and a given increment in body size has the same effect on both sexes (see Fig. 3A), it will also evolve if variation in reproductive success is similar in both sexes but size has a greater influence on success in males (Fig.

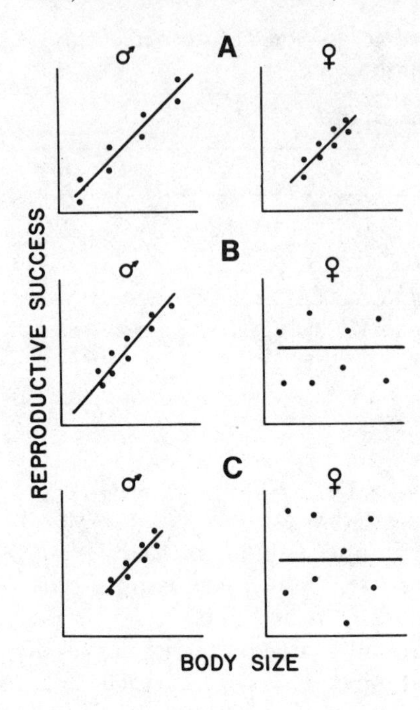

Fig. 3. Three hypothetical relationships between body size and reproductive success in males and females likely to produce size differences between the sexes.

3B) or even if variation in success is greater in females but the effects of size are stronger among males (Fig. 3C). Conversely, sexual dimorphism in body size is unlikely to evolve where the effects of size on breeding success are similar in the two sexes, even if male breeding success varies more widely overall.

Of course, there is no reason to suppose that relationships between size and reproductive success will be linear in contemporary populations or that they will follow a similar pattern in both sexes or in different species. Indeed, where size differences are heritable and stabilizing selection is operating, there is every reason to suppose that relationships between size and reproductive success will be more complex. However, it will still be the comparative effects of body size on breeding success in males and females and not the extent to which success varies that is likely to determine selection pressures affecting sexual dimorphism.

Polygyny and Sexual Dimorphism

Emphasis on the importance of the comparative effects of size on breeding success in males and females provides a framework within which exceptions to the general relationship between polygyny and size dimorphism can be easily accommodated. Since the costs and benefits of increasing body size need not be similar among the males and females of different polygynous species, there is no reason to suppose that species with similar breeding systems should necessarily show similar development of dimorphism. Species in which size dimorphism is especially pronounced are presumably those where the effects of size on male success are particularly strong or where the upper limits of female size are most firmly constrained. Species showing relatively little size dimorphism are likely to be those where increasing body size has little effect on male breeding success—or has similar effects on females. In addition, size dimorphism may arise in some species as a consequence of sex differences in feeding behavior (Selander, 1972) or energy requirements (Downhower, 1976).

This framework suggests why it is that sexual dimorphism in body size is not always closely correlated with sexual dimorphism in coat coloration, display organs, or weaponry (Leutenegger and Kelly, 1977; Leutenegger, 1982; Lowther, 1975). Since there is no reason to predict that coloration or weaponry will always have the strongest effect on male success relative to female success in those species where male size is most important, there are no firm grounds for predicting that the distribution of these traits will be closely related to each other.

These arguments have some bearing on recent arguments concerning the common tendency for sexual dimorphism to increase in larger species (Huxley, 1932; Rensch, 1959; Clutton-Brock *et al.*, 1977, 1980; Leutenegger, 1978, 1982). Huxley (1932), Rensch (1959), and Maynard Smith (1978) have all suggested that the association between body size and sexual dimorphism may be a nonadaptive consequence of linkage between genetic factors affecting body size in males and females, while Leutenegger (1978, 1982) has suggested that the association may occur because large body size favors the development of polygyny. Neither explanation is satisfactory. The first is unlikely because the strong selection pressures that operate on body size would be likely to disrupt any linkage between the genetic determinants of absolute and relative size, and because the association between body size and sexual dimorphism is not a particularly close one (Clutton-Brock *et al.*, 1977; Leutenegger, 1982), while Leutenegger's explanation is unsatisfactory because there is no reason why an increase in the average size of males should promote polygyny. Although it is possible to suggest an almost infinite number of hypotheses, the answer may merely be that male size is more important in determining the outcome of fights in larger species because such fights involve a greater amount of pushing, tussling, and prolonged physical combat and strength and size may play a large part in their outcome (Geist, 1966; Janis, 1982). In smaller species, which commonly rely more on the use of weapons and less on pushing, maneuverability may be at a premium and size less important.

The question we should perhaps be asking is not why sexual dimorphism in body size is not always well correlated with the degree of polygyny, but why the two measures are correlated at all. The probable answer is that the effects of body size on the breeding success of males and females are generally most similar in monogamous species and most different in polygynous ones (Clutton-Brock, 1983). Unfortunately, we do not yet have any measures of the factors affecting lifetime breeding success in males and females in any monogamous vertebrate. While it is clear that these will not be identical in monogamous species [see McGregor *et al.* (1981) for one example], mate quality and territory quality will probably exert an important influence on the breeding success of both sexes and selection is likely to lead to greater similarity between the sexes. In contrast, female breeding success in polygynous species is likely to depend principally on access to food supplies, while male success will depend principally on access to females (Trivers, 1972). Where the distribution of food supplies precludes effective defense (Clutton-Brock and Harvey, 1978) while access to females depends on fighting ability (e.g., Sussman and Richard, 1974; Packer, 1979*a,b*), selection for traits that favor fighting ability will be stronger in males than females and, if these include body size, sexual dimorphism in size is likely to evolve.

Summary

1. The widespread association between sexual dimorphism in body size and polygyny is commonly explained as a consequence of increased variation in reproductive success in males in polygynous species, leading to increased intermale competition and selection pressures favoring size dimorphism.
2. However, variation in the lifetime breeding success of males may not be substantially greater in polygynous species than in monogamous ones; there is no firm evidence that, judged over the whole year, fighting between males is more frequent or more intense in polygynous species; and it is not variation in breeding success but the comparative effects of body size on breeding success in males and females that will determine selection pressures favoring dimorphism.
3. The common association between polygyny and dimorphism probably occurs because the effects of body size on the reproductive success of males and females are most similar in monogamous species and most different in polygynous ones.

ACKNOWLEDGMENTS

I am very grateful to Michael Reiss, Paul Harvey, John Coulson, Callum Thomas, and John Maynard Smith for their comments; Anthony Arak, Dafila Scott, Jon Seeger, Steve Albon, Joanne Reiter, Meg McVey, and Burney Le Boeuf for helpful discussion; and to G. P. Baerends for criticism of an earlier draft of this manuscript.

References

Alexander, R. D., Hoogland, J. L., Howard, R. D., Noonan, M., and Sherman, P. W. 1979. Sexual dimorphisms, and breeding systems in pinnipeds, ungulates, primates and humans, in: *Evolutionary Biology and Human Social Behavior: An Anthropological Perspective* (N. A. Chagnon and W. Irons, eds.), pp. 402–604, Duxbury Press, North Scituate, Massachusetts.

Bateman, A. J. 1948. Intrasexual selection in *Drosophila. Heredity* **2:**349–368.

Charles-Dominique, P. 1974. Aggression and territoriality in nocturnal prosimians, in: *Primate Aggression, Territoriality and Xenophobia; A Comprehensive Perspective* (R. L. Holloway, ed.), pp. 31–48, Academic Press, New York.

Clutton-Brock, T. H. 1983. Selection in relation to sex, in: *Evolution from Molecules to Men* (Proceedings of the Darwin Centennial Conference, Cambridge) (J. S. Bendall, ed.), pp. 457–482, Cambridge University Press, Cambridge.

Clutton-Brock, T. H., and Harvey, P. H. 1978. Mammals, resources and reproductive strategies. *Nature* **273:**191–195.

Clutton-Brock, T. H., Harvey, P. H., and Rudder, B. 1977. Sexual dimorphism, socionomic sex ratio and body weight in primates. *Nature* **269:**797–800.

Clutton-Brock, T. H., Albon, S. D., Gibson, R. M., and Guinness, F. E. 1979. The logical stag: Adaptive aspects of fighting in red deer (*Cervus elaphus* L.). *Anim. Behav.* **27:**211–225.

Clutton-Brock, T. H., Albon, S. D., and Harvey, P. H. 1980. Antlers, body size and breeding systems in the Cervidae. *Nature* **285:**565–567.

Clutton-Brock, T. H., Guinness, F. E., and Albon, S. D. 1982. *Red Deer: Behavior and Ecology of Two Sexes,* University of Chicago Press, Chicago.

Darwin, C. 1871. *The Descent of Man and Selection in Relation to Sex,* Murray, London.

Downhower, J. F. 1976. Darwin's finches and the evolution of sexual dimorphism in body size. *Nature* **263:**558–563.

Gautier-Hion, A. 1975. Dimorphisme sexuel et organisation sociale chez les cercopithécinés africains. *Mammalia* **39:**365–374.

Gautier-Hion, A. 1980. Seasonal variations of diet related to species and sex in a community of *Cercopithecus* monkeys. *J. Anim. Ecol.* **49:**237–269.

Geist, V. 1966. The evolution of horn-like organs. *Behaviour* **27:**175–214.

Geist, V. 1971. *Mountain Sheep: A Study in Behavior and Evolution,* University of Chicago Press, Chicago.

Grafen, A. 1982. How not to measure inclusive fitness. *Nature* **298:**419–420.

Hrdy, S. B. 1977. *The Langurs of Abu: Female and Male Strategies of Reproduction,* Harvard University Press, Cambridge.

Huxley, J. S. 1932. *Problems of Relative Growth,* Methuen, London.

Janis, C. 1982. Evolution of horns in ungulates: Ecology and paleoecology. *Biol. Rev.* **57:**261–318.

Kleiman, D. G. 1977. Monogamy in mammals. *Q. Rev. Biol.* **52:** 39–69.

Klingel, H. 1972. Social behaviour of African equidae. *Zool. Africana* **7:**175–185.

Kruuk, H. 1972. *The Spotted Hyena: A Study of Predation and Social Behavior,* University of Chicago Press, Chicago.

Lack, D. 1968. *Ecological Adaptations for Breeding in Birds,* Methuen, London.

Leutenegger, W. 1978. Scaling of sexual dimorphism in body size and breeding system in primates. *Nature* **272:**610–611.

Leutenegger, W. 1982. Scaling of sexual dimorphism in body weight and canine size in primates. *Folia Primatol.* **37:**163–176.

Leutenegger, W., and Kelly, J. T. 1977. Relationship of sexual dimorphism in canine size and body size to social behavioral and ecological correlates in anthropoid primates. *Primates* **18:**117–136.

Lowther, P. 1975. Geographic and ecological variation in the family Icteridae. *Wilson Bull.* **87:**481–495.

McGregor, P. K., Krebs, J. R., and Perrins, C. M. 1981. Song repertoires and lifetime reproductive success in the great tit, *Parus major. Am. Nat.* **118:**149–159.

Moreno-Black, G., and Maples, W. R. 1977. Differential habitat utilization of four Cercopithecidae in a Kenyan forest. *Folia Primatol.* **27:**85–107.

Nagel, U., and Kummer, H. 1974. Variation in cercopithecoid aggressive behavior, in: *Primate Aggression, Territoriality and Xenophobia; A Comprehensive Perspective* (R. L. Holloway, ed.), pp. 159–184, Academic Press, New York.

Packer, C. 1979*a*. Inter-troop trnsfer and inbreeding avoidance in *Papio anubis. Anim. Behav.* **27:**1–36.

Packer, C. 1979*b*. Male dominance and reproductive activity in *Papio anubis. Anim. Behav.* **27:**37–45.

Ralls, K. 1976. Mammals in which the female is larger than the male. *Q. Rev. Biol.* **51:**245–272.

Ralls, K. 1977. Sexual dimorphism in mammals: Avian models and unanswered questions. *Am. Nat.* **111:**917–938.

Rensch, B. 1959. *Evolution above the Species Level,* Methuen, London.

Selander, R. K. 1972. Sexual selection and dimorphism in birds, in: *Sexual Selection and the Descent of Man,* 1871–1971 (B. Campbell, ed.), pp. 180–230, Aldine, Chicago.

Shine, R. 1979. Sexual selection and sexual dimorphism in the Amphibia. *Copeia* **2:**297–306.

Smith, J. M. 1958. *The Theory of Evolution,* Penguin, Hardmondsworth.

Smith, J. M. 1978. *The Evolution of Sex,* Cambridge University Press, Cambridge.

Sorenson, M. W. 1974. A review of aggressive behavior in the tree shrews, in: *Primate Aggression, Territoriality and Xenophobia; A Comprehensive Perspective* (R. L. Holloway, ed.), pp. 13–30, Academic Press, New York.

Sussman, R. W., and Richard, A. 1974. The role of aggression among diurnal prosimians, in: *Primate Aggression, Territoriality and Xenophobia; A Comprehensive Perspective* (R. L. Holloway, ed.), pp. 49–76, Academic Press, New York.

Trivers, R. L. 1972. Parental investment and sexual selection, in: *Sexual Selection and the Descent of Man, 1871–1971* (B. Campbell, ed.), pp. 136–179, Aldine, Chicago.

Wade, M. J. 1979. Sexual selection and variance in reproductive success. *Am. Nat.* **114:**742–746.

Wade, M. J., and Arnold, S. J. 1980. The intensity of sexual selection in relation to male sexual behavior, female choice, and sperm precedence. *Anim. Behav.* **28:**446–461.

Zuckerman, S. 1932. *The Social Life of Monkeys and Apes,* Routledge, Kegan & Paul, London.

Gastrointestinal Allometry in Primates and Other Mammals

5

R. D. MARTIN, D. J. CHIVERS,
A. M. MACLARNON, AND
C. M. HLADIK

Introduction

There has been a considerable upsurge of interest recently in the concept of interspecific allometry (Gould, 1966), which relates to the scaling of individual characteristics to match body size in different species. Two particularly important developments have been taking place in theoretical aspects of allometry (e.g., see other contributions to this volume) and in the extension of allometric analysis to new fields of enquiry. One example of the application of allometric principles to an entirely new area of research is provided by the work conducted by Chivers and Hladik (1980) on the morphology of the gastrointestinal tract in primates and other mammals. Further development of this particular approach forms the subject of this chapter.

It is now common practice to use the empirical allometric formula $Y = bX^k$ in interspecific comparisons, where X is a measure of body size (usually body weight) and Y is the character under consideration (in this chapter,

---

R. D. MARTIN AND A. M. MacLARNON • Department of Anthropology, University College, London WC1E 6BT, England. D. J. CHIVERS • Sub-department of Veterinary Anatomy, University of Cambridge, Cambridge CB2 1QS, England. C. M. HLADIK • C.N.R.S., Muséum National d'Histoire Naturelle, Laboratoire d'Ecologie Générale, 91800 Brunoy, France.

surface area of compartments of the gastrointestinal tract). This equation becomes linear in its logarithmic form, $\log Y = k \log X + \log b$, and it is now standard practice to fit lines to plots of logarithmically transformed data. When this is done, the slope of the best fit line yields the exponent k of the allometric formula, while the intercept, $\log b$, reveals the value of the allometric coefficient b. In a typical case where a line is fitted to real biological data, with the individual points representing average conditions for separate species, the overall trend will reflect some systematic scaling principle (e.g., the surface:volume law), while the positions of specific points above or below the best fit line will reflect the special adaptations of individual species ["grade differences"; see Martin (1980)].

Allometric analysis can be used, therefore, both for the recognition of general scaling principles governing the adjustment of characteristics (e.g., stomach surface area) to body size *per se,* and for the identification of special adaptations (e.g., increase in stomach surface area, relative to body size, in folivorous mammals exhibiting foregut fermentation). In any comparison of individual species covering a range of body sizes, it is essential to take into account any systematic influence of body size exerted on the characteristics considered, and allometric analysis now provides a well-established procedure for achieving this. In their previous consideration of gastrointestinal morphology of mammals, Chivers and Hladik (1980) set out to identify scaling principles for the three main compartments of the gastrointestinal tract (stomach, small intestine, cecum plus colon) and investigated ways of taking such scaling into account in order to identify residual differences between groups of species. They were able to demonstrate grade differences between three broad dietary groups—folivores, frugivores, and faunivores—on the basis of distinctions in gastrointestinal surface areas and volumes relative to body size.

The present chapter is designed to go beyond the original paper published by Chivers and Hladik (1980) in the following respects:

1. Additional data have been collected for a number of species. In particular, data are now available for six *Homo sapiens,* permitting inclusion of average values for our own species in the comparisons and hence raising the possibility of inferring some basic adaptation of the human gastrointestinal tract to suit a broad dietary category.

2. Use of overall scaling patterns in a predictive manner, for example, with respect to *Homo sapiens,* requires explicit consideration of the linked problems of *intraspecific variability* and of *modification of the gut by a changed diet* (e.g., in captivity).

3. All comparisons discussed below have been conducted with respect to body weight, whereas most of the allometric formulas reported by Chivers and Hladik (1980) relate to the cube of body length. While the use of the cube of body length avoids certain problems (e.g., obesity in individual specimens), it introduces others (e.g., confusion of grade differences in body length between mammal groups with grade differences in gut morphology), and in any

case body weight has been far more widely used as the reference standard in mammalian allometric comparisons generally.

4. Scaling of gastrointestinal compartments is obviously related in some direct way to the metabolic requirements of individual mammal species. Hence, explicit reference to metabolic scaling is advisable both in the analysis and in the interpretation of gastrointestinal allometry. (This also provides an accessory argument for the use of body weight as the measure of body size, since metabolic data have been traditionally reported with respect to body weight.)

5. In parallel to the now widespread practice of calculating *encephalization quotients* to express relative brain size in mammals (and other vertebrates), it is possible to calculate *gastrointestinal quotients.* These quotients take into account the allometric scaling principles that have been recognized and indicate for each species the degree to which the surface area of any individual gut compartment is bigger or smaller than expected for the body size of that species.

6. A recurring problem encountered in the comparative quantitative study of primates and other mammals with respect to dietary adaptations is that of *dietary classification.* (A similar problem is encountered with *locomotor classifications* when considering quantitative aspects of locomotor adaptations in primates and other mammals.) It might always be argued that a particular basic classification of dietary categories (e.g., folivores versus frugivores versus faunivores) is bedeviled by so many intermediate forms and special cases that any analysis based on such *a priori* categories is suspect. This extreme view is countered by the recognition of fairly clear-cut grades in the allometric analysis of gut morphology (Chivers and Hladik, 1980), but the fact remains that any classification must impose arbitrary boundaries in some instances at least. For this reason, the analyses discussed in this chapter have been conducted initially without any prior classification into dietary categories, and the gastrointestinal quotients determined for each species have been subjected to both bivariate and multivariate clustering techniques (bivariate plots of quotients; construction of dendrograms; multidimensional scaling) to allow the data, so to speak, to fall into natural groupings of their own accord with no prompting from the investigator.

Methods

The manner of collection of specimens and the techniques used in measuring parameters of the gastrointestinal tract have already been described in full by Chivers and Hladik (1980). The procedures used, therefore, are summarized only briefly here:

1. Animals collected were weighed intact wherever possible (though this was not always practical in the field), and their body lengths were meaured (1) from bregma to ischial tuberosity and (2) from tip of nose

to base of tail (the latter measure was not used, because of the distortion introduced by varying lengths of muzzle, especially when contrasting primates with other mammals).

2. The guts of all specimens included in this analysis were typically examined and measured in the fresh state (although zoo specimens were usually deep frozen or placed in saturated saline until collection).
3. Many specimens were examined, drawn, and photographed with the guts *in situ* and/or displayed under water in a large dissecting tray (Chivers and Hladik, 1980).
4. The dimensions of each region were then measured, for calculations of area and volume, and subsequently weighed after the removal of excess moisture.

Techniques were standardized throughout between two of the authors (DJC and CMH), on occasions when they worked together, so as to obtain comparable accuracy. Several hours were allowed to elapse after the death of the individual to permit relaxation of muscle tone in the gut wall. Measurements of length and breadth of stomach, small intestine, cecum, and colon were then made without stretching, after opening and flattening the gut wall, usually under water in a dissecting tray (except for the larger specimens). Remeasuring specimens on the moist surface of a dissecting trolley produced no significant discrepancies. Because different parts of the gut can be fully contracted or relaxed simultaneously or sequentially, this seems to be the best compromise in functional terms for measuring what is a very malleable system for comparative purposes.

The surface areas of small and large intestines were calculated from lengths and a series of breadths; sometimes it was more appropriate to treat the cecum as a triangle rather than an elongated rectangle. The irregular shape of the stomach required summing the area of its parts, usually arranged to cover the different compartments or division into fundus and pylorus. The areas of such nontubular parts were also measured by cutting pieces of aluminum foil or polythene to the exact shape of the part(s) immersed in water and then weighing to calculate from the weight of unit area; this provided a means for checking the accuracy of length and breadth measurements. Errors resulting from the different methods, or from repeated measurements, amounted to less than 5%. The additional area afforded by gross structures such as folds and papillae was measured, but the further contribution of villi to overall surface area awaits further study.

The raw data assembled in this way for 168 specimens of 73 species (see Table 1) have been analyzed as follows:

1. For each species, average values were calculated for the surface area of each gut compartment (stomach, small intestine, cecum, colon), with separate values determined for wild and captive specimens where applicable.
2. The average values for the various species for a given gut compartment were then plotted against body weight on logarithmic coordi-

Table 1. Key to the Species Included in the Dendrograms and the Multidimensional Scaling Diagrams[a]

No.	Species	No.	Species
1	*Arctocebus calabarensis* (5)	37	*Hylobates pileatus* (1)
2	*Avahi laniger* (3)	38	*Hylobates syndactylus* (3)
3	*Cheirogaleus major* (1)	39	*Pongo pygmaeus* (2)
4	*Euoticus elegantulus* (4)	40	*Pan troglodytes* (1)
5	*Galago alleni* (4)	41	*Gorilla gorilla* (1)
6	*Galago demidovii* (4)	42	*Homo sapiens* (6)
7	*Lepilemur mustelinus* (1)	43	*Felis domestica* (6)
8	*Lepilemur leucopus* (1)	44	*Canis familiaris* (13)
9	*Loris tardigradus* (1)	45	*Mustela nivalis* (1)
10	*Microcebus murinus* (1)	46	*Vulpes vulpes* (5)
11	*Perodicticus potto* (1)	47	*Atilax paludinosus* (1)
12	*Saguinus geoffroyi* (2)	48	*Nandinia binotata* (3)
13	*Aotus trivirgatus* (1)	49	*Poiana richardsoni* (1)
14	*Ateles belzebuth* (1)	50	*Genetta servalina* (1)
15	*Saimiri oerstedii* (3)[b]	51	*Mustela* sp. (1)
16	*Cebus capucinus* (1)	52	*Ailurus fulgens* (1)
17	*Alouatta palliata* (3)	53	*Nasua narica* (1)
18	*Lagothrix lagotricha* (2)	54	*Genetta* sp. (1)
19	*Miopithecus talapoin* (3)	55	*Panthera tigris* (1)
20	*Cercopithecus cephus* (4)	56	*Sus scrofa* (3)
21	*Cercopithecus neglectus* (2)	57	*Capra hircus* (2)
22	*Cercopithecus nictitans* (4)	58	*Ovis aries* (3)
23	*Cercocebus albigena* (2)	59	*Cervus elaphus* (11)
24	*Macaca sylvana* (1)	60	*Equus caballus* (1)
25	*Macaca sinica* (1)	61	*Halichoerus grypus* (1)
26	*Macaca fascicularis* (3)	62	*Phocoena phocoena* (1/3)
27	*Papio sphinx* (3)	63	*Tursiops truncata* (1)
28	*Erythrocebus patas* (1)	64	*Sciurus vulgaris* (1)
29	*Colobus polykomos* (2)	65	*Epixerus ebii* (2)
30	*Presbytis entellus* (2)	66	*Heliosciurus rufobrachium* (1)
31	*Presbytis cristata* (1)	67	*Sciurus carolinensis* (1)
32	*Presbytis obscura* (2)	68	*Oryctolagus cuniculus* (1)
33	*Presbytis melalophos* (5)	69	*Potamogale velox* (1)
34	*Presbytis rubicunda* (1)	70	*Manis tricuspis* (1)
35	*Nasalis larvatus* (1)	71	*Dendrohyrax dorsalis* (4)
36	*Pygathrix nemaeus* (2)	72	*Bradypus tridactylus* (1)
		73	*Macropus rufus* (1)

[a]Number of specimens per species is given in parentheses.
[b]Data on small intestine and colon surface areas of 80 wild *Saimiri sciureus* from Middleton and Rosal (1972) are included in the relevant single gut compartment analyses (Figs. 2 and 4), giving $N = 74$ in these cases.

nates (using data for wild-caught individuals only, wherever available, but including captive data in the few cases where no other material was available for a given species). Best fit lines were then determined for each plot.

There is still some uncertainty over the relative appropriateness of the alternative line fitting techniques available for this task (e.g., regression, reduced major axis, major axis). Chivers and Hladik (1980) used regressions for their analysis of allometric relationships (having checked that major axes gave very similar results), but there are good reasons for avoiding the use of regressions with biological data where no clear distinction can be drawn between a dependent variable and an independent variable (Kermack and Haldane, 1950; Harvey and Mace, 1982). Reduced major axis and major axis techniques require no such distinction to be made, and there are a number of additional reasons for preferring them (e.g., there is no inherent assumption that the *X* variable—body weight in this context—is measured without error). Accordingly, the major axis technique has been used throughout for the fitting of lines to bivariate plots of logarithmically transformed data. Such calculation of empirical best fit lines, however, is only a preliminary step.

In any biological comparison, the departures of individual points from a best fit line are just as interesting as the parameters of the line itself, and if a set of data contains a number of different "grades," an overall best fit line may not really be meaningful. For instance, it is well known that simian primates (monkeys and apes) have bigger brains, relative to body size, than do prosimian primates (lemurs, lorises, and tarsiers). It also happens to be the case that simian primates are generally larger in body size than prosimian primates. An overall best fit line through all data points (prosimians and simians) therefore has an artificially high slope, and it is obviously more meaningful to plot *separate* best fit lines for prosimians and simians. It is essential, therefore, to examine any bivariate plot to see whether distinct grades are likely to be present before attempting to define allometric relationships.

It is also useful, whenever possible, if a line of *fixed* slope can be fitted to the data where good theoretical reasons exist for proposing a particular value for the allometric exponent in advance. For instance, it is now well established that basal metabolic rate scales to body weight in mammals with an exponent value of 0.75 ["Kleiber's law" (Kleiber, 1961; Hemmingsen, 1950, 1960; Schmidt-Nielsen, 1972)], and there is new evidence to indicate that active metabolic rate (i.e., total metabolic turnover in a standard time) also scales to body size with a comparable exponent value (Mace and Harvey, 1982). Hence, one might expect any organ in the body that is directly concerned with metabolic turnover to scale to body size in accordance with Kleiber's law. Compartments of the gastrointestinal tract are concerned with metabolic turnover both with respect to secretory activity and with respect to absorption, so it is quite reasonable to fit a line with a fixed slope of 0.75 to the bivariate plots of gut compartment surface areas against body weight as a standard procedure. This has been done for all four gut compartments, and in addition the empirical slopes determined have been examined to see to what extent they do in fact accord with Kleiber's law (see p. 87).

Either on the basis of empirically fitted lines or on the basis of lines of fixed slope fitted to bivariate plots, it is possible to calculate *quotients* for each species by dividing the observed value for each species by the value expected for the body weight of that species from the best fit line. (This is equivalent to taking the antilog of the distance of each species point from the line on the log–log bivariate plots.) Since the concept of scaling of gut compartment surface areas in accordance with Kleiber's law in fact proved to be generally compatible with the data (see p. 87), quotients were calculated with respect to best fit lines with a fixed slope of 0.75. It was possible, therefore, to calculate four quotients for each species to express the relative surface area of the main gut compartments (gastric, intestinal, cecal, and colonic quotients). For any given gut compartment, a quotient value of one indicates that the surface area of that compartment is as expected for a typical mammal of that body size, while values greater or smaller than one indicate that the surface area is larger or smaller than expected, respectively.

Having calculated values for the various gut quotients, thus effectively eliminating the effects of body size differences between the species under comparison, it is possible to proceed to a detailed analysis of interspecific contrasts and of shared patterns of quotient values. This can be accomplished to some extent by plotting one quotient against another (e.g., gastric quotient versus intestinal quotient) in order to identify trends and to test individual hypotheses [e.g., that mammals with large stomachs have relatively reduced small intestines (Böker, 1932)]. Such bivariate comparisons, however, are limited in scope and some form of multivariate comparison is required, ideally leading to some clustering of the data. Bauchot (1982) tackled this problem with respect to individual brain components of insectivores and primates by calculating indices (quotients) for the 22 separate structures (using a best fit line with a fixed scope of 0.67 in each case) and then calculating the Euclidean distances between species for these indices. From the matrix of Euclidean distances between pairs of species, he proceeded to use a hierarchical clustering technique based on Ward (1963) to generate a dendrogram. In a comparable approach, the four gut quotient values calculated for each species in the present study were used to generate a matrix of Euclidean distances and then two separate clustering techniques were applied, one producing a dendrogram (as in Bauchot's approach), and the other providing a two-dimensional representation of distances between species [multidimensional scaling (Kruskal, 1964*a,b*)].

Results

The logarithmic plots of the raw surface area data (species averages) against body weight are shown in Figs. 1–4. The plot of stomach surface area S against body weight W (Fig. 1) shows a clear grade distinction in that the colobine primates and certain nonprimate mammals (notably artiodactyls, *Bradypus* and *Macropus*) have markedly larger stomachs than other mammals of the same body size. When a major axis is fitted to the data without taking

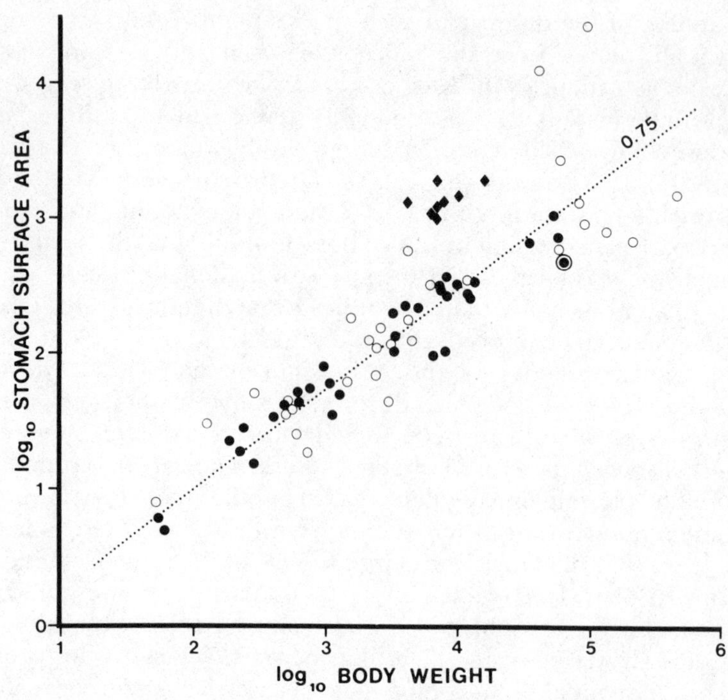

Fig. 1. Logarithmic plot of stomach surface area against body weight for 73 mammal species (primates shown in black; nonprimates in white; ringed black circle is *Homo sapiens*). Note that the colobine primates, indicated by black diamonds, have conspicuously large stomachs both absolutely and relatively compared to other primates. The dotted line is the best-fit line with a fixed slope of 0.75.

this grade shift into account, the following formula is obtained (S in cm^2, W in g):

$$\log_{10} S = 0.88 \log_{10} W - 0.85 \qquad (r = 0.86)$$
$$(N = 73;\ 95\% \text{ confidence limits on slope: } 0.77\text{–}1.00)$$

However, when these larger stomached mammals are ignored in the calculation of the major axis, the value of the slope is decreased and there is a notable improvement in the value of the correlation coefficient r:

$$\log_{10} S = 0.71 \log_{10} W - 0.38 \qquad (r = 0.92)$$
$$(N = 61;\ 95\% \text{ confidence limits on slope: } 0.64\text{–}0.79)$$

In this case the allometric relationship is clearly negative (i.e., $k < 1$) and the empirical value determined for the slope (0.71) is compatible with the value of 0.75 expected on the basis of Kleiber's law. Hence it is justifiable to fit a line of fixed slope 0.75 to the data yielding the following formula:

$$\log_{10} S = 0.75 \log_{10} W - 0.38 \qquad (N = 73)$$

This formula has been used for the calculation of *gastric quotient* (GQ) values for the species concerned.

With the plot of surface area I of the small intestine against body weight (Fig. 2), the empirical best fit line (major axis) is in fact in perfect accord with Kleiber's law (empirical slope value 0.75). Although there is some scatter of points around the best fit line, there is a clearly defined negative allometric trend with increasing body size, and the correlation coefficient for all data ($r = 0.95$) is higher than for any of the other gut compartments. The empirically determined allometric formula is (I in cm^2, W in g)

$$\log_{10} I = 0.75 \log_{10} W + 0.19$$

$$(N = 74;\ 95\%\ \text{confidence limits on slope: } 0.70\text{–}0.81)$$

This formula has therefore been used directly as the basis for calculating *intestinal quotient* (IQ) values for the individual species.

The logarithmic plot of surface area C of the cecum against body weight shows considerably more scatter. Indeed, some insectivorous and carnivorous mammals have no cecum at all (Chivers and Hladik, 1980), which accounts for

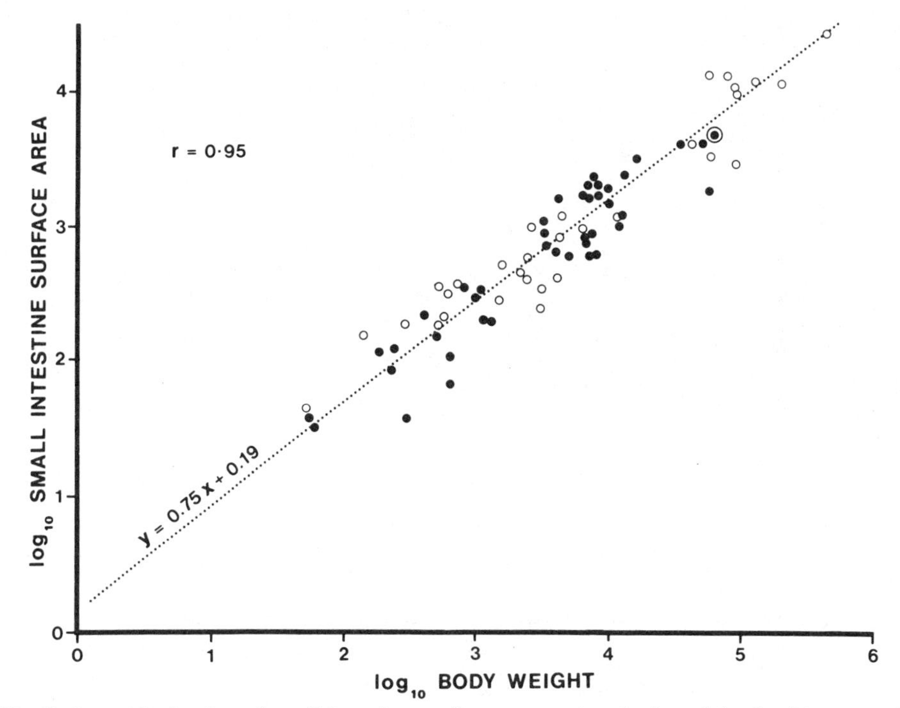

Fig. 2. Logarithmic plot of small intestine surface area against body weight for 74 mammal species (key to symbols as for Fig. 1). The dotted line is the empirical best fit line (major axis).

the reduced sample size shown in Fig. 3. Despite the great scatter (and the correspondingly relatively low value of the correlation coefficient: $r = 0.65$), the value of the slope of the empirical best fit line (major axis) is close to that predicted by Kleiber's law (C in cm^2, W in g):

$$\log_{10} C = 0.77 \log_{10} W - 1.07$$

$$(N = 63;\ 95\%\ \text{confidence limits on slope: } 0.57\text{–}1.03)$$

It is justifiable, therefore, to fit a standard line of fixed slope 0.75 to the data, yielding the following formula:

$$\log_{10} C = 0.75 \log_{10} W - 0.98 \qquad (N = 63)$$

This formula has been used for calculating cecal quotients (CQ) for the various species involved, and species lacking a cecum altogether have been given a quotient value of zero.

The logarithmic plot of surface area L of the colon or large intestine against body weight also shows a considerable degree of scatter (Fig. 4). The empirical best fit line (major axis) yields a formula that is only just compatible with Kleiber's law (L in cm^2, W in g):

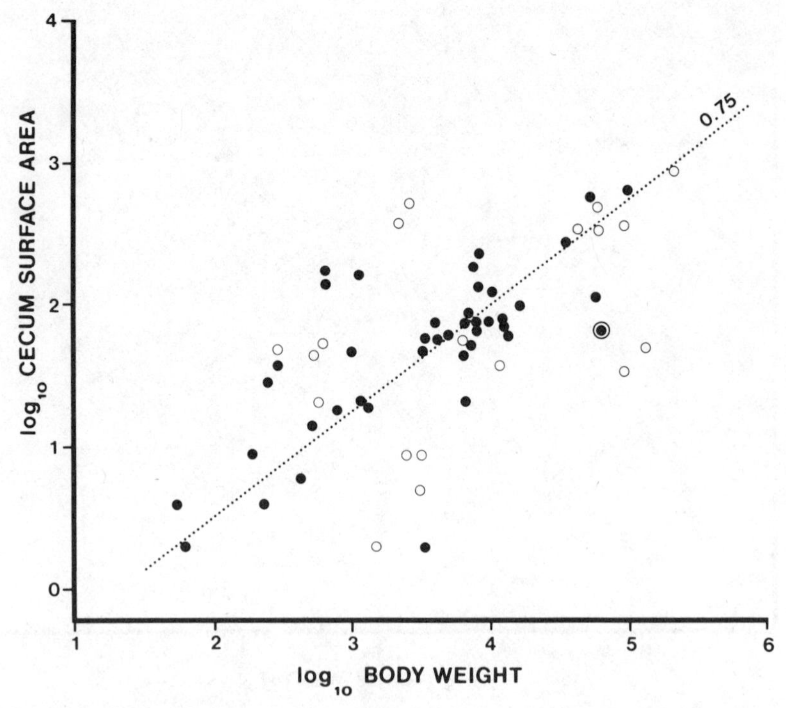

Fig. 3. Logarithmic plot of cecum surface area against body weight for 63 mammal species (key to symbols as for Fig. 1). The dotted line is the best fit line with a fixed slope of 0.75.

$$\log_{10} L = 0.91 \log_{10} W - 0.84 \qquad (r = 0.75)$$

$$(N = 74;\ 95\%\ \text{confidence limits on slope: } 0.74\text{–}1.11)$$

As with the plot of stomach surface area against body weight, the allometric relationships involved are obscured to some extent by grade distinctions.

This applies in particular to a number of mammalian species that are essentially faunivorous in habit (insectivores, carnivores, toothed cetaceans) and that have realtively small colons, and to the horse (*Equus*), which has a conspicuously well-developed colon. The point for the capuchin monkey (*Cebus capucinus*), referring to a single specimen, is also aberrant in that the colon is very small for the body size concerned. When all of these outliers are removed from the calculation, the correlation coefficient improves markedly (from $r = 0.75$ to $r = 0.95$), and the allometric formula becomes

$$\log_{10} L = 0.85 \log_{10} W - 0.39$$

$$(N = 55;\ 95\%\ \text{confidence limits on slope: } 0.78\text{–}0.92)$$

The relationship is clearly negatively allometric (i.e., $k < 1$), but the lower confidence limit just excludes compatibility with Kleiber's law (slope 0.75). Since the other three gut compartments all exhibit scaling compatible with

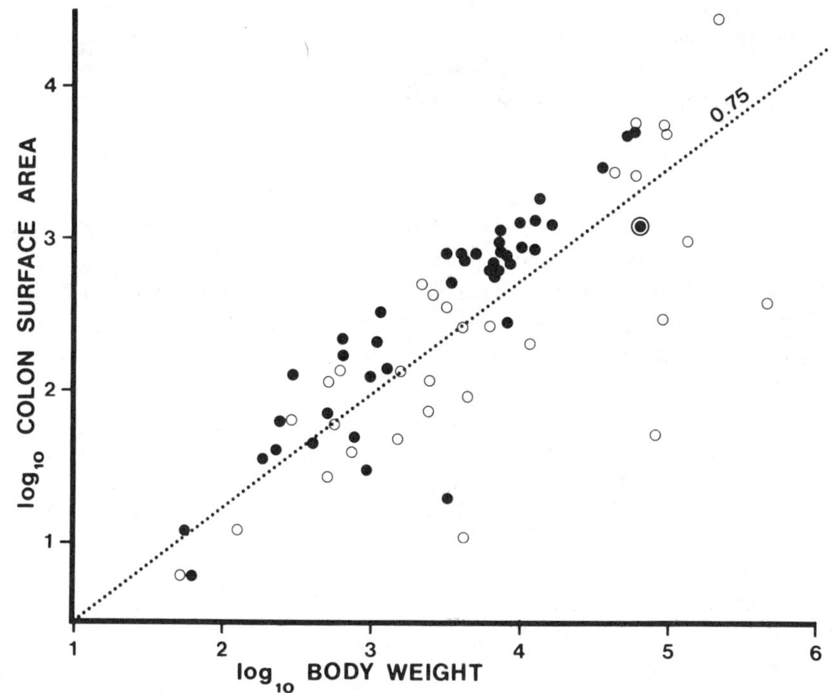

Fig. 4. Logarithmic plot of colon surface area against body weight for 74 mammal species (key to symbols as for Fig. 1). The dotted line is the best fit line with a fixed slope of 0.75.

Kleiber's law, a fixed slope of 0.75 was also applied in the case of the colon, yielding the following formula for calculating the colonic quotients (LQ):

$$\log_{10} L = 0.75 \log_{10} W - 0.26 \qquad (N = 74)$$

For the sake of consistency, quotients have thus been calculated for all four gut compartments using the formulas generated with best fit lines of slope 0.75, though the scaling of the colon surface area may in fact represent a special case.

In principle, calculation of gut quotients should very largely eliminate the influence of body size, so that direct and meaningful comparisons can be made between species of widely differing body sizes. This, of course, can only be true if the best fit lines used to calculate the quotients have the appropriate slope, so that calculation of colonic quotient values (in particular) may be to some extent inappropriate. Only detailed analysis using an improved data set can resolve this point. As a first approximation, however, it can be argued that calculation of gut quotients for all four gut compartments with respect to a fixed slope of 0.75 renders direct comparisons between species more meaningful in metabolic terms. Certainly, use of quotients that take into account the negatively allometric scaling of all gut compartments relative to body weight represents a distinct improvement over the use of simple proportions (e.g., a ratio of gut surface area to body weight).

Calculation of gut quotient values has particular interest in the case of the four average surface areas of the gut compartments determined for six *Homo sapiens*. It can be seen from Figs. 1–4 that man has values of less than one for all four gut compartments, most notably with respect to the cecum:

$$\text{GQ} = 0.31; \qquad \text{IQ} = 0.76; \qquad \text{CQ} = 0.16; \qquad \text{LQ} = 0.58$$

This is a pattern shared with a number of mammals relying heavily on animal food ["faunivores" (Chivers and Hladik, 1980)]. Such comparisons can be carried out for other single species, but it is very difficult to obtain a clear overall picture of interspecific differences in this way. Some systematic information can be obtained by observing overall patterns in bivariate plots of two gut quotients, however, and certain hypotheses regarding the overall adaptation of the mammalian gastrointestinal tract can be tested in this way.

One notion prevalent in the literature is that mammals that have a relatively large stomach are likely to have a relatively restricted small intestine, either because more *digestion* can take place in the stomach (Böker, 1932), or because more *absorption* can take place there. Chivers and Hladik (1980) found that the best separation between best fit lines for faunivores, frugivores, and folivores was found when they calculated a "potential area of absorption" by adding half the combined areas of the stomach, cecum, and colon to the surface of the small intestine and plotting this total area against body weight. This was done on the grounds that some absorption is known to take place in parts of the gastrointestinal tract other than the small intestine, at least in certain mammal species. Regardless of whether it is suggested that more *digestion* or more *absorption* takes place in the stomach when this organ is relatively enlarged, there is the same implied prediction that any mammal species with a relatively large stomach surface area should require a correspondingly smaller surface area in the small intestine. This can be tested by plotting the gastric quotient GQ against the intestinal quotient IQ to see whether there is the expected negative correlation (Fig. 5). In fact, it emerges that there is a slight *positive* correlation between these two quotients ($r = 0.28$), though the slope of the line is extremely shallow:

$$\text{GQ} = 0.03\ \text{IQ} + 1.08$$

(95% confidence limits on slope: −0.02 to +0.08)

Thus, there is no evidence that the surface area of the intestine is reduced when the surface area of the stomach is increased, once body size has been taken into account.

Another prediction that emerges from the literature (e.g., Janis, 1976; Chivers and Hladik, 1980) is that two divergent strategies may be exhibited by mammal species specialized for food types that require extensive digestion (e.g., leaves). The required expansion of the gastrointestinal tract (to house symbiotic bacteria in the case of folivorous species) can take place *either* in the stomach or in the midgut (cecum and proximal part of the colon), but not significantly in both. This prediction is fully confirmed by plotting (Fig. 6) the gastric quotient GQ against the cecal quotient CQ. Specialized folivorous species may have a high GQ value combine with a low CQ value or *vice versa*, but no species has high values for both quotients. This clear evidence of a definite

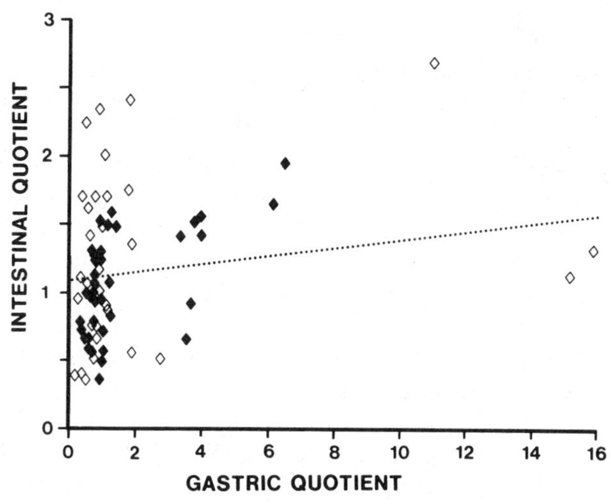

Fig. 5. Plot of intestinal quotient against gastric quotient for 73 mammal species (primates shown in black; nonprimates in white). The dotted line is the major axis.

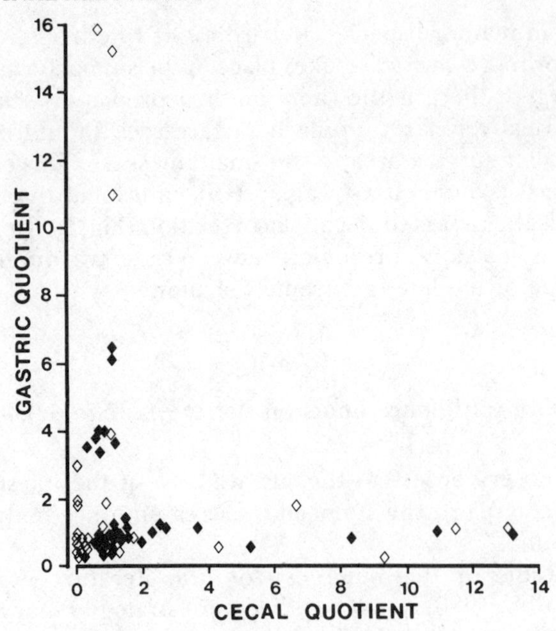

Fig. 6. Plot of gastric quotient against cecal quotient for 73 mammal species (key to symbols as for Fig. 5).

dichotomy between foregut- and midgut-fermenting species suggests, incidentally, that the common ancestral mammalian condition would have been typified by the absence of any commitment to either foregut or midgut specialization. Further, as has already been suggested by Hill and Rowell (1954), it is likely that the cecum was present in the common ancestral stock of mammals. This gut compartment has probably been lost as a secondary feature in lipotyphlous insectivores, some carnivores, edentates, cetaceans, and certain other mammals that have diets composed predominantly of animal food, and also in some folivorous mammals (e.g., artiodactyls) that have undergone specialization of the foregut at the expense of the cecum. The possession of a cecum in the ancestral mammalian condition doubtless indicates that *some* plant food was usually consumed, and the common blanket reference to ancestral placental mammals as "insectivores" is therefore probably misleading.

The development of the cecum can also be examined with respect to that of the colon, since there is evidence that in some mammals at least the two compartments of the gut are closely linked. In the horse, for example, enlargement of the colon is accompanied by expansion and considerable specialization of the proximal part of the colon (primitive right colon). A plot of the colonic quotient LQ against the cecal quotient CQ shows that there is some degree of positive correlation between enlargement of the colon and of the cecum, relative to body size (Fig. 7). The weak but positive correlation coefficient ($r = 0.51$) and the slight upward trend of the major axis suggest a weak link between these two compartments in the mammals overall. The formula indicated by the major axis is:

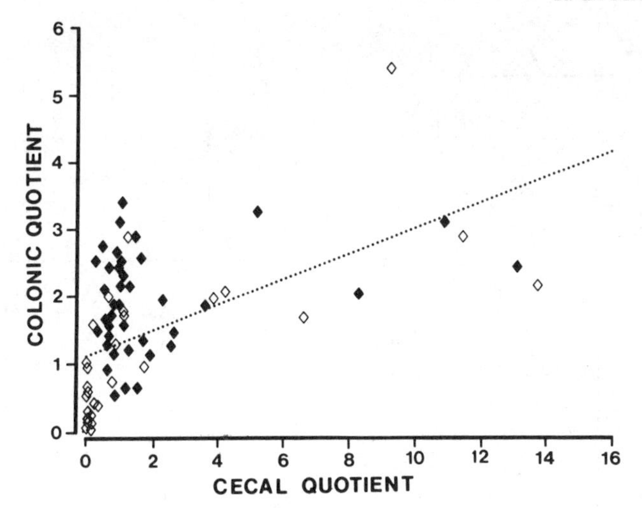

Fig. 7. Plot of colonic quotient against cecal quotient for 73 mammal species (key to symbols as for Fig. 5). The dotted line is the major axis.

$$LQ = 0.18\ CQ + 1.20$$

(95% confidence limits on slope: 0.11–0.25)

Examination of the plot of LQ against CQ (Fig. 7) shows that the correlation is relatively weak because some mammal species may have an enlarged colon (relative to body size) without any marked enlargement of the cecum. This indicates that at least part of the colon, presumably the primitive right colon, is enlarged as part of the adaptation for midgut fermentation, providing support for the contention of Chivers and Hladik (1980) that this region of the colon may be more important than the cecum in such fermentation. These results suggest that there are two different categories of colonic enlargement in mammals, only one of which is associated with cecal enlargement. Numerous primate species, in particular, exhibit an elevated LQ value in the absence of any marked elevation in CQ ($CQ < 2$).

Bivariate comparison of the various gut quotients, therefore, has some value, but a multivariate approach is required if some kind of overall picture is to be obtained. This has been attempted, both through the construction of dendrograms and through multidimensional scaling. Since there is evidence that maintenance of mammals in captivity may modify dimensions of the gut (see p. 84), the first dendrogram was constructed using only data for wild-caught primates (Fig. 8). It can be seen from the dendrogram that five major groups emerge (A–E), with three of them forming a well-defined cluster at a higher level (A–C). The two peripheral groups are composed of species known to exhibit specialization either of the foregut (D: colobine monkeys) or of the midgut (E: *Avahi, Euoticus, Lepilemur*). Group D species can be generally regarded as folivores, whereas group E species include a gum-feeding form

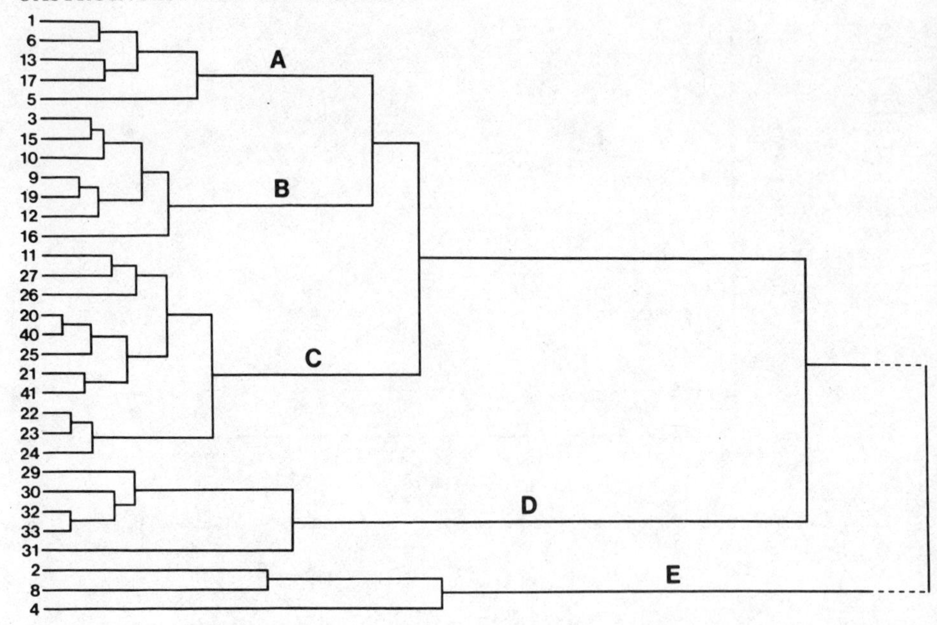

Fig. 8. Dendrogram constructed from the Euclidean distance matrix defined by the four gut quotient values for each of 31 primate species with data for wild-caught specimens. (See Table 1 for identification of individual species according to the numbers shown.) Capital letters indicate major groupings discussed in the text.

(*Euoticus elegantulus*) and a folivorous form (*Lepilemur leucopus*); *Avahi* has yet to be subjected to a detailed field study, but there is a clear indication from the dendrogram that this lemur is adapted for midgut fermentation of some kind. The other major groups (A–C) are less clear-cut in terms of dietary implications. Group B could be described as a general frugivore–insectivore category (*Cheirogaleus, Loris, Microcebus, Saguinus, Saimiri, Cebus, Miopithecus*), but groups A and C are both somewhat heterogeneous in composition. Group C forms could perhaps be described as frugivore–omnivores (*Perodicticus;* various cercopithecine monkey species; *Pan, Gorilla*), though this would not fit the general interpretation of the gorilla as a folivore. Group A forms are even more heterogeneous: *Arctocebus, Galago alleni, G. demidovii, Aotus, Alouatta.* One would certainly not infer from this grouping that *Alouatta* is the most folivorous of the New World monkeys (Milton, 1980)! Thus, the dendrogram has successfully separated out the foregut and midgut fermenters, but the remaining primate species (which usually include some fruit in their diets) are not clearly divided up into meaningful dietary categories. The dendrogram does indicate one thing quite clearly, however: the prior calculation of gut quotients has effectively eliminated body size as a factor in the grouping procedure, and there is no evidence that species have been grouped according to size rather than to differential adaptation of the gastrointestinal tract.

Having examined the situation with respect to wild-caught primates only,

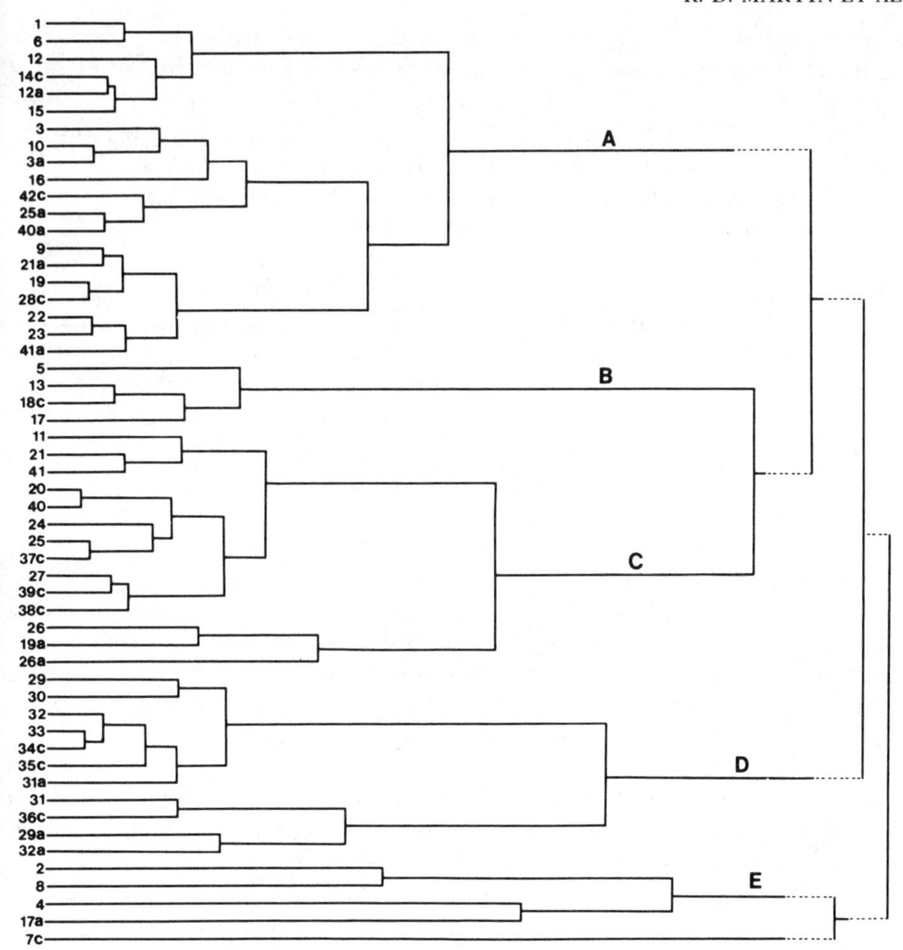

Fig. 9. Dendrogram constructed for 42 primate species (including *Homo sapiens*) with data for wild-caught and captive specimens. Species represented only by captive specimens are indicated by **c**; captive specimens of species for which wild-caught specimens were also available are indicated by **a**. (See Table 1 for key to species.) Capital letters indicate major groupings discussed in the text.

one can proceed to include captive specimens to see where they appear in the dendrogram, and whether they modify the overall configuration. For certain species, only captive specimens were available. For others, both wild-caught and captive specimens were available, so these were separated in the calculation of gut quotients and then entered separately into the dendrogram as a direct test of detectable differences brought about by captive diets. It can be seen from the dendrogram (Fig. 9) that there are five major groups, as before. Groups D and E retain their integrity. Captive specimens of *Colobus polykomos, Presbytis cristata, P. obscura, P. rubicunda, Pygathrix,* and *Nasalis* cluster with the other colobine monkeys in group D, while the captive *Lepilemur mustelinus* clusters in the midgut-fermenting group E with the wild-caught *Lepilemur leucopus*. The only anomaly is that the captive specimen of *Alouatta palliata* is grouped with the midgut fermenters (group E), suggesting a gut adapted for folivory, while the wild-caught specimen of this species falls in with group B. The reason in this case seems to be that the body weight of 1.47 kg given for the captive *Alouatta palliata* specimen (Chivers and Hladik, 1980) is well below the normal adult body weight of about 7 kg for this species. Hence, all gut compartments are larger than expected for body weight, though the general pattern is the same as for adults of normal body weight. This highlights the fact that results for captive specimens may be influenced not only by the effects of artificial diets but also by any deviations in body weight from the normal adult condition. *Alouatta palliata* is the most extreme case of difference in body weight between wild and captive specimens in the present sample.

The other groups (A–C) show rather more reorganization as a result of the inculsion of captive specimens. The groups themselves show some changes in composition, and there are a number of cases where captive specimens are separated from wild-caught specimens of the same species within the dendrogram. Apart from the problem of *Alouatta palliata* discussed above, five species appear in different categories when wild and captive specimens are compared (*Cercopithecus neglectus, Gorilla gorilla, Macaca sinica, Miopithecus talapoin,* and *Pan troglodytes*). Four of these species shift from group C to group A, while the remaining species (*Miopithecus talapoin*) shifts in the opposite direction for captive versus wild specimens. For the remaining three species, however, there is no difference between wild and captive specimens with respect to clustering into the major groups of the dendrogram (*Cheirogaleus major, Macaca fascicularis, Saguinus geoffroyi*). Once again, body weight differences between wild and captive specimens seem to be a contributory factor. Contrary to expectation, captive specimens may be either heavier or lighter than wild specimens, but the difference is only of major significance among these species in the case of *Gorilla*. (The captive gorilla specimen, a male weighing 236 kg, was undoubtedly obese.) In any event, groups A–C are relatively labile and do not clearly indicate dietary categorization, so shifts between these groups are not of major significance. In view of this, the inclusion of *Homo sapiens* within the subcluster A does not provide a very specific indication of dietary adaptation of gut compartments.

At this point, it is useful to turn to the diagram produced by the multidimensional scaling technique for primate species (Fig. 10). In this analysis, the species documented by wild specimens represented in the dendrogram of Fig. 8 (N = 31) have been supplemented with 11 species for which only captive data are available (including *Homo sapiens* as a "captive" species), giving a total sample of 42 primate species. It is clear from the diagram that the foregut-fermenting colobine species (numbers 29–36 inclusive) form a distinct outlying cluster (bottom right) quite separate from another outlying cluster

→

Fig. 10. Multidimensional scaling diagram for 42 primate species. Wherever possible, only data for wild-caught specimens have been used; species represented only by captive specimens are indicated by **c**. (See Table 1 for key to species.)

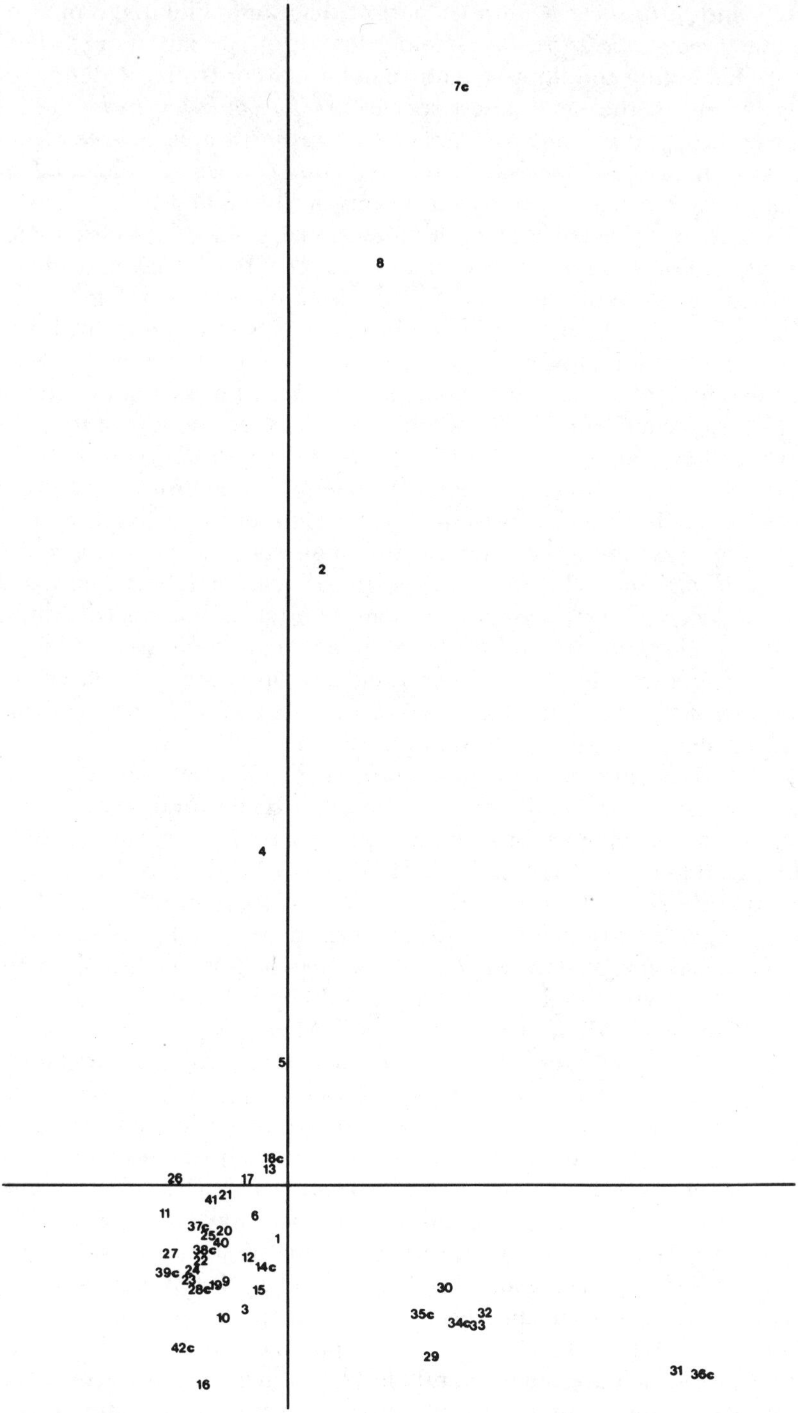
7c
8
2
4
5
18c
13
26
17
41
21
11
6
37c
25
20
40
1
27
38c
22
12
14c
39c
24
23
28c
19
9
15
10
3
30
35c
32
34c
33
42c
29
31
36c
16

formed by primates characterized by midgut specialization (*Avahi,* 2; *Lepilemur,* 7 and 8; *Euoticus,* 4—upper part of diagram). The diagram also suggests that *Galago alleni* (5) shows some degree of specialization toward the midgut-fermenting condition, a feature not apparent from the dendrograms in Figs. 8 and 9; the next closest species are *Lagothrix lagotricha* (18), *Aotus trivirgatus* (13), and *Alouatta palliata* (17), suggesting a mild degree of midgut specialization in these species as well. All four species are included in the anomalous category A of the dendrograms in Figs. 8 and 9.

Thus it would seem that multidimensional scaling provides somewhat different, and perhaps more accurate, information on clustering primate species according to gut quotient values. It is also apparent from Fig. 10 that the remaining primates form a fairly dense central cluster of points with no very clear differentiation. However, a number of general comments can be made. First, the gorilla (41) shows some possible affinity with the midgut specialization cluster. Second, those wild-caught primates that are known to include a fair proportion of insects in the diet (e.g., *Arctocebus,* 1; *Cheirogaleus,* 3; *Galago demidovii,* 6; *Loris,* 9; *Microcebus murinus,* 10; *Saguinus geoffroyi,* 12; *Saimiri,* 15; *Miopithecus,* 19) form a cluster to the bottom right of the generalized primate group. This positioning of the cluster suggests that primates with frugivorous–insectivorous habits may have guts that are mildly specialized in the direction associated with foregut fermentation (cf. points for colobines). Finally, *Cebus capucinus* (16) and *Homo sapiens* (42c) are quite separate from the main, more central cluster of generalized primates, being displaced toward the bottom of the diagram. This, too, represents a clear departure from the picture presented by the dendrogram in Fig. 9.

Comparisons limited to primates have the drawback that only a restricted perspective on dietary adaptation is permitted. Inclusion of the other mammal species in the comparisons should therefore permit a more detailed analysis of gut dimensions. As Chivers and Hladik (1980) argue, it is these other mammals that show the greatest specializations for faunivory and folivory, while primates have remained relatively unspecialized and flexible. Data for a total of 73 mammal species were available for analysis, including 51 species represented by wild-caught specimens and 22 by only captive specimens (42 species of primate, 31 of other mammals). When the data are processed to produce a dendrogram, as before, it is possible to recognize five major groups of species (Fig. 11, groups A–E). Group E, as before, contains species characterized by midgut specialization. In addition to the prosimian primates previously recognized in this group, there are four nonprimates from four different orders (*Dendrohyrax, Equus, Heliosciurus, Oryctolagus*) similarly characterized by midgut specialization. The inclusion of *Heliosciurus* in this group is perhaps somewhat surprising, since this West African sun squirrel is supposed to have a very generalized diet (Walker, 1968) rather like North American and European squirrels (*Sciurus*), yet the latter cluster quite separately in group B (Fig. 11). Group D is composed of three artiodactyls (*Capra, Cervus, Ovis*), which are only distantly linked to memebers of group C (colobine monkeys, sloth, kangaroo, and pangolin), yet groups C and D together are characterized by foregut fermentation.

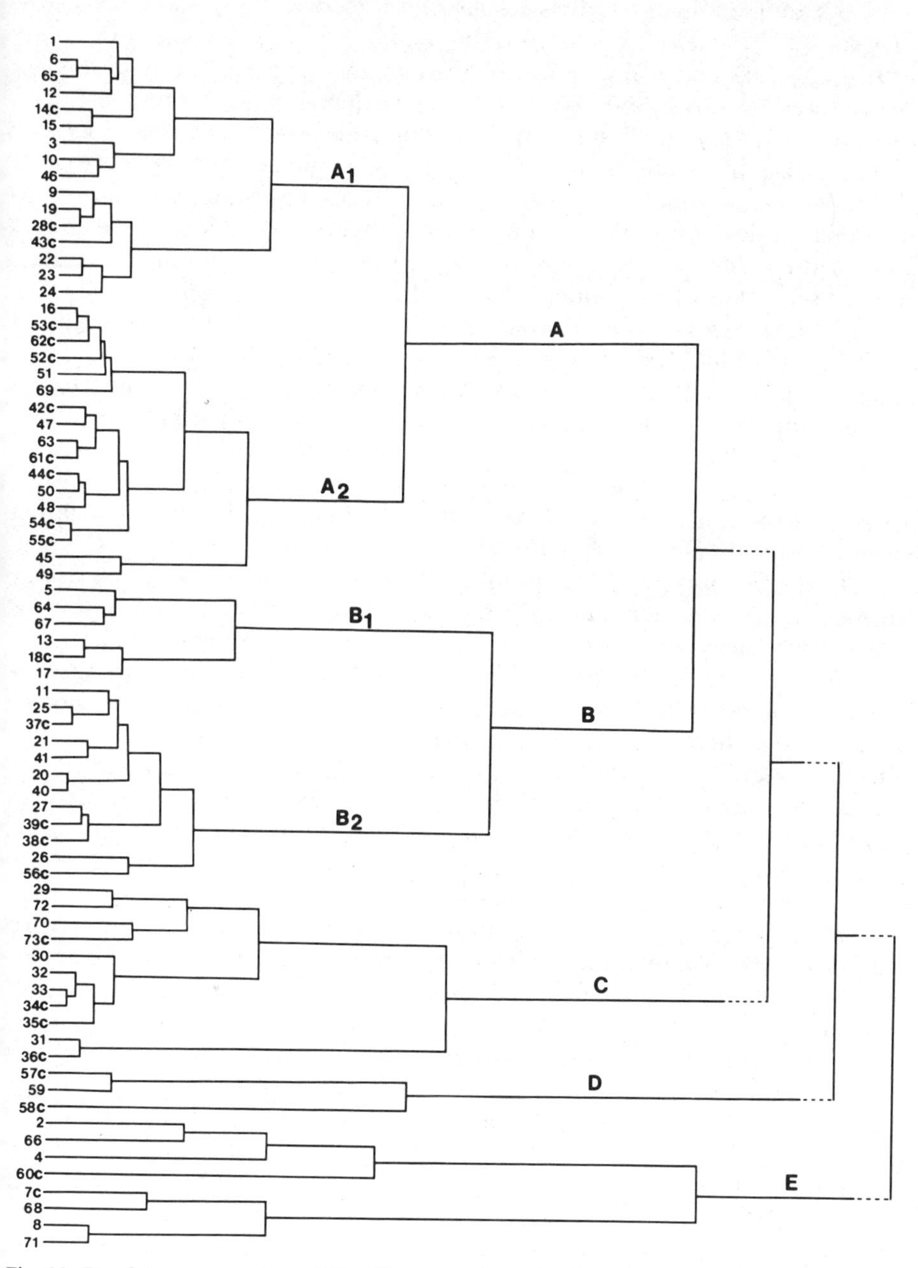

Fig. 11. Dendrogram constructed for 73 mammal species (including *Homo sapiens*). Wherever possible, only data for wild-caught specimens have been used; species represented only by captive specimens are indicated by **c**. Capital letters indicate major groupings discussed in the text.

Group B contains a number of primate species that include fairly large amounts of fruit in their diets (*Galago alleni, Perodicticus, Aotus, Lagothrix, Alouatta, Hylobates, Cercopithecus neglectus, Pan, Pongo, Macaca fasicularis*) along with the gorilla and a few nonprimates that might loosely be described as "omnivores" (*Sciurus, Sus*). Further, there is clear separation into two subcategories (B_1, B_2), the first of which contains the primates mentioned above (on the basis of multidimensional scaling) as exhibiting a mild degree of midgut specialization. Group A can be characterized as containing numerous mammal species (primates and nonprimates) that include at least some animal food in their diets. Again, there is a separation into two subcategories (A_1, A_2), the second of which contains most of the mammalian carnivores and only two primate species—*Cebus capucinus* and *Homo sapiens*. Group A_1 contains, in addition to primate frugivore–insectivores (see above), a number of "omnivorous" primates (wild-caught *Cercopithecus nictitans, Cercocebus,* and *Macaca sylvana,* along with captive *Erythrocebus* and *Ateles*), two generalist carnivores (*Vulpes, Felis*), and a rodent (*Epixerus*). Unfortunately, nothing seems to be known of the natural diet of the African palm squirrel, *Epixerus* (Walker, 1968), but there is a clear prediction from the dendrogram that this rodent relies to some extent on animal food.

Once again, it is possible to examine the data in a somewhat different way through multidimensional scaling (Fig. 12). As with Fig. 10, the most obvious trends are indicated by species with either foregut specialization (bottom left of diagram) or midgut specialization (top). The diagram thus clearly associates the foregut-fermenting forms into a single cluster, which was not so obvious from the dendrogram (Fig. 11). Once again, the colobines are distinctly separated from the other primates, and the trend they indicate is carried to an extreme in the artiodactyls (*Capra, Ovis, Cervus*). The sloth lies quite close to tbe colobine assemblage, in line with its specialized folivorous habits. It is interesting to note that, whereas the kangaroo also lies relatively close to the colobines, so does the myrmecophagous pangolin. This reinforces the suggestion derived from the multidimensional scaling diagram of primates alone that adaptation to include arthropods in the diet may be associated with mild modification of the gut in the direction found with foregut fermenters.

Among the species with midgut specialization, included in the outlying cluster running obliquely to the top left of the diagram (Fig. 12), there are all of the species included in group E of the dendrogram (Fig. 11). In addition, it can be seen that the species included in group B_1 of that dendrogram lie at the origin of this trend, along with *Perodicticus* and *Cercopithecus neglectus.* For some reason, *Mustela nivalis* is plotted quite close to the two *Sciurus* species, suggesting that this carnivore may have some degree of midgut specialization.

As before, there is a fairly dense, more central cluster of points formed by the remaining species, in the bottom right quadrant of the diagram (Fig. 12). There is some differentiation within this cluster, however. First, there is a fairly well-demarcated group at the extreme bottom right of the diagram, which corresponds to group A_2 of the dendrogram in Fig. 11, with the exclu-

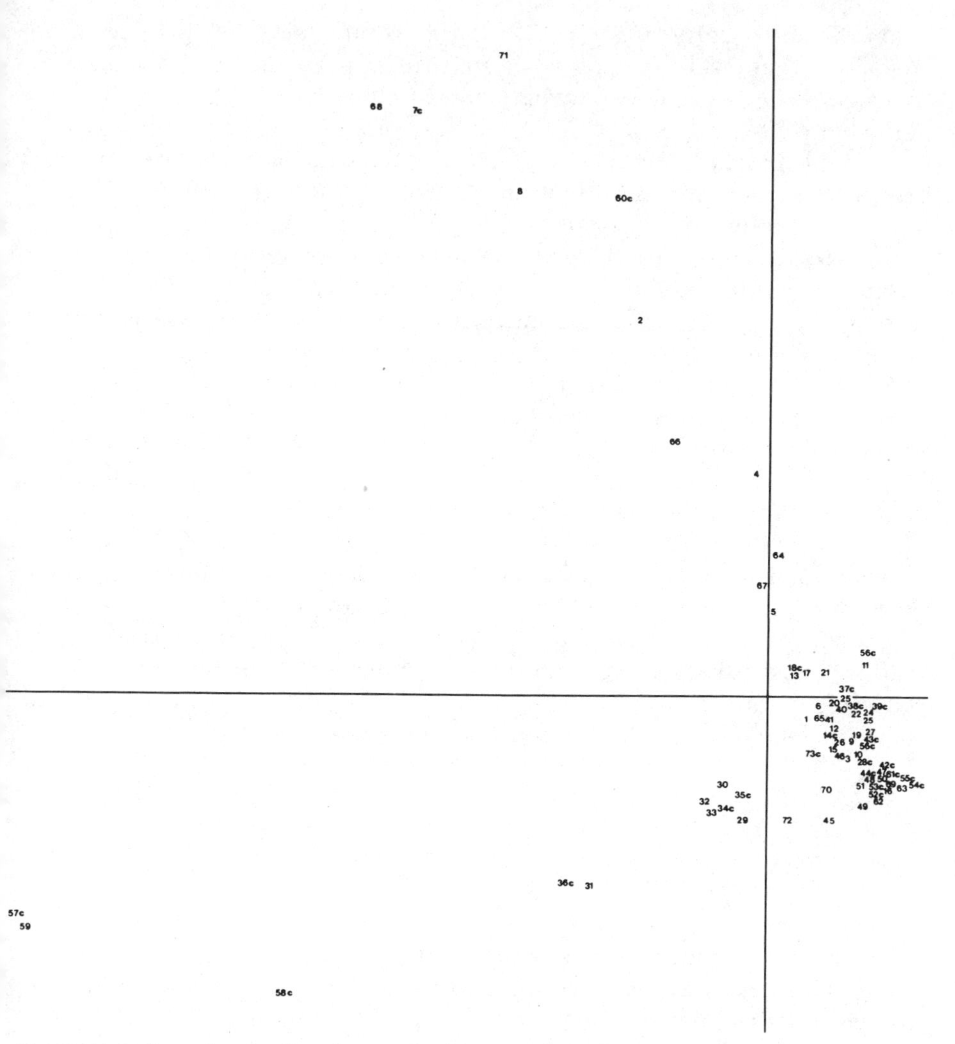

Fig. 12. Multidimensional scaling diagram for 73 mammal species (42 primates; 31 nonprimates). Wherever possible, only data for wild-caught specimens have been used; species represented only by captive specimens are indicated by **c**. (See Table 1 for key to species.)

sion of *Mustela nivalis* (see above). This group contains most of the species from the order Carnivora, two faunivorous toothed cetaceans, the faunivorous African water shrew (*Potomogale*), and just two primate species—*Cebus capucinus* and *Homo sapiens*. This provides, therefore, some evidence that both the capuchin monkey and man exhibit some gut adaptations for faunivory; since the former were wild-caught, this indication is fairly reliable, but in man we may be dealing with direct gut modifications through particular dietary habits, particularly since our own species (42c) is at the periphery of the faunivorous assemblage. Nevertheless, there is some suggestion here of adaptation of the modern human gut for a diet containing a fair proportion of animal food, and this merits further study (e.g., using data from the guts of humans with a wide variety of dietary habits). The remaining cluster of points in the bottom right quadrant of the diagram can also be divided broadly into two roughly parallel bands running obliquely from left to right. The left-hand band is composed primarily of primate frugivore–insectivores, but also includes three other wild primates (*Erythrocebus, Gorilla, Macaca fascicularis*), the captive *Ateles,* a carnivore (*Vulpes*), and a rodent (*Epixerus*). This corresponds to some extent to the subgroup A_1 in the dendrogram (Fig. 11), but differs both in the exclusion of the domestic cat (*Felis*) and three cercopithecine monkeys (*Cercocebus, Cercopithecus nictitans,* and *Macaca sylvanus*), and in the inclusion of *Gorilla* and *Macaca fascicularis* (part of subgroup B_2 in Fig. 11). The right-hand band is composed largely of primates, most of which include some fruit in their diets (*Cercopithecus cephus, C. nictitans, Hylobates, Pan, Pongo, Macaca sylvanus, M. sinica, Papio sphinx*), and also includes two domestic mammals—the cat and the pig. Thus, as in the case of the analysis of primates alone, the multidimensional scaling diagram does differ in some respects from the dendrogram, and there are a few noteworthy differences in the clustering patterns produced with the two techniques.

Finally, it is worth noting that all of these multivariate approaches have been repeated using simple ratios of gastrointestinal compartment areas to body weight, rather than quotient values. As expected, the results are virtually meaningless and are still strongly influenced by body weight effects.

Discussion

One of the main concerns in attempting to interpret the results of the analyses presented above is the extent to which the dimensions of the gastrointestinal tract may be modified by maintenance of the species concerned on an artificial diet. This problem is particularly acute when attempting to infer any kind of adaptation of the human gut for a particular diet, since the data obtained may simply reflect the diet of the individuals concerned, rather than any species-specific pattern. Evidence has already presented above for the modification of the gut in some captive primate species. Inclusion of wild and captive data for 12 primate species as separate entries into the dendrogram shown in Fig. 9 demonstrated that significant modification of the resulting position in the dendrogram occurred in half of them. With the colobine species the captive specimens clustered as expected with their wild counterparts in the distinct group of foregut fermenters. There were three other primate species whose position in the dendrogram was virtually the same for both wild and captive specimens (*Cheirogaleus major, Saguinus geoffroyi,* and *Macaca fascicularis*), suggesting that for many species the gut may be modified very little by artificial diets. Hladik (1967) quoted evidence that surface areas of gut compartments may be modified in both directions by diets in captivity. An individual *Cercopithecus nictitans* was found to exhibit considerably smaller gut surface areas compared to the average condition for three wild-caught conspecifics, whereas an individual *Miopithecus talapoin* was found to exhibit considerably larger gut surface areas compared to the average condition for three wild-caught conspecifics. Since all gut compartment surface areas, however, show negative allometric variation with respect to body weight, the figures provided are difficult to interpret without reference to body weight scaling effects. One useful comparison, with relatively larger sample sizes, is provided by data on gut compartment *lengths* in the squirrel monkey (*Saimiri sciureus*), which are available both for wild-caught animals (Fooden, 1964) and for animals maintained for a number of years in captivity (Beischer and Furry, 1964). Direct comparison of the data indicates that the captive specimens do indeed have shorter stomachs, shorter small intestines, and shorter colons, though the cecum shows no difference from the wild specimens (Table 2). Even this comparison, however, must be treated with caution, since the squirrel monkeys concerned may have come from quite different sources (Fooden's animals were from Surinam; those of Beischer and Furry were simply referred to as "*Saimiri sciureus*" with no indication of their country of origin). Further, there is a mild difference in average body weight between the two groups of squirrel monkeys (704 ± 90 g for wild-caught; 644 ± 130 g for captive), which might account to some extent for smaller gut dimensions in the captive animals examined by Beischer and Furry. Nevertheless, it can be concluded that there is some evidence overall for modification of gut dimensions in captive primates, notably in certain folivorous species.

Accordingly, data from wild-caught specimens should be used wherever possible in conducting comparative quantitative studies of the mammalian gastrointestinal tract. It is interesting to note, in this connection, that birds also exhibit differences in gut dimensions when maintained in captivity. Both barnacle geese and red grouse have been shown to exhibit shorter intestines and ceca in captivity as compared with the condition found in wild-living

Table 2. Comparison of Gastrointestinal Tract Dimensions for Wild and Captive Squirrel Monkeys

	Gut compartment lengths, mm			
Source	Stomach	Small intestine	Cecum	Colon
Wild squirrel monkeys, from Surinam; *N* = 11 (Fooden, 1964)	84 ± 10	1149 ± 91	39 ± 6	159 ± 15
Captive squirrel monkeys, source unspecified; *N* = 10 (Beischer and Furry, 1964)	63 ± 16	966 ± 151	39 ± 10	83 ± 20
Percent reduction in captive specimens	25%	16%	0%	48%

conspecifics (Sibly, 1981). It has even been demonstrated that gut dimensions may change in the course of the annual cycle in various wild bird species [see Sibly (1981) for references]. In the case of starlings (*Sturnus vulgaris*), the minimal length of the intestine and cecum has been shown to coincide with the time of year when plant food represents the smallest proportion of the diet (Al-Jaborae, 1980). There is also experimental evidence that gut dimensions *increase* when certain bird species are maintained on poor diets (Sibly, 1981).

The question of modification of gut dimensions through change in diet is just one part of the overall phenomenon of *intraspecific variation*. It is important to know what the range of variation in gut dimensions may be, even under broadly similar conditions, in order to assess (for example) what sample size of a single species might be required in order to determine typical values. In the present analysis, most species have been represented by only a few individuals and there are only nine species for which sample sizes are five or more. Coefficients of variation have been calculated for the surface area of each gut compartment for each of these nine species (Table 3). From this analysis, it emerges that the gut compartments all exhibit considerable intraspecific variation in surface area, with the overall coefficient of variation approximating 25%. In other words, the 95% range of variation for the surface area of a given gut compartment in any species is approximately 50%, and the 99% range of variation is 75%. This degree of intraspecific variation is markedly greater than for most other organs of the body that have been subjected to allometric analysis (e.g., brain, heart, lungs), and provides yet another indication of the relative lability of gut dimensions. Clearly, confident interpretation of the gut surface area data (e.g., for predictive purposes with

Table 3. Coefficients of Variation for Surface Areas of Gastrointestinal Compartments (Intraspecific Variation)

Species	Number	Coefficients of variation, %			
		Stomach	Small intestine	Cecum	Colon
Homo sapiens	6	27.5	15.8	24.5	16.7
Arctocebus calabarensis	5	36.8	18.5	18.9	15.6
Cheirogaleus major	5	28.4	28.6	35.9	35.5
Presbytis melalophos	6	27.8	21.8	35.1	26.9
Felis domestica	6	12.8	15.1	21.2	16.9
Canis familiaris	9	35.1	33.3	21.4	33.1
Vulpes vulpes	5	16.6	6.1	19.8	17.5
Dendrohyrax dorsalis	5	9.8	27.9	20.2	21.0
Cervus elaphus	16	31.9	26.7	30.5	28.6
Overall averages	—	28.5	21.5	24.9	23.5

species for which the natural diet is unknown) requires fairly large sample sizes, which must be collected under clearly defined conditions. In the case of *Homo sapiens,* there is a tentative indication that our own species is somewhat aberrant in comparison with nonfolivorous primates (Fig. 10), and that our affinities lie with faunivorous mammals when the basis for comparison is widened (Figs. 11 and 12). More detailed work is required, however, with larger samples and covering a defined range of dietary habits before this indication can be taken really seriously.

Even given the limitations of the data set analyzed here, a number of firm conclusions can be drawn. It is, for instance, quite clear that surface areas of gut compartments scale in a negative allometric fashion with body size in mammals overall. Further, this scaling is compatible with Kleiber's law (slope value $\approx$ 0.75) in the case of the stomach, small intestine, and cecum, though some doubt exists about the colon, where the slope may be significantly in excess of 0.75 (though still negatively allometric). Calculation of gut compartment quotients (GQ, IQ, CQ, LQ) according to a fixed slope of 0.75 is justified, at least as a first approximation, and it is successful in the sense that species clustered on the basis of overall gut quotient values do seem to group according to dietary adaptation and not simply as a direct reflection of body size. By contrast, simple ratios of gut compartment surface area to body weight do not group in this way. The excellent fit of the empirical allometric formula for the scaling of the surface area of the small intestine against body weight to the expectation from Kleiber's law (slope 0.75) is particularly encouraging for the basic hypothesis that the gastrointestinal tract is essentially scaled according to metabolic requirements. Nevertheless, it must be emphasized that this very simple approach to allometry is essentially static, and that *gut transit times* should ideally be incorporated into any overall model based on a direct link between metabolic requirements and gut allometry.

Overall, analysis of the gut quotients calculated for the 73 mammal species investigated (42 primates, 31 nonprimates) has been quite productive. Bivariate plots of certain quotients have demonstrated that:

1. The surface area of the small intestine is *not* consistently smaller in species with enlarged stomachs.
2. Mammals may exhibit *either* an enlarged stomach *or* an enlarged cecum (and primitive right colon) relative to body size, but no mammal species investigated showed enlargement of both these quotients.
3. There is a link between the cecum and colon in that mammal species with an enlarged cecum normally exhibit enlargement of the colon, but there are several mammal species that exhibit colonic enlargement in the absence of any enlargement of the cecum.

Multivariate clustering techniques (dendrogram, multidimensional scaling), based on the matrix of Euclidean distances derived from the four gut quotient values calculated for each species, have also proved to be generally useful. Mammals (both primates and nonprimates) with specialization of either the foregut or midgut for a folivorous diet are clearly separated out with both clustering techniques. There is, however, considerable lability among mammals species exhibiting less specialized guts, and the multidimensional scaling technique proved to be somewhat more useful for these species, correcting certain anomalies arising with the dendrogram technique.

As a final note, it should be emphasized particularly that special care must be exercised in the interpretation of dendrograms and multidimensional scaling diagrams with respect to *phylogenetic* relationships. Multivariate clustering techniques of this kind provide nothing more than an overall assessment of similarity, and it is now well recognized that for numerous reasons similarity alone does not provide an adequate basis for inferring phylogenetic relationships. For this reason, no attempt has been made here to draw any conclusions regarding phylogeny; indeed, it is fairly obvious from the results obtained that direct functional adaptation of the gut for different dietary habits is a far more potent factor than phylogenetic affinity in determining the patterns observed.

ACKNOWLEDGMENTS

Special thanks are expressed to F. Brett for his invaluable advice and many hours of direct assistance in the computation methods used for multivariate clustering. Thanks also go to Dr. M. Hills and Dr. P. Harvey for many extremely useful discussions of techniques of allometric analysis. All basic computations of data were carried out by A. M. MacLarnon, with the support of a Research Assistantship from University College, London. We thank Professor A. Gresham for access to the human material, Miss M. Brancker, FRCVS, for most of the captive primate material, the Zoological Society of London for other captive primate material, and Professor R. Harrison for the marine mammals.

References

Al-Jaborae, F. F. 1980. The Influence of Diet on the Gut Morphology of the Starling (*Sturnus vulgaris* L 1758), D. Phil. Thesis, University of Oxford.

Bauchot, R. 1982. Brain organization and taxonomic relationships in Insectivora and Primates, in: *Primate Brain Evolution: Methods and Concepts* (E. Armstrong and D. Falk, eds.), pp. 163–175, Plenum Press, New York.

Beischer, D. E., and Furry, D. E. 1964. *Saimiri sciureus* as an experimental animal. *Anat. Rec.* **148:**615–624.

Böker, H. 1932. Beobachtungen und Untersuchungen an Säugetieren während einer biologisch-anatomischen Forschungsreise nach Brasilien im Jahre 1928. *Morphol. Jahrb.* **70:**1–66.

Chivers, D. J., and Hladik, C. M. 1980. Morphology of the gastrointestinal tract in primates: Comparisons with other mammals in relation to diet. *J. Morphol.* **166:**337–386.

Fooden, J. 1964. Stomach contents and gastro-intestinal proportions in wild shot Guianan monkeys. *Am. J. Phys. Anthropol.* **22:**227–232.

Gould, S. J. 1966. Allometry and size in ontogeny and phylogeny. *Biol. Rev.* **41:**587–640.

Harvey, P. H., and Mace, G. M. 1982. Comparisons between taxa and adaptive trends: Problems of methodology, in: *Current Problems in Sociobiology* (King's College Research Group, ed.), pp. 343–361, Cambridge University Press, Cambridge.

Hemmingsen, A. M. 1950. The relation of standard (basal) energy metabolism to total fresh weight of living organisms. *Rep. Steno. Mem. Hosp.* **4:**7–58.

Hemmingsen, A. M. 1960. Energy metabolism as related to body size and respiratory surfaces, and its evolution. *Rep. Steno. Mem. Hosp.* **9:**1–110.

Hill, W. C. O., and Rewell, R. E. 1954. The caecum of Monotremata and Marsupialia. *Trans. Zool. Soc. Lond.* **28:**185–240.

Hladik, C. M. 1967. Surface rélative du tractus digestif de quelques primates, morphologie des villosités intestinales et correlations avec le régime alimentaire. *Mammalia* **31:**120–147.

Janis, C. 1976. The evolutionary strategy of the Equidae and the origins of rumen and caecal digestion. *Evolution* **30:**757–774.

Kermack, K. A., and Haldane, J. B. S. 1950. Organic correlation and allometry. *Biometrika* **37:**30–41.

Kleiber, M. 1961. *The Fire of Life: An Introduction to Animal Energetics,* Wiley, New York.

Kruskal, J. B. 1964*a*. Multidimensional scaling by optimising goodness of fit to a nonmetric hypothesis. *Psychometrika* **29:**1–27.

Kruskal, J. B. 1964*b*. Nonmetric multidimensional scaling: A numerical method. *Psychometrika* **29:**115–129.

Mace, G. M., and Harvey, P. H. 1982. Energetic constraints on home range size. *Am. Nat.* **121:**120–132.

Martin, R. D. 1980. Adaptation and body size in primates. *Z. Morphol. Anthropol.* **71:**115–124.

Middleton, C. C., and Rosal, J. 1972. Weights and measurements of normal squirrel monkeys (*Saimiri sciureus*). *Lab Anim. Sci.* **22:**583–586.

Milton, K. 1980. *The Foraging Strategy of Howler Monkeys,* Columbia University Press, New York.

Schmidt-Nielsen, K. 1972. *How Animals Work,* Cambridge University Press, Cambridge.

Sibly, R. M. 1981. Strategies of digestion and defecation, in: *Physiological Ecology: An Evolutionary Approach to Resource Use* (C. R. Townsend and P. Catlow, eds.), pp. 109–139, Blackwell Scientific Publications.

Walker, E. P. 1968. *Mammals of the World,* 2nd ed., John Hopkins Press, Baltimore.

Ward, J. H. 1963. Hierarchical grouping to optimize an objective function. *J. Am. Stat. Assoc.* **58:**236.

Organ Weight Scaling in Primates

6

SUSAN G. LARSON

Introduction

Analysis of organ weight allometry in primates has been largely confined to examination of brain weight scaling [see Gould (1975) and references therein]. This emphasis reflects our quite natural interest in the unique degree of cerebral development in primates in general and in our species in particular. The scaling of the other internal organs (e.g., heart, lungs, kidneys) has received far less attention, however. Only a handful of studies have addressed the question of primate soft tissue allometry. These include Stahl (1965) and Stahl and Gummerson (1967) on interspecific organ weight allometry; Gest and Siegel (1983) and Larson (1978) on intraspecific organ weight scaling among adult olive baboons and stumptail macaques, respectively; and Snow and Vice (1965) and Stahl *et al.* (1968) on ontogenetic organ weight allometry in olive baboons and howler monkeys, respectively. Though all represent valuable contributions, to answer general questions about organ weight allometry, such as whether all primate species share a common developmental pattern, or if some species possess unique specializations in terms of the size of certain organs, we require a broad, comparative data base in order to discern regular patterns and recognize special cases. The present study represents an effort to supply such a data base. It examines the ontogenetic scaling of nine internal organs in each of six primate species. The importance of studying ontogentic trends has been emphasized in several recent papers (e.g., Alberch *et al.*, 1979; Jungers and Fleagle, 1980; Shea, 1981). These authors have shown that understanding the developmental patterns of related species can be an extremely useful aid in determining their phylogenetic relationships

SUSAN G. LARSON • Department of Anatomical Sciences, School of Medicine, State University of New York, Stony Brook, New York 11794.

and the evolutionary mechanisms behind their differentiation. This study will focus on the developmental origin of adult intersexual and interspecific size differences.

Materials and Methods

Experimental Subjects

The experimental subjects of this study include five macaque species (*Macaca fascicularis, M. radiata, M. mulatta, M. nemestrina, M. arctoides*) and one baboon species (*Papio cynocephalus**). There are a number of advantages to this choice of species. Delson (1980) has described *Macaca* as a conservative genus, retaining numerous morphological features that characterize the early cercopithecines. This retention of ancestral features is no doubt a consequence of their position near the generalist end of an adaptive strategy spectrum. Though different macaque species tend to prefer particular habitats, all divide their time between the trees and the ground, and all are omnivorous, opportunistically exploiting a wide variety of resources, including cultivated fields (Roonwal and Mohnot, 1977). This unspecialized and flexible adaptive strategy is reflected in their rather generalized body form. Interspecies differences in organ sizes, then, will be due in large part to differences in absolute size, thus making macaques ideal subjects for a study of the allometric relations between organ sizes and body size. In addition, this generalized body form suggests that the organ weight scaling patterns displayed by these species will be useful as a comparative base for other, more specialized species.

The baboon species *P. cynocephalus* was included as a larger body-sized comparison to the macaques. Though baboons are somewhat more specialized to a terrestrial habitat than the macaques, nearly all spend some time in the trees for feeding and sleeping (Napier and Napier, 1967). Moreover, like macaques, their diet is omnivorous and somewhat opportunistic. In addition, baboons and macaques are closely related, sharing a common ancestor as little as 6–7 million years ago (Delson, 1975, 1980; Cronin *et al.*, 1980). In terms of internal design, baboon morphology, then, is unlikely to be very different from that of the macaques except in those features that reflect their differences in overall size.

Data Sources

The organ and body weight data for this study were obtained from the necropsy records of the following research institutions: the California, Wash-

*The sample for this species is made up of hybrids of the *cynocephalus* and *anubis* groups (A. M. Coelho, personal communication). Though perhaps more correctly called *P. cynocephalus anubis*, the baboons examined in this study will be referred to simply as *P. cynocephalus*.

ington, Wisconsin, and Yerkes Regional Primate Research Centers; the Southwest Foundation for Research and Education; and The Johns Hopkins University. All records were screened in order to collect samples of ordinary, healthy individuals. Animals considered ideal were those sacrificed as controls for other studies or for normal tissue samples. Animals used in projects potentially affecting normal growth and development or in projects affecting total body weight were not included, nor were individuals that were obese or emaciated, or generally debilitated for any reason. [For a more complete discussion of necropsy screening and sample collection procedures, see Larson (1982).]

Animals of all ages were collected, and for each individual total body weight and the weights of its heart, lungs, kidneys, liver, pancreas, spleen, adrenals, thyroid, and gonads were recorded.* In addition, estimates of age at time of death were recorded when available (for *M. mulatta, M. nemestrina, M. arctoides,* and *P. cynocephalus* only).

Data Analysis

Organ weight variation was examined relative to age and to total body weight. The analysis by age involved two phases: first, a year by year growth profile was developed for the rhesus monkey sample to give a general picture of a macaque growth pattern. Second, the body and organ weight data for each species were regressed against age to produce an interspecies comparison of growth rates. This was done on those portions of the samples up to early adulthood in age (0 to 9 years in males and 0 to 6 years in females). Although the relation between size increase and age through the entire life history of these animals is curvilinear, during this initial period, when the major portion of growth is completed, size increase is essentially a linear function of age. The increase in size during adulthood proper is minor in comparison and most likely due to increased fat deposition. Least squares regression was used to derive these growth rate equations, since age was assumed to be a measurement error-free variable (Sacher, 1970). For each regression a correlation coefficient and a 95% confidence interval for the slope estimate were computed.

In the body to organ weight analysis, each species was examined with the sexes combined and each sex separately. After logarithmic transformation†, allometric equations of the form

$$\log Y = k \log X + \log b$$

*In many cases one or more organ weights were missing in the necropsy report. Therefore, sample sizes for any particular organ will differ from the total sample size for a species.

†See Larson (1982) for a discussion of the applicability of logarithmic transformation in this study.

where Y is organ weight and X is body weight, were fit to the data using the major axis technique (Jolicoeur and Heusner, 1971). Least squares regression, the traditional line fitting technique in allometric analysis, was not used here due to the objections raised by many workers to its underlying assumption of error-free measurement of X (e.g., Kuhry and Marcus, 1977; Jolicoeur and Heusner, 1971; Sacher, 1970; Pilbeam and Gould, 1974). For each bivariate relation, a correlation coefficient was computed and a 95% confidence interval was calculated for the slope estimate. As an indicator of the general relationship between an organ weight and body weight, the arithmetic mean of the individual species' slopes was computed for each organ.

Results

Analysis by Age

The analysis-by-age results are summarized in Fig. 1 and Table 1. Figure 1 displays the average year-to-year growth profile for total body weight and each organ weight for the rhesus monkey. Part A presents the curve for body weight, parts B–J for the various organs. The points along the curves represent the mean body or organ weight values for each one year age group, and the vertical lines reflect one standard deviation on either side of the means.[5] In most cases the curves display an initial phase of rapid weight increase, which gradually levels off in early adulthood. Thus among adults there is little or no increase in weight with increasing age (though there appear to be somewhat higher levels of variability). A major distinction between the sexes emerges in the age at which this leveling off occurs. Among females growth ceases soon after attainment of sexual maturity (approximately 4 years of age). Males, however, continue to grow until 7 or 8 years of age.

The gonads display a pattern of increase quite unlike that of the other internal organs. In males, the testes show little weight change in the first years of life, but undergo a marked enlargement beginning at approximately 3 years of age. This size increase no doubt corresponds to the histological alterations associated with the onset of spermatogenesis. The ovaries, however, do not show a similar change in association with menarche. In fact, ovary weight does not display any apparent correlation to increasing age.

Table 1 presents the equations for body and organ weights regressed against age. These equations cover the initial period of growth among juveniles and young adults. The slope of each equation represents the growth rate, and the y intercept represents the initial size of the organ. Also included in the

*Recently Cupp and Uemura (1981) have taken very similar data and derived exponential functions to fit weights plotted against age. Their equations agree well with the mean weights in Fig. 1 and are recommended to anyone wishing to predict a weight from an animal's age.

table is the correlation coefficient for each regression, a 95% confidence interval for each slope, and the probability of sex differences in slope values.

In all four species the growth rate for the body as a whole is significantly higher among males. Among the individual organs, however, there are few consistent sex differences in growth rates. Only the heart parallels the body as a whole, with higher growth rates among males for all species (though not all differences are statistically significant). Sex differences, then, appear both in the rate and duration of growth.

Interspecies comparison of the duration of growth is difficult due to small sizes and uneven distributions of two of the samples. But visual inspection of the point scatters of weights plotted against age does not suggest any major timing differences between the species. So, unlike intersexual size dimorphism, alterations in the length of the growth period do not contribute to interspecies size differences. Comparison of the regression slopes among the four species, however, reveals a gradient of increasing growth rates corresponding to the order of increasing mean adult sizes (smallest to largest: *M. mulatta, M. nemestrina, M. arctoides, P. cynocephalus*). This gradient is fairly consistent throughout all of the regressions. The association of higher growth rates with larger size is especially apparent from comparison of baboon growth rates to those of the macaque species. Baboon adults are approximately three times the size of macaque adults both in overall size and in organ weights. Accordingly, their rates of body and organ weight growth are roughly three times those of an average macaque. In terms of rates, then, interspecies differences parallel intersexual differences: larger species, like larger males, grow at faster rates. But whereas the intersexual differences in rates applied only to body and heart weights, the interspecies differences are reflected in the body and nearly all the organs.

Allometric Analysis

Table 2 contains the allometric results for each species, presenting the equations for the various organs. For each species, males (♂) and females (♀) are analyzed separately and together (T). Also included in the table are the correlation coefficients, 95% confidence intervals for the slopes, tests for significant sex differences in slope values, and the arithmetic averages of the six species slopes.

The correlation coefficients between organs and body weight were all positive and significantly different from zero, most at a probability level of less than 0.001. This clearly indicates coordination between the size of the body and each of the organs during ontogeny. However, the magnitudes of the coefficients vary in a systematic way, reflecting variation in the strength of these relationships. Summarized briefly, the highest values were found for the major organs, i.e., heart, lungs, kidneys, and liver, with an average coefficient of 0.96; the correlation coefficients for the spleen, pancreas, and testes

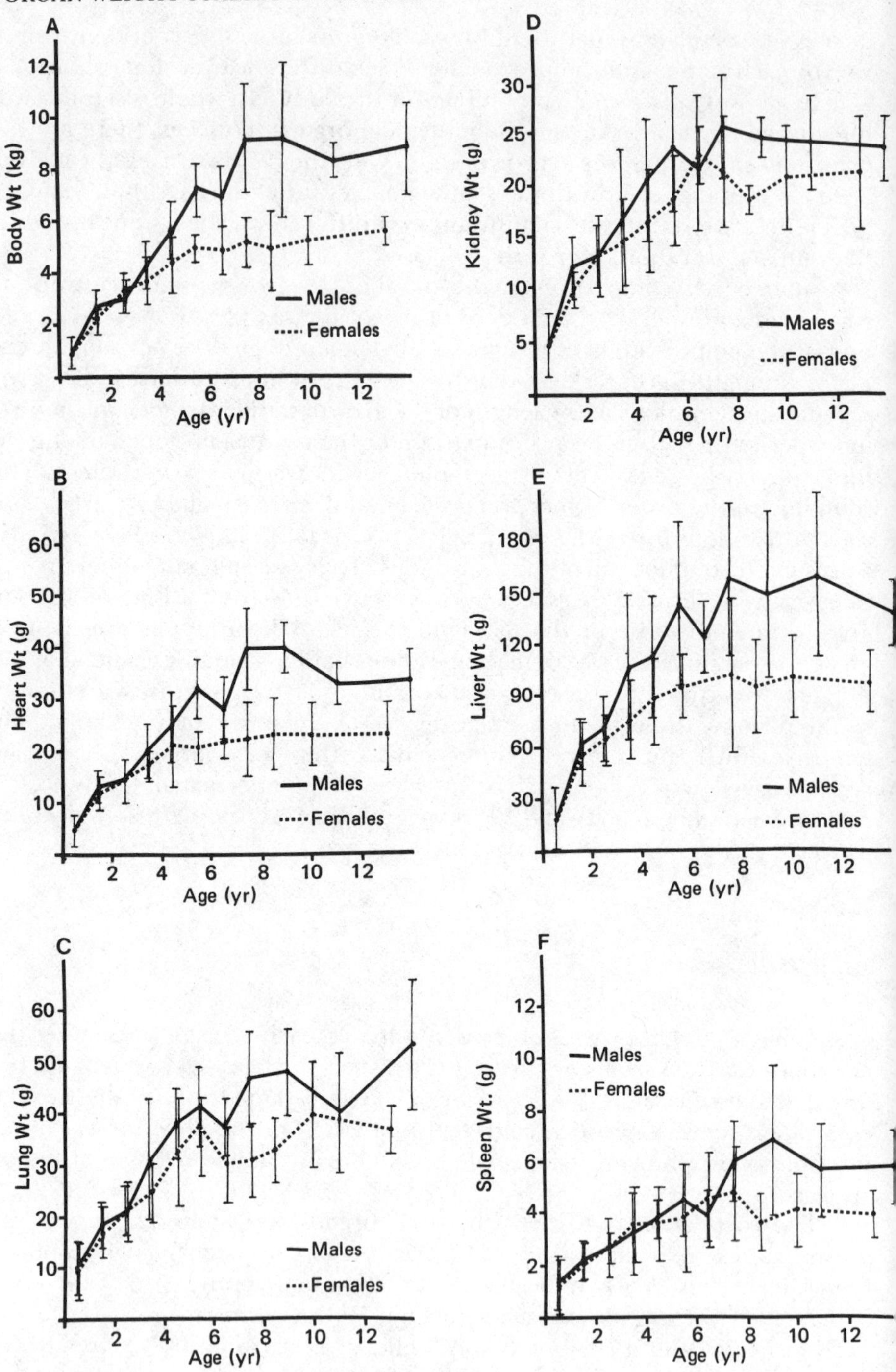

Fig. 1. Age-grouped mean weight data for *M. mulatta* plotted against age. Points represent the mean weight values for each one-year age group and the vertical lines reflect one standard deviation on either side of the means. See text for discussion.

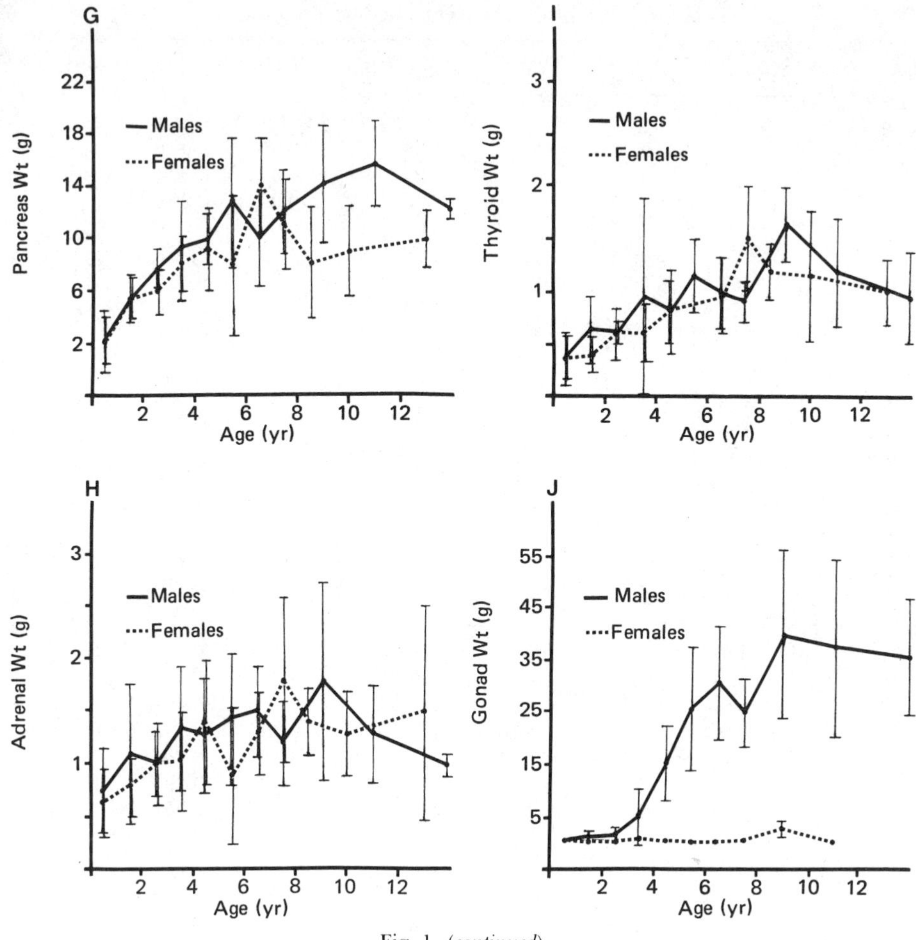

Fig. 1. (*continued*)

were slightly lower, averaging around 0.90; and the lowest and most variable correlation coefficients were those for the adrenals, thyroid, and ovaries.

It is clear from even a brief scan of the major axis slopes in Table 2 that negative allometry (k less than 1.0) is the predominate relationship between organ weights and body weight. That is, in general, the body as a whole is increasing at a faster rate than the individual organs. The only consistent exceptions are the pancreas and the gonads, which show positive allometry with respect to body weight. (In addition, a few cases of isometry ($k = 1.0$) are found in *M. fascicularis* and *M. radiata.*)

A general impression of the magnitudes of the slopes can be gained from examination of the average slopes for each organ presented at the bottom of each part of Table 2. Among the major organs, the heart, kidneys, and liver all scale at roughly the same power of body weight (0.84, 0.86, and 0.86, respectively). The average slope for the lungs, however, is somewhat lower, at

Table 1. Weight Data Regressed against Age[a]

Species[b]		N	B	a	95% CI(a)	r	Prob(r)	Prob(sd)
Body weight								
M.m.	♂	220	0.58	1.10	1.04–1.15	0.94	0.001	0.05
	♀	155	0.71	0.87	0.82–0.93	0.93	0.001	
M.n.	♂	22	0.33	1.23	1.06–1.40	0.96	0.001	0.05
	♀	20	0.57	0.75	0.65–0.86	0.96	0.001	
M.a.	♂	79	0.54	1.30	1.18–1.41	0.93	0.001	0.05
	♀	62	0.61	1.05	0.91–1.19	0.89	0.001	
P.c.	♂	29	0.86	3.53	2.94–4.12	0.92	0.001	0.05
	♀	14	0.51	2.32	1.71–2.93	0.92	0.001	
Heart weight								
M.m.	♂	195	4.25	4.43	4.16–4.70	0.92	0.001	0.05
	♀	138	4.29	3.79	3.44–4.14	0.87	0.001	
M.n.	♂	21	2.71	3.79	3.18–4.40	0.95	0.001	0.05
	♀	19	2.76	3.13	2.63–3.63	0.95	0.001	
M.a.	♂	67	4.68	4.15	3.62–4.67	0.89	0.001	NS
	♀	53	4.63	3.94	3.27–4.60	0.86	0.001	
P.c.	♂	29	9.73	13.96	11.48–16.44	0.91	0.001	NS
	♀	13	3.30	11.62	8.44–14.81	0.92	0.001	
Lung weight								
M.m.	♂	172	8.89	5.38	4.94–5.82	0.88	0.001	NS
	♀	125	7.70	5.57	4.94–6.21	0.84	0.001	
M.n.	♂	16	4.54	6.83	5.86–7.80	0.97	0.001	0.05
	♀	14	6.66	3.18	2.58–3.77	0.96	0.001	
M.a.	♂	37	7.70	7.34	6.44–8.23	0.94	0.001	NS
	♀	32	7.47	8.18	6.11–10.25	0.83	0.001	
P.c.	♂	22	26.10	18.26	14.72–21.81	0.92	0.001	NS
	♀	11	6.44	19.53	11.75–27.31	0.88	0.001	
Kidney weight								
M.m.	♂	206	5.38	3.11	2.84–3.38	0.85	0.001	NS
	♀	145	4.35	3.17	2.86–3.48	0.86	0.001	
M.n.	♂	20	3.23	3.91	3.68–4.13	0.99	0.001	NS
	♀	19	4.10	3.22	2.32–4.13	0.88	0.001	
M.a.	♂	72	7.07	4.60	4.14–5.06	0.92	0.001	0.05
	♀	60	5.49	5.65	4.87–6.43	0.88	0.001	
P.c.	♂	28	11.16	7.68	6.07–9.30	0.89	0.001	0.05
	♀	14	4.13	12.02	9.65–14.39	0.95	0.001	
Liver weight								
M.m.	♂	186	13.18	20.74	18.96–22.52	0.86	0.001	0.05
	♀	135	28.61	18.49	16.44–20.53	0.84	0.001	
M.n.	♂	21	24.30	20.96	18.01–23.92	0.96	0.001	NS
	♀	20	19.25	23.18	17.91–23.44	0.91	0.001	

Table 1 *(Continued)*

Species[b]		N	B	a	95% CI(a)	r	Prob(r)	Prob(sd)
M.a.	♂	38	21.54	31.23	25.47–37.01	0.88	0.001	NS
	♀	30	17.34	33.47	26.58–40.36	0.88	0.001	
P.c.	♂	24	49.10	38.99	32.41–45.57	0.93	0.001	NS
	♀	10	26.80	57.52	37.98–77.05	0.92	0.001	
Spleen weight								
M.m.	♂	199	1.23	0.56	0.50–0.63	0.77	0.001	NS
	♀	139	1.22	0.60	0.50–0.70	0.71	0.001	
M.n.	♂	20	2.29	1.24	0.87–1.61	0.86	0.001	—
	♀	19	3.40	0.65	—	0.42	NS	
M.a.	♂	60	2.30	1.59	1.31–1.88	0.83	0.001	NS
	♀	54	2.12	1.76	1.27–2.25	0.71	0.001	
P.c.	♂	26	1.79	4.53	3.34–5.71	0.85	0.001	0.05
	♀	12	3.20	1.36	0.84–1.88	0.88	0.001	
Pancreas weight								
M.m.	♂	94	2.32	1.64	1.36–1.93	0.77	0.001	NS
	♀	88	1.76	1.69	1.40–1.97	0.79	0.001	
M.n.	♂	0	—	—	—	—	—	—
	♀	0	—	—	—	—	—	
M.a.	♂	43	3.28	1.35	1.04–1.67	0.80	0.001	NS
	♀	35	2.49	1.60	1.07–2.14	0.73	0.001	
P.c.	♂	25	1.21	3.32	2.82–3.82	0.94	0.001	NS
	♀	9	1.85	2.85	0.71–4.99	0.77	0.02	
Adrenal weight								
M.m.	♂	182	0.73	0.12	0.08–0.15	0.47	0.001	NS
	♀	134	0.57	0.16	0.11–0.20	0.53	0.001	
M.n.	♂	20	0.43	0.25	0.18–0.32	0.86	0.001	NS
	♀	18	0.38	0.22	0.13–0.31	0.79	0.001	
M.a.	♂	67	0.71	0.23	0.18–0.28	0.76	0.001	0.05
	♀	53	0.62	0.34	0.26–0.42	0.78	0.001	
P.c.	♂	26	1.32	0.18	0.03–0.32	0.45	0.02	NS
	♀	11	0.68	0.52	0.12–0.91	0.70	0.02	
Thyroid weight								
M.m.	♂	167	0.37	0.11	0.09–0.14	0.57	0.001	NS
	♀	117	0.31	0.10	0.07–0.12	0.59	0.001	
M.n.	♂	6	−0.20	0.66	0.36–0.95	0.95	0.003	—
	♀	4	0.12	0.09	—	0.90	NS	
M.a.	♂	51	0.31	0.12	0.08–0.16	0.65	0.001	NS
	♀	46	0.20	0.18	0.12–0.23	0.71	0.001	
P.c.	♂	20	0.22	0.31	0.05–0.57	0.51	0.02	—
	♀	7	0.22	0.26	—	0.74	NS	

[a]Regression equation: weight = $a \times$ age + B; a is the slope, B the y intercept. Here 95% CI(a) is the 95% confidence interval for the slope estimate a, Prob(r) is the probability that r is different from zero, and Prob(sd) is the probability of sex differences in slope values a. NS: significance level greater than 0.05.

[b]Species abbreviations: *M.m.*, *M. mulatta; M.n.*, *M. nemestrina; M.a.*, *M. arctoides;* and *P.c.*, *P. cynocephalus.*

Table 2. Allometric Equations for Each Sample[a]

Species[b]		N	log b	k	95% CI(k)	r	Prob(r)	Prob(sd)
Log Heart weight								
M.f.	♂	32	0.67	0.83	0.76–0.92	0.97	0.001	
	♀	53	0.67	0.84	0.76–0.92	0.95	0.001	NS
	T	85	0.67	0.84	0.78–0.89	0.96	0.001	
M.r.	♂	28	0.71	0.82	0.77–0.87	0.99	0.001	
	♀	40	0.73	0.82	0.75–0.89	0.96	0.001	NS
	T	68	0.72	0.82	0.77–0.86	0.97	0.001	
M.m.	♂	236	0.75	0.85	0.83–0.88	0.97	0.001	
	♀	208	0.74	0.87	0.84–0.90	0.97	0.001	NS
	T	444	0.75	0.86	0.84–0.88	0.97	0.001	
M.n.	♂	36	0.68	0.84	0.77–0.92	0.97	0.001	
	♀	72	0.69	0.85	0.79–0.92	0.95	0.001	NS
	T	108	0.69	0.85	0.80–0.89	0.96	0.001	
M.a.	♂	91	0.76	0.80	0.76–0.84	0.97	0.001	
	♀	78	0.78	0.78	0.74–0.83	0.97	0.001	NS
	T	169	0.77	0.79	0.76–0.82	0.97	0.001	
P.c.	♂	35	0.88	0.82	0.77–0.87	0.89	0.001	
	♀	33	0.75	0.95	0.85–1.06	0.96	0.001	0.05
	T	68	0.83	0.86	0.81–0.92	0.97	0.001	
			♂	0.83	0.80–0.85			
Average slopes:			♀	0.85	0.79–0.91			NS
			T	0.84	0.81–0.86			
Log Lung weight								
M.f.	♂	15	0.95	0.82	0.65–1.03	0.94	0.001	
	♀	23	0.93	0.78	0.67–0.89	0.95	0.001	NS
	T	38	0.94	0.79	0.70–0.89	0.94	0.001	
M.r.	♂	0	—	—	—	—	—	
	♀	0	—	—	—	—	—	—
	T	0	—	—	—	—	—	
M.m.	♂	204	1.01	0.71	0.67–0.74	0.95	0.001	
	♀	185	1.01	0.73	0.69–0.78	0.93	0.001	NS
	T	389	1.01	0.72	0.69–0.74	0.94	0.001	
M.n.	♂	19	0.98	0.69	0.61–0.78	0.97	0.001	
	♀	33	0.98	0.82	0.71–0.94	0.93	0.001	0.05
	T	52	0.99	0.75	0.68–0.83	0.95	0.001	
M.a.	♂	47	1.03	0.78	0.73–0.84	0.97	0.001	
	♀	47	1.04	0.80	0.74–0.87	0.97	0.001	NS
	T	94	1.03	0.79	0.75–0.83	0.97	0.001	
P.c.	♂	26	1.20	0.70	0.63–0.76	0.97	0.001	
	♀	24	1.05	0.93	0.84–1.04	0.97	0.001	0.05
	T	50	1.15	0.77	0.70–0.85	0.96	0.001	
			♂	0.74	0.67–0.81			
Average slopes:			♀	0.81	0.72–0.90			NS
			T	0.76	0.72–0.80			
Log Kidney weight								
M.f.	♂	33	0.77	0.90	0.82–0.99	0.97	0.001	
	♀	51	0.74	1.03	0.95–1.13	0.95	0.001	0.05
	T	84	0.75	0.97	0.91–1.03	0.96	0.001	

Table 2 (*Continued*)

Species[b]		N	log b	k	95% CI(k)	r	Prob(r)	Prob(sd)
M.r.	♂	39	0.78	0.89	0.82–0.97	0.97	0.001	
	♀	54	0.79	1.03	0.97–1.10	0.97	0.001	0.05
	T	93	0.79	0.97	0.92–1.02	0.97	0.001	
M.m.	♂	247	0.75	0.75	0.72–0.79	0.95	0.001	
	♀	215	0.73	0.84	0.81–0.87	0.96	0.001	0.05
	T	462	0.74	0.79	0.77–0.82	0.95	0.001	
M.n.	♂	39	0.75	0.78	0.73–0.85	0.97	0.001	
	♀	77	0.74	0.90	0.83–0.97	0.94	0.001	0.05
	T	116	0.76	0.84	0.79–0.88	0.95	0.001	
M.a.	♂	93	0.84	0.86	0.81–0.91	0.96	0.001	
	♀	81	0.80	0.97	0.91–1.03	0.96	0.001	0.05
	T	174	0.82	0.90	0.86–0.94	0.96	0.001	
P.c.	♂	34	0.86	0.69	0.62–0.75	0.97	0.001	
	♀	35	0.89	0.80	0.71–0.90	0.95	0.001	0.05
	T	69	0.91	0.71	0.65–0.77	0.94	0.001	
			♂	0.81	0.72–0.90			
Average slopes:			♀	0.93	0.83–1.03			0.05
			T	0.86	0.75–0.97			
Log Liver weight								
M.f.	♂	31	1.52	0.88	0.80–0.95	0.98	0.001	
	♀	45	1.53	0.87	0.80–0.95	0.96	0.001	NS
	T	76	1.52	0.87	0.82–0.93	0.97	0.001	
M.r.	♂	35	1.50	0.94	0.87–1.01	0.98	0.001	
	♀	42	1.54	0.99	0.93–1.07	0.98	0.001	NS
	T	77	1.52	0.97	0.92–1.02	0.97	0.001	
M.m.	♂	213	1.52	0.79	0.75–0.82	0.95	0.001	
	♀	188	1.50	0.82	0.78–0.86	0.95	0.001	NS
	T	401	1.51	0.80	0.78–0.83	0.95	0.001	
M.n.	♂	38	1.54	0.76	0.71–0.83	0.97	0.001	
	♀	67	1.50	0.94	0.86–1.02	0.95	0.001	0.05
	T	105	1.53	0.84	0.79–0.89	0.96	0.001	
M.a.	♂	44	1.49	0.91	0.84–0.97	0.97	0.001	
	♀	43	1.51	0.95	0.89–1.00	0.98	0.001	NS
	T	87	1.51	0.93	0.89–0.97	0.98	0.001	
P.c.	♂	28	1.54	0.69	0.63–0.75	0.98	0.001	
	♀	27	1.52	0.85	0.76–0.96	0.96	0.001	0.05
	T	55	1.56	0.73	0.67–0.79	0.95	0.001	
			♂	0.83	0.73–0.93			
Average slopes:			♀	0.90	0.83–0.97			NS
			T	0.86	0.77–0.95			
Log Spleen weight								
M.f.	♂	25	0.23	1.04	0.77–1.41	0.82	0.001	
	♀	43	0.22	1.20	0.98–1.48	0.84	0.001	NS
	T	68	0.23	1.13	0.96–1.33	0.83	0.001	
M.r.	♂	39	0.31	0.79	0.69–0.89	0.94	0.001	
	♀	49	0.33	0.88	0.78–0.99	0.92	0.001	NS
	T	88	0.32	0.83	0.76–0.91	0.93	0.001	
M.m.	♂	240	0.08	0.72	0.67–0.77	0.87	0.001	

(*continued*)

Table 2 (*Continued*)

Species[b]		N	log b	k	95% CI(k)	r	Prob(r)	Prob(sd)
	♀	208	0.13	0.71	0.66–0.76	0.88	0.001	NS
	T	448	0.09	0.71	0.67–0.74	0.87	0.001	
M.n.	♂	39	0.43	0.73	0.63–0.85	0.92	0.001	
	♀	74	0.43	0.81	0.69–0.96	0.82	0.001	NS
	T	113	0.44	0.77	0.69–0.85	0.87	0.001	
M.a.	♂	75	0.35	0.84	0.76–0.92	0.92	0.001	
	♀	70	0.33	0.90	0.79–1.04	0.87	0.001	NS
	T	145	0.34	0.86	0.80–0.93	0.90	0.001	
P.c.	♂	33	0.14	0.98	0.90–1.06	0.98	0.001	
	♀	31	0.31	0.82	0.67–1.00	0.89	0.001	NS
	T	64	0.21	0.93	0.86–1.01	0.95	0.001	
	♂			0.85	0.71–0.99			
Average slopes:	♀			0.89	0.71–1.06			NS
	T			0.87	0.72–1.03			
Log Pancreas weight								
M.f.	♂	0	—	—	—	—	—	
	♀	0	—	—	—	—	—	—
	T	0	—	—	—	—	—	
M.r.	♂	0	—	—	—	—	—	
	♀	0	—	—	—	—	—	—
	T	0	—	—	—	—	—	
M.m.	♂	118	0.06	1.28	1.18–1.40	0.90	0.001	
	♀	130	0.06	0.43	1.30–1.57	0.88	0.001	0.05
	T	248	0.07	1.33	1.25–1.42	0.89	0.001	
M.n.	♂	0	—	—	—	—	—	
	♀	0	—	—	—	—	—	—
	T	0	—	—	—	—	—	
M.a.	♂	61	0.21	1.05	0.91–1.22	0.88	0.001	0.05
	♀	47	0.09	1.28	1.08–1.54	0.86	0.001	
	T	108	0.17	1.12	1.01–1.25	0.87	0.001	
P.c.	♂	29	−0.07	1.04	0.95–1.14	0.97	0.001	
	♀	26	0.20	0.92	0.77–1.10	0.92	0.001	NS
	T	55	0.06	0.98	0.89–1.08	0.95	0.001	
	♂			1.13	0.79–1.47			
Average slopes:	♀			1.21	0.56–1.86			NS
	T			1.15	0.71–1.59			
Log Adrenal weight								
M.f.	♂	33	−0.46	0.86	0.58–1.25	0.70	0.001	
	♀	45	−0.28	0.55	0.37–0.77	0.65	0.001	0.05
	T	78	−0.37	0.72	0.56–0.93	0.67	0.001	
M.r.	♂	40	−0.32	0.43	0.32–0.54	0.80	0.001	
	♀	50	−0.32	0.60	0.44–0.78	0.71	0.001	0.05
	T	90	−0.32	0.51	0.41–0.61	0.74	0.001	
M.m.	♂	221	−0.20	0.40	0.33–0.46	0.62	0.001	
	♀	201	−0.25	0.51	0.43–0.60	0.65	0.001	0.05
	T	422	−0.14	0.45	0.40–0.50	0.64	0.001	
M.n.	♂	39	−0.26	0.56	0.45–0.68	0.84	0.001	
	♀	77	−0.33	0.85	0.69–1.03	0.76	0.001	0.05
	T	116	−0.27	0.69	0.59–0.80	0.78	0.001	
M.a.	♂	89	−0.19	0.60	0.53–0.69	0.85	0.001	

Table 2 (*Continued*)

Species[b]		N	log b	k	95% CI(k)	r	Prob(r)	Prob(sd)
	♀	76	−0.14	0.60	0.51–0.70	0.83	0.001	NS
	T	165	−0.17	0.61	0.55–0.67	0.84	0.001	
P.c.	♂	33	−0.08	0.33	0.14–0.53	0.54	0.002	
	♀	28	−0.23	0.62	0.37–0.95	0.68	0.001	0.05
	T	61	−0.11	0.40	0.25–0.56	0.56	0.001	
			♂	0.53	0.33–0.73			
Average slopes:			♀	0.62	0.50–0.75			NS
			T	0.56	0.42–0.70			
Log Thyroid weight								
M.f.	♂	13	−0.69	0.60	0.00–1.09	0.73	0.005	
	♀	13	−0.89	1.25	0.78–2.11	0.84	0.001	0.05
	T	26	−0.80	0.95	0.64–1.38	0.76	0.001	
M.r.	♂	0	—	—	—	—	—	
	♀	0	—	—	—	—	—	—
	T	0	—	—	—	—	—	
M.m.	♂	202	−0.53	0.69	0.60–0.78	0.73	0.001	
	♀	179	−0.49	0.68	0.60–0.76	0.73	0.001	NS
	T	381	−0.51	0.68	0.62–0.74	0.75	0.001	
M.n.	♂	11	−0.60	0.87	0.42–1.68	0.78	0.004	
	♀	10	−0.76	0.83	0.42–1.52	0.82	0.004	NS
	T	21	−0.67	0.83	0.54–1.24	0.77	0.001	
M.a.	♂	68	−0.64	0.76	0.64–0.90	0.83	0.001	
	♀	63	−0.67	0.87	0.72–1.04	0.81	0.001	NS
	T	131	−0.65	0.80	0.71–0.90	0.82	0.001	
P.c.	♂	25	−0.68	0.71	0.36–1.24	0.62	0.002	
	♀	23	−0.84	0.91	0.63–1.29	0.80	0.001	NS
	T	48	−0.75	0.78	0.59–1.03	0.74	0.001	
			♂	0.73	0.60–0.85			
Average slopes:			♀	0.90	0.64–1.17			NS
			T	0.81	0.69–0.93			
Log Gonad weight								
M.f.	♂	16	−3.01	7.88	4.71–23.01	0.66	0.006	—
	♀	7	−1.08	2.01	0.27–12.04	0.80	0.05	
M.r.	♂	21	−0.44	2.06	1.13–5.53	0.56	0.005	—
	♀	2	—	—	—	—	—	
M.m.	♂	187	−0.49	2.07	1.94–2.22	0.91	0.001	—
	♀	74	−0.95	1.25	1.02–1.55	0.46	0.001	
M.n.	♂	34	−0.12	1.62	0.34–2.00	0.91	0.001	—
	♀	19	−0.95	1.51	1.00–2.45	0.77	0.001	
M.a.	♂	72	−0.27	1.83	1.64–2.05	0.91	0.001	—
	♀	14	−1.03	1.20	0.41–4.71	0.60	0.02	
P.c.	♂	25	−0.73	1.75	1.55–1.99	0.96	0.001	—
	♀	11	−2.58	2.45	1.72–3.91	0.89	0.001	
			♂	2.87	0.27–5.45			
Average slopes:			♀	1.58	0.63–2.53			

[a]Allometric equation: log organ weight = $k \times$ log body weight + log b; k is the slope, log b the y intercept. Here 95% CI(k) is the 95% confidence interval for slope estimate k, Prob(r) is the probability that r is different from zero, and Prob(sd) is the probability of sex differences in slope values k. NS: significance level greater than 0.05.

[b]Species abbreviations: *M.f.*, *M. fascicularis*; *M.r.*, *M. radiata*; *M.m.*, *M. mulatta*; *M.n.*, *M. nemestrina*; *M.a.*, *M. arctoides*; *P.c.*, *P. cynocephalus*.

0.76. The spleen scales at a power of body weight quite close to those of the heart, kidneys, and liver: 0.87. The positive allometric slopes of the pancreas, ovaries, and testes are 1.15, 1.58, and 2.88, respectively. Finally, the average slope for the adrenals is 0.56 and for the thyroid is 0.81. (Average slopes vary slightly for males and females considered separately.)

Comparison of male and female slopes points to a single consistent pattern throughout all of the organs: higher slopes for females. (The very different growth patterns of the gonads of the two sexes have already been described in some detail and will not be treated further here.) In every case the average female slope exceeds the average male slope. Within the allometric results for the individual species, there are a few isolated exceptions, but overall the pattern is quite evident. The actual amount of difference between the slopes, however, varies from one organ to the next, e.g., it is very small for the heart but quite marked for the kidneys, and in many cases the differences are not statistically significant.

Comparison of the allometric results for the individual species suggests considerable interspecies variation. Some slopes are higher than the average, others are lower. To more easily visualize these differences, the slope values for each species are plotted in Fig. 2 to display their relative magnitudes. The average slopes for the six species are also included in the figure for use as reference points. It can be readily seen from this display that those organs having the highest correlations to body weight within the species (i.e., heart, lungs, kidneys, and liver) also show the greatest uniformity in allometric relations across species. The small variation in slope values for the heart and lungs is especially noteworthy. Those organs that expressed weaker relations to body weight within each species show less uniformity in the interspecies comparisons, though it is often the case that the wide dispersion of points is mainly due to one or two outliers. The sizes of the major organs, then, seem genuinely dependent on total body size, since all species display approximately the same allometric relations despite their differences in growth rates and differences in ultimate size. For the remaining organs (e.g., spleen or pancreas), the relationship to body weight seems less consistent and other factors that influence their size apparently have varying effects across the species.

Discussion

Correlation Values

The very high correlation values between body weight and the weights of the heart, lungs, kidneys, and liver clearly indicate the existence of strong relationships in each case. Body size variation explains over 90% of their size variation. This correspondence between body weight and these organ weights

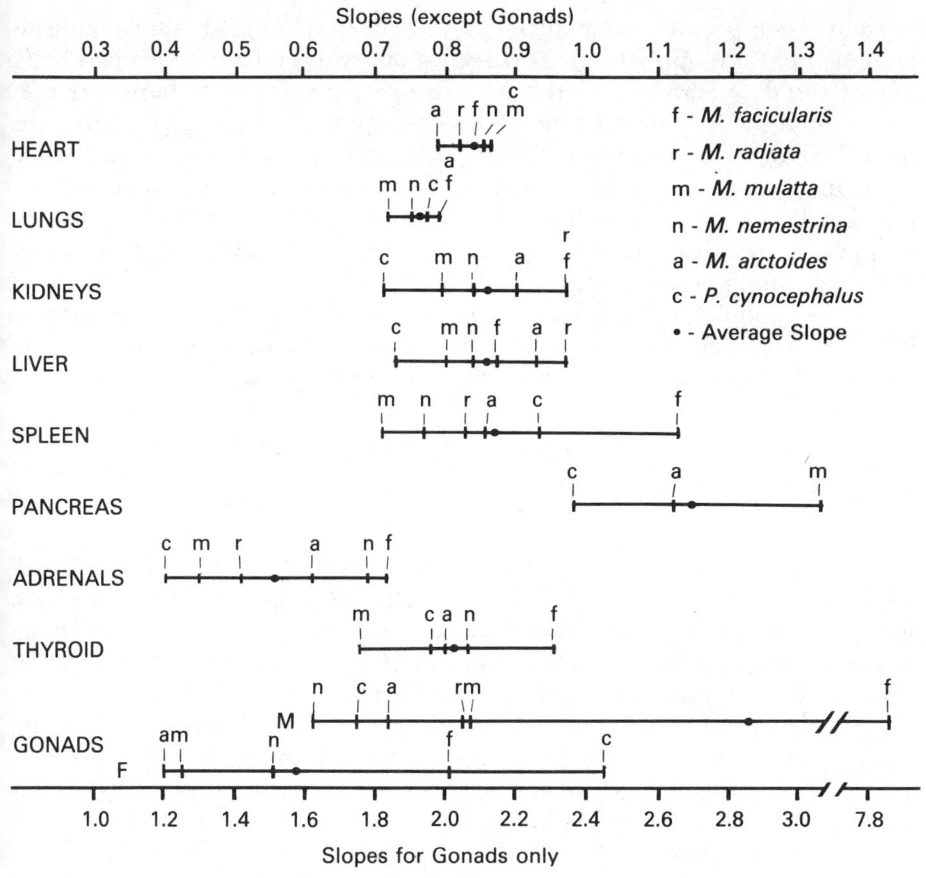

Fig. 2. Interspecies comparison of allometric slopes. Lines reflect dispersion of the allometric slopes for the six species from the highest to lowest values, with the position of the individual species along the line as indicated. The average of the six species' slopes is also plotted as a reference point.

suggests that functional requirements demand fairly exact relative weight relationships in these cases. The slightly lower correlations for the spleen, testes, and pancreas indicate less strict or perhaps indirect relations to body weight. Generally higher variability for spleen size has been observed in various laboratory animals by Webster and co-workers (Webster *et al.*, 1947; Webster and Liljegren, 1949, 1955) and they have shown that, to some degree, conditions prior to death and manner of death can affect spleen size. In the case of the testes, the reason for their imperfect correspondence to body weight increase is clear from the analysis-by-age results: while the body is growing in an essentially linear fashion, the testes are following a sigmoidal growth pattern. Pancreas weight may be affected by dietary factors, independently of any effect on overall body size. It is possible that functional efficiency in this organ can be maintained within fairly broad size limits or perhaps "weight" is simply an inadequate measure of its development. Similar considerations may also apply to the occurrence of low correlations between body weight and the weights of the thyroid, adrenals, or ovaries. Perhaps variation in an animal's physiological state has a more significant effect on these organs than do changes in overall size. Snow and Vice (1965) have suggested that the small increments in the hormonal products of these glands that are demanded by increased body size may be met by small increases in cellular output or by relatively minor increases in glandular mass. Finally, the difficulties in obtaining accurate weight measurements for these very small organs cannot be eliminated as a factor contributing to their variability. In any event, it is apparent that variation in body size does not account for much of the observed variation in adrenal, thyroid, or ovary weights.

Average Slope Values

The similarity of the average slopes for the heart, kidneys, liver, and spleen indicates that these organs keep roughly the same pace of increase with increasing body weight. This coordination of growth suggests some functional interrelations between them. One element common to the functioning of at least the first three is the processing of blood. Very simply, the heart pumps it, the liver processes metabolites in it, and the kidneys filter it. In a resting human adult, the digestive tract (including the liver) receives approximately 24% and the kidneys approximately 20% of the total cardiac output (Vander *et al.,* 1970). Edwards (1975) found that throughout mammals renal blood flow represents about 26% of total cardiac output and hepatic portal blood flow about 24% (the latter increases slightly with increasing body size). Clearly coordination of the size of the heart, liver, and kidneys would be functionally adaptive in maximizing efficiency in their interlocking activities. This coordination could be brought about by genetic control of growth, or simple accommodation to the physical demands of total blood volume, or a combination of both.

As primarily a member of the lymphatic system, the role of the spleen in this internal design is not entirely obvious. However, the spleen also performs some of the same blood-vascular functions as the liver. One of these is removal of damaged red blood cells (RBCs). As blood flows through the splenic sinusoids, damaged RBCs are filtered out and destroyed by phagocytes, and the iron released from their hemoglobin is stored in the spleen (Hollinshead, 1974). By virtue of this filtering action the spleen becomes a storage site for RBCs. During trauma involving blood loss, the spleen can assist in maintaining constant blood pressure by contracting and expelling its RBCs into the blood stream. In humans the spleen is relatively small and its contribution in this area is considered to be minor. But in other mammals its assistance in countering a drop in blood pressure takes on greater importance and the splenic capsule contains smooth muscle fibers to assist in contraction (Hollinshead, 1974). Thus its roles as a blood filter and a RBC storage site make the coordination of splenic growth with that of the heart, liver, and kidneys more understandable. In addition, the association of trauma with splenic contraction and expulsion of RBCs has been suggested as a major contributor to the variability in splenic size when measured at autopsy (Barcroft *et al.*, 1925; Barcroft and Stephens, 1927; Webster *et al.*, 1971).

The lungs scale at the 0.76 power of body weight, which is intriguingly close to the so-called "metabolic constant" of 0.75. This constant is actually the slope at which metabolic rate (usually measured as oxygen consumption) scales to total body weight (Kleiber, 1961; McMahon, 1973; Schmidt-Nielsen, 1972, 1975). This similarity in slopes appears to suggest that the growth rate of the lungs is directly related to metabolic rate as overall body size increases. However, the metabolic constant reflects the relationship between body size and metabolic rate among adults of various species; it is not clear it would take on the same value when examined ontogenetically. Kleiber (1947) finds no evidence against applying the "¾ power rule" to intraspecific data, but Brody (1945) reports that growth and maturation variously affect the relationship within a species. Few data are available on the ontogentic scaling of metabolic rate in primates and further clarification of the relationship between metabolic rate and lung weight scaling (if one exists) will have to await their compilation.

The extremely high average slope for scaling of testes weight is not unexpected given the degree of enlargement these organs undergo at sexual maturity. Similarly, the low slope for the adrenal glands could be predicted from their low growth rate relative to age. But the relatively high slopes for the ovaries and thyroid are rather surprising given the analysis-by-age results, which indicated very little growth in these organs over time. But perhaps their small size and high variability obscure what is actually significant weight change. The size range of these organs is from less than 1 g to 3–4 g. An increase in thyroid weight, for instance, from 0.3 g to 0.6 g seems minor when compared to, say, a change from 50.0 g to 80.0 g in liver weight corresponding to the same increase in body size. But in terms of the initial size of the organs, the latter represents only a 60% increase, while the former reflects a doubling of thyroid weight. In any case, it appears that, despite their high levels of variability, there is significant corresponding size adjustment in thyroid and ovary weights to increase in body weight.

A few comparable all-ages-combined allometric slopes for organ weights are available in the literature. Von Bertalanffy and Pirozynsky (1952) studied ontogenetic scaling in the albino rat and found slopes remarkably close to those reported here: heart, 0.82; kidneys, 0.80; lungs, 0.73 (liver growth was divided into two phases and thus their slope estimates are not comparable for this organ). Hall and MacGregor (1937) report an average slope of 0.86 for kidney growth in cats. Mixner *et al.* (1943) give slopes for thyroid scaling in guinea pigs of 0.73 for males and 0.83 for females. All of these values are similar to those observed here. The only study on primate organ weight allometry in the literature that treats animals of all ages is the work of Stahl *et al.* (1968) on the howler monkey. With the exception of a rather high slope for heart weight in males ($k = 0.995$), by and large their results agree well with those reported here. Snow and Vice (1965) included only immature animals in their study on organ weight scaling in the olive baboon, but most of their slope values do not differ significantly from the present all-ages-combined slopes. This remarkable similarity in allometric slopes for organ weight scaling in these different species suggests that the ontogenetic development of many mammalian species may follow very similar patterns.

Male–Female Differences

The analysis-by-age results indicate that males grow for a longer period of time than females. This supports earlier reports of prolongation of growth in males by Van Wagenen and Catchpole (1956) for rhesus macaques and Snow (1967) for olive baboons. Geist (1971), studying bighorn sheep, and Estes (1974), studying African buffalo, have suggested a link between delayed somatic maturation in males and the establishment and maintenance of a dominance hierarchy. Females among these species typically cease growth at sexual maturity, but males continue to grow, not only in body size but also in armaments, such as horns. This long period of postreproductive change establishes an extensive series of distinguishable forms among males and allows them to maintain a dominance hierarchy on largely visual criteria. It seems highly likely that the prolongation of male growth observed here serves a very similar function. But though it is clearly a major factor in the developmental origin of the sexual size dimorphism characterizing adult macaques and baboons, the regression of weight on age results indicate that males also grow at a faster rate than females. Thus the size dimorphism between the sexes observed here is a product of differences in both the rate and duration of growth.

In the allometric analysis, females typically displayed higher slopes than males. This implies that they have higher rates of organ weight increase, relative to body weight increase, than do males. The traditional interpretation of such allometric results (e.g., Gray, 1945; Stahl *et al.*, 1968) would be that females have larger organs and/or that these organs grow at faster rates than in males. But such an interpretation is at odds with both the mean organ weight data reported in the literature (e.g., Kerr *et al.*, 1974; Larson, 1978) and the analysis-by-age results discussed above. The difficulty here is with our usual methods of interpreting allometric slopes. Generally one assesses differences between slopes by comparing rates of organ size increase relative to a common rate of body size increase. But for most organs growth rates do not differ in males and females. The key to understanding this dimorphism in allometric slopes lies not in organ growth rates, but in the difference between the sexes in rates of body weight growth. The higher slopes in females are actually a result of their slower rate of body weight growth, and conversely, the lower slopes in males reflect their higher body weight growth rate. The fact that the growth rates for the individual organs are approximately equal in the two sexes indicates that their internal designs are quite similar. The higher rate of overall body weight growth in males is perhaps due to a differential development of muscle mass and/or skeletal dimensions.

There is, however, one notable instance of intersexual differences in organ growth rates—that of the heart. The age-related results indicate that the rate of heart weight growth, like the rate of body rate growth, is higher among males. Thus heart weight increase seems to keep pace with body weight increase in the two sexes. This can be seen from the allometric results for heart weight scaling, which show very little difference between slopes for males versus females. The higher rate of heart weight increase in males coupled with their higher rate of body weight increase produces allometric slopes nearly identical to those of females. It appears that the allometric line in this case reflects a necessary relationship, which is maintained despite differences in overall growth rates.

Allometric differences between males and females, then, reduce to principally a difference in overall growth rates. It was suggested earlier that the source of the male size differential was larger muscle mass or skeletal dimensions. Discussion of factors that affect the degree of sexual dimorphism in body weight is beyond the scope of this chapter (see Leutenegger and Kelly, 1977; Leutenegger and Cheverud, 1982, and this volume), but it seems clear that the selective forces producing size dimorphism have acted primarily on those features that make males "look" larger.[6] Their internal design has not apparently been affected to the same degree. While it is true that mean organ weights are larger in adult males, this is mainly because they grow for a longer period of time. At a given body weight females have slightly larger organs. I can only speculate as to how or why this is the case. Perhaps the selective pressure for larger appearance in males has "uncoupled" the development of the musculoskeletal system from the internal organs to some degree, accelerating the growth of the former. Conversely, perhaps the extra physiological demands of pregnancy and lactation in females require slightly larger organs relative to body weight, resulting in organ weight growth rates equal those of larger males. Stahl *et al.* (1968) note that during pregnancy the ovaries and adrenals increase in weight. A combination of these two alternatives could be stated as follows: within the general developmental design of these monkeys, selection has operated to maintain the maximum potential physiological capacity of the female's internal organs to accommodate periodic increased demands, while in males it has utilized the broad limits of their physiological efficiency to allow larger overall body size.

Interspecies Comparisons

Comparison of allometric results across the six species (summarized in Fig. 2) revealed close similarity in the slopes for heart and lung weight scaling,

*It is interesting to note that in environments where dietary factors limit the degree of weight dimorphism in males, e.g., *P. hamadryas,* the appearance of size dimorphism is maintained through the development of shoulder capes and manes (Leutenegger and Kelly, 1977).

slightly greater differences in the slopes for kidney and liver weight scaling, and wide variation in the slopes for pancreas, spleen, adrenal, thyroid, and gonad weight scaling. The existence of a close relation between heart and body weight has already been noted in the context of male–female comparisons. The observation here of nearly identical allomeric slopes for heart weight scaling across all species is therefore not surprising. Although no special relation was observed between lung weight and body weight in this context, the similarity of lung weight allometric slopes across the species is intuitively reasonable, especially if the growth of the lungs is in fact related to a constant like metabolic rate. In any event, the slopes for the six species are similar enough in these two organs to permit comparison of their *y*-intercept values. Scanning the log *b* values in Table 2, we see a clear gradient of increasing values in both organs corresponding to increasing mean adult size (smallest to largest: *M. fascicularis* to *P. cynocephalus*). Only *M. nemestrina* is out of place. This comparison of log *b* values indicates that the allometric relations for the six species arrange themselves into a series of parallel lines. When such a situation arises, the allometric lines are said to be shifted or "transposed" in relation to each other. Such transpositions occur when size has been altered but not shape (Gould, 1971). In other words, for the heart and lungs each species displays the same relationship between the rates of organ weight and body weight growth during ontogeny, but they do so at different overall sizes.

The species slopes for kidney and liver weight scaling, though similar, are somewhat more widely dispersed than was the case for the heart and lungs. However, the position of each relative to the average follows a fairly regular pattern: slopes for *M. fascicularis, M. radiata,* and *M. arctoides* tend to fall above the average, and those for *M. mulatta, M. nemestrina,* and *P. cynocephalus* fall below. Ignoring *M. arctoides* for the moment, the variation in slope values seems to be inversely related to absolute size differences, that is, the small species have higher slopes and the large species have lower ones. This distinction between the species parallels the sex differences in slopes observed within each species, and perhaps a similar explanation for the differences can be applied: the high slopes among *M. fascicularis* and *M. radiata* are due to their relatively low overall growth rates, and the low slopes for *M. mulatta, M. nemestrina,* and *P. cynocephalus* are a reflection of their higher overall growth rates. The fact that the slopes for *P. cynocephalus* are usually the lowest of the six species supports this interpretation, given its very high rate of body weight growth (see Table 1). For the kidneys and liver, then, though there are clearly high correlations between their sizes and total body size, the relationships are not the same in all species. It does not appear to be physiologically necessary that the same proportions be maintained at different body sizes as it is for the heart and lungs. Larger species "get by" with relatively smaller kidneys and liver. Perhaps the increased demands placed on them by larger body size can be met to some degree by internal reorganization, or possibly these demands are not a simple linear function of the increase in body size.

Given the above explanation for the interspecies differences in the slopes for kidney and liver weight scaling, the high slopes for *M. arctoides* are anomalous. Since it is the largest macaque species in this study, its slopes should in fact be among the lowest. But in other features, such as morphology of the external genitalia (Fooden, 1980) or gene frequencies for polymorphic protein systems (Weiss *et al.*, 1973), *M. arctoides* appears to be physically distinct among the macaques. The fact that they have relatively larger organs than would be expected for their body size might be another feature pointing to this distinction.

There is considerable variability in the species' slope values for the remaining organs, though to some degree those for the spleen and pancreas follow a pattern similar to that observed for the kidneys and liver. For the adrenals, thyroid, and gonads there are no consistent patterns.

Across the six species, then, the heart and lungs display the same allometric slopes despite differences in overall size; the scaling of the kidneys and liver is very similar but influenced slightly by size differences; the spleen and pancreas are also affected by differences in size but probably by other factors as well, producing their wider allometric variability; and the adrenals, thyroid, and gonads display wide variation in their scaling relationships, reflecting their low correlations to body weight.

Summary

This study has involved the comparison of ontogenetic scaling of organ weights in six species of Old World monkeys. It has shown that by and large the scaling patterns are quite similar across all species and may reflect a common mammalian developmental pattern. The observed differences can in part be related to differences in growth rates for total body weight, as in the case of the kidneys, liver, pancreas, or spleen. Across all species, the similarity of the allometric slopes for heart, kidney, liver, and spleen weight scaling suggests a coordination in the ontogenetic development of these organs, possibly related to their blood-vascular functions. Male and female scaling differences were found to be primarily a reflection of differences in body weight growth rates. This difference in growth rates coupled with delayed somatic maturation in males produces the sexual size dimorphism so characteristic of the adults of these species.

The allometric equations derived here represent the ontogenetic relationships between organs and body weight among ordinary healthy individuals of each of these species. Besides displaying aspects of the primate developmental pattern as described above, it is hoped that they will prove useful to other primate biologists as predictors of normal organ weights for comparison to their experimental groups.

Acknowledgments

I would like to thank the following individuals for their assistance in the data collection for this study: Judy Johnson and Susan Hemingway of the

Washington Regional Primate Research Center, Larry Jacobsen and Peter Goy of the Wisconsin Regional Primate Research Center, Harold McClure of Yerkes Regional Primate Center, Bernice Brenner of the California Regional Primate Research Center, Thomas Kuehl of the Southwest Foundation for Research and Education, and John Strandberg of Johns Hopkins University. I also wish to thank Walter Leutenegger for comments and criticisms on this manuscript.

References

Alberch, P., Gould, S. J., Oster, G. F., and Wake, D. B. 1979. Size and shape in ontogeny and phylogeny. *Paleobiology* **5:**296–317.

Barcroft, J., and Stephens, J. G. 1927. Observations upon the size of the spleen. *J. Physiol.* **64:**1–22.

Barcroft, J., Harris, H. A., Orahovats, D., and Weiss, R. 1925. A contribution to the physiology of the spleen. *J. Physiol.* **60:**443–456.

Brody, S. 1945. *Bioenergetics and Growth.* Van Nostrand Reinhold, New York.

Cronin, J., Cann, R., and Sarich, V. M. 1980. Molecular evolution and systematics of the genus *Macaca,* in: *The Macaques* (D. G. Lindburg, ed.), pp. 31–51, Van Nostrand Reinhold, New York.

Cupp, C. J., and Uemura, E. 1981. Body and organ weights in relation to age and sex in *Macaca mulatta. J. Med. Primatol.* **10:**110–123.

Delson, E. 1975. Evolutionary history of the Cercopithecidae, in: *Approaches to Primate Paleobiology* (Contrib. Primatol., Vol. 5, F. Szalay, ed.), pp. 167–217, S. Karger, Basel.

Delson, E. 1980. Fossil macaques, phyletic relationships and a scenario of development, in: *The Macaques* (D. G. Lindburg, ed.), pp. 10–30, Van Nostrand Reinhold, New York.

Edwards, N. A. 1975. Scaling of renal functions in mammals. *Comp. Biochem. Physiol.* **52A:**63–66.

Estes, R. D. 1974. Social organization of the African Bovidae, in: *The Behavior of Ungulates and its Relation to Management,* (V. Geist and F. Walther, eds.), pp. 166–205, IVCN, Morges, Switzerland.

Fooden, J. 1980. Classification and distribution of living macaques (*Macaca* Lacepede, 1799), in: *The Macaques* (D. G. Lindburg, ed.), pp. 1–9, Van Nostrand Reinhold, New York.

Gest, T. R., and Siegel, M. I. 1983. The relationship between organ weights and body weights, facial dimensions, and dental dimensions in a population of olive baboons (*Papio cynocephalus anubis*). *Am. J. Phys. Anthropol.* **61:**189–196.

Geist, V. 1971. *Mountain Sheep: A Study in Behavior and Evolution,* University of Chicago Press, Chicago.

Gould, S. J. 1971. Geometric similarity in allometric growth: A contribution to the problem of scaling in the evolution of size. *Am. Nat.* **105:**113–136.

Gould, S. J. 1975. Allometry in primates, with emphasis on scaling and the evolution of the brain, in: *Approaches to Primate Paleobiology* (Contrib. Primatol., Vol. 5, F. Szalay, ed.), pp. 244–292, S. Karger, Basel.

Gray, H. 1945. Heart weight and body weight in rodents. *J. Mammal.* **26:**285–299.

Hall, V. E., and MacGregor, W. W. 1937. Relation of kidney weight to body weight in the cat. *Anat. Rec.* **69:**319–331.

Hollinshead, W. 1974. *Textbook of Human Anatomy,* Harper and Row, New York.

Jolicoeur, P., and Heusner, A. A. 1971. The allometry equation in the analysis of the standard oxygen consumption and body weight of the white rat. *Biometrics* **27:**841–855.

Jungers, W. L., and Fleagle, J. G. 1980. Postnatal growth allometry of the extremities in *Cebus albifrons* and *Cebus apella:* A longitudinal and comparative study. *Am. J. Phys. Anthropol.* **53:**471–478.

Kerr, G. R., Allen, J. R., Scheffler, G., and Conture, J. 1974. Fetal and postnatal growth of rhesus monkeys (*M. mulatta*). *J. Med. Primatol.* **3:**221–235.

Kleiber, M. 1947. Body size and metabolic rate. *Physiol. Rev.* **27:**511–541.

Klebier, M. 1961. *The Fire of Life: An Introduction to Animal Energetics,* Wiley, New York.

Kuhry, B., and Marcus, L. F. 1977. Bivariate linear models in biometry. *Syst. Zool.* **26:**201–209.

Larson, S. G. 1978. Scaling of organ weights in *Macaca arctoides. Am. J. Phys. Anthropol.* **49:**95–102.

Larson, S. G. 1982. The Scaling of Organ Weights in Six Old World Monkey Species, Ph.D. Diss., University of Wisconsin at Madison.

Leutenegger, W., and Cheverud, J. 1982. Correlates of sexual dimorphism in primates: Ecological and size variables. *Int. J. Primatol.* **3:**387–402.

Leutenegger, W., and Kelly, J. T. 1977. Relationship of sexual dimorphism in canine size and body size to social, behavioral, and ecological correlates in anthropoid primates. *Primates* **18:**117–136.

McMahon, T. 1973. Size and shape in biology. *Science* **179:**1201–1204.

Mixner, J. P., Bergman, A. J., and Turner, C. W. 1943. Relation of certain endocrine glands to body weight in growing and mature guinea pigs. *Endocrinology* **32:**298–304.

Napier, J. R., and Napier, P. H. 1967. *A Handbook of Living Primates,* Academic Press, New York.

Pilbeam, D., and Gould, S. J. 1974. Size and scaling in human evolution. *Science* **186:**892–902.

Roonwal, M. L., and Mohnot, S. M. 1977. *Primates of South Asia,* Harvard University Press, Cambridge.

Sacher, G. A. 1970. Allometric and factorial analysis of brain structure in insectivores and primates, in: *The Primate Brain* (Adv. Primatol., Vol. 1, C. R. Noback and W. Montagna, eds.), pp. 245–287, Appleton-Century-Crofts, New York.

Schmidt-Nielsen, K. 1972. *How Animals Work,* Cambridge University Press, Cambridge.

Schmidt-Nielsen, K. 1975. Scaling in biology: The consequences of size. *J. Exp. Zool.* **194:**287–308.

Shea, B. T. 1981. Relative growth of the limbs and trunk in the African apes. *Am. J. Phys. Antrhopol.* **56:**179–201.

Snow, C. C. 1967. Some observations on the growth and development of the baboon, in: *The Baboon in Medical Research* (H. Vagtborg, ed.), Vol. II, pp. 187–199, University of Texas Press, Austin.

Snow, C. C., and Vice, T. 1965. Organ weight allometry and sexual dimorphism in the olive baboon, *Papio anubis,* in: *The Baboon in Medical Research* (H. Vagtborg, ed.), Vol. I, pp. 151–163, University of Texas Press, Austin.

Stahl, W. R. 1965. Organ weights in primates and other mammals. *Science* **150:**1039–1042.

Stahl, W. R., and Gummerson, J. Y. 1967. Systematic allometry in five species of adult primates. *Growth* **31:**21–34.

Stahl, W. R., Malinow, M. R., Maruffo, C. A., Pope, B. L., and DePaoli, R. 1968. Growth and age estimation of howler monkeys, in: *Biology of the Howler Monkey* (*Allouatta caraya*) (Bibliotheca Primatol., No. 7), pp. 59–80, S. Karger, Basel.

Vander, A. J., Sherman, J. H., and Luciano, D. S. 1970. *Human Physiology: The Mechanisms of Body Function,* McGraw-Hill, New York.

Van Wagenen, G., and Catchpole, H. R. 1956. Physical growth of the rhesus monkey (*Macaca mulatta*). *Am. J. Phys. Anthropol.* **14:**245–273.

Von Bertalanffy, L., and Pirozynski, W. J. 1952. Ontogenetic and evolutionary allometry. *Evolution* **6:**387–392.

Webster, S. H., and Liljegren, E. J. 1949. Organ:body-weight ratios for certain organs of laboratory animals. II. Guinea pig. *Am. J. Anat.* **85:**199–230.

Webster, S. H., and Liljegren, E. J. 1955. Organ–body weight ratios for certain organs of laboratory animals. III. White Swiss mouse. *Am. J. Anat.* **97:**129–153.

Webster, S. H., Liljegren, E. J., and Zimmer, D. J. 1947. Organ weight ratios for liver, kidneys, and spleen of laboratory animals. I. Albino rat. *Am. J. Anat.* **81:**477–513.

Weiss, M. L., Goodman, M., Prychodko, W., Moore, G. W., and Tanaka, T. 1973. An analysis of macaque systematics using gene frequency data. *J. Hum. Evol.* **2:**213–226.

Allometric Considerations of the Adult Mammalian Brain, with Special Emphasis on Primates

7

ESTE ARMSTRONG

Introduction

Allometric studies of the brain investigate differences in the size of the total brain or its subdivisions and associate those differences with the size of the organism or, for its parts, with the size of the brain. Two quantitative features are examined in these studies: (1) the intercepts, or how big a part is in relation to the whole, and (2) the slope, or how the two features scale together. Over the past 100 years comparisons of adult vertebrates have demonstrated that taxonomic groups differ according to the amount of brain per body weight and that brain weights do not show as much enlargement as do body weights (negative allometry). Although many studies have covered all vertebrates, only mammalian data will be discussed in this review.

Much of the early impetus for the study of relative brain size came from a desire to study and rank the intelligence of animals (e.g., Dubois, 1897; Brummelkamp, 1937; Jerison, 1973). A major assumption for this approach is that increases in brain size not directly tied to the body are associated with increased neural interactions and thus "intelligence." Many other behaviors,

ESTE ARMSTRONG • Department of Anatomy, Louisiana State University Medical Center, New Orleans, Louisiana 70112.

however, such as differing levels of arousal, memory, and social repertoires, including forms of communication, social control, and parenting, need to be tested for their association with relative brain size. Behaviors (such as enlarged auditory structures of bats, visual structures of primates, or structures regulating diving capacities of marine mammals) that appear more directly connected to the body also need to be explored for their role in enlarging parts of the brain and thus whole brains independent of body mass changes.

The association of behavior with neural structures is difficult, yet must precede interpretations associating relative brain size with particular behaviors. Physiological, chemical, and connection studies are paramount for the establishment of function with structure. The association of a complex behavior like intelligence with the brain is neither simple nor known. Correlations of intelligence with brain size are well reviewed elsewhere (Radinsky, 1982; MacPhail, 1982).

Among adult mammalian species, body mass enlarges more than brain mass size does. Different studies have shown different scalings between these two variables. The earliest brain–body associations of 5⁄9 (Dubois, 1897, 1914) were superceded by values close to ⅔ (von Bonin, 1937; Bauchot and Stephan, 1969; Jerison, 1955, 1973). Most recent studies show an association close to ¾ (Eisenberg, 1981; Martin, 1981, 1982; Armstrong, 1982*b,c*, 1983). The discrepancies may stem from sampling different numbers of animals with differing relative brain sizes (L. Radinsky, personal communication; Hofman, 1982). The scaling values have influenced the interpretations. The early values of ⅔ were associated with a presumed role for surface effectors and receptors in brain size (e.g., Jerison, 1955, 1973). A ¾ slope was noted by Martin (1981, 1982) to be equivalent to the metabolic expenditure (kcal) of the organism. While most of the work has tried to explain relative brain size on the basis of what the brain can do for the organism (in terms of coordination of sensory information and sensory–motor interactions), it is also necessary to examine what the body can do for the brain. Maintaining an adequate nutritional supply is one suggestion (Armstrong, 1982*b,c*, 1983).

Allometry is a necessary and powerful tool for investigation differences in sizes, but, like all methods, it has limitations. Allometric studies of the brain can tell us which brains are relatively large or small or which particular parts have enlarged, but these studies cannot show how the outcome of the brain's activities, behavior, has changed. The situation is similar to many other allometric studies, including that of the skeleton. It is well known that increases in body mass are associated with increases in skeletal mass. The mass of an organism increases as a volume (or the third power of the linear dimension), so the cross sections of the bones must increase at a rate greater than the square of the limb diameters (keeping a linear proportion) or the animal would be incapable of supporting itself (Gould, 1966). Indeed, when the weight of skeletal mass was correlated with body weight among seven mammals, the skeletal weight had a positive allometric relation with body weight (1.13); that is, it increased more than body weight did (Kayser and Heusner, 1964). (The data were from two different species of rodents, a rabbit, two carnivore species, man, and elephant.) The allometric study cannot tell us anything about the function of the skeletal mass beyond its support function. No indication of which animals are quadrupedal or bipedal, how many vertebrae they have, whether they have relatively short carpals or tarsals, are claviculate or have nails or claws can be gleaned from this analysis. All of these morphological characteristics are important for understanding skeletal form and function of the animal as well as the animal's biology, but these features cannot be determined from skeletal-to-body allometric analyses. The allometric analysis, however, can make important statements about mechanical support. Similarly, studies of brain allometry are imperative for determining how brain scales with body mass and how the sizes of different brain regions scale with overall size.

Historical Development

Initial interest in relative brain size was closely associated with the desire to rank unstudied species by intelligence. The intelligence of a particular species was "known" by an investigator and the index of relative brain size then judged according to whether the index matched the preconceived (and as yet) untested assumptions. Different investigators have had different prejudices: for Darwin (1871), it was clear that hymenoptera are brainier than beetles; for Count (1947), it was obvious that gorillas are smarter than lions; and for Hemmer (1971), that apes are smarter than monkeys. A major assumption by all such proponents is that human values must be highest. Although this assumption is intuitively correct, its subjective bias has an undetermined effect on the analyses.

While earlier scientists studied differences in brain sizes in hopes of elucidating factors about intelligence, absolute brain size was rejected as an inadequate parameter because humans do not possess the largest brains. Around the turn of the century it was realized that relative brain size based on a simple ratio of brain to body weight was also not a suitable parameter because the human and mouse ratios were similar (Dubois, 1897, 1914). Searching for a better way to separate the portions of the brain involved in intelligence from those involved in mere bodily functions, Dubois (1897, 1914) paired closely related animals of different body sizes. He assumed that closely related animals were likely to be of equal intelligence, and thus he could quantitatively determine for the selected pairs how differences in brain and body weights scaled when intelligence was held constant. It was thought that the association could then be extrapolated to additional animals of widely divergent phylogeny and intelligence. Whatever the faults of Dubois' analysis (Sholl, 1948; Gould, 1975), he recognized the problem of interspecific comparisons in matters of intelligence and tried to control for this problem by choosing his pairs from closely related animals.

At that time, the expectation was that brains would scale isometrically with the body's surface area, the repository of neural effectors and receptors. Thus brains were thought to scale at about ⅔ with body weight (Brandt, 1867; Snell, 1892). The average of Dubois' series of paired animals, however, came to 0.56 and he assumed this slope represented increases in brain size due to increases in body weight and that increases in excess to this were due to increases in intelligence (the psychencephalon).

From these beginnings, studies of brain and body weights have shown that increases in brain weights are regularly associated with increases in body weights, but the enlargements in brain weights do not keep pace with those of body weights. The scaling is best described by the familiar equation

$$\log Y = \log b + k \cdot \log X$$

where Y is brain weight, X is body weight, $\log b$ is a constant or the y intercept of brain weight at a body weight of one, and k = the rate (slope) at which changes in Y are associated with X. In many allometric studies $Y = E$ and $X = S$ (e.g., Jerison, 1973). This equation is, of course, the equation for relative growth or allometry (Huxley, 1932; Gould, 1966). While the initial interest in relative brain size may have stemmed from a desire to understand "intelligence" or the "psyche," it has emerged as an interesting puzzle in and of itself.

Lapique (1912) extended Dubois' investigations to birds, and by using log–log plots, he reconfirmed Manouvrier's work (1885) that intraspecific scaling of brain to body weights was much lower than interspecific scaling, being about 2/9. Additional studies confirm that in many, but not all, mammalian taxa, intraspecific brain weights have a much smaller increase in relation to body weight than do cross-species comparisons (e.g., Bronson, 1979; Bauchot and Diagne, 1973).

Brummelkamp (1939) continued to study interspecific relations by fitting Dubois' interspecific scaling exponents to a series of mammals. A series of parallel lines was necessary to account for the values of the various species and he thought the parallel lines were separated by $\sqrt{2}$, a distance he attributed to differing proportions of nuclear and cytoplasmic masses during division (Brummelkamp, 1939). About the same time, von Bonin (1937) critiqued the simplistic association of intelligence and brain size. He also determined a brain-to-body scaling of 0.66 among 115 mammals. This was the scaling that had been predicted on the basis of surface area 50 years earlier (Snell, 1892; Brandt, 1867).

A little later Crile (1941), who is well known for his and Quiring's collection of brain and body weights (Crile and Quiring, 1940), wrote a now almost forgotten book hypothesizing that the brain and thyroid gland must have evolved together. He saw the metabolically active brain as incapable of survival without an adequately large thyroid gland, an organ that modulates the metabolic rate. Much earlier, Brandt (1867) also concluded that metabolism might influence brain volume.

Count (1947) returned to the classical analysis of brain and body weights.

He criticized a one-to-one association of relative brain size with intelligence, yet thought relative brain size somehow must be associated with intelligence because gorillas are much smarter than lions and have relatively larger brains. Although a major emphasis of his monograph dealt with ontogenetic development, Count avoided considering any preconceived ideas of mind–brain problems, and instead proposed "straightening" the somewhat curved line describing the interspecific brain-to-body association with an additional factor. Count's interspecific brain–body equation was

$$\log Y = \log b + 0.56X - cX^2$$

where c is a constant. The $-cX^2$ term had no biological meaning, so with the exception of Bauchot (1978), it has not been investigated further. Bauchot (1978) concluded that such a factor is important for large mammals and produces estimates of relative brain sizes that better conform to preconceived ideas about their behavior and learning abilities.

Sholl (1948), whose major contribution to comparative neurobiology is his development of quantitative Golgi techniques, criticized Count's (1947) use of a biologically undefined parameter in straightening the curve. Although he collected brain and body weights for his critique, he thought that only after neural homologies had been established would size comparisons be meaningful (Sholl, 1948).

In the 1950s several investigators began their research and greatly extended our knowledge of brain and body relationships. In Europe, Stephan (1956) and later Bauchot began several series of studies separately and together as well as in collaboration with Andy and others (e.g., Stephan and Andy, 1964). In the U.S., Jerison (1955) also began his well-known work.

On the continent, many scientists examined brain and body relationships in a wide variety of mammals, and most interspecific values of brain to body were reported as scaling between 0.5 and 0.6 (Frick, 1957; von Röhrs, 1966). Stephan, Bauchot, and Andy analyzed brain and body relationships primarily among insectivores, primates, and bats (Stephan, 1960, 1972; Stephan and Andy, 1964; Stephan *et al.*, 1970). They compared different mammalian species to the most primitive mammalian group they could identify, the basal insectivores. Basal insectivores were defined as the insectivores with an unspecialized cerebral pattern (Stephan, 1972; Stephan *et al.*, 1970). The descriptive equation $\log E = 1.632 + 0.63 \log S$ of the brain-to-body relationships in this small group was then used to generate progression indices (IP) among living mammals. The progression index is the difference between the observed brain weight and the brain weight expected of a basal insectivore of equivalent body weight. They found a wide range of primate values, with the human IP value being highest. While ape and monkey values are higher than but overlap prosimian values, the latter IPs resemble those of many other mammalian groups. *Daubentonia* was found to have an extremely high IP and *Gorilla* a very low IP. In addition, the scientists have emphasized that toothed whales and seals overlap the nonhuman anthropoid IP values, whereas baleen whales and sirenians have low values that overlap rodent IPs (Stephan, 1972; Bauchot and Stephan, 1966, 1969). Stephan *et al.* (1970) also collected and analyzed volumetric divisions of the brain, as will be discussed below (pp. 131–137).

Like others before him, Jerison, looking for an anatomical correlate of intelligence, chose relative brain size. Using Count's (1947) data, he independently confirmed von Bonin's slope of 0.66. Interestingly, when primates were excluded, the mammalian brain-to-body association had a higher correlation and a slope of 0.73 (Jerison, 1955), a slope that some scientists now think may represent a more accurate interspecific brain-to-body relationship and equation (see pp. 126–127). After deriving best fit empirical slopes for taxonomic groups, he compared *y* intercepts, which he called indices of cephalization (Jerison, 1955). Later he derived an interspecific equation ($E = 0.125S^{2/3}$) and used it to compare observed brain weights with the brain weight expected for an average mammal of that particular body mass (Jerison, 1973). He termed the ratios of the expected and observed brain weights encephalization quotients (EQs). Among extant mammals he found that anthropoids have larger relative brain sizes than do other mammals (Jerison, 1955, 1973), and the enlargement is more pronounced in apes than in monkeys (Jerison, 1973) [but not when the index of cephalization is used (Jerison, 1955)], whereas humans have the highest deviation (Jerison, 1955, 1973). Ungulates and carnivores have intermediate values, and rodents the smallest relative brain sizes (Jerison, 1955, 1973). He has also attempted to estimate numbers of neurons, cortical volume, and sizes of cortical neurons and their constituent elements (e.g., dendritic widths) from EQs (Jerison, 1973, 1979, 1982). These features will be reviewed below (pp. 130–131). Jerison also examined relative brain sizes among fossil vertebrates (Jerison, 1973). Jerison's application and interpretation of his various formulas have engendered a lively debate (e.g., Radinsky, 1977; Holloway, 1979), which will undoubtedly help clarify our knowledge of mammalian brain evolution.

Radinsky (1975, 1978, 1981, 1982) and others (Jerison, 1973; McHenry, 1975; Gurche, 1982) have analyzed mammalian fossil endocast materials allometrically. Where external morphologies distinguish divisions of the brain (e.g., cerebellum, frontal lobes), parts of the brain have been allometrically analyzed (see pp. 135–138). The estimation of body weight is a difficult problem for fossil analyses (see Smith, this volume; Stendel, this volume). The estimates are derived by measuring the dimensions of various fossilized structures (which are themselves frequently fragmentary) and then choosing an appropriate living animal to model the expected body weight for a given structure length. The choice of models has produced the most controversy (Radinsky, 1981, 1982; Holloway and Post, 1982; Gurche, 1982).

Over the past decade allometric analyses of fossil brain–body relationships have been made. Different mammalian orders appear to have independently evolved relatively larger brains (Jerison, 1973; Radinsky, 1978). Modern carnivore and ungulate grades of relative brain size may have been achieved by the Oligocene or Miocene (Radinsky, 1978, 1981). Although it is not yet known when modern prosimian and anthropoid relative brain sizes were attained (Radinsky, 1979), early Eocene fossil prosimians appear to have smaller relative brain sizes than do modern species (Radinsky, 1975, 1977, 1979, 1982; Gurche, 1982). Differences in estimating body sizes led Jerison (1979) to suggest an earlier attainment of the modern prosimian grade of relative brain size. The modern anthropoid level may have been reached after the Oligocene, since the anthropoid *Aegyptopithecus* appears to have a relatively small brain (Radinsky, 1979).

During the last few years most of the extant brain-to-body allometric analyses have considered larger data bases or analyzed groups according to dietary or different taxonomic criteria. The larger data bases among living mammals have determined an interspecific brain-to-body slope close to 0.75 (Martin, 1981, 1982; Eisenberg, 1981; Hofman, 1982). Most of the additional species have been small mammals with relatively small brains and this disproportionate addition could produce the steeper slope (L. Radinsky, personal communication; Hofman, 1982). Relationships of major taxa within the new and old data sets, however, remain consistent [e.g., anthropoids and toothed whales have always had relatively large brains (Martin, 1981, 1982; Stephan, 1972; Radinsky, 1975)].

Recently, dietary groups have also been determined to differ in the relative sizes of their brains (Pirlot and Stephan, 1970; Eisenberg and Wilson, 1978; Mace *et al.*, 1981). While the dietary differences among bats have been shown to be statistically significant (Pirlot and Stephan, 1970; Eisenberg and Wilson, 1978), differences in primates are not (Clutton-Brock and Harvey, 1980). The differences between the two orders may stem from the fact that primate diets are preferential, rather than obligatory.

Over the past 100 years, then, a regular and well-established pattern of brain and body wieght relationships have been established. Brain and body weights increase in a regular and negatively allometric (less than 1.0) fashion, with interspecific correlations having a steeper alignment than intraspecific ones. When body sizes are controlled, humans have consistently been found to have the largest brains. Nonhuman anthropoids, seals, and odontocetes have high and overlapping brain to body mass proportions. The relative brain sizes of many prosimians overlap those of other mammals. Frugivorous bats have bigger brains per body weight than do insectivorous bats. While differences in relative brain size have frequently been interpreted as stemming from differences in a complex behavior, intelligence, an association of intelligence with relative brain size is neither clear nor simple (Radinsky, 1979, 1982; MacPhail, 1982).

Causes of Brain-to-Body Allometry

Although much of the regularity associating brain and body weights has been worked out, little is known about the underlying causes or mechanisms of this association. The major theoretical underpinning for expecting a 0.67

slope among extant mammals (or vertebrates) is that the brain receives information from and sends information to the body's surface receptors and effectors (Brandt, 1867; Rensch, 1960; Jerison, 1955, 1973). Thus, as the body expands, its weight would increase as a cube while its surface area would increase as a square: assuming no changes in shape, the relationship between brain and body weights should slope at 0.67. The mechanisms of how the numbers of incoming and outgoing fibers would control size development of the brain have not been investigated.

Evidence to support or refute this theory for the 0.67 slope minimally depends on two major pieces of information. Surface area to body weight must scale at approximately 0.67 and the density of surface effectors and receptors must remain constant in different body sizes.

The first requirement appears to be met. The surface area of most mammals appears to scale to body size at 0.64, a figure that is almost identical to the expected 0.67 slope (Brody, 1945; Stahl, 1967; Falk, 1982*b*; Falk and Waide, 1982). For the 0.67 scaling of surface area to body weight to affect the size of the brain by the same magnitude, information flowing to and from the brain must also scale at 0.67. Few studies have examined the density of surface effectors and receptors. Although density of endplates is not known, striated muscle fibers may not increase in number as body size increases (Szarski, 1980). Two studies have examined the densities of morphological structures that are associated with nerve endings. Sensory epithelial pads (Straile, 1969) and hair follicles from the mid-dorsum (HF) (Carter, 1965) decrease in density as body size increases [Armstrong (1982*c*); body weights from Eisenberg (1981)]. The data so far available do not support the assumption of a constant density among surface effectors and receptors (Armstrong, 1982*c*).

Numbers of nerve fibers within afferent and efferent tracts would also need to scale at 0.67 if surface effectors and receptors are to control brain size as predicted. Both the mammalian pyramidal tract (PT) and optic nerve (Ot) have been studied quantitatively at the light-microscopic level—a level which may have precluded observations of the smallest fibers.

Towe (1973) reviewed the literature concerning the number of fibers in the PT among 21 mammals. When all mammals ($N = 21$) were included, the number of PT fibers scaled at 0.30, with the 95% confidence limits of the slope (cls) being 0.16–0.43 ($r = 0.74$). A stronger association of number of fibers with body weight came from dividing the mammals into two groups. The relationship among carnivore, primate, and rodent ($N = 14$) number of PT fibers with body weight is $\mathrm{PT} = 1.27 \times 10^5 S^{0.44}$ with 95% cls being 0.36–0.52, $r = 96$. The equation for the ungulates and marsupials ($N = 7$) is $\mathrm{PT} = 3.10 \times 10^4 S^{0.47}$, 95% cls being 0.396–0.54. In no case do the number of fibers in this major tract overlap the expected 0.67 slope (Armstrong, 1982*c*).

A major afferent mammalian tract is the optic nerve. One early quantitative study was made among 11 mammals (Bruesch and Arey, 1942). In this case the number of optic nerves (Ot) is associated with body size [Armstrong (1982*c*); body weight data from Eisenberg (1981)], $r = 0.854$, $p < 0.01$. Although the number of fibers scales at less than 0.67 ($\log \mathrm{Ot} = 3.2 + 0.583 \log$

S), there is overlap with the theoretical 0.67 slope (95% cls:0.36–0.81). The optic nerve, the output of which comes from a very restricted surface area, the retina, supplies the only support for tracts to or from the brain scaling at 0.67 (Armstrong, 1982*c*).

A second hypothesis associating brain weight and surface area suggests that these two features are linked because they both develop from the ectoderm (Falk and Waide, 1982; Falk, 1982*b*). An understanding of the ontogenetic relationship between these two ectodermal derivatives should help clarify which factors are important for relative brain size. Neonatal brain–body associations have had some allometric analyses (Sacher, 1982; Leutenegger, 1982).

While the brain is related to surface effectors and receptors and is derived from ectoderm, the adult brain also has other important associations with the body. Its never-ceasing changes in electrical potentials and manufacture of chemical neurotransmitters and structural proteins are energetically expensive. Consequently the brain requires a constant and large supply of oxygen and glucose (Sokoloff, 1981). The brain consumes a much higher proportion of O_2 than brain to body weight ratios predict (Table 1). The brain uses the oxygen for the aerobic oxidation of glucose, and unlike other organs, the brain appears to have little choice for substrate (Sokoloff, 1981). During long periods of starvation, however, the proportion of ketone bodies in the blood rises and they are oxidized by the brain (Sokoloff, 1981). While some anaerobic glycolysis, as determined by lactate production, normally occurs in the brain (Lajtha *et al.*, 1981), it increases markedly after the onset of total cerebral ischemia (Duffy and Plum, 1981). Primates differ in the proportions of their kinds of cerebral lactate dehydrogenase (Goodman *et al.*, 1969), but the significance of this fact is difficult to interpret.

The central nervous system's demand for energy is mainly due to the

Table 1. Cerebral Metabolic Rates[a]

Species	CMR, cm^3 O_2 $(100\ g)^{-1}\ min^{-1}$	$\frac{\text{Brain weight}}{\text{body weight}} \times 100$	$\frac{\text{CMR} \times \text{brain weight}}{\text{body weight} \times \text{BMR}}$
Rat	7.6[b]	0.83	4.76
Cat	4.5[c]	1.03	6.15
Dog	3.4[d]	0.72	4.12
Rhesus monkey	3.7[e]	2.57	12.42[e]
Human	3.3[f]	2.34	20.2

[a]Data collected from Mink *et al.* (1981).
[b]Original source Nilsson and Siesjo (1976).
[c]Original source Geiger and Magnes (1947).
[d]Original source Gilboe and Betz (1973).
[e]Original source, Schmidt *et al.* (1948). The proportions listed in the table are based on a body weight of 3627 g. The use of a more typical adult body weight (6000 g) and adjusted BMR gives a brain/body ratio of 1.55% and (CMR × brain weight)/(BMR × body weight) percentage of 9.0%.
[f]Kety (1957).

active transport of ions required for the maintenance and restitution of membrane potentials (Sokoloff, 1981). Because no respite of electrical activity occurs during sleep, O_2 consumption is not diminished (Mangold *et al.*, 1955). Indeed, during rapid-eye-movement (REM) sleep, which is characterized by low-voltage, high-frequency electroencephalogram (EEG) wave patterns, the brain may increase its metabolic rate (Townsend *et al.*, 1973).

A possible exception to a continual high cerebral metabolic rate may occur during hibernation. Hibernation is a normal physiological state for some mammals and they appear to enter hibernation from slow-wave sleep, a segment of the sleep cycle characterized by high-voltage, low-frequency EEG patterns (Heller, 1979). Eventually in deep hibernation all spontaneous cortical activity is lost, whereas some limbic and reticular formation activity remain (Walker *et al.*, 1977; Beckman and Stanton, 1976). The loss of electrical activity by a large part of the brain suggests that the metabolic rate of the hibernating mammal would also be diminished; however, these brains maintain an ability to restore ionic balances after perturbations faster than can nonhibernating individuals and this fact suggests that the probable decrease in cerebral metabolic rate may not be so large. Our understanding of the relationship between hibernation and metabolic rates will only be clarified when the cerebral metabolic rate has been measured in the various stages of hibernation. Thus for all living adult mammals, with the possible exception of those in hibernation, no respite for a need for O_2 and glucose is found. Depressions in cerebral metabolic rates are associated with pathological states such as stupor and coma (Duffy and Plum, 1981).

Although mammals have no apparent release from meeting the energy demands of their brains, neither do these demands appear normally to escalate. REM sleep may cause the only normal increase in O_2 consumption (Townsend *et al.*, 1973) and this needs further investigation (Siesjo, 1978). Increases in mental work, determined by having the human subjects perform arithmetic problems, produced no noticeable effect on overall cerebral O_2 consumption (Sokoloff *et al.*, 1955). Recent evidence suggests that changing behaviors (e.g., solving arithemetic problems) are associated with shifting patterns of energy use, so that particular regions increase their cerebral blood flow, and presumably O_2 consumption, and other regions have lessened activity and O_2 consumption (Risberg and Ingvar, 1973). While particular populations of neurons increase their activities and O_2 consumption, the overall cerebral metabolic rate remains unchanged. Except during REM sleep, the only increases in cerebral metabolism occur during pathological states, in this case, seizures (Duffy and Plum, 1981).

The constantly high metabolic rate of the adult brain means that the brain must be constantly supplied with O_2 and glucose. Because the mammalian brain stores almost no substrate (glycogen) or O_2, severe hypoglycemia leads to stupor (Duffy and Plum, 1981). Neurons are irreversibly injured within minutes of the onset of hypoglycemic coma (Ghajar *et al.*, 1982).

Various safety factors protect the brain from a lack of O_2 and glucose. The neurons receive a continuous supply of blood despite fluctuations in systemic pressures through many autoregulatory mechanisms, of which the vasoactive effect of CO_2, K^+, P_{O_2}, and neural effectors on the cerebral vascular walls are among the better known (Sokoloff, 1981; Kuschinsky and Wahl, 1978; Lund-Andersen, 1979). The brain is also supplied with more O_2 than it uses (Buchweitz *et al.*, 1980) and a dense network of capillaries whose density may correlate with average regional cerebral blood flow [compare von Lierse (1963) with Sokoloff (1973)].

Although these safety factors protect an organism's brain from small perturbations, the brain's incessant requirements for metabolic substrates and O_2 make it vulnerable. Thus, a likely design feature of the brain is for it to be provided with a sufficiently large supply of O_2 and glucose for its maintenance. One may hypothesize from this that if selection pressures favored a larger brain, a concomitant increase in energy supply to the brain must also have been favored. Several possible mechanisms to do this exist. An increase in body size would provide the brain with more O_2 and a larger cardiovascular delivery system (Weibel, 1979; Weibel *et al.*, 1981). A higher basal (standard) metabolic rate (BMR) would provide the organism with a faster turnover rate of O_2. Although BMR decreases as animals increase in size (Kleiber, 1961), animals of equivalent body sizes also vary in their BMR. Insectivorous mammals are thought to have lower BMR than their frugivorous counterparts (McNab, 1978), domesticated mammals less than their wild counterparts (Schmidt-Nielsen, 1975), and seals and toothed whales higher than terrestrial mammals (Slijper, 1962). Thus animals that have higher BMR may be able to supply their brains with sufficient O_2 and glucose with a smaller body (Armstrong, 1982*b,c,* 1983).

In addition, mammals differ in the proportion of the total O_2 and energy reserves that they send to their brains (see Table 1). Finally, mammals could differ in the degree of safety factors, hemoglobin carrying capacities of O_2, transport of glucose, diffusion distances for O_2, and other histological and biochemical features. These additional features have not been analyzed among a comparative series of mammals, although some interesting beginnings have been made (Dhindsa *et al.*, 1982).

A first approximation looking at the interrelationships among brain weight, body weight, and standard (basal) metabolic rate has been made (Armstrong, 1982*b,c,* 1983). Data were collected from the literature, transformed logarithmically, and analyzed allometrically (Bruhn, 1934; Compoint–Monmignaut, 1973; Count, 1947; Crile and Quiring, 1940; Dawson and Hulbert, 1970; Eisenberg, 1981; Eisenberg and Wilson, 1978; Goffart, 1977; Hart, 1971; Hart and Irving, 1959; Herreid and Schmidt-Nielsen, 1966; Hildwein, 1972; Hildwein and Goffart, 1975; Hudson and Brower, 1974; Irving and Hart, 1957; Irving *et al.*, 1941; Jerison, 1973; Kamau and Maloiy, 1981; Karandeeva *et al.*, 1973; Kraus and Pilleri, 1969; MacMillen and Nelson, 1969; McNab, 1969; Milton *et al.*, 1979; Mink *et al.*, 1981; Muller, 1975; Nakayama *et al.*, 1971; Nelson and Asling, 1962; Pilleri, 1959; Pirlot and Stephan, 1970; Proppe and Gale, 1970; Scholander and Irving, 1941; Scholander *et al.*, 1950; Stephan *et al.*, 1970; von Bonin, 1937). Among the sample of mammals ($N = 93$), brain weight scales to body weight at 0.76 with 95% cls being 0.743–0.779 (Armstrong, 1983). Because a disproportionate number of the mammals are small and may have relatively small brains, a steeper slope than an "accurate" mammalian slope might be generated (L. Radinsky, personal communication; Hofman, 1982). To control for this, values within each of the 16 mammalian orders were averaged and then analyzed. In this case, the brain-to-body slope scales slightly lower, at 0.72, but these values overlap the slope generated from species' values: 95% cls 0.623–0.816 (Armstrong, 1982*c*, and Fig. 1). These slopes are close to the recently

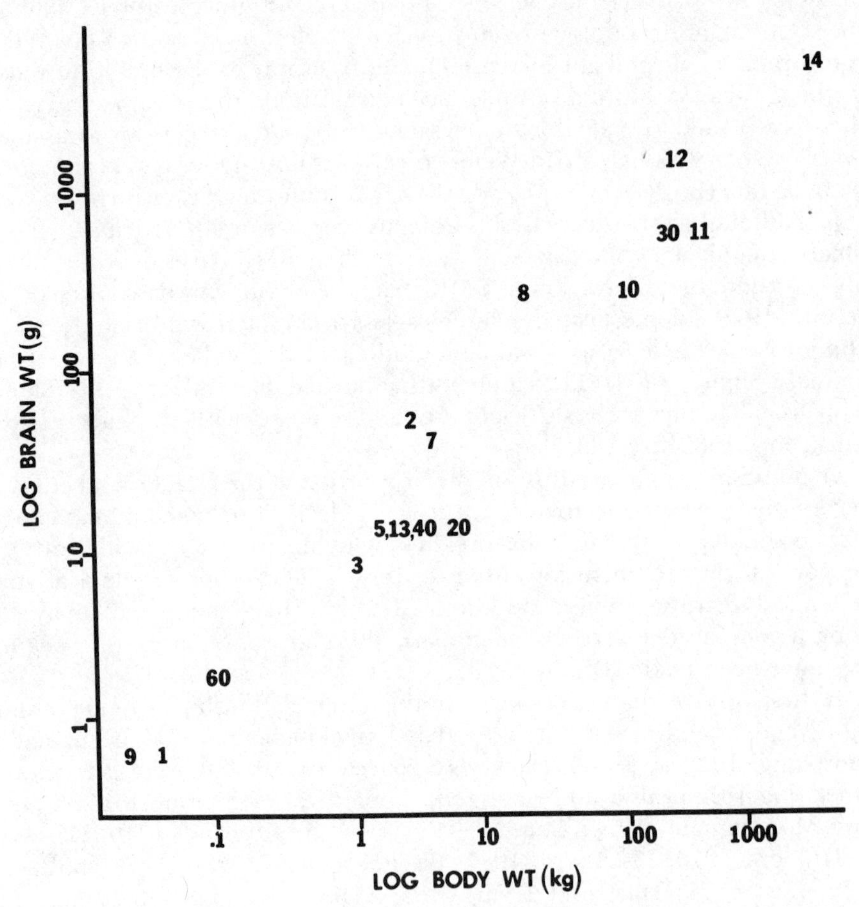

Fig. 1. Logarithmic plot of brain weight against body weight for the averages of 16 mammalian orders. The slope for the sample is 0.72. The plotted numbers identify the order; the numbers in parentheses refer to the number of species studied: 1, Insectivores (6); 2, primates (15); 3, marsupials (3); 5, lagomorphs (2); 7, carnivores (3); 8, pinnipeds (2); 9, bats (19); 10, artiodactyls (7); 11, perissodactyls (3), 12, cetaceans (5); 13, hyrax (1), 14, elephant (1); 20, monotremes (2); 30, manatee (1); 40, edentates (2); 60, rodents (21). [From Armstrong (1982*c*), courtesy of Elsevier Press.]

determined brain-to-body slopes of Eisenberg (1981) and Martin (1981, 1982).

The basal metabolic rate (cm^3 O_2 100 $g^{-1} \cdot min^{-1}$) diminishes as body weight enlarges. The rate among these 93 species is −0.269, 95% cls −0.292 to −0.248 (Fig. 2) (Armstrong, 1983). The BMR slope for the averages of the 16 orders is slightly lower, −0.22, with 95% cls −0.33 to −0.12 (Armstrong, 1982*c*). The unexplained variance of BMR-to-body weight (r^2 of species = 0.86; r^2 of orders = 0.36) is much higher than the brain weight-to-body weight variance (r^2 of species = 0.95; r^2 of orders = 0.90). The increase in variance may represent the difficulties in measuring BMR or a more labile relationship with body weight (Armstrong, 1982*c*, 1983).

The multiplication of body weight times BMR produces a figure that represents the amount of available energy for the organism. It parallels, but is lower by a factor of 4.8, than an animal's kilocaloric expenditure. Thus, if brain weight is constrained by the amount of energy available, one would predict that brain weight and the energy supply from body size and BMR

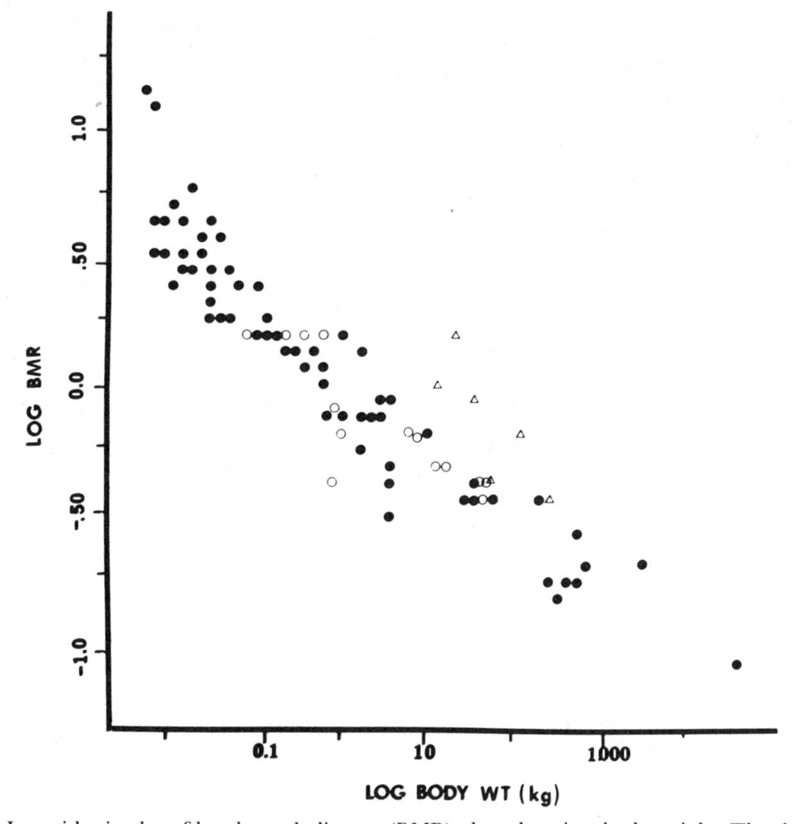

Fig. 2. Logarithmic plot of basal metabolic rate (BMR) plotted against body weight. The slope is −0.269; r^2 = 0.95. (●) mammals, (○) primates, (△) odontocetes and pinnipeds.

would scale isometrically, and that is what has been determined (Armstrong, 1982*b,c,* 1983). Using species, $\log E = -2.11 + 1.026 \log(S \cdot \mathrm{BMR})$, with 95% cls 0.97–1.08 (Armstrong, 1983); averaged values for orders show a similar relationship, $\log E = -1.86 + 0.95 \log(S \cdot \mathrm{BMR})$, with 95% cls 0.84–1.055 (Fig. 3) (Armstrong, 1982*c*), where E is brain weight and S is body weight. Although the BMR-to-body weight relationship has a much bigger amount of unexplained variance, once BMR is combined with body weight the unexplained variance is not intermediate between BMR-to-body and brain-to-body, but rather the variance is equivalent to that of brain-to-body (species r^2 E-to-$S \cdot$BMR = 0.94; r^2 by order is 0.93). The data suggest that among extant adult mammals, brain weight scales isometrically with its supply of energy.

Recent attention has focused on differences in the relative brain sizes of bats and primates that feed on fruits as opposed to those that eat predominantly insects. The division is statistically significant in bats (Pirlot and Stephan, 1970; Eisenberg and Wilson, 1978). Similar dietary divisions of primates are not statistically significant (Clutton-Brock and Harvey, 1980), perhaps

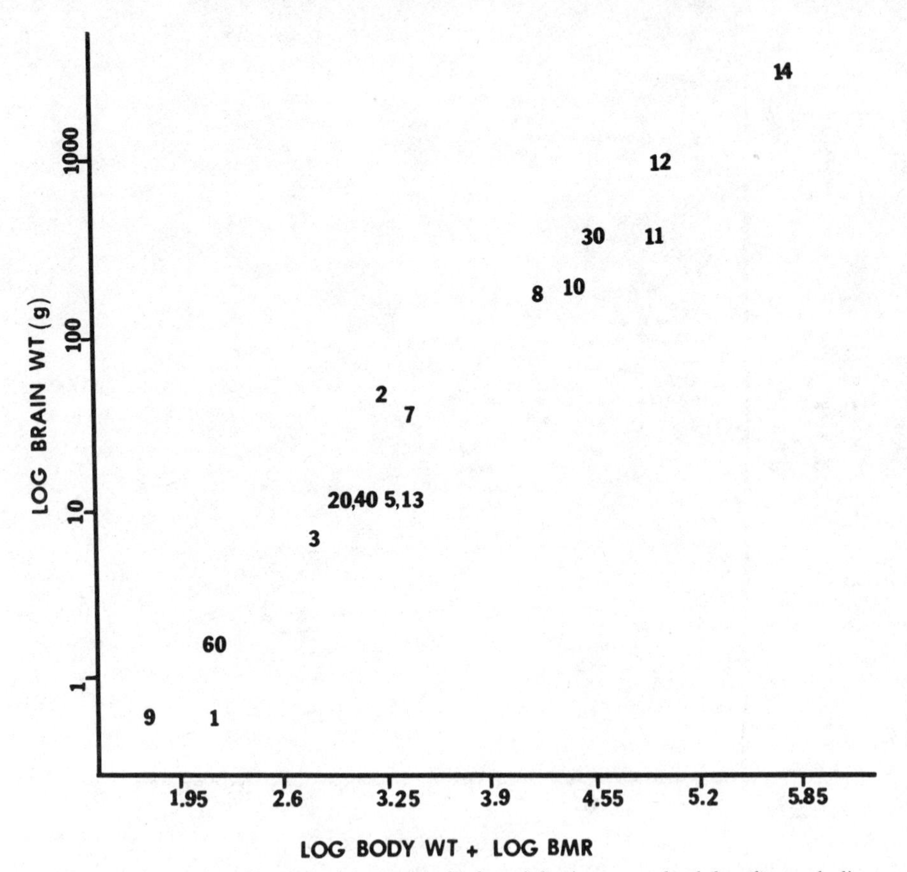

Fig. 3. Logarithmic plot of brain weight against body weight times standard (basal) metabolic rate. In this case brain weight scales isometrically with the adjusted body weight (slope ~1.0). Numbers as in Fig. 1. [From Armstrong (1982*c*), courtesy of Elsevier Press.]

reflecting the more eclectic diets of primates (Kay, 1975; Clutton-Brock and Harvey, 1980). My analysis of bat data corroborates the noted differences in relative brain size between frugivorous and insectivorous bats ($p < 0.01$) (Armstrong, 1983). The frugivorous bats also have higher BMR than insectivorous ones ($p < 0.03$) (Armstrong, 1983). Although the two groups can be separated according to the amount of BMR per body weight, the slopes associating BMR to body weight differ, although not statistically significantly so, between the two dietary groups. When one looks at how brain weight scales to the available sources of energy ($S \cdot$BMR), the dietary groups show an overlap of relative brain size, but also follow slopes that are statistically different, $p < 0.05$. To analyze the brain-to-body $S \cdot$BMR relationship in this case, the values from the mean of log($S \cdot$BMR) instead of the intercepts were compared and, in addition, the intercepts from the mean slopes of both groups were compared. In neither case are statistically significant differences found between the bat dietary groups; in both cases $p > 0.01$ (Armstrong, 1983).

Differences among the larger terrestrial mammals and seals and toothed whales also shift. Brain to body size ratios of these marine mammals are higher than the terrestrial ungulates ($p = 0.03$) and overlap extensively with anthropoid values ($p = 0.78$). After accounting for their high BMR, however, (Fig. 2) one finds that the adjusted relative brain sizes overlap these of ungulates ($p > 0.05$) and anthropoids ($p = 0.13$) (Armstrong, 1983).

These data suggest that brain size is related to the systems delivering O_2 and glucose. Indeed relative brain size (E/S) is highly correlated with BMR according to a Spearman rank correlation ($r_s = 0.785$, $p < 0.0001$) (Armstrong, 1983). Primates, however, retain a relatively larger brain per body ratio even after correction for BMR. If primate data are removed from the mammalian data, the correlation of relative brain size (E/S) to body size increases, $r_s = 0.868$, $p << 0.0001$ (Armstrong, 1983).

How does the primate brain maintain a large size relative to the total body's energy supply? One possibility is that it receives a larger proportion of the total energy supplies. Data about the energy demands of the brain (cerebral metabolic rate times brain weight) from the literature [Mink *et al.* (1981); collected data from Nilsson and Siesjo (1976), Geiger and Magnes (1947), Gilboe and Betz (1973), Kety (1957), Schmidt *et al.* (1945)] show that the rhesus monkey and human brains use a larger proportion of their body's energy reserves than do the brains of rat, cat, or dog (Table 1). The proportions correlate significantly with the residuals of brain-to-body and brain-to-body times BMR. That is, after the effects of body size and energy supply ($S \cdot$BMR) are removed from brain size, the residual amount of brain size strongly correlates with the amount of cerebral metabolism per body reserves ($p < 0.01$ for both body weight and body weight times BMR) (Armstrong, 1983).

The data suggest that brain size scales isometrically with its energy reserves. Relatively large brains, such as primate brains, must have mechanisms for directing more energy reserves to the brain. To clarify the brain-to-body relationship, future studies will need to analyze the energetics of the brain.

Allometric Considerations of Internal Divisions of the Brain

The total brain size may become enlarged via increases in size among different regions. On the assumption that regions of the brain that are large for an organism reflect which neuronal populations and functions are important for that taxon (Holloway, 1968; Stephan *et al.*, 1970; Welker, 1973), relative size differences are interpreted as indicating functional differences. Changes in circuitry, neurochemical transmitters, and physiological functions, however, may profoundly affect function without making any changes in size (Welker, 1976; Galaburda and Pandya, 1982). Such forms of differentiation are beyond the scope of this essay.

Several forms of differentiation affect the sizes of internal brain regions. Currently the best studied taxa are primates, and the most common parameter used for allometric studies are volumes. Analyses of volumes cannot indicate which feature has caused the size change, but volumetric studies can indicate how various neuronal populations scale vis-à-vis each other. Neuronal populations may increase by an addition in redundant circuitry, by an addition of specialized circuitry while maintaining a general functional constancy, by an addition of a new neuronal population with new functions, and by an addition to individual neuronal elements (enlarged dendritic or axonal arborizations) or changes in nonneural elements.

Increases in redundancy have been documented for cerebral and cerebellar cortices. Both of these cortices have anatomically and physiologically defined repetitive units—columns in the cerebral cortex (Hubel and Wiesel, 1963; Mountcastle, 1957; Szenthagothai, 1978) and repetitive Purkinje cell trees in the cerebellar cortex (Braitenberg, 1977; Palkovits *et al.*, 1971). In most mammals a constant number of neurons is found between the surface and the medullary substance of neo- and allocortical regions, that is, a constant number of neurons exists within a column (Bok *et al.*, 1939; Rockell *et al.*, 1974, 1980; West and Andersen, 1980; West, 1981). Many evolutionary changes in cortical size may have occurred through an addition of columns (Rockell *et al.*, 1980; Jerison, 1982). At least one cortical region, the primate visual cortex, does not follow this pattern and has more neurons per column than does a typical mammalian column [Rockell *et al.* (1980); see below, p. 137]. The cellular organization within noncortical structures is not known in sufficient detail to detect whether changes at these levels also involve repetitive units.

Specialized circuitry may be added to regions in which the overall functions remain constant. In both the somatosensory cortex and ventrobasal complex, mammals that have increased tactile areas have enlarged sensory regions (Welker, 1973). Presumably a parallel situation exists in motor regions. Such increases in specialized regions maintain the functions of proprioception and touch discrimination.

Specialized circuitry with new functions has also been found. An increased number of neuronal populations has been documented in primate visual system (Allman, 1982; Zilles *et al.*, 1982). The number and organization of tertiary visual regions distinguish primates from other mammals and also separate various primate families from each other (Allman, 1982). A particular subset of pulvinar neurons may exist in the human, but not the rhesus, pulvinar (Ogren, 1982). If this finding is substantiated it would document a species-specific or hominoid-specific addition of a neuronal population.

Although it is frequently stated that larger mammalian brains have larger dendritic and/or axonal branches (Jerison, 1973, 1980, 1982), very little work has actually been done to document this statement. Bok (1959) found no correlation of dendritic size with body size. Haug (1972) concluded [Shariff's (1953) and Tower's (1954) data] that primate neurons are miniaturized and more densely arrayed than are those of other mammals. Differences in neuronal densities may also emerge through changes in other systems, such as capillary densities, and number or size of glial elements.

Diminished neuronal densities may permit larger neurons (Jerison, 1973, 1979), but whether one should talk in terms of "a cortical neuron" when over 64 different types of cortical neurons have been identified is not clear (Carpenter, 1976). Rather, changes in neuronal sizes may be the result of increased differentiation (Herrick, 1926). Statements that argue for a particular change in neuronal size without using any procedure to visualize the neurons directly (such as a Golgi technique or HRP-filled neurons) cannot be seriously considered. Statements concerning subneuronal elements (e.g., "average" dendritic widths) on the basis of overall size changes in the brain are not considered except to note that pyramidal cell dendrites commence from the perikarya at widths up to 7–8 μm and taper from there (Shepherd, 1974).

Changes in size may stem from changes in nonneural elements. Different regions vary in their glial and capillary densities (e.g., von Lierse, 1963; Diemer, 1978). How these different features will scale in different parts of the brain is not yet known. Only in the cerebral and cerebellar cortex has any attempt been made to study glial–neuronal ratios quantitatively in mammals of different sizes (Friede, 1954, 1963; Tower and Young, 1973). In these studies larger brained mammals have more glia per neuron. The possibilities of changes in glial sizes and/or differentiation have not been examined. Allometric considerations of various parts of the brain provide an introduction to one aspect of evolutionary change within the brain. These regional studies must be accompanied by detailed comparative anatomic work for a solid interpretation.

Medulla Oblongata and Pons

The medulla, pons, and cerebellum are embryologically derived from the rhombencephalon (Carpenter, 1976). The cerebellum will be discussed separately. Most of the morphometric data listed as medulla oblongata [Bauchot (1982) and Stephan *et al.* (1970, 1981) and the analyses of these data, Jerison

(1973) and Sacher (1970)] include the tegmentum of the medulla, the pons, and parts of the mesencephalon (Stephan *et al.*, 1970).

The medulla, *sensu stricto*, is the most caudal part of the brain, extending from the spinal cord to the pons. A large part of its mass is composed of ascending and descending tracts. Several large and important nuclei are also located within this portion of the brain, including all or parts of cranial nerve nuclei 5 and 8–12, respiratory centers, the inferior olivary nucleus, and part of the reticular formation (Carpenter, 1976).

Little morphometric work exists for the medulla *per se*. Tilney's (1928) data (one specimen of *Tarsius*, *Macaca*, baboon, gibbon, orang, chimp, gorilla, and *Homo*) suggest that the pyramids occupy a large proportion of the hominoid medulla than they do in the monkey or prosimian medulla. In contrast, the number of corticospinal tract fibers in the mammalian medullary pyramids scales at 0.30 with body weight [see above; Towe (1973)]. The principal inferior olivary nucleus is claimed to be larger in more "advanced" mammalian forms (Sarnat and Netsky, 1981). Comparisons of cross-sectional size of the inferior olive to the medulla show a trend among primates, with the prosimian inferior olive occupying the smallest ratio of the medulla, the human olive the largest (Tilney, 1928). How variable the regression of the inferior olive is among primates and how primates compare to other mammals cannot be stated at this time. Whales, for example, have a relatively small inferior olive and a much larger development of the medial accessory olive, perhaps reflecting their differential development of trunk and tail muscles (Sarnat and Netsky, 1981). Whether these different proportions would be reflected in total medullary morphometric data is not clear. Hofman (1982) found that his medulla–body weight correlations deviated for the whale. This finding could reflect the problems of estimating a true body weight with so much blubber present (Hofman, 1982) or a different organization within the medulla. In addition to altered olivary proportions, whales and other diving mammals may be expected to have changes in their respiratory centers to regulate their diving capacities and this might change the size of the medulla.

Several scientists have suggested that the size of the medulla may be used to generate estimates of body sizes (Hofman, 1982; Jerison, 1973; Gould, 1975). It should be reiterated that estimates of medulla oblongata size from Stephan's data include other parts of the brainstem. Hofman (1982) also collected some medullary lengths from an atlas, but did not define its rostral border. While the "medulla oblongata" (*sensu lato*) is very strongly correlated with body weight, it is even more highly correlated with brain weight (Stephan *et al.*, 1970; Sacher, 1970; Bauchot, 1982). Generating body weights from the hind brain must be done with caution.

The pons, the rostral part of the rhombencephalon, is composed of the basis pontis, pontine nuclei, cranial nerve nuclei 8–4, part of the reticular formation, and cardiovascular control centers. Except for Tilney's study (1928) of the pontine nuclei (they occupy a slightly larger proportion of the pons in hominoids than in monkeys or prosimians), the pons has not been studied separately. Rather, the tegmentum (grey matter) has formed part of a unit labeled medulla oblongata, whereas the nuclei of the basis pontis are included in the cerebellum. Based on the data of Stephan *et al.*, the human medullary-pontino complex (medulla) was found to be about the same size predicted for a nonhuman primate of our body mass (Passingham, 1975), and less than encephalization quotients would predict (Holloway, 1979).

Mesencephalon

The mesencephalon extends from the pons to the diencephalon. In addition to tracts, of which the cerebral peduncles are most prominent, it includes the superior and inferior colliculi (the tectum), the red nucleus, cranial nerve nuclei 3 and part of 5, and part of the reticular formation. Both the superior and inferior colliculi appear to be relatively enlarged parts of the mesencephalon in prosimians and decline as proportion to the rostral mesencephalon when they are compared with monkey, ape, and human proportions (Tilney, 1928). The red nucleus occupies an increasingly larger proportion of the mesencephalon in the same series of primates when data from the small-brained prosimians are compared with monkeys, apes, and humans (Tilney, 1928).

Taking the mesencephalon as a whole [the data of Stephan *et al.* (1970)], one finds that the human volume is about twice what one would predict for a nonhuman primate of equivalent body weight (Passingham, 1975), but less than encephalization quotients predict (Holloway, 1979). In another insectivore–primate series, the human mesencephalon was only slightly larger than expected (Bauchot, 1982).

Cerebellum

The cerebellum, which also develops from the rhombencephelon, includes both a tightly packed cerebellar cortex and deep nuclei. In addition, the cerebellar data of Stephan *et al.* (1970) include the basis pontis (the middle cerebellar peduncle and pontine nuclei) and the inferior and superior cerebellar peduncles. The data of Stephan *et al.* (1970), analyzed by others, show the following. Insectivore and primate cerebellar volumes correlate very strongly with brain weights (Sacher, 1970). The human cerebellum is almost three times as large as one would predict for a primate of our body weight (Passingham, 1975), but as large as expected for a basal insectivore or primate of our brain size (Radinsky, 1975; Gurche, 1982) and thus predictable by EQs (Holloway, 1979). Cerebellar volumes are highly correlated with diencephalic, striate, and subthalamic volumes (Bauchot, 1982).

Volumetric data on the dentate nucleus indicate that among nonhuman primates the size of the dentate nucleus is allometrically negative to body weight, and isometric with brain weight. The human dentate nucleus is more than twice as large as primate body data predict, but may be slightly smaller than brain weight predicts [dentate volume of ten primates from Solnitsky (1945); brain/body data from Stephan *et al.* (1970)]. More precise statements will come from using the same specimens for all the measurements.

Diencephalon

The mammalian diencephalon extends from the mesencephalon to the telencephalon. It includes four major regions (thalamus, hypothalamus, epithalamus, subthalamus), which are themselves conglomerates of neuronal populations and functions (Carpenter, 1976). Several of these populations have recently been studied allometrically (Baron, 1979; Stephan *et al.*, 1981; Bauchot, 1982). Treated as one unit, the primate and insectivore diencephalons increase in size slightly more than body weights predict, but less than the overall brain weights do. The human diencephalon is about 1.5 times as large as nonhuman body weights predict [the data of Stephan *et al.* (1970); Passingham (1975)] and less than EQs predict (Holloway, 1979). It correlates more strongly with brain size than body size (Bauchot, 1982; Stephan *et al.*, 1970). The major components within the diencephalon—the thalamus, hypothalamus, subthalamus, and epithalamus—scale along different slopes as brain and body weights change, and species differ in the amount of deviation from the expected size of their nuclei. That is, a particular species with a high encephalic index for the hypothalamus may have a low or high value for the subthalamus (Bauchot, 1982; Baron, 1979). When more differentiated regions are used, such as individual thalamic nuclei, correlations among functionally related structures begin to appear (Bauchot, 1982). Using a progression index based on *Tenrec,* Baron (1979) determined that among insectivores and primates, human subthalamic, thalamic, hypothalamic, and epithalamic volumes were larger than expected, with the largest deviation being found for the subthalamus, the smallest for the epithelamus. *Lagothrix lagothrix* had a larger subthalamic deviation than did *Homo* and several monkeys had larger epithalamic deviations than did *Homo* (Baron, 1979). Slightly different patterns of species deviations were determined by Bauchot (1982), but looking at *Homo* and considering the four major diencephalic regions, both scientists found the human subthalamus to have the largest positive deviation.

Comparative work on neuronal populations small enough to permit conjectures about functions have been done on thalamic nuclei. Bauchot (1982) found that compared to his insectivore–primate series, the human dorsal tier of the lateral thalamic nuclei had the largest deviation, followed in order by the medial dorsal complex, the ventral tier, the anterior complex, midline nuclei, and the lateral and medial geniculate bodies. The human anterior principal and lateral dorsal nuclei clearly have many more neurons than ape data (Armstrong, 1980*a*) or primate data (Armstrong and St. Onge, 1981) predict on the basis of brain or body size. The medial dorsal nucleus has slightly more neurons than ape brain data predict (Armstrong, 1982*a*), the pulvinar complex as many as brain mass predicts (Armstrong, 1981), and the motor and sensory nuclei fewer neurons than would be predicted (Armstrong, 1979, 1980*b*, 1982*a*).

The kinds of differentiation that have occurred at the thalamic level are not yet established. Until the patterns of thalamic circuitry, particularly in reference to the cortex, are worked out, detailed species comparisons of the forms of differentiation are precluded. Certain regions, however, have been studied.

The ventrobasal complex contains a mammunculus and, for *Ateles* with its tactile tail pad, it is relatively enlarged (Welker, 1973; Pubols and Pubols, 1972). This type of differentiation, an increase in a neuronal population while maintaining an overall function (in this case maintenance of the mammunculus), is possibly a common comparative feature in motor and sensory regions throughout the neuraxis.

Neuronal populations that differ in proportions of constituent cell sizes are likely to differ in function. Studies comparing thalamic perikaryal sizes among hominoids tentatively suggest that the sensory, motor, and limbic nuclei have stabilized or increased their neuronal numbers in the human brain without changing characteristics of the neuronal pool (Armstrong, 1979, 1980*a,b*, 1982*a*). On the other hand, as the human pulvinar increased in size it apparently differentiated its neuronal population into two populations (Armstrong, 1980*b*, 1982*a*). Whether this finding is related to the development of a telencephalic ganglionic eminence (Rakic and Sidman, 1969; Ogren, 1982) remains to be determined. The human mediodorsal nucleus may also share this pattern of differentiation (Armstrong, 1982*a*).

Telencephalon

The telencephalon includes both subcortical nuclei, such as the striatum, amygdala, and septum, and the cerebral cortex. Because the telencephalon is the largest component of the human brain (Sacher, 1970; Passingham, 1975; Bauchot, 1982), and because its surface, the cerebral cortex, is well studied both anatomically and physiologically, and leaves traces of its phylogenetic history in the form of endocasts, the telencephalon has been studied more extensively than have other parts of the brain.

Recent advances in computerized image analysis suggest that a more precise quantification of neuronal populations is possible (Schleicher *et al.*, 1978). This advance may represent a powerful new tool for the extension of allometric analyses to neuronal populations. To date the method has been used to investigate the cerebral cortex (e.g., Zilles and Schleicher, 1980; Zilles *et al.*, 1978, 1982).

The power of the allometric method for neurobiologists was perhaps most dramatically shown when Stephan and Andy showed that the human septum not only had not diminished in size, but is as large as primate body weights predict. Previously, the relatively large size of other neuronal structures had obscured this relationship (Andy and Stephan, 1968).

Other human limbic structures have not been so conservative. In the diencephalon the thalamic anterior and lateral dorsal nuclei are relatively enlarged for a primate brain of human dimensions (Armstrong, 1980*a*, 1982*a*). The telencephalic amygdala, particularly the corticobasolateral portions, is also relatively large in the human brain. The primate, including human, centromedial group of amygdalar nuclei scale according to body weight, and only the nucleus of the lateral olfactory tract is reduced in size in humans (Stephan and Andy, 1977).

Although the limbic and olfactory systems are traditionally considered together, the two systems scale independently. Olfactory bulbs and related structures are small in primates (Radinsky, 1970, 1975; Gurche, 1982; Stephan and Andy, 1977), but many limbic structures are relatively large.

The "limbic" cortex also shows this independence. The sizes of the primate and insectivore schizo- and allocortices (entorhinal and hippocampal regions) correlate strongly with each other, but not with the neocortex (Stephan and Andy, 1970). Neither the schizo- nor the hippocampal cortex increase as much as brain weight does [Sacher (1970), from data of Stephan and Andy (1970)]. For their given body weights, the human and the aye-aye hippocampi are the most encephalized (Stephan and Andy, 1970), although Bauchot (1982) found the highly encephalized human hippocampus to be within the range of monkeys. The hippocampus also scales isometrically with the amygdala (Stephan and Andy, 1977). In rodents estimates of the size of the hippocampal and entorhinal cortices can provide accurate estimates of the numbers of neurons because the number of neurons through the depth of these cortices is stable (West and Andersen, 1980; West, 1981). Given the ubiquitous nature of columns, this may be true for other mammals as well.

The neocortical volumes of Stephan *et al.* (1970) include the corona radiata. Analysis of these data shows that primates have relatively more neocortex than do basal insectivores (Radinsky, 1975). Additional analyses show that the size of the human neocortex is about three times larger than expected for a primate of our body weight, but predictable on the basis of primate brain weights (Radinsky, 1975; Passingham, 1973).

In all mammals studied, the cortex increases in size as the brain enlarges [the data of Stephan *et al.* (1970); Passingham (1975) and Jerison (1982)], but the enlargement is bigger than predicted by the scaling of surface area to a sphere. To accommodate its size, the relatively enlarged cortical surface area of large-brained species is folded into gyri. The length of the gyri appears directly proportional to brain size among extant mammals [Elias and Schwartz (1971); reanalyzed by Jerison (1982)], although mechanical intracortical forces may be important (Richman *et al.*, 1975). Allometric studies have helped identify some of the geometric regularities among brains of different sizes.

Allometric analyses of intracortical regions, demarcated by cytoarchitectural differences, suggest that eulaminate (late maturing, "association") cortex and agranular (motor) cortex occupy a larger proportion of the human brain than they do in other primates. Granular (sensory) cortex, on the other hand, has maintained a similar proportion among all primates [Shariff's (1953) data; Passingham (1973, 1975)]. This suggests an increase in the amount of eulaminate to granular cortex (Holloway, 1968; Passingham, 1975). The classical correlation of eulaminate cortex with the function of association is no longer as clear as was once thought (Diamond, 1979), and so the meaning of the changed proportions is more problematic.

The primary visual or striate cortex is homologous among primates and mammals in general (Allman, 1982; Zilles *et al.*, 1982). At the same time the primate, including human, striate cortex is distinct in that it contains more neurons per column than do the visual columns of other mammals (Rockell *et al.*, 1980). Other primate neocortical sensory regions do not have more neurons. The human striate cortical area is smaller than primate brain weight predicts (Stephan, 1969; Passingham, 1975). Diurnal prosimians have wider supragranular striate regions than do nocturnal prosimians (Zilles *et al.*, 1982), suggesting that retinal input is more extensively processed in day-active primates than it is in nocturnal forms.

Since the lunate sulcus delimits the rostral boundary of the striate cortex, the relative size of lateral aspects of the striate cortex can possibly be determined from fossil endocasts. The Oligocene anthropoid *Aegyptopithecus* has a larger postlunate surface area than prosimians (Radinsky, 1975, 1979). In Australopithecines, the determination of the lunate sulcus is not clear and the interpretations of relative visual cortical sizes are thus in dispute (Holloway, 1975, 1981; Radinsky, 1979; Falk, 1980, 1983*b*).

While primate secondary visual cortices are probably homologous with those of other mammals, homologies of tertiary regions are very difficult to determine (Allman, 1982; Kaas, 1978; Campbell, 1982). Although quantitative cytoarchitectonic divisions are possible (Schleicher *et al.*, 1978; Zilles and Schleicher, 1980; Zilles *et al.*, 1978, 1982), allometric analyses have not yet been carried out on tertiary visual structures.

Little morphometric work has been done on the somatosensory cortex. The anthropoid central sulcus as differentiated from the prosimian coronolateral sulcus is present in *Aegytopithecus,* and so is at least 27 million years old (Radinsky, 1974, 1975, 1979). Both fossil and living prosimians have a horizontal coronolateral sulcus separating sensory and motor regions from forelimb regions. The earlier prosimian coronolateral sulcus is located closer to the midline than it is in extant prosimians. The changed proportions may indicate an increased representation for the extremities (Radinsky, 1975; Gurche, 1982).

Other changes in somatosensory-motor cortex that stem from increases of particular body regions have been documented among extant mammals (Pubols and Pubols, 1972; Welker, 1973). An apparently enlarged sensory-motor face region and increased sensitivity in the Pleistocene *Archaeolemur* (Radinsky, 1970) and a baboon (Falk, 1981, 1982*a*) have been so interpreted.

Eulaminate cortex is the well-ordered, distinctly six-layered homotypical cortex (Carpenter, 1976). The vast majority of the primate and human cortex is composed of variants of this kind of cortex. Recently, it has been determined that in this cortex topography and connection patterns do not necessarily covary (Galaburda and Pandya, 1982). Part of the human cortex involved in language, area 44 or Broca's area, for example, is found in a distinctly different region of the human frontal cortex than is its cytoarchitecturally identified homologue in rhesus monkeys (Galaburda and Pandya, 1982). For topographically similar regions, different structures would be compared. Allometric analyses are most useful after homologies based on cytoarchitecture, neuronal morphology, and connections have been established.

Whole lobes have been analyzed morphometrically and allometrically. The temporal lobes in early primates appear to be relatively large compared with those of insectivores (Radinsky, 1970, 1975, 1979), probably reflecting a role in visual functions (Radinsky, 1970; Allman, 1982). Large temporal lobes in extinct notoungulates, on the other hand, may reflect specializations in auditory functions (Radinsky, 1981). Size and shape of the temporal poles and inferior temporal surfaces may separate humans from apes (Holloway, 1972, 1975).

The frontal lobe has also been measured as an entity. Taking the end of the Sylvian fissure as the posterior boundary of the frontal lobe (and thus including some of the parietal lobe), Radinsky (1970, 1975, 1979) found a relative increase in frontal lobe size in later (late Eocene and Oligocene) fossil prosimians compared with those of early and middle Eocene forms. Prosimian frontal lobe development occurred later than occipital and temporal lobe expansion (Radinsky, 1975, 1979). The relative frontal lobe size of Oligocene fossil primates is not as large as that of modern prosimians (Radinsky, 1970, 1979). The Oligocene anthropoid *Aegyptopithecus* may also have had a relatively small frontal lobe compared with those of modern species (Radinsky, 1974, 1975, 1979). Although a large frontal lobe has been viewed as a hallmark of the human brain, morphometric analyses challenge this assumption (Holloway, 1968). The shape of the human orbito frontal surface differs from that of apes (Holloway, 1975) and the human configuration may have been attained after the Australopithecines (Falk, 1980, 1983*a*).

Conclusions

1. Although brain–body relationships were first examined as a way to investigate behavior (intelligence), research showed that brain–body relationships present an interesting puzzle in itself. Anthropoids, seals, and toothed whales have large brains for their body weights. Prosimian brain–body relationships appear equivalent to those of most mammals. Frugivorous bats have larger brains per body weight than do insectivorous bats.

2a. Causal explanations of brain–body relationships have been few. A relationship stemming directly from the number of incoming sensory afferents or exiting motor efferents does not withstand quantitative analyses. Possibilities of an allometric relationship of the brain with another major derivative of the ectodermal anlage—the epidermis—needs further investigation.

2b. Data on mammalian energetics, and brain and body sizes, suggest that adult mammalian brains scale isometrically with their energy supply. The relatively enlarged brains of seals and toothed whales are associated with relatively high metabolic rates. High anthropoid brain–body ratios and superhigh human brain–body ratios appear to be supported by a higher proportion of the body's energetics being used by the brain than in other mammals.

3a. During the evolution of different mammalian lineages, brains increased in size both absolutely and relatively. Changes in different aspects of neural tissue underlie enlargements in brain size mass. Insufficient knowledge prohibits the establishment of "laws of neural evolution," but data from the well-studied mammalian visual system permit some generalizations. Although cortical tissue is most likely to increase by increasing the number of repetitive columns, specialized regions such as the primate striate cortex may increase the number of neurons per column. The primate secondary and tertiary cortical areas show interesting panprimate trends as well as intraorder differences. Detailed analyses of these regions may help provide trenchant insights into primate phylogeny as well as primate brain evolution.

3b. Not all regions of the brain have differentiated at the same rate. The hindbrain has been the object of few comparative studies, perhaps because it appears to be evolutionarily conservative. The conservatism may reflect its involvement in basic mammalian regulatory mechanisms, e.g., the modulation of cardiovascular and respiratory functions. The cerebellum, an outgrowth of the hindbrain anlage, integrates many motor functions, and in primates, enlarges at the same rate as the overall brain.

3c. Although the primate diencephalon has not increased in size as much as the whole brain, certain regions appear particularly enlarged in the human brain. The subthalamus (a motor region) and nuclei involved in "Papez circuit" (limbic regions) are relatively enlarged in human brains.

3d. Telencephalic structures also vary in their morphometric divergences. Primates have more neocortex per brain weight than do insectivores. Visual areas are relatively large and differentiated and these changes appeared early in primate evolution. While frontal lobes are large in modern primates, the enlargement was attained relatively late in both prosimian and anthropoid lineages.

Acknowledgments

I thank L. Radinsky, D. Falk, W. Leutenegger, G. Conroy, and K. Zilles for critically reviewing this paper. This research was supported by the Harry

F. Guggenheim Foundation and by National Science Foundation grant BNS-82-04480.

References

Allman, J. 1982. Reconstructing the evolution of the brain in primates through the use of comparative neurophysiological and neuroanatomical data, in: *Primate Brain Evolution: Methods and Concepts* (E. Armstrong and D. Falk, eds., pp. 13–28, Plenum Press, New York.

Andy, O. J., and Stephan, M. 1968. The septum in the human brain. *J. Comp. Neurol.* **133:**388–409.

Armstrong, E. 1979. A quantitative comparison of the hominoid thalamus: I. Specific sensory relay nuclei. *Am. J. Phys. Anthropol.* **52:**405–419.

Armstrong, E. 1980*a*. A quantitative comparison of the hominoid thalamus. II. Limbic nuclei anterior principalis and lateralis dorsalis. *Am. J. Phys. Anthropol.* **52:**43–54.

Armstrong, E. 1980*b*. A quantitative comparison of the hominoid thalamus. III. A motor substrate—The ventrolateral complex. *Am. J. Phys. Anthropol.* **52:**405–419.

Armstrong, E. 1981. A quantitative comparison of the hominoid thalamus. IV. The pulvinar and lateral posterior complex. *Am. J. Phys. Anthropol.* **55:**369–383.

Armstrong, E. 1982*a*. Mosaic evolution in the primate brain: Differences and similarities in the hominoid thalamus, in: *Primate Brain Evolution: Methods and Concepts* (E. Armstrong and D. Falk, eds.), pp. 131–161, Plenum Press, New York.

Armstrong, E. 1982*b*. An analysis of brain allometry: Consideration of the cerebral metabolic demand. *Am. J. Phys. Anthropol.* **57:**167–168.

Armstrong, E. 1982*c*. A look at relative brain size in mammals. *Neurosci. Lett.* **34:**101–104.

Armstrong, E. 1983. Metabolism and relative brain size. *Science* **220:**1302–1304.

Armstrong, E. and St. Onge, M. 1981. Evolution of the human anterior thalamic complex: Results of morphometric and allometric analyses. *Soc. Neurosci.* **7:**755.

Baron, G. 1979. Quantitative changes in the fundamental structural pattern of the diencephalon among primates and insectivores. *Folia Primatol.* **31:**74–105.

Bauchot, R. 1978. Encephalization in vertebrates. *Brain Behav. Evol.* **15:**1–18.

Bauchot, R. 1982. Brain organization and taxonomic relationships in insectivora and primates, in: *Primate Brain Evolution: Methods and Concepts* (E. Armstrong and D. Falk, eds.), pp. 163–175, Plenum Press, New York.

Bauchot, R., and Diagne, M. 1973. La croissance encéphalique chez *Hemicentetes semispinosus* (Insectivora, Tenrecidae). *Mammalia* **37:**468–477.

Bauchot, R., and Stephan, H. 1966. Données nouvelles sur l'encéphalisation des insectivores et des prosimiens. *Mammalia* **30:**160–196.

Bauchot, R., and Stephan, H. 1969. Encéphalisation et niveau évolutif chez les simiens. *Mammalia* **33:**225–275.

Beckman, A. L., and Stanton, T. L. 1976. Changes in CNS responsiveness during hibernation. *Am. J. Physiol.* **231:**810–816.

Bok, S. T. 1959. *Histonomy of the Cerebral Cortex,* Elsevier, Amsterdam.

Bok, S. T., Kip, M. J., and Taalman, V. E. 1939. The size of the body and the size and number of the nerve cells in the cerebral cortex. *Acta Neerl. Morphol. Norm. Pathol.* **3:**1–22.

Braitenberg, V. 1977. *On the Texture of Brains,* Springer-Verlag, New York.

Brandt, A. 1867. Sur le rapport du poids du cerveau à celui dur corps chez différents animaux. *Bull. Soc. Imp. Nat. Moscow* **40:**525–543.

Brody, S. 1945. *Bioenergetics and Growth,* Hafner, New York.

Bronson, R. T. 1979. Brain weight–body weight scaling in breeds of dogs and cats. *Brain Behav. Evol.* **16:**227–236.

Bruesch, S. R., and Arey, L. B. 1942. The number of myelinated and unmyelinated fibers in the optic nerve of vertebrates. *J. Comp. Neurol.***77:**631–665.

Bruhn, J. M. 1934. The respiratory metabolism of infrahuman primates. *Am. J. Physiol.* **110:**477–484.

Brummelkamp, R. 1939. Das sprungweise wachstum der kernmasse. *Acta Neerl. Morphol. Norm. Pathol.* **2:**177–188.

Buchweitz, E., Sinha, A. K., and Weiss, H. R. 1980. Cerebral regional oxygen consumption and supply in anesthetized cat. *Science* **209:**499–501.

Campbell, C. B. G. 1982. Some questions and problems related to homology, in: *Primate Brain Evolution: Methods and Concepts* (E. Armstrong and D. Falk, eds.), pp. 1–11, Plenum Press, New York.

Carpenter, M. B. 1976. *Human Neuroanatomy,* Williams and Wilkins, Baltimore.

Carter, H. B. 1965. Variation in the hair follicle population of the mammalian skin, in: *Biology of the Skin and Hair Growth* (A. G. Lyne and B. F. Short, eds.), pp. 25–33, Elsevier, New York.

Clutton-Brock, T. H., and Harvey, P. H. 1980. Primates, brains and ecology. *J. Zool. (Lond.)* **190:**309–323.

Compoint-Monmignaut, C. 1973. Anatomie comparée: L'encéphalisation chez les rongeurs. *C. R. Acad. Sci. Paris* **277:**861–863.

Count, E. W. 1947. Brain and body weight in man: Their antecedents in growth and evolution. *Ann. N.Y. Acad. Sci.* **46:**993–1122.

Crile, G. 1941. *Intelligence, Power and Personality,* McGraw-Hill, New York.

Crile, G. W., and Quiring, D. P. 1940. A record of the body weight and certain organ and gland weights of 3690 animals. *Ohio J. Sci.* **40:**219–259.

Darwin, C. 1871. *The Descent of Man and Selection in Relation to Sex,* Murray, London.

Dawson, T. J., and Hulbert, A. J. 1970. Standard metabolism, body temperature and surface areas of Australian marsupials. *Am. J. Physiol.* **218:**1233–1238.

Dhindsa, D. S., Hoversland, A. S., and Metcalfe, J. 1982. Comparative studies of the respiratory functions of mammalian blood. XII. Black galago (*Galago crassicaudatus argintatus*) and brown galago (*Galago crassicaudatus crassicaudatus*). *Respir. Physiol.* **47:**313–323.

Diamond, I. T. 1979. The subdivision of neocortex: A proposal to revise the traditional view of sensory, motor and association areas. *Prog. Psychobiol. Physiol. Psychol.* **8:**81–151.

Diemer, N. H. 1978. Glial and neuronal changes in experimental hepatic encephalopathy: A quantitative morphological investigation. *Acta Neurol. Scand.* **58**(Suppl. 71):1–144.

Dubois, E. 1897. Sur le rapport du poids de l'encéphale avec la grandeur du corps chez le mammifères. *Bull. Soc. Anthropol.* **8:**337–376.

Dubois, E. 1914. Die gesetzmässige Beziehung von Gehirnmasse zu Körpergrösse bei den Wirbeltieren. *Z. Morphol. Anthropol.* **18:**323–350.

Duffy, T. E., and Plum, F. 1981. Seizures, coma and major metabolic encephalopathies, in: *Basic Neurochemistry* (G. J. Siegel, R. W. Albers, B. W. Agranoff, and R. Katzman, eds.), pp. 681–718, Little, Brown, Boston.

Eisenberg, J. F. 1981. *The Mammalian Radiations,* University of Chicago Press, Chicago.

Eisenberg, J. F., and Wilson, D. 1978. Relative brain size and feeding strategies in the Chiroptera. *Evolution* **32:**740–751.

Elias, H., and Schwartz, D. 1971. Cerebrocortical surface areas, volumes, lengths of gyri and their interdependence in mammals, including man. *Z. Saeugetierkd.* **36:**147–163.

Falk, D. 1980. A reanalysis of the South African Australopithecine natural endocasts. *Am. J. Phys. Anthropol.* **53:**525–539.

Falk, D. 1981. Sulcal patterns of fossil *Theropithecus* baboons: Phylogenetic and functional implications. *Int. J. Primatol.* **2:**187.

Falk, D. 1982*a*. Mapping fossil endocasts, in: *Primate Brain Evolution: Methods and Concepts* (E. Armstrong and D. Falk, eds.), pp. 217–226, Plenum Press, New York.

Falk, D. 1982*b*. Allometry: Scaling of brain size, body surface area and body shapes in primates. *Int. J. Primatol.* **3:**281.

Falk, D. 1983*a*. Cerebral cortices of East Asian hominids. *Science* **221:**1072–1074.

Falk, D. 1983*b*. The Taung endocast: A reply to Holloway. *Am. J. Phys. Anthropol.* **60:**479–489.

Falk, D., and Waide, R. 1982. Allometry: Body shape as a key factor in brain evolution. *Am. J. Phys. Anthropol.* **57:**186.

Frick, H. 1957. Betrachtungen über die Beziehungen zwischen körpergewicht und organgewicht. *Z. Saugetierkd.* **22:**193–207.

Friede, R. L. 1954. Der quantitative Anteil der Glia an der Cortexentwicklung. *Acta Anat.* **20:**290–296.

Friede, R. L. 1963. The relationship of body size, nerve cell size, axon length and glial density in the cerebellum. *Proc. Natl. Acad. Sci. U.S.A.* **49:**187–193.

Galaburda, A. M., and Pandya, D. N. 1982. Role of architectonics and connections in the study of primate brain evolution, in: *Primate Brain Evolution: Methods and Concepts* (E. Armstrong and D. Falk, eds.), pp. 203–216, Plenum Press, New York.

Geiger, A., and Magnes, J. 1947. The isolation of the cerebral circulation and the perfusion of the brain in the living cat. *Am. J. Physiol.* **149:**517–537.

Ghajar, J. B. G., Plum, F., and Duffy, T. E. 1982. Cerebral oxidative metabolism and blood flow during acute hypoglycemia and recovery in unanesthetized rats. *J. Neurochem.* **38:**397–409.

Gilboe, D. D., and Betz, A. G. 1973. Oxygen uptake in the isolated canine brain. *Am. J. Physiol.* **224:**588–595.

Goffart, M. 1977. Hypométabolisme chez *Aotus trivirgatus.* (Primates Platyrhini, Cebidae). *C. R. Séances Soc. Belge. Biol.* **171:**1149–1152.

Goodman, M., Snyder, F. N., Stimson, C. W., and Rankin, J. J. 1969. Phylogenetic changes in the proportions of two kinds of lactate dehydrogenase in primate brain regions. *Brain Res.* **14:**447–459.

Gould, S. J. 1966. Allometry and size in ontogeny and phylogeny. *Biol. Rev.* **41:**587–640.

Gould, S. J. 1975. Allometry in primates, with emphasis on scaling and the evolution of the brain, in: *Approaches to Primate Paleobiology* (Contrib. Primatol., Vol. 5, F. Szalay, ed.), pp. 244–292, S. Karger, Basel.

Gurche, J. A. 1982. Early primate brain evolution, in: *Primate Brain Evolution: Methods and Concepts,* (E. Armstrong and D. Falk, eds.), pp. 227–246, Plenum Press, New York.

Hart, J. S. 1971. Rodents, in: *Comparative Physiology of Thermoregulation* (G. G. Whittow, ed.), p. 1–149, Academic Press, New York.

Hart, J. S., and Irving, L. 1959. The energetics of harbor seals in air and in water with special consideration of seasonal changes. *Can. J. Zool.* **37:**447–57.

Haug, H. 1972. Stereological methods in the analysis of neuronal parameters in the central nervous system. *J. Microsc.* **95:**165–180.

Heller, H. C. 1979. Hibernation: Neural aspects. *Ann. Rev. Physiol.* **41:**305–321.

Hemmer, H. 1971. Beitrag zur Erfassung der Progressiven Cephalisation bei Primaten, in: *Proceedings of the 3rd International Congress of Primatology,* Vol. 1 (H. Hemmer, H. Biegert, and W. Leutenegger, eds.), pp. 99–107, S. Karger, Basel.

Herreid, C. F., and Schmidt-Nielsen, K. 1966. Oxygen consumption, temperature and water loss in bats from different environments. *Am. J. Physiol.* **211:**1108–1112.

Herrick, C. J. 1926. *Brains of Rats and Men,* University of Chicago Press, Chicago.

Hildwein, G. 1972. Métabolisme énergétique de quelques mammifères et oiseaux de la forêt équatioriale. *Arch. Sci. Physiol.* **26:**379–385.

Hildwein, G., and Goffart, M. 1975. Standard metabolism and thermoregulation in a prosimian *Perodicticus potto. Comp. Biochem. Physiol.* **50A:**201–212.

Hofman, M. A. 1982. Encephalization in mammals in relation to the size of the cerebral cortex. *Brain Behav. Evol.* **20:**24–96.

Holloway, R. L., Jr. 1968. The evolution of the primate brain: Some aspects of quantitative relations. *Brain Res.* **7:**121–172.

Holloway, R. L. 1972. New Australopithecine endocast SK 1585 from Swartkrans, S. Africa. *Am. J. Phys. Anthropol.* **37:**173–186.

Holloway, R. L., Jr. 1975. *The Role of Human Social Behavior in the Evolution of the Brain* (43rd James Arthur Lecture), American Museum of Natural History, New York.

Holloway, R. L., Jr. 1979. Brain size, allometry and reorganization: Toward a synthesis, in: *Development and Evolution of Brain Size: Behavioral Implications* (M. E. Hahn, C. Jensen, and B. C. Dudek, eds., pp. 59–88, Academic Press, New York.

Holloway, R. L., Jr. 1981. Revisiting the South African Taung Australopithecine endocast: The position of the lunate sulcus as determined by the stereo polotting technique. *Am. J. Phys. Anthropol.* **56:**43–58.

Holloway, R. L., and Post, D. G. 1982. The relativity of relative brain measure and hominid mosaic evolution, in: *Primate Brain Evolution: Methods and Concepts* (E. Armstrong and D. Falk, eds.), pp. 57–76, Plenum Press, New York.

Hubel, D. H., and Wiesel, T. N. 1963. Shape and arrangement of columns in cat's striate cortex. *J. Physiol.* **165:**559–568.

Hudson, J. W., and Brower, J. E. 1974. Oxygen consumption: Vertebrates, in: *Biology Data Book III* (P. L. Altman and D. S. Dittimer, eds.), pp. 1613–1616, Federation of American Societies for Experimental Biology, Washington, D.C.

Huxley, J. S. 1932. *Problems of Relative Growth,* Methuen, London.

Irving, L., and Hart, J. S. 1957. The metabolism and insulation of seals as bare-skinned mammals in cold water. *Can. J. Zool.* **35:**497–511.

Irving, L., Scholander, P. F., and Grinnell, S. W. 1941. The respiration of the porpoise, *Tursiops truncates. J. Cell. Comp. Physiol.* **17:**145–168.

Jerison, H. J. 1955. Brain to body ratios and the evolution of intelligence. *Science* **121:**447–449.

Jerison, H. J. 1973. *Evolution of the Brain and Intelligence,* Academic Press, New York.

Jerison, H. J. 1979. The evolution of diversity in brain size, in: *Development and Evolution of Brain Size: Behavioral Implications* (M. E. Hahn, C. Jensen, and B. C. Dudek, eds.), pp. 30–57, Academic Press, New York.

Jerison, H. J. 1982. Allometry brain size, cortical surface and convolutedness, in: *Primate Brain Evolution: Methods and Concepts* (E. Armstrong and D. Falk, eds.), pp. 77–84, Plenum Press, New York.

Kaas, J. H. 1978. The organization of visual cortex in primates, in: *Sensory Systems of Primates* (C. R. Noback, ed.), pp. 151–179, Plenum Press, New York.

Kamau, J. M. Z., and Maloiy, G. M. D. 1981. The fasting metabolism of a small East African antelope, the dik-dik. *J. Physiol.* **319:**50–51p.

Karandeeva, O. G., Matisheua, S. K., and Shapunov, V. M. 1973. Features of external respiration in the Delphinidae, in: *Morphology and Ecology of Marine Mammals* (K. K. Chapskii and V. E. Sokolov, eds.), pp. 196–206, Wiley, New York.

Kay, R. F. 1975. The functional adaptations of primate molar teeth. *Am. J. Phys. Anthropol.* **43:**195–215.

Kayser, C., and Heusner, H. 1964. Etude comparative du métabolisme énergétique dans la série animale. *J. Physiol.* **56:**489–524.

Kety, S. S. 1957. The general metabolism of the brain *in vivo,* in: *Metabolism of the Nervous System* (D. Richter, ed.), pp. 221–237, Pergamon, New York.

Kleiber, M. 1961. *The Fire of Life: An Introduction to Animal Energetics,* Wiley, New York.

Kraus, C., and Pilleri, G. 1969. Quantitative Untersuchunaen uber die Grosshirnrinde der Cetaceen, in: *Investigations on Cetacea* (G. Pilleri, ed.), pp. 127–150, Waldau, Berne.

Kuschinsky, W., and Wahl, M. 1978. Local chemical and neurogenic regulation of cerebral vascular resistance. *Physiol. Rev.* **58:**656–689.

Lajtha, A. L., Maker, H. S., and Clarke, D. D. 1981. Metabolism and transport of carbohydrates and amino acids, in: *Basic Neurochemistry* (G. J. Siegel, R. W. Albers, B. W. Agranoff, and R. Katzman, eds., pp. 329–353, Little, Brown, Boston.

Lapicque, L. 1912. Le poids du cerveau et la grandeur du corps. *Biologica* **21:**257–265.

Leutenegger, W. 1982. Encephalization and obstetrics in primates with particular reference to human evolution, in: *Primate Brain Evolution: Methods and Concepts* (E. Armstrong and D. Falk, eds.), pp. 85–95, Plenum Press, New York.

Lund-Andersen, H. 1979. Transport of glucose from blood to brain. *Physiol. Rev.* **59:**305–352.

Mace, G. M., Harvey, P. H., and Clutton-Brock, T. H. 1981. Brain size and ecology in small animals. *J. Zool.* **193:**333–354.

McHenry, H. M. 1975. Fossils and the mosaic nature of human evolution. *Science* **190:**425–431.

MacMillen, R. E., and Nelson, J. E. 1969. Bioenergetics and body size in dasyurid marsupials. *Am. J. Physiol.* **217:**1246–1251.

McNab, B. K. 1969. The economics of temperature regulation in neotropical bats. *Comp. Biochem. Physiol.* **31:**227–268.

McNab, B. K. 1978. Energetics of arboreal folivores: Physiological problems and ecological consequences of feeding on an ubiquitous food supply, in: *The Ecology of Arboreal Folivores* (G. G. Montgomery, ed.), pp. 153–162, Smithsonian Institution, Washington, D.C.

MacPhail, E. 1982. *Brain and Intelligence in Vertebrates,* Oxford University Press, New York.

Mangold, R., Sokoloff, L., Conner, E., Kleinerman, J., Therman, P. G., and Kety, S. S. 1955. The effects of sleep and lack of sleep on the cerebral metabolism of normal young men. *J. Clin. Invest.* **34:**1092–1100.

Manouvrier, L. 1885. Sur l'interprétation de la quantité dans l'encéphale et dans le cerveau en particulier. *Bull. Soc. Anthropol. (Paris)* **3:**137–323.

Martin, R. D. 1981. Relative brain size and basal metabolic rate in terrestrial vertebrates. *Nature* **293:**57–60.

Martin, R. D. 1982. Allometric approaches to the evolution of the primate nervous system, in: *Primate Brain Evolution: Methods and Concepts* (E. Armstrong and D. Falk, eds.), pp. 39–56, Plenum Press, New York.

Milton, K., Casey, T. M., and Casey, K. K. 1979. The basal metabolism of mantled howler monkeys (*Alouatta palliata*). *J. Mammal.* **60:**373–376.

Mink, J. W., Blumenschine, R. J., and Adams, D. B. 1981. Ratio of central nervous system to body metabolism in vertebrates: Its constancy and functional basis. *Am. J. Physiol.* **241:**R203–R212.

Mountcastle, V. B. 1957. Modality and topographic properties of single neurons of cat's somatic sensory cortex. *J. Neurophysiol.* **20:**408–434.

Muller, E. 1975. Temperature regulation in the slow loris. *Naturwissenschaften* **62:**140–141.

Nakayama, T., Hori, T., Nagasaka, T., Tokura, H., and Tadaki, E. 1971. Thermal and metabolic responses in the Japanese monkey at temperatures of 5–38°C. *J. Appl. Physiol.* **31:**332–337.

Nelson, L. E., and Asling, C. W. 1962. Metabolic rate of tree-shrews (*Urogale everetti*). *Proc. Soc. Exp. Biol. Med.* **109:**602–604.

Nilsson, B., and Siesjo, B. K. 1976. A method for determining blood flow and oxygen consumption in the rat brain. *Acta Physiol. Scand.* **96:**72–82.

Ogren, M. P. 1982. The development of the primate pulvinar, in: *Primate Brain Evolution: Methods and Concepts* (E. Armstrong and D. Falk, eds.), pp. 113–129, Plenum Press, New York.

Palkovits, M., Magyar, P., and Szentagothai, J. 1971. Quantitative histological analysis of the cerebellar cortex in the cat. I. Number and arrangement in space of the Purkinje cells. *Brain Res.* **32:**1–13.

Passingham, R. E. 1973. Anatomical differences between the neocortex of man and other primates. *Brain Behav. Evol.* **7:**337–359.

Passingham, R. E. 1975. Changes in the size and organization of the brain in man and his ancestors. *Brain Behav. Evol.* **11:**73–90.

Pilleri, G. 1959. Beitrage zur vergleichenden Morphologie des Nagetiergehirns. *Acta Anat.* **39**(Suppl.)**:**1–124.

Pirlot, P., and Stephan, H. 1970. Encephalization in Chiroptera. *Can. J. Zool.* **48:**433–444.

Proppe, D. W., and Gale, C. C. 1970. Endocrine thermoregulatory responses to local hypothalamic warming in unanesthetized baboons. *Am. J. Physiol.* **219:**202–207.

Pubols, B. H., and Pubols, C. M. 1972. Neural organization of somatic sensory representation in the spider monkey. *Brain Behav. Evol.* **5:**342–366.

Radinsky, L. 1970. The fossil evidence of prosimian brain evolution, in: *The Primate Brain* (C. R. Noback and W. Montagna, eds.), pp. 209–224, Appleton-Century-Crofts, New York.

Radinsky, L. 1974. The fossil evidence of anthropoid brain evolution. *Am. J. Phys. Anthropol.* **41:**15–27.

Radinsky, L. 1975. Primate brain evolution. *Am. Sci.* **63:**656–663.

Radinsky, L. 1977. Early primate brains: Facts and fiction. *J. Hum. Evol.* **6:**79–86.

Radinsky, L. 1978. Evolution of brain size in carnivores and ungulates. *Am. Nat.* **112:**815–831.

Radinsky, L. 1979. *The Fossil Record of Primate Brain Evolution* (49th James Arthur Lecture on the Evolution of the Human Brain), American Museum of Natural History, New York.

Radinsky, L. 1981. Brain evolution in extinct South American ungulates. *Brain Behav. Evol.* **18:**169–187.

Radinsky, L. 1982. Some cautionary notes on making inferences about relative brain size, in: *Primate Brain Evolution: Methods and Concepts* (E. Armstrong and D. Falk, eds.), pp. 29–37, Plenum Press, New York.

Rakic, P., and Sidman, R. L. 1969. Telencephalic origin of pulvinar neurons in the fetal human brain. *Z. Anat. Entwicklungsgesch.* **129:**53–82.

Rensch, B. 1960. *Evolution above the Species Level,* Columbia University Press, New York.

Richman, D. P., Stewart, R. M., Hutchinson, J. W., and Caviness, V. S., Jr. 1975. Mechanical model of brain convolutional development. *Science* **189:**18–21.

Risberg, J., and Ingvar, D. H. 1973. Patterns of activation in the grey matter of the dominant hemisphere during memorization and reasoning. A study of regional cerebral blood flow changes during psychological testing. *Brain* **96:**737–756.

Rockell, A. J., Hiorns, R. W., and Powell, T. P. S. 1974. Numbers of neurons through full depth of neocortex. *A. Anat.* **118:**371.

Rockell, A. J., Hiorns, R. W., and Powell, T. P. S. 1980. The basic uniformity in structure of the neocortex. *Brain* **103:**221–244.

Sacher, G. A. 1970. Allometric and functional analysis of brain structure in insectivores and primates, in: *The Primate Brain* (C. R. Noback and W. Montagna, eds.), pp. 245–287, Appleton-Century-Crofts, New York.

Sacher, G. A. 1982. The role of brain maturation in the evolution of the primates, in: *Primate Brain Evolution: Methods and Concepts* (E. Armstrong and D. Falk, eds.), pp. 97–112, Plenum Press, New York.

Sarnat, H. B., and Netsky, M. G. 1981. *Evolution of the Nervous System.* Oxford University Press, New York.

Schleicher, A., Zilles, K. V., and Kretschman, H. Z. 1978. Automatische Registrierung und Auswertung eines Grauwertindex in histogischen Schnitten. *Anat. Anz.* **144:**413–415.

Schmidt, C. F., Ketz, S. S., and Pennes, H. H. 1945. The gaseous metabolism of the brain of the monkey. *Am. J. Physiol.* **143:**33–52.

Schmidt-Nielsen, K. 1975. Scaling in biology: The consequences of size. *J. Exp. Zool.* **194:**287–308.

Scholander, P. F., and Irving, L. 1941. Experimental investigations on the respiration and diving of the Florida manatee. *J. Cell. Comp. Physiol.* **17:**169–191.

Scholander, P. F., Hock, R., Walters, V., Johnson, F., and Irving, L. 1950. Heat regulation in some arctic and tropical mammals. *Biol. Bull.* **99:**237–258.

Shariff, G. A. 1953. Cell counts in the primate cerebral cortex. *J. Comp. Neurol.* **98:**381–400.

Shepherd, G. M. 1974. *The Synaptic Organization of the Brain,* Oxford University Press, New York.

Sholl, D. A. 1948. The quantitative investigation of the vertebrate brain and the applicability of allometric formulae to its study. *Proc. R. Soc. B* **135:**243–258.

Siesjo, B. K. 1978. *Brain Energy Metabolism,* Wiley, New York.

Slijper, E. J. 1962. *Whales,* Hutchinson, London.

Snell, O. 1892. Die Abhängigkeit des Hirngewichts von dem Körpergewicht und den geistigen Fahigkeiten. *Arch. Psychiatr.* **23:**436–446.

Sokoloff, L. 1973. The (^{14}C) deoxyglucose method: Four years later. *Acta Neurol. Scand. Suppl.* **72:**640–649.

Sokoloff, L. 1981. Circulation and energy metabolism of the brain, in: *Basic Neurochemistry* (G. T. Siegel, R. W. Albers, B. W. Agranoff, and R. Katzman, eds.), pp. 471–495, Little, Brown, Boston.

Sokoloff, L., Mangold, R., Wechsler, R. L., Kennedy, C., and Kety, S. S. 1955. The effect of mental arithmetic on cerebral circulation and metabolism. *J. Clin. Invest.* **34:**1101–1106

Solnitsky, O. 1945. Volumetric and reconstruction studies of the primate cerebellar nuclei. *Anat. Rec.* **91:**300.

Stahl, W. R. 1967. Scaling of respiratory variables in mammals. *J. Appl. Physiol.* **219:**1104–1107.

Stephan, H. 1956. Vergleichend-anatomische Untersuchungen an Insektivorengehirnen. *Morphol. Jahrb.* **97:**77–122.

Stephan, H. 1960. Methodische Studien über den quantitativen Vergleich architektonischer Struktureinheiten des Gehirns. *Z. Wiss. Zool.* **164:**143–172.

Stephan, H. 1969. Quantitative investigations on visual structures in primate brains, in: *Proceedings of the 2nd International Congress of Primatology,* Vol. 3 (H. O. Hofer, ed.), pp. 34–42, S. Karger, Basel.

Stephan, H. 1972. Evolution of primate brains: A comparative anatomical investigation, in: *The Functional and Evolutionary Biology of Primates* (R. Tuttle, ed.), pp. 155–174, Aldine, Chicago.

Stephan, H. 1975. Allocortex, in: *Handbuch der mikroskopischen Anatomie des Menschen: IV/9* (W. Bargmann, ed.), Springer, New York.

Stephan, H., and Andy, O. J. 1964. Quantitative comparisons of brain structures from insectivores to primates. *Am. Zool.* **4:**59–74.

Stephan, H., and Andy, O. J. 1970. The allocortex in primates, in: *The Primate Brain* (C. R. Noback and W. Montagna, eds.), pp. 109–135, Appleton-Century-Crofts, New York.

Stephan, H., and Andy, O. J. 1977. Quantitative comparison of the amygdala in insectivores and primates. *Acta Anat.* **98:**130–153.

Stephan, H., Bauchot, R., and Andy, O. J. 1970. Data on size of the brain and various brain parts in insectivores and primates, in: *The Primate Brain* (C. R. Noback and W. Montagna, eds.), pp. 289–297, Appleton-Century-Crofts, New York.

Stephan, H., Frahm, and Baron, G. 1981. New and revised data on volumes of brain structures in insectivores and primates. *Folia Primatol.* **35:**1–29.

Straile, W. E. 1969. Encapsulated nerve end-organs in the rabbit, mouse, sheep and man. *J. Comp. Neurol.* **136:**317–336.

Szarski, H. 1980. A functional and evolutionary interpretation of brain size in vertebrates, in: *Evolutionary Biology,* Vol. 12 (M. Hecht, W. Steere, and B. Wallace, eds.), pp. 149–174, Plenum Press, New York.

Szenthagothai, J. 1978. The neuron network of the cerebral cortex: a functional interpretation. The Ferrier Lecture, 1977. *Proc. Roy. Soc. Lond. B* **201:**219–248.

Tilney, F. 1928. *The Brain from Ape to Man,* Hoeber, New York.

Towe, A. L. 1973. Relative numbers of pyramidal tract neurons in mammals of different sizes. *Brain Behav. Evol.* **7:**1–17.

Tower, D. B. 1954. Structural and functional organization of the mammalian cerebral cortex. The correlation of neuron density with brain size. *J. Comp. Neurol.* **101:**14–53.

Tower, D. B., and Young, O. M. 1973. The activities of butyrylcholinesterase and carbonic anhydrase, the rate of anaerobic glycolysis, and the question of constant density of glial cells in cerebral cortices of mammalian species from mouse to whale. *J. Neurochem.* **20:**269–278.

Townsend, R. E., Prinz, P. N., and Obrest, W. D. 1973. Human cerebral blood flow during sleep and waking. *J. Appl. Physiol.* **35:**620–625.

Von Bonin, G. 1937. Brain-weight and body-weight of mammals. *J. Gen. Psychol.* **16:**379–389.

Von Lierse, W. 1963. Die Kapillarlichte im Wirbeltiergehirn. *Acta Anat.* **54:**1–31.

Von Röhrs, M. 1966. Vergleichende Untersuchungen zur Evolution der Gehirne von Edentaten. I. Hirngewicht-Körpergewicht. *Z. Zool. Syst. Evolutionsforsch.* **4:**196–207.

Walker, J. M., Glotzbach, S. F., Berger, R. J., and Heller, H. C. 1977. Sleep and hibernation in ground squirrels (*Citellus* spp.): Electrophysological observations. *Am. J. Physiol.* **233:**R213–21.

Weibel, E. R. 1979. *Stereological Methods,* Vol. 1, Academic Press, New York.

Weibel, E. R., Taylor, C. R., Gehr, P., Hoppeler, H., Mathiew, O., and Maloiy, G. M. O. 1981. Design of the mammalian respiratory system. *Respir. Physiol.* **44:**151–164.

Welker, W. I. 1973. Principles of organization of the ventrobasal complex in mammals. *Brain Behavior. Evol.* **7:**253–336.

Welker, W. I. 1976. Brain evolution in mammals: A review of concepts, problems and methods, in: *Evolution of Brain and Behavior in Vertebrates* (B. Masterton, M. E. Bitterman, C. B. G. Campbell, and N. Hotton, eds.), pp. 251–344, Lawrence Erlbaum, Hillsdale, New Jersey.

West, M. J. 1981. The constant number of granule cells per unit surface area of the fascia dentatae of three different species. *Soc. Neurosci.* **7:**465.

West, M. J., and Andersen, A. H. 1980. An allometric study of the area dentata in the rat and mouse. *Brain Res. Rev.* **2:**317–348.

Zilles, K., and Schleicher, A. 1980. Similarities and differences in the cortical areal patterns of *Galago demidovii* (E. Geoffroy, 1796), (Lorisidae, primates) and *Microcebus murinus* (E. Geoffroy, 1828), (Lemuridae, primates). *Folia Primatol.* **33:**161–171.

Zilles, K., Schleicher, A., and Kretschmann, H. J. 1978. A quantitative approach to cytoarchitectonics. I. The areal pattern of the cortex of *Tupaia belangeri. Anat. Embryol.* **153:**195–212.

Zilles, K., Stephan, H., and Schleicher, A. 1982. Quantitative cytoarchitectonics of the cerebral cortices of several prosimian species, in: *Primate Brain Evolution: Methods and Concepts,* (E. Armstrong and D. Falk, eds.), pp. 177–201, Plenum Press, New York.

Brain Size Allometry 8

Ontogeny and Phylogeny

R. D. MARTIN AND PAUL H. HARVEY

Introduction

Interspecific allometric analysis rests upon the assumption that individual characteristics typically change in a regular, predictable fashion with body size according to some recognizable *scaling principle,* unless special adaptation has led to fundamental reorganization identifiable as a *grade shift* (see Martin, 1980). The widely used empirical allometric formula relating an individual character dimension Y to some measure of body size X can be said to reflect the scaling principle through the value of the allometric exponent k and grade through the value of the allometric coefficient b:

$$Y = bX^k \tag{1}$$

or, in its (linear) logarithmic form

$$\log Y = k \log X + \log b \tag{2}$$

When best fit lines are determined for logarithmically transformed data sets relating to *interspecific* comparisons, it is common to find that a number of parallel lines are identifiable. The common slope of the lines k reflects the scaling principle, while the vertical separation of the lines reflects grade dif-

R. D. MARTIN • Department of Anthropology, University College, London WC1E 6BT, England. PAUL H. HARVEY • School of Biological Sciences, University of Sussex, Falmer, Brighton BN1 9QG, Sussex, England.

ferences between groups of species. This approach has been widely applied in the analysis of relative brain size in primates and a variety of other vertebrate groups (Bauchot and Stephan, 1966, 1969; Stephan, 1972; Jerison, 1973; Harvey *et al.*, 1980).

There is, however, a major problem involved in the empirical application of this model of interspecific allometric analysis to data on brain size in mammals and other vertebrates. Since Lapicque's (1907) pioneering study, it has been generally accepted that the value of the allometric exponent k determined from best fit lines can change quite markedly with the taxonomic level at which allometric analysis is conducted. Typically, interspecific exponent values are highest when high-level taxonomic categories are examined (e.g., class Mammalia; order Primates) and lowest for relatively small taxonomic units (e.g., family, subfamily, or genus). The limiting case is *intraspecific* scaling, involving adults of a given species, for which the lowest exponent values of all have been reported. Lapicque (1907) and Brody (1945) gave a typical exponent value of 0.25 for intraspecific scaling of brain weight in mammals, and several subsequent authors have accepted an intraspecific scaling exponent value of 0.2–0.4 (Count, 1947; von Bertalanffy and Pirozynski, 1952; Klemmt, 1960; Bauchot and Stephan, 1964; Gould, 1975; Lande, 1979; Shea, 1983). One possible interpretation of this information is that exponent values gradually increase from the intraspecific level through successively higher taxonomic groups until a given "true" value (Jerison, 1973) is attained that reflects some optimal scaling principle. Jerison, for instance, took the "true" value to be 0.67 and specifically linked this to a surface–volume relationship (see also Lande, 1979). It could, for example, be argued that both sensors and effectors relate more to surfaces of the body than to volumes and that brain size might therefore be expected to keep pace with body surface area (k = 0.67), rather than with body volume (k = 1.0). However, there is no longer any compelling reason to accept 0.67 as the "true" value for the exponent governing brain–body scaling in mammals. Four separate studies involving allometric analysis of brain–body scaling in larger samples of mammal species ($N > 240$) have shown that the exponent value is closer to 0.75 (Bauchot, 1978; Martin, 1981; Eisenberg, 1981; Hofman, 1982). This empirical finding, along with other evidence, led Martin (1981) to suggest that brain size is in some way linked to basal metabolic rate, which also scales to body size with an exponent value close to 0.75. The specific hypothesis suggested was that the metabolic turnover of a mammalian mother sets constraints (directly or indirectly) on development of the fetal brain, such that neonatal brain size broadly reflects maternal metabolic capacity. Since there is relatively little growth in brain size after a mammal infant has become independent of its mother (see below, p. 165), this relationship is maintained such that adult brain size still scales in the same way to adult (= maternal) body size.

It has been shown for a sample of 309 placental mammal species that the empirically determined brain size scaling exponent value (0.76) is significantly greater than Jerison's expected value of 0.67 (Martin, 1981). However, it can be argued that the exponent value determined may be biased by the particular collection of species examined (e.g., overrepresentation of a particular group of mammals). There were, indeed, in Martin's sample a disproportionate number of rodent species, which have relatively small brains overall and might perhaps have tended to increase the slope of the line. On the other hand, the largest mammals in the sample (baleen whales) also have relatively small brains and should have influenced the slope of the line in the opposite direction. In an attempt to exclude overrepresentation of particular mammal types and disproportionate influence exerted by extreme outliers as possible sources of bias, Martin (1983) took the averaged logarithmic values for brain and body weight for each of the 10 placental mammal orders represented in the original sample. The best fit line through the 10 ordinal average points gives an exponent value of 0.78 with 95% confidence limits of 0.72–0.84 and a very high correlation (r = 0.993), which rules out even more emphatically a value of 0.67 as being too low. But even this approach might be said to be biased at a lower level, since the ordinal logarithmic average values determined obviously depend upon the representation of species within each order. For this reason, a more detailed analysis is ideally required to determine whether an exponent value close to 0.75 (and significantly higher than 0.67) is indeed characteristic at the level of the class Mammalia.

Even if it can be established that an exponent value of approximately 0.75 applies to the viviparous mammals as an overall group, there remains the problem that exponent values appear to be lower when allometric analysis is restricted to groups at lower taxonomic levels. In attempting to tackle this problem, the first step is to define the pattern of change in exponent values with changing taxonomic level, in order to see whether any regularity can be detected. For instance, the analysis conducted by Bauchot (1978) shows that for 12 placental mammal orders (with sample sizes ranging from four to 147 species per order and a total sample of 653 species) the average brain size scaling exponent is 0.595 ± 0.155. Although the average exponent value is therefore well below 0.75, the 95% confidence limits are extremely wide (0.285–0.905) and do not exclude either surface scaling or the metabolic scaling hypothesis. (Inclusion of the marsupials as an order in the comparison makes very little difference to the result, with an overall average exponent value of 0.596 ± 0.148 for the 13 orders and 95% confidence limits of 0.300–0.892.) It is noteworthy that there is a strong influence exerted by the number of species in each order. For the six placental mammal orders with the largest number of species included (N = 34–147 per order), the average exponent value is 0.708 ± 0.090, whereas with the six placental orders with the smallest number of species represented (N = 4–25 per order), the average exponent value is 0.482 ± 0.119. It is hence clear that the actual pattern of change in exponent value with changing taxonomic level within each order of mammals requires detailed investigation. Somewhat surprisingly, there has been no comprehensive analysis to date of changes in exponent with taxonomic level within the mammals. Given previously available evidence, it may be that lowered exponents are only found with very small taxonomic groups (e.g., among species within genera from orders containing large numbers of species), or slopes might actually increase with taxonomic group size up to the level of the class. Alternatively, changes of scaling with taxonomic group size might vary between taxonomic groups so that equivalent analyses within different mammalian orders would reveal different patterns. Hypotheses attempting to explain either an overall mammalian exponent or changes of exponent with taxonomic level must accommodate the available data. One of the principal aims of this chapter is to describe the patterns revealed at different taxonomic levels by a much larger sample of interspecific mammalian data than has hitherto been used.

The limiting case for variation in brain scaling exponent values with taxonomic group size is obviously the individual species, and at this point *interspecific* allometric comparisons meet up with the phenomenon of *intraspecific* variability in brain–body size relationships. Since it is this variability that provides the starting point for the evolution of relative brain size, as has been explicitly recognized by Lande (1979) in his seminal paper on genetic aspects of brain–body size allometry, it is reasonable to expect that an explanation of changing exponent value with changing taxonomic rank should involve some understanding of intraspecific allometric relationships. However, although it is widely accepted that intraspecific brain scaling exponent values are low (0.2–0.4, as noted above), the evidence has not been effectively reviewed, particularly since there has been continuing confusion about different kinds of intraspecific brain size allometry. It is, above all, necessary to separate *ontogenetic* allometric relationships (which refer to brain–body size relations as the brain and body are still growing to reach their typical adult values) from *adult* intraspecific allometric relationships (which concern variation in the plateau values attained at the end of individual growth trajectories). Accordingly, the other principal aims of this chapter are to analyze brain–body size relationships at the *intraspecific* level and to integrate any conclusions with information on *interspecific* brain size allometry in an attempt to explain the significance of changes in exponent value with changing taxonomic level.

Methods

Interspecific Comparisons

Brain and body weights of 883 mammal species from 15 orders were extracted from the literature. Data sources and the taxonomic distribution of the sample are given in Table 1. The classification used follows Corbet and Hill (1980), which is adapted from Simpson (1945) to take into account more recent work. Tests for changes in slope with taxonomic level were performed within orders to minimize effects due to groups with different phylogenetic status having similar taxonomic status: for example, the criteria used to de-

Table 1. The Number of Species in Each Mammalian Order for Which Data on Body and Brain Weight Were Used[a]

Order	Number of species	Body weight, g	Brain weight, g	Reference[b]
Artiodactyla	72	85,220	196.0	3, 4, 10–13
Carnivora	168	8,920	50.2	2
Cetacea	11	3,506,048	2219.4	1, 3, 4
Chiroptera	180	29	0.7	6
Edentata	8	3,512	19.4	3, 4
Hyracoidea	2	1,620	15.3	3
Insectivora	49	127	1.5	5, 7
Lagomorpha	14	418	4.7	5
Marsupialia	24	1,028	7.9	1, 4, 9
Perissodactyla	9	232,117	378.4	3
Pholidota	1	3,498	11.0	4
Pinnipedia	9	248,699	558.9	1, 3
Primates	114	1,151	18.0	8
Proboscidea	2	3,269,017	4505.3	3
Rodentia	220	412	4.9	5

[a]Average body and brain weights given are calculated from family estimates to reduce effects resulting from unequal representation at lower taxonomic levels within the sample (see text).
[b]1, Sacher and Staffeldt (1974); 2, Gittleman (1983); 3, Crile and Quirling (1940); 4, von Bonin (1937); 5, Mace *et al.* (1981); 6, Stephan *et al.* (1981); 7, Bauchot and Stephan (1966); 8, Clutton-Brock and Harvey (1980); 9, Eisenberg and Wilson (1981); 10, Kruska (1973); 11, Oboussier and Schliemann (1966); 12, Ronnefeld (1970); 13, Tyska (1966).

note family status may differ *between* rodents and carnivores, while the criteria and thus the status *within* each of those orders is more nearly equivalent. Because of varying sample sizes and the need to allocate degrees of freedom as evenly as possible across each order, taxonomic levels of analysis varied according to the order being considered.

Hierarchical analyses of covariance using both regression (Snedecor and Cochran, 1967) and major axis (Sokal and Rohlf, 1969) models were performed on each order. Data were logarithmically transformed before analysis so that slopes are estimates, using formula (2), of the exponent in formula (1) above.

The lowest level of each analysis is among species within genera, and only genera containing more than two species provide additional degrees of freedom for analyses of covariance. However, even monospecific genera are used in higher level analyses. For example, if a family contains 15 species with seven in each of two genera and one in a third, then the two larger genera are used to calculate "common" (or average) major axis and regression generic slopes. But three data points are used for calculating the appropriate family slope—the averages of log-transformed brain and body weights for each of the two seven-species genera and the single-species point for the monospecific genus. Analyses were performed in this way at each increasing taxonomic level, so that each common slope was calculated using only data from the taxonomic level immediately below it (species for genus, genera for family, and so on).

The usual variance-ratio test (Snedecor and Cochran, 1967) was used to test for heterogeneity of regression estimates which together make the common slope at each taxonomic level. A maximum likelihood ratio with an associated χ^2 statistic (Harvey and Mace, 1982) was employed in an equivalent test for major axis slopes.

To produce a standardized analysis at the class level, we have compared orders within the class, using a single point for each mammalian order. Order points were calculated as the geometric mean of family points, which were themselves derived from the generic points described above. This procedure was designed to reduce bias resulting from unequal representation of the different subtaxa within each order.

Intraspecific Comparisons

Actual data for intraspecific brain–body size relationships were taken from the literature wherever possible and reanalyzed to allow for (1) clear separation of ontogenetic allometry from interadult allometry, and (2) determination of exponent values from three different best fit lines (regression; reduced major axis; major axis) because of continuing uncertainty about the most appropriate method (Harvey and Mace, 1982).

In several cases, however, the raw data were not published. Some authors provided only average brain weight values for different size-classes, while others only provided the results of their analyses (often based exclusively on regressions, which may not be appropriate). Such information has been included, despite its limitations, where no more comprehensive material was available to establish particular points, and this is specifically indicated in the text where applicable. Certain relevant points are also drawn from other analyses already conducted on ontogenetic aspects of brain allometry in primates and other mammals (Martin, 1983).

Unless otherwise mentioned in the text, best fit lines for reanalysis of intraspecific data used to infer scaling exponents are major axes.

Results

Interspecific Comparisons

The interorder comparison across the whole class gives a major axis exponent of 0.72 with 95% confidence limits between 0.68 and 0.77 (see Fig. 1; values of brain and body weight for the different orders are given in Table 1). This analysis therefore confirms the view that for viviparous mammals as a

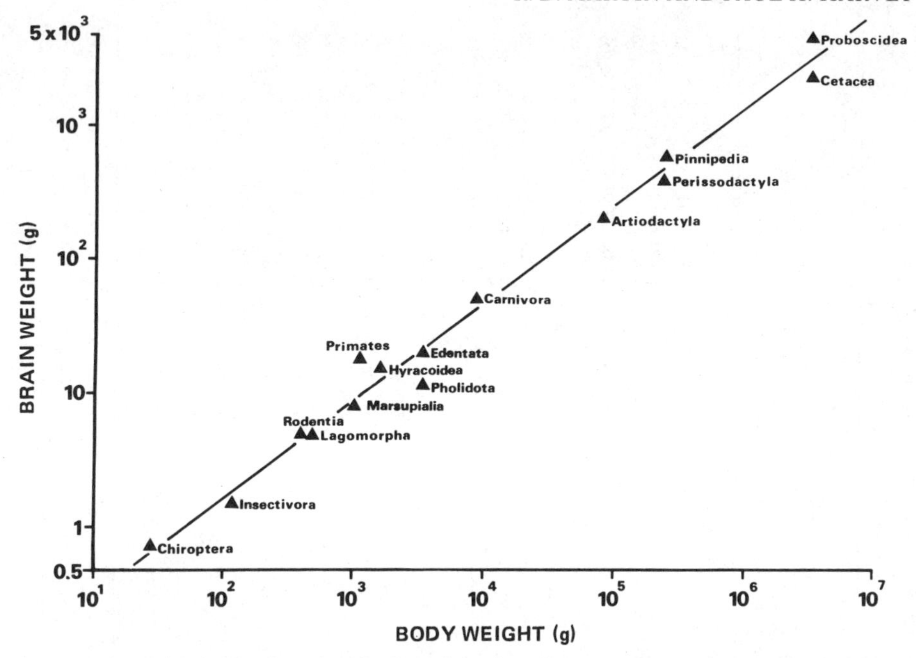

Fig. 1. Brain weight plotted against body weight for 15 mammalian orders. Body and brain weight estimates for each order are given in Table 1, and are derived from family averages (see text). The best-fit major axis line has a slope of 0.72 (95% confidence limits 0.68–0.77).

whole, brain size scales to body size with an exponent value close to 0.75, and the 95% confidence limits still exclude the value of 0.67 required by the surface scaling hypothesis of brain allometry in mammals.

Sample sizes were large enough to test for changes of slope with taxonomic level in six orders: Artiodactyla, Carnivora, Chiroptera, Insectivora, Primates, and Rodentia. Results of the analyses are presented in Table 2, with major axis slope and 95% confidence limits plotted in Fig. 2. Reduced major axis slopes, defined as the ratio of the standard deviations of brain size to that of body size, may be estimated by dividing the regression slopes by their appropriate correlation coefficients. Although regression analyses invariably give the shallowest slopes, patterns revealed by the data are relatively robust and apply whether regression, reduced major axis, or major axis models are used to estimate slopes. We accordingly restrict our discussion to the major axis analyses.

The most striking finding is that none of the 21 major axis slopes calculated at the order level and below is significantly steeper than 0.75, whereas in two cases the 95% confidence limits clearly exclude a value of 0.67 as being too low. In the primates, bats (Chiroptera), and artiodactyls, slopes increase progressively from the generic to the infraorder, suborder, and order level, respectively. A less consistent pattern of increase is seen for the rodents and the carnivores. No consistent pattern is evident among the insectivores, al-

Table 2. Results of Statistical Analyses Used to Test for Changes with Taxonomic Level of the Exponent Relating Brain to Body Size[a]

Order	Taxonomic level	Number of taxa		Number of subtaxa	Common correlation	Common regression	F ratio (d.f.)		Common major axis	95% confidence limits	χ^2 (d.f.)	
		Total	$N > 2$									
Artiodactyla	Genus	16	7	52	0.96	0.52	0.16	(6,20)	0.53	0.48–0.59	1.0	(6)
	Family	3	2	30	0.98	0.56	1.70	(1,24)	0.57	0.53–0.61	1.8	(1)
	Order	1	1	9	0.97	0.57	—	—	0.58	0.49–0.69	—	—
Carnivora	Genus	25	12	95	0.88	0.52	0.52	(11,45)	0.56	0.49–0.63	10.3	(11)
	*Family	8	7	98	0.93	0.61	6.38	(6,28)	0.64	0.59–0.69	27.7	(6)
	Superfamily	2	2	8	0.99	0.64	—	—	0.64	0.60–0.69	0.04	(1)
	Order	1	0	2	—	—	—	—	—	—	—	—
Chiroptera	Genus	32	16	124	0.90	0.54	10.70	(15,60)	0.57	0.52–0.63	22.2	(15)
	Family	11	10	78	0.97	0.69	1.48	(9,56)	0.71	0.67–0.75	12.8	(9)
	Suborder	1	1	18	0.90	0.76	—	—	0.82	0.65–1.02	—	—
	Order	1	0	2	—	—	—	—	—	—	—	—
Insectivora	Genus	6	2	25	0.85	0.57	0.10	(1,13)	0.60	0.43–0.80	0.9	(1)
	*Family	5	3	27	0.90	0.49	6.40	(2,17)	0.51	0.42–0.62	12.2	(2)
	Superfamily	2	1	5	1.00	0.75	—	—	0.75	0.70–0.79	—	—
	Order	1	1	5	0.94	0.58	—	—	0.59	0.42–0.80	—	—
Primates	Genus	19	13	86	0.77	0.40	1.63	(12,49)	0.45	0.36–0.53	15.9	(12)
	Subfamily	11	9	40	0.94	0.57	1.42	(8,18)	0.59	0.52–0.66	6.7	(8)
	Infraorder	4	3	17	0.94	0.73	0.06	(2,9)	0.76	0.63–0.91	0.04	(2)
	Order	1	1	5	0.99	0.82	—	—	0.82	0.73–0.93	—	—
Rodentia	*Genus	39	19	159	0.87	0.35	1.85	(18,81)	0.37	0.33–0.40	31.4	(18)
	Subfamily	14	7	84	0.94	0.40	0.62	(6,56)	0.40	0.37–0.44	4.3	(6)
	Family	3	2	14	0.89	0.31	0.04	(1,8)	0.32	0.23–0.41	0.03	(1)
	Order	1	1	17	0.98	0.57	—	—	0.58	0.52–0.63	—	—

[a]Sufficient data were available for analyses of six mammalian orders. When, for example, the family is the level for analysis then generic points were used to calculate best fit lines using logarithmically transformed data (see text). Correlation coefficients are high, so that regression and major axis models give similar results. Asterisked analyses indicate that the common regression and major axis slopes were estimated from statistically heterogeneous component slopes (revealed by significance, $p < 0.05$, F-ratio, and χ^2 texts). d.f., Degrees of freedom.

though the data would not reject a hypothesis of increased slope with taxonomic level.

Generic common slopes differ among the orders, ranging from 0.37 in the rodents to 0.60 in the insectivores. Similar variation across the orders is apparent in analyses performed at higher taxonomic levels. Fourteen of the common slopes are derived from more than two estimates, and in three of those cases the constitutent slopes are significantly heterogeneous (see Table 2); in those cases, at least, the confidence limits should be treated only as approximate guides.

Intraspecific Comparisons

A search of the literature followed by preliminary analysis of data revealed that there are numerous special problems involved in the determination of intraspecific brain–body size scaling exponents:

1. Sample sizes were often extremely small (less than a dozen specimens in many cases). Although this problem is also encountered in interspecific analyses, the difficulties are exacerbated at the intraspecific level because correlation coefficients tend to be markedly lower and confidence limits on exponent values are consequently much wider.
2. It is ideally necessary to distinguish wild-caught from captive individuals, since conditions in captivity are likely to modify body weights in both directions (debilitation, obesity) and since it is well established that in captive mammals, especially following extensive breeding in captivity, brain size can be reduced by a factor of as much as 25% [Lapicque (1907); see Herre, (1959) for a review]. Again, the distinction between captive and wild-caught specimens may influence interspecific comparisons as well, but because the range of body sizes involved is typically far greater, the overall scaling relationship is less drastically affected. Further, it is obviously difficult to obtain data on wild-caught specimens for large samples of individual species, and most of the available data relate to animals kept in captivity. Finally, because artificial selection in captivity has often produced an unusually large range of body sizes (e.g., for dogs), intraspecific data obtained in such cases may introduce an entirely new element into the coadaptation of brain size and body size (see p. 167).
3. It is known (e.g., for humans) that brain size may actually decline with senescence and obviously senescent animals should therefore be excluded from interadult comparisons.
4. There may be an effect of sex on intraspecific allometry (Lapicque, 1907; Röhrs, 1959). In cases where sexual dimorphism in body size is present, males and females may have different typical adult values for brain size as well. If, for example, males typically have larger brains and larger bodies than females as well as having a larger allometric coefficient, then fitting a line to

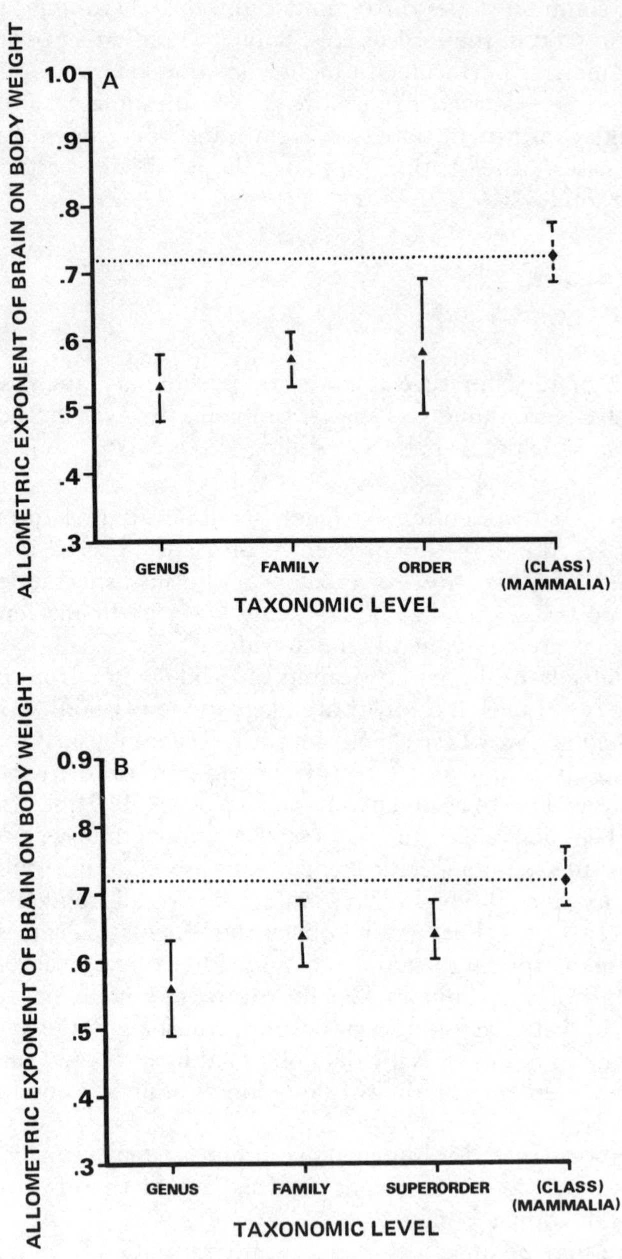

Fig. 2. Common major axis slopes and 95% confidence limits at different taxonomic levels for six mammalian orders. The dotted line and class estimate shown are from the interorder comparison

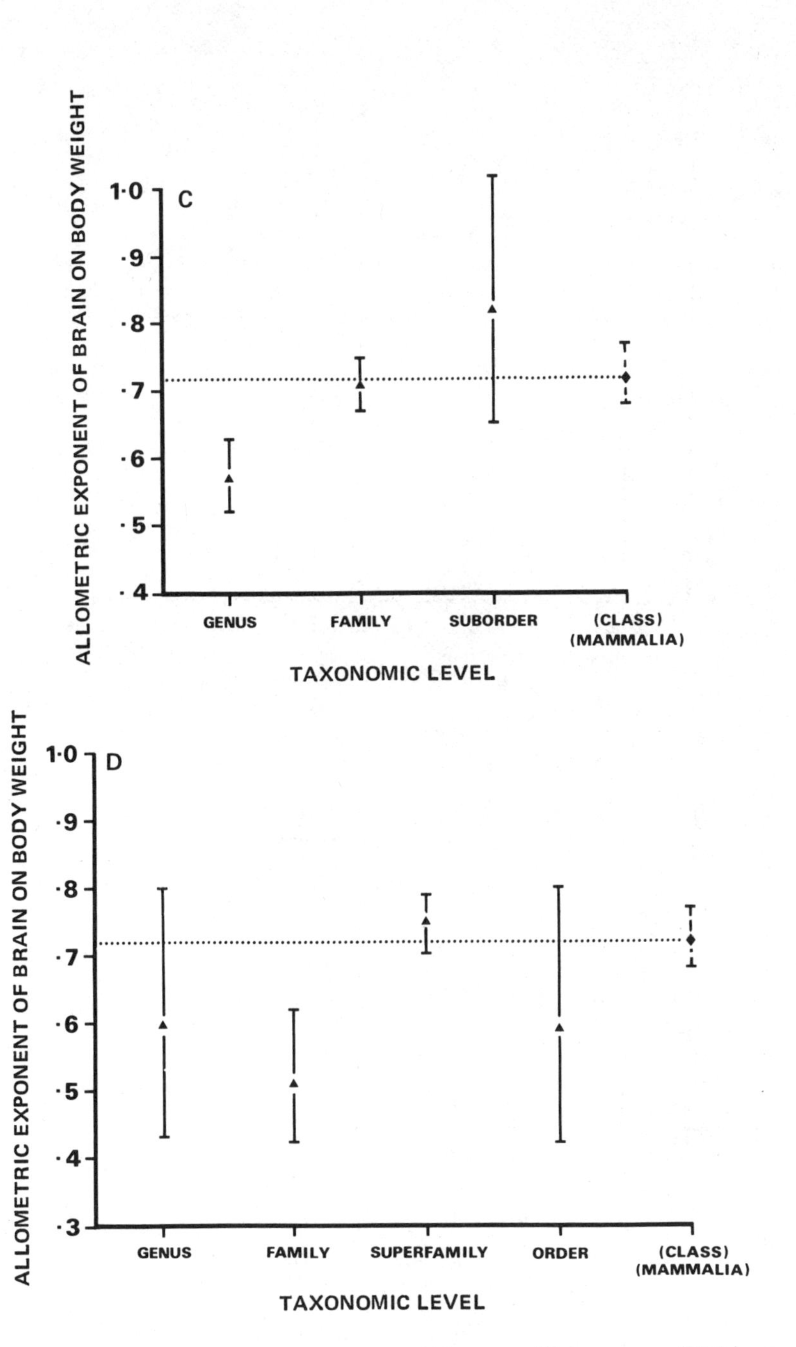

plotted in Fig. 1. (A) Artiodactyla, (B) Carnivora, (C) Chiroptera, (D) Insectivora, (E) Primates, (F) Rodentia.

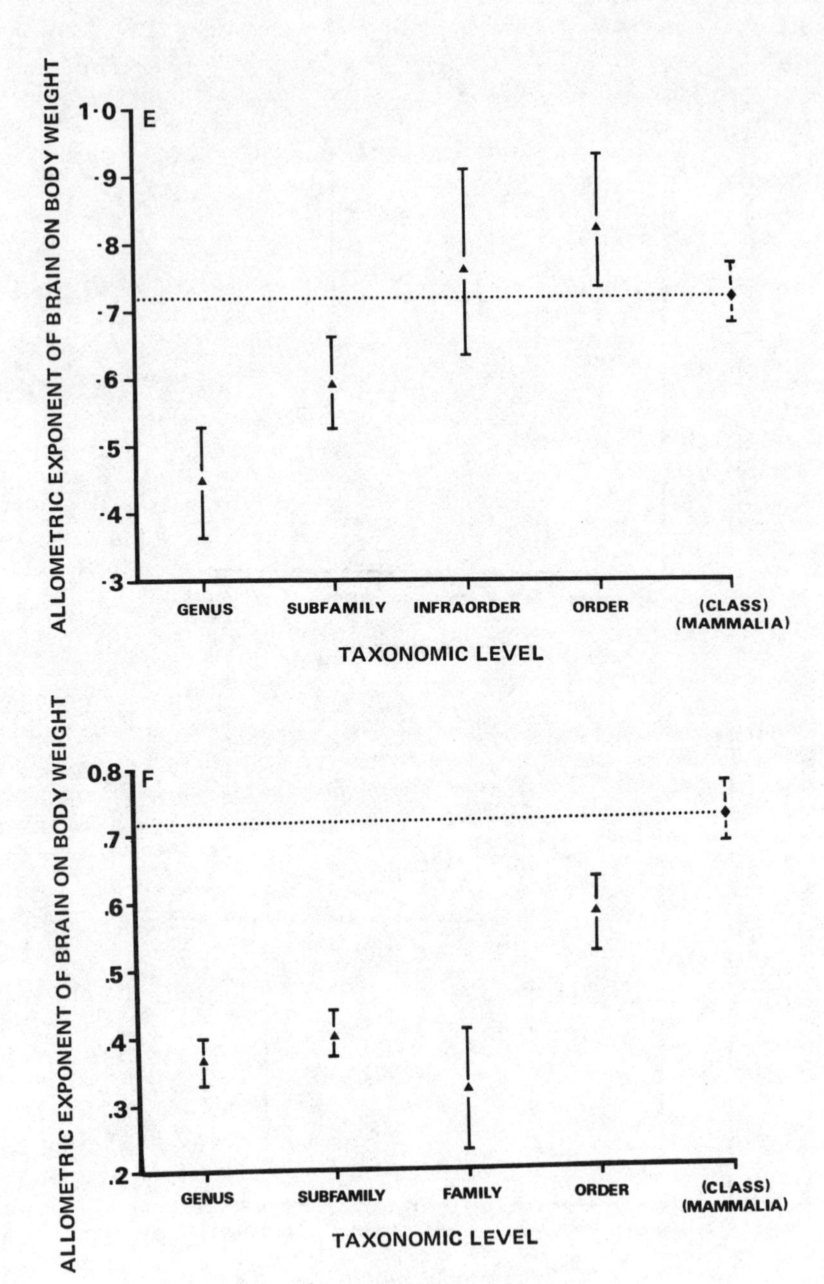

Fig. 2. (*continued*)

combined male and female data will produce an allometric exponent value higher than for males and females alone [see Holloway (1980) for *Homo sapiens* and a number of other primate species]. Indeed, such an effect may be found in the absence of overall body size sexual dimorphism if the sexes differ in brain size and in the *distribution* of body weights within each sex.

5. If earlier ontogenetic stages and sexually mature adults differ in scaling exponents, combination of the two in analyses will confuse ontogenetic scaling with interadult scaling. For example, Bauchot and Stephan (1964) fitted lines to data on all postnatal stages together in determining intraspecific brain scaling exponents for Madagascar insectivores. Their average "intraspecific" exponent value of 0.24 (range: 0.18–0.34) for five species may therefore be misleading if it is interpreted as an *interadult* scaling exponent. In fact, it is true of most published studies that "intraspecific scaling" actually refers to scaling through all postnatal stages, rather than to either preadult or interadult scaling alone. Accordingly, the widely held view that values of 0.2–0.4 are generally applicable to intraspecific brain–body size scaling *among adults* is questionable. It is also worth noting that Sholl's study (1948) of scaling exponents, which is widely quoted as throwing doubt on the regularity of brain–body size scaling at the intraspecific level, indiscriminately considered young and adult animals together as well as combining males and females. Hardly any of the much-cited macaques were, in fact, adults.

6. Problems involved in the choice of an appropriate statistical technique for fitting of lines to logarithmically transformed data are considerably magnified by a combination of small sample size with low scaling exponent values. With the relatively high exponent values typically found in interspecific comparisons (e.g., 0.4–0.75; see Table 2), the major axis and reduced major axis yield closely comparable results, while the regression systematically yields a lower exponent value (though the discrepancy is important only when the correlation coefficient is relatively weak). With the low exponent value typical of intraspecific comparisons, however, it emerges that the major axis and regression yield similar results, while the reduced major axis yields a higher exponent value. Since, with exponent values, correlation is typically relatively low, and since the reduced major axis exponent is equivalent to the regression exponent divided by the correlation coefficient, considerable discrepancies are found between values determined from the major axis or regression on the one hand and the value determined from the reduced major axis on the other. This being the case, if there were in fact good reasons for using the reduced major axis instead of the regression or major axis (e.g., Kermack and Haldane, 1950), the very low exponent values reported for certain intraspecific comparisons might be to some extent an artifact. Alternatively, since there is fairly good agreement between the major axis and the regression with low exponent values, this observation might be cited as evidence for the unsuitability of the reduced major axis for allometric analysis. In any event, it is important to note that the reliability of allometric analyses (and resulting discussion of the significance of differences in exponent values) depends on

the suitability of available statistical techniques for determining best fit lines, and that low scaling exponents pose particular problems in this context.

The best available data set for the analysis of intraspecific brain–body size scaling proved to be that provided by Wingert (1969) for the albino mouse (*Mus musculus*), which covers the entire period from conception to adulthood and provides brain and body weights for animals of known ages. Wingert showed that postnatal brain development in the albino mouse can be clearly divided into two separate phases with respect to brain–body size scaling exponents. There is an initial phase, terminating at about 12 days after birth, in which the exponent value is relatively high, followed by a second phase with a much lower exponent value. The point of inflection in the brain–body size relationship at a postnatal age of about 12 days is very obvious. But laboratory mice do not become sexually mature until at least 30 days of age (Asdell, 1964), so the line fitted by Wingert to all postnatal stages from 12 days of age onward combines data from the preadult growth stage with data from sexually mature adults of different body weights. When Wingert's data are reanalyzed, it emerges that *four* different phases can be recognized, with different scaling exponents, calculated from the double logarithmic graph of brain size against body size from conception through to the largest bodied, sexually mature adults:

1. Fetal stage: scaling exponent 0.57 (95% confidence limits 0.54–0.60; $N = 57$; $r = 0.98$).
2. Early postnatal stage (0–12 days): scaling exponent 0.78 (95% confidence limits 0.74–0.83; $N = 57$; $r = 0.98$).
3. Intermediate postnatal stage (13–31 days): scaling exponent 0.13 (95% confidence limits 0.08–0.18; $N = 41$; $r = 0.62$).
4. Mature adult stage: scaling exponent −0.02 (95% confidence limits −0.10–0.07; $N = 42$; $r = -0.07$).

The three postnatal stages identified are illustrated in Fig. 3, showing that the early postnatal stage is characterized by relatively rapid expansion of brain size (relative to body size), while brain size scales with significantly lower exponent values in the second and third stages. Indeed, for the mature adult stage of the albino mouse there is virtually no correlation between brain size and body size, so "intraspecific" brain scaling in this case is entirely confined to ontogenetic stages. This raises the possibility that the accepted range of 0.2–0.4 for exponent values of "intraspecific" brain scaling in mammals generally may apply largely or exclusively to ontogenetic processes rather than to variation among mature adults. This certainly seems to be true of a number of primate species for which adequate data are available for analysis of mature adults alone [Table 3; see also Holloway (1980)]. For the five primate species studied by Holloway, regression analysis alone was used and no confidence limits on the expected values determined were provided, but great care was taken to exclude immature individuals from the samples used. It is clear for all 11 species in Table 3 that, if males and females are analyzed separately,

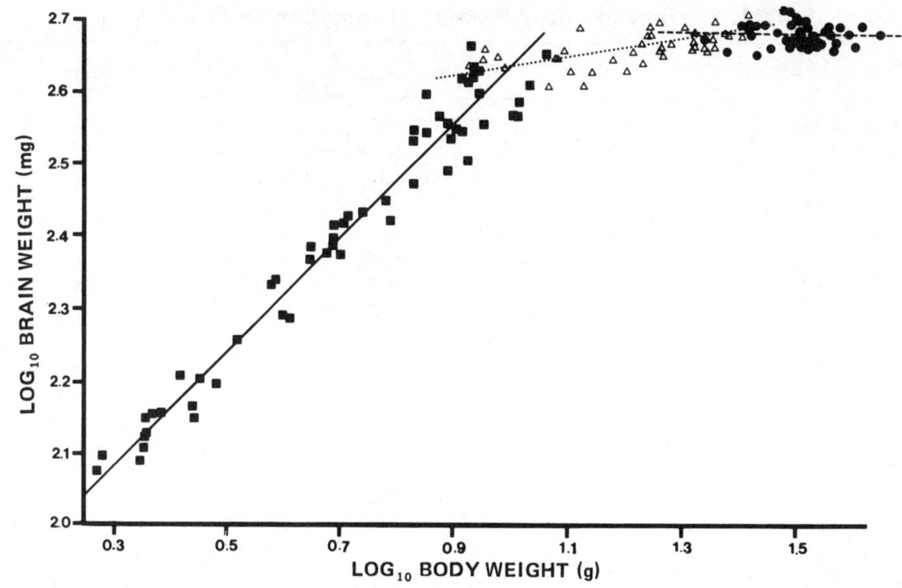

Fig. 3. Double logarithmic plot of brain weight (mg) against body weight (g) for postnatal stages of the albino mouse, based on data from Wingert (1969). Best fit lines (major axes) have been fitted to the following successive stages of postnatal development: (1) early postnatal stage (0–12 days): black squares; (2) intermediate postnatal stage (13–31 days): open triangles; (3) mature adult stage (32–170 days): black circles.

interadult scaling exponent values determined from regressions or major axes are low and in some cases even negative (average 0.05 ± 0.07; range −0.17 to 0.19) and that there is no obvious difference in this respect between wild-caught and captive samples. However, spuriously high values for the exponent are commonly found where males and females are combined in the analysis for the nine sexually dimorphic species, excluding *Callithrix jacchus* and *Ateles paniscus* (average 0.18 ± 0.09; range 0.10–0.35). As expected (see above), the highest exponent values are found with three species showing extreme sexual dimorphism (*Macaca mulatta; Gorilla gorilla; Pongo pygmaeus*). Unfortunately, insufficient data are available for an analysis of ontogenetic allometry of brain size in individual primate species, but Snow and Vice (1965) indicate for a sample of wild-caught immature plains baboons (*Papio anubis*) that there is a positive scaling exponent. They divided their animals into size classes and fitted regressions through the means of the size classes to obtain exponent values of 0.12 for males (N = 46) and 0.14 for females (N = 36), with no clear distinction between the sexes (overall exponent value 0.13). This result is compatible with the expectation that postnatal ontogenetic allometry in brain size exhibits a higher exponent value than interadult allometry, where the value is not significantly different from zero in primates for single-sex samples; but more ontogenetic data from primates are required to resolve this point clearly.

Low exponent values for brain–body size scaling in fully adult animals

Table 3. Intraspecific Allometry of Brain Size in Primates[a]

Species	N	r	k_{reg}	k_{RMA}	k_{MA}	Source
Homo sapiens ♂ + ♀	667	0.31	0.17	(0.55)	—	Holloway (1980)[b]
♂	502	0.20	0.10	(0.50)	—	
♀	165	0.05	0.03	(0.60)	—	
Pan troglodytes ♂ + ♀	30	0.60	0.14	(0.23)	—	Holloway (1980)
♂	14	0.68	0.16	(0.24)	—	
♀	16	0.34	0.08	(0.24)	—	
Gorilla gorilla ♂ + ♀	20	0.87	0.29	(0.33)	—	Holloway (1980)
♂	10	0.61	0.19	(0.31)	—	
♀	10	−0.42	−0.10	(−0.24)	—	
Pongo pygmaeus ♂ + ♀	40	0.70	0.22	(0.31)	—	Holloway (1980)
♂	16	0.03	0.05	(1.67)	—	
♀	24	0.02	0.04	(2.00)	—	
Macaca mulatta ♂ + ♀	29	0.38	0.35	(0.92)	—	Holloway (1980)
♂	37	0.40	0.10	(0.25)	—	
♀	24	0.09	0.05	(0.56)	—	
Papio papio ♂ + ♀	30	0.67	0.15	0.22	0.15	Riese and Riese (1952)[c]
♂	16	0.10	0.02	0.14	0.02	
♀	14	0.75	0.15	0.20	0.15	
Saimiri sciureus ♂ + ♀	38	0.40	0.12	(0.30)	—	Holloway (1980)
♂	16	0.25	0.09	(0.36)	—	
♀	22	−0.10	−0.03	(−0.30)	—	
Callithrix jacchus ♂ + ♀	23	0.04	0.02	0.53	0.03	Bauchot and Stephan (1969)
♂	—	—	—	—	—	
♀	—	—	—	—	—	
Ateles paniscus ♂ + ♀	30	−0.06	−0.03	(−0.50)	—	J. M. Ayres (personal communication)[d]
♂	9	0.40	0.18	(0.45)	—	
♀	21	−0.30	−0.17	(−0.57)	—	
Cebus apella ♂ + ♀	29	0.24	0.11	(0.46)	—	Ayres (personal communication)[d]
♂	13	−0.003	−0.001	(−0.33)	—	
♀	16	0.19	0.10	(0.53)	—	
Alouatta seniculus ♂ + ♀	33	0.26	0.10	(0.38)	—	Ayres (personal communication)[d]
♂	13	0.11	0.08	(0.73)	—	
♀	20	−0.05	−0.03	(−0.60)	—	

[a]Analyses based on brain weight for most species, but conducted on cranial capacities for *Ateles*, *Cebus*, and *Alouatta*. Here k_{reg}, k_{RMA}, and k_{MA} are determined from regression, reduced major axis, and major axis, respectively. Values of k_{RMA} in parentheses have been inferred from k_{reg} and r by using the formula $k_{RMA} = k_{reg}/r$, with the sign of the slope determined by the sign of r.
[b]Data derived from Pakkenberg and Voigt (1964).
[c]Males older than 4 years and females older than 3 years taken as adult. (N.B.: These captive specimens varied greatly in physical condition.)
[d]Body weight from wild-caught animals; cranial capacities measured on prepared skulls.

are also found with a number of other mammal species, though sample sizes are usually small for the studies that have been conducted. For example, in the shrew *Crocidura orientalis*, data collected for 23 adults by Stephan (1958) yield an overall scaling exponent of 0.09 (95% confidence limits 0.01–0.18; $r = 0.42$) and this result is not greatly affected if the sexes are analyzed separately (exponent for 18 males, 0.13: exponent for five females, −0.02). Similarly, data for 13 adult specimens of the sexually monomorphic *Lama guanocoë* collected by Herre and Thiede (1965) yield an overall scaling exponent of −0.01 (95% confidence limits −0.25 to 0.23; r = −0.03). Finally, it is clear from the graph of brain weight against body weight for three species of *Peromyscus* provided by King (1965) that in this genus there is little change in brain size after adulthood has been reached. Thus, it can be concluded that there are several mammal species, at least, that do not exhibit a significant increase in brain size with increasing body size among sexually mature adults. Nevertheless, there are also a number of mammal species that *do* show some increase in brain size with increasing body size among adults.

A large body of data on postnatal brain and body weights was collected by Klemmt (1960) for wild-caught field mice (*Apodemus sylvaticus*). Klemmt's regression analysis of the entire sample showed the following relationships between brain weight and body weight (net of contents of gastrointestinal tract and bladder):

Males (N = 143): scaling exponent 0.16 (r = 0.63)
Females (N = 98): scaling exponent 0.18 (r = 0.77)

As expected (see above), in Klemmt's analysis as reported reduced major axes yielded rather higher values for these exponents (males 0.26; females 0.24). Klemmt then cites this result as confirming the widespread view that "values for the intraspecific brain exponents thus lie between 0.15 and 0.33 for the species investigated" (our translation). However, it is clear that the data set includes animals of different sexual status. Klemmt's data on testis size show that males have relatively small testes (0.03–0.37 g) at body weights up to 19 g and that there is a sharp increase in testis weight (to 0.80 g) at body weights of between 19 and 20 g. Thereafter, testis weight increases only slowly, to reach 1.3 g at the maximum male body weight of just over 30 g. A body weight of 20 g can therefore be taken as the lower limit for fully adult male field mice. No comparable criterion is available for attainment of adulthood in female field mice from Klemmt's data, though it is known that sexual maturity is reached at an age of 80–90 days (Asdell, 1964). Since female field mice are generally smaller than males, a weight of 16 g can be taken as the provisional lower limit for full adulthood in females. Reanalysis of Klemmt's data with the samples divided according to these cutoff points reveals the following scaling relationships:

1. Preadult field mice:
   Males (10.7–19.9 g; N = 71): scaling exponent 0.24
   (95% confidence limits 0.17–0.31; r = 0.62)
   Females (7.5–15.6 g; N = 44): scaling exponent 0.19
   (95% confidence limits 0.11–0.28; r = 0.55)
2. Adult field mice:
   Males (20.0–32.4 g; N = 72): scaling exponent 0.26
   (95% confidence limits 0.13–0.39; r = 0.42)
   Females (16.0–29.9 g; N = 54): scaling exponent 0.18

(95% confidence limits 0.10–0.25; r = 0.54)

In this case it emerges that there is no difference in scaling exponent values between preadult and adult stages, though there may be a difference between the sexes in exponent values. Instead, brain size seems to increase steadily with body size over the entire range of body weights represented, with an average exponent value of 0.22 when the sexes are combined.

A similar result is obtained when Kretschmann's data (1966, 1968) on the spiny mice *Acomys cahirinus dimidiatus* and *A. c. minous* are reanalyzed with the samples divided into "young" and "adult" stages (assuming that sexual maturity is reached at about 45 days of age):

1. Preadult *Acomys c. dimidiatus* (Kretschmann, 1966):
   Males (N = 10): scaling exponent 0.18
   (95% confidence limits 0.10–0.26; r = 0.81)
   Females (N = 5): scaling exponent 0.17
   (95% confidence limits −0.15 to 0.52; r = 0.42)
   Adult *Acomys c. dimidiatus*
   Males (N = 16): scaling exponent 0.19
   (95% confidence limits 0.09–0.29; r = 0.66)
   Females (N = 12): scaling exponent 0.13
   (95% confidence limits 0.001–0.26; r = 0.49)
   Overall average scaling for adults, 0.16
2. Preadult *Acomys c. minous* (Kretschmann, 1968):
   Males (N = 19): scaling exponent 0.32
   (95% confidence limits 0.29–0.36; r = 0.97)
   Females (N = 14): scaling exponent 0.29
   (95% confidence limits 0.26–0.32; r = 0.98)
   Adult *Acomys c. minous*
   Males (N = 30): scaling exponent 0.18
   (95% confidence limits 0.14–0.22; r = 0.86)
   Females (N = 6): scaling exponent 0.12
   (95% confidence limits 0.07–0.17; r = 0.88)
   Overall average scaling for adults, 0.15

The results for both subspecies agree in indicating that there is a definite increase in brain size with increasing body size among adults, even when the sexes are analyzed separately, with a scaling exponent of about 0.15. The situation for the preadult stage is less clear, since at first sight there seems to be a difference in exponent values between the two subspecies. However, this can be explained by the fact that Kretschmann's sample (1966) for *Acomys c. dimidiatus* does not contain individuals younger than 20 days old, whereas most of the preadult animals in his sample (1968) for *A. c. minous* are less than 20 days old. If the preadult individuals from the latter sample are divided into those of ages between 0 and 20 days old and those older than 20 days [which was, in fact, the criterion used by Kretschmann (1968) for separating "young" from "adults"], it emerges that the relatively high exponent values for preadult stages is entirely due to relative increase in brain size up to 20 days of age. From that age onward, brain size increases with the same exponent value as for individuals older than 45 days (the age of sexual maturity assumed above). Thus, it would seem that in both *Acomys* species there is an initial stage from 0 to 20 days of age in which brain size increases moderately rapidly with body size (exponent value approximately 0.30), followed by a stage that continues through adulthood with brain size increasing less rapidly with body size (exponent value approximately 0.15). Nevertheless, for both *Apodemus* and *Acomys* there is a definite increase in brain size with increasing body size among sexually mature adults, with males and females analyzed separately, in contrast to the situation with several primate species (including man) and a number of other mammals. This distinction in interadult relationships between brain size and body size requires some explanation if an overall interpretation of interspecific brain scaling is to be attained.

Discussion

Contrary to the impression given in the literature, intraspecific scaling of brain size to body size does not exhibit a typical exponent value of 0.25 when interadult allometry is clearly separated from ontogenetic allometry. In some mammal species, adults of a single sex do exhibit scaling exponents approaching this level (0.15 in *Acomys;* 0.22 in *Apodemus*), but in others the scaling exponent approaches zero, and may even become negative (e.g., with the primate species listed in Table 3), if the effects of sexual dimorphism are excluded. Explanation of this variation in interadult brain scaling requires consideration of the entire ontogenetic sequence through which adult brain–body size relationships are attained. It is known (Martin, 1983; Sacher, 1982) that the initial, fetal stage of brain development is characterized by relatively high exponent values (close to unity in primates and approximately 0.80 for most other mammals). The subsequent pattern of change in brain size with increasing body size depends upon the distinction between altricial and precocial mammals (Portmann, 1962). In altricial mammals, a high exponent value for brain–body scaling persists for a short period after birth, as is the case with laboratory mice [Wingert (1969); see Fig. 3]. Thereafter, there is a point of inflection in the logarithmic plot of brain size against body size, with the exponent value decreasing markedly to 0.25 or below. Subsequently, there may be another point of inflection in the plot when sexual maturity is reached and (as with Wingert's data on laboratory mice) there may be no further increase in brain size with increasing body size among adults. With precocial mammals, on the other hand, the early part of the pattern is typically different [though *Homo sapiens* is an exception specifically discussed by Martin (1983)]; the point of inflection marking the transition from the high exponent value characterizing early brain development coincides closely with the time of birth, and *all* postnatal brain development typically involves a low exponent value of 0.25 or less. In primates (including man), the evidence so far available indicates that in sexually mature adults of each sex there is virtually no further increase in brain size with increasing body size (exponent value close to zero; see Table 3). This contrast between the patterns for altricial and precocial mammals is also found in comparing the data for the precocial rodent *Acomys* (Kretschmann, 1966, 1968) with those for the altricial laboratory mouse (Wingert, 1969). Even for the earliest stages of postnatal brain development in *Acomys* [see above analysis of Kretschmann's (1968) data], the brain–body size scaling exponent is close to 0.30, whereas the earliest postnatal stage in the laboratory mouse exhibits a scaling exponent of 0.78. [Although *Apodemus sylvaticus* is also altricial, Klemmt's data (1960) do not cover very early ontogenetic stages; but it can be predicted that the earliest postnatal stage of development of the brain in this species would be characterized by a high scaling exponent value as in the albino mouse.]

In all cases, the relatively low exponent values found with later ontogenetic stages reflect the fact that the brain develops relatively rapidly compared to most other organ systems (see also von Bertalanffy and Pirozynski, 1952). This can be illustrated by taking Schönheit's (1970) data for different postnatal age groups of albino laboratory mice (Fig. 4). Adult brain size is attained by approximately 20 days of age, whereas adult body weight is attained markedly later (though probably somewhat before the 60-day-old sample that represents the next step in Schönheit's data). Because the brain develops relatively rapidly initially, scaling exponent values tend to be high, whereas at a later stage the brain develops relatively little while body size

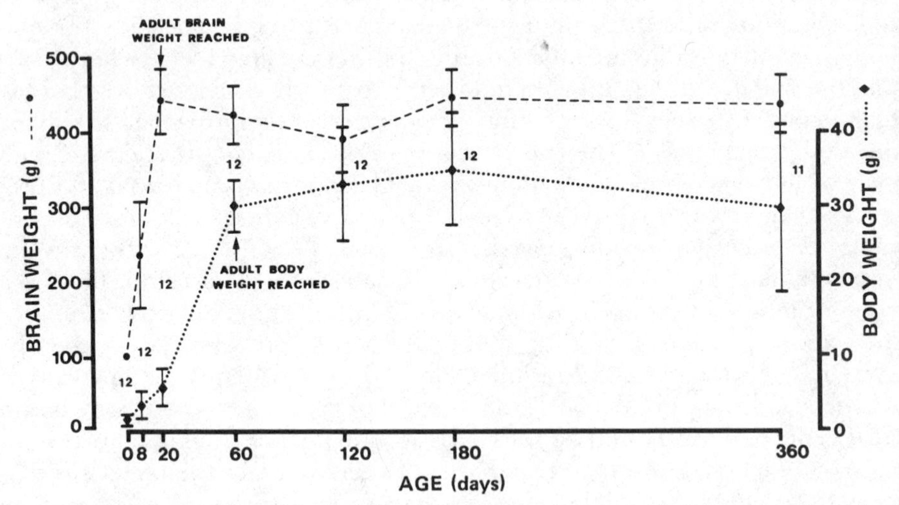

Fig. 4. Plot of brain and body weight for successive age groups of albino mice [data from Schönheit (1970)]. Vertical bars represent standard deviations. Note that adult brain weight values are reached much earlier than adult body weight values.

continues to increase, thus generating low exponent values. In the laboratory mouse, the exponent value of 0.13 characterizing the intermediate postnatal stage (see above) thus represents a combination of the concluding phase of brain ontogeny with continued quite rapid body growth preceding the attainment of sexual maturity.

As has been demonstrated above, in some mammal species there is no further increase in brain size with increasing body size among sexually mature adults, whereas in other species there is a small but nevertheless positive scaling exponent for brain size–body size relationships among mature adults. There are at least two possible explanations for this latter observation. It is conceivable that in these species there is continued growth in brain size with increasing adult body size. However, it is also possible that the interadult scaling exponent merely reflects variability in the body weight attained at sexual maturity (with accompanying variation in mature brain weight, following the ontogenetic scaling process). In the absence of detailed data, it is difficult to decide which of these explanations is correct, but the general occurrence of early maturation of the brain in mammals suggests that the second explanation is more likely to be true and that apparent interadult scaling of the brain in certain mammals (e.g., *Apodemus, Acomys*) simply reflects variation in the endpoints attained by ontogenetic processes, rather than the actual matching of brain size to body size through continued growth of the brain during adult life. It is important to note here [as has been cogently argued by Shea (1983)] that static allometric analysis of immature stages may not clearly reflect the allometric relationships of individual ontogeny, which can only be established through longitudinal studies.

It is in this context that scaling of brain size to body size among adults in *domesticated species* is of particular interest. Lapicque (1907) was particularly influenced by data on the domestic dog suggesting that an exponent value of 0.25 is characteristic of intraspecific scaling of brain size to body size. He cites two sets of data on brain and body size in adult dogs, both grouped into body size classes for purposes of analysis. The first data set (obtained from Richet's laboratory) was derived from 188 laboratory mongrels, which Lapicque divided into 10 size classes. A regression fitted to the means for brain and body weights in these size classes yields an exponent value of 0.23 ($r = 0.995$). The second data set, from Lapicque's own studies, included 47 dogs divided into seven size classes. A regression fitted to the means for brain and body weights yields an exponent value of 0.25 ($r = 0.987$). An exponent value of approximately 0.25 has recently been confirmed by Bronson (1979), who analyzed data on average brain and body size for 26 domestic breeds of dogs (total sample size 1037) and obtained an empirical exponent value of 0.27 ($r = 0.95$) from a regression fitted to the mean of breed values. As was pointed out by Lapicque, the dog is particularly notable among domestic animals for the large range of body sizes generated through artificial selection. In this case, it is undoubtedly true that diversity in adult body weight is largely a reflection of differential body weights attained at sexual maturity. Accordingly, scaling of brain size to body size among breeds of dogs is almost certainly generated largely by differences in ontogenetic patterns, rather than by actual scaling of brain size to body size among adults. In other words, brain size scaling in domestic dogs (and doubtless among other domestic species exhibiting a large body size range) is a direct reflection of ontogenetic scaling of brain size to body size during individual growth.

If this interpretation is correct, the data on intraspecific brain scaling among adults generally actually reflect early ontogenetic processes. This accords well with the hypothesis (Martin, 1981) that neonatal brain weight is largely determined by the mother's metabolic turnover and that postnatal development of the brain follows a very predictable course, which may be essentially "ballistic," rather than reflecting continued fine adjustment between brain and body weight in the individual developing mammal. Further support for this view is provided by a recent analysis of a large body of data on brain and body development in the laboratory rat, which attempts to separate genetic from environmental correlations between brain and body weight (Atchley *et al.*, 1983). Among other things, this study indicates that the genetic correlation between brain size and body size in laboratory rats drops from about 0.5 to zero with increasing age (in a sample covering ages from 14 to 189 days) and that the decline in correlation begins at a postnatal age of approximately 60 days, which would quite closely correspond to the cessation of the main ontogenetic period.

Lande (1979) has discussed in some detail the possible genetic background to changes in brain scaling exponent values with taxonomic level, referring to data from selection experiments on brain and body weights in mice. He concluded that short-term differentiation of brain and body sizes in closely related mammal species might result largely from directional selection acting principally on body size (or perhaps from genetic drift), whereas over the longer term there is more directional selection acting on brain size itself. However, this conclusion is based on the assumption that *interadult* brain scaling within species is governed by an exponent value of between 0.2 and 0.4, and (as demonstrated above) this is not always the case. Instead, one might argue that in the short term selection accompanying diversification of closely related species acts primarily on the body size attained at sexual maturity and that in the first instance brain size in intraspecific comparisons would increase along the *ontogenetic* trajectory, with a typical scaling exponent of 0.25 or somewhat less. Subsequent selection for brain size itself might then lead to higher exponent values for interspecific scaling as individual species become slowly adjusted to reach the upper limits imposed by maternal metabolic constraints. This minor modification of Lande's hypothesis could account for all of the patterns in both intraspecific and interspecific brain–body size scaling as presented above.

Lande cites experiments involving artificial selection of laboratory mice for either body size (Falconer, 1973) or brain size (Roderick *et al.*, 1976). It is noteworthy that when *brain size* was subjected to artificial selection the average slope of lines fitted to logarithmic plots of brain and body weights for the two separate experimental groups (selection for large brain size versus selection for small brain size) was found to be quite high, 0.77. However, it is not clear how artificial selection actually exerts an effect on adult brain–body size relationships in such experiments. Selection might act on any of the factors involved in brain development (e.g., maternal metabolic rate, gestation period, rate of fetal brain development, neonatal brain weight, rate of postnatal brain development, time of weaning, time of attainment of sexual maturity) and it is accordingly difficult to predict a particular outcome of selection of adult brain size as such. Nevertheless, it is clear from such experiments that selection for increased adult brain size entails an accompanying increase in adult body size and it is perhaps significant that the slope value determined by Lande is very close to the value of 0.75 that would be expected from a link with the scaling of basal metabolic rate. (Lande himself suggested that the empirical value of 0.77 was compatible with the value of 0.67 expected from Jerison's concept of brain scaling according to a surface law, but the agreement is by no means as good.) In fact it could be argued that selection for increased brain size would be subject to maternal metabolic constraints, such that a scaling exponent value of 0.75 would normally be attained but not greatly exceeded. However, the validity of this argument can only be tested by examining in detail how selection for increased adult brain size actually affects the individual parameters involved.

The hypothesis invoking maternal metabolic rate as a major factor in the determination of brain–body size relationships (Martin, 1981) entails an additional source of complexity when *intraspecific* scaling is considered. It is commonly taken for granted that Kleiber's law applies both interspecifically and intraspecifically, i.e., that in both cases basal metabolic rate scales to body weight with an exponent value of 0.75. But there is considerable evidence that *intraspecific* scaling of basal metabolic rate among adults does not in fact exhibit this exponent value (Brody, 1945; Thonney *et al.*, 1976). Thonney *et al.* provide details for six mammalian species (rat, rabbit, dog, sheep, human, cattle), taking the sexes separately, which reveal an average intraspecific scaling exponent of 0.64. For humans, the intraspecific scaling exponent is particularly low, with a combined average of 0.44 for male and female values. Given the fact that basal metabolic rate seems to scale with an exponent value of less than 0.75 within species, one might expect an indirect influence of intraspecific metabolic scaling on intraspecific brain scaling, tending to produce a lower exponent value for the latter as well. Such an effect would be indirect in that the relationship between the reduced exponent value for maternal metabolic scaling and the reduced exponent value for brain size scaling in adults would depend upon the heritability of body size from mothers to offspring.

Finally, it must be reiterated that the analysis of brain size scaling at low taxonomic levels, particularly at the intraspecific level, can be greatly influenced by the choice of technique for determining best fit lines. When both exponent values and correlation coefficients are low, there is a marked discrepancy between results inferred from the reduced major axis on the one hand and those inferred from the major axis or regression on the other. In the above discussion, figures have generally been provided either on the basis of regressions (when quoting from published sources) or on the basis of major axes (from our own data analyses), which agree fairly closely. However, if the reduced major axis had been used throughout as the basis for analysis, considerably higher exponent values would have been inferred in many intraspecific analyses (e.g., see Table 3). Thus, until a biologically convincing case has been established for the choice of a particular procedure for determining best fit lines in allometric analysis, there must be a residual suspicion that low exponent values recorded at the intraspecific level might to some extent be a statistical artifact. Nevertheless, some low exponent values determined at low taxonomic levels are accompanied by relatively high correlation coefficients [e.g., see above analysis of Kretschmann's (1968) data for *Acomys cahirinus minous*], so there is undoubtedly some real decrease in exponent values at low taxonomic levels.

It therefore seems reasonable to draw the provisional conclusion that the "taxon-level problem" in brain–body size scaling probably arises from differences in the pattern of natural selection typifying different stages of phylogenetic diversification. In the early stages of evolution, where closely related species are diversifying, selection may act primarily on body size, and differences in brain size between species of different average body weights would accordingly reflect different degrees of progression along a common track of postnatal brain and body development, characterized by a low scaling exponent value (see also Shea, 1983). Subsequently, long-term selection acting on various parts of the ontogenetic process would permit brain size to change relative to body size to match more closely the constraints set by maternal metabolic turnover. Since a number of interlocking parameters are involved in the overall ontogeny of the brain (e.g., gestation period; maternal fetal exchange across the placenta; neonatal brain weight; neonatal body weight; age of cessation of brain growth; age of cessation of body growth; lactation period; milk composition), it is perhaps understandable that long periods of evolutionary time might elapse before initial ontogenetic scaling exponents became modified to match the "optimal" interspecific exponent value of about 0.75. Variation in times of divergence and in different degrees of selection (as well as different rates of response to selection) for brain size versus body size would explain why different orders of mammals have come to exhibit different patterns of increase in brain–body size scaling exponents with increasing taxonomic level (Fig. 2). This being the case, variation in exponent values with taxonomic level is a factor that must obviously be taken into account in any allometric analysis. Since the actual pattern of variation varies from one group of animals to another, it would seem that the best approach to adopt is a pragmatic one, with the choice of taxonomic level for any given allometric analysis determined by the taxonomic level of the particular biological questions involved.

Acknowledgments

Thanks go to Ann MacLarnon for assistance in data analysis and for numerous helpful discussions in the development of the work covered in this chapter. Thanks also go to Dr. J. Marcio Ayres for measuring cranial capacities on the skulls of *Ateles paniscus, Alouatta seniculus,* and *Cebus apella* and for providing data on the body weights of the wild-caught specimens (cf. Table 3). Dr. L. Aiello, Dr. M. Hills, and Dr. G. M. Mace have contributed through helpful advice and comments at various stages.

References

Asdell, S. A. 1964. *Patterns of Mammalian Reproduction,* 2nd ed., Cornell University Press, Ithaca, New York.

Atchley, W. R., Leamy, L., and Riska, B. 1983. Genetics of brain and body size associations: Data from the rat (unpublished manuscript).

Bauchot, R. 1978. Encephalization in vertebrates: A new mode of calculation for allometry coefficients and isoponderal indices. *Brain Behav. Evol.* **15:**1–18.

Bauchot, R., and Stephan, H. 1964. Le poids encéphalique chez les insectivores malgaches. *Acta Zool.* **45:**63–75.

Bauchot, R., and Stephan, H. 1966. Données nouvelles sur l'encéphalisation des Insectivores et des Prosimiens. *Mammalia* **30:**160–196.

Bauchot, R., and Stephan, H. 1969. Encéphalization et niveau evolutif chez les simiens. *Mammalia* **33:**225–275.

Brody, S. 1945. *Bioenergetics and Growth.* Rheinhold, New York.

Bronson, R. T. 1979. Brain weight–body weight scaling in breeds of dogs and cats. *Brain Behav. Evol.* **16:**227–236.

Clutton-Brock, T. H., and Harvey, P. H. 1980. Primates, brains and ecology. *J. Zool.* **190:**309–324.

Corbet, G. B., and Hill, J. E. 1980. *A World List of Mammalian Species,* British Museum (Natural History), London.

Count, E. W. 1947. Brain and body weight in man: Their antecedents in growth and evolution. *Ann. N.Y. Acad. Sci.* **46:**993–1122.

Crile, G. W., and Quiring, D. P. 1940. A record of the body weight and certain organ and gland weights of 3690 animals. *Ohio J. Sci.* **40:**219–259.

Eisenberg, J. F. 1981. *The Mammalian Radiations: A Study in Evolution and Adaptation,* University of Chicago Press, Chicago.

Eisenberg, J. F., and Wilson, D. E. 1981. Relative brain size and demographic strategies in didelphid marsupials. *Am. Nat.* **118:**1–15.

Falconer, D. S. 1973. Replicated selection for body weight in mice. *Genet. Res.* **22:**291–321.

Gittleman, J. L. 1983. The Behavioural Ecology of Carnivores, D. Phil. Thesis, University of Sussex, England.

Gould, S. J. 1975. Allometry in primates, with emphasis on scaling and the evolution of the brain, in: *Approaches to Primate Paleobiology* (Contrib. Primatol., Vol. 5, F. Szalay, ed.), pp. 244–292, S. Karger, Basel.

Harvey, P. H., and Mace, G. M. 1982. Comparisons between taxa and adaptive trends: Problems of methodology, in: *Current Problems in Sociobiology* (King's College Research Group, ed.), pp. 343–361, Cambridge University Press, Cambridge.

Harvey, P. H., Clutton-Brock, T. H., and Mace, G. M. 1980. Brain size and ecology in small mammals and primates. *Proc. Natl. Acad. Sci. USA* **77:**4387–4389.

Herre, W. 1959. Domestikation und Stammesgeschichte, in: *Die Evolution der Organismen* (G. Herberer, ed.), pp. 801–856, Gustav Fischer, Stuttgart.

Herre, W., and Thiede, U. 1965. Studien an Gehirnen südamerikanischer Tylopoden. *Zool. Jahrb. Anat.* **82:**155–176.

Hofman, M. A. 1982. Encephalization in mammals in relation to the size of the cerebral cortex. *Brain Behav. Evol.* **20:**84–96.

Holloway, R. L. 1980. Within-species brain–body weight variability: A reexamination of the Danish data and other primate species. *Am. J. Phys. Anthropol.* **53:**109–121.

Jerison, H. J. 1973. *Evolution of the Brain and Intelligence,* Academic Press, New York.

Kermack, K. A., and Haldane, J. B. S. 1950. Organic correlation and allometry. *Biometrika* **37:**30–41.

King, J. A. 1965. Body, brain and lens weight of *Peromyscus. Zool. Jahrb. Anat.* **82:**177–188.

Klemmt, L. 1960. Quantitative Untersuchungen an *Apodemus sylvaticus* (L. 1758). *Zool. Anz.* **165:**249–275.

Kretschmann, H.-J. 1966. Über die Cerebralisation eines Nestflüchters (*Acomys cahirinus dimidiatus* [Cretzschmar 1826]) im Vergleich mit Nesthockern (Albinomaus, *Apodemus sylvaticus* [Linnaeus 1758] und Albinoratte). *Morphol. Jahrb.* **109:**376–410.

Kretschmann, H.-J. 1968. Über die Cerebralisation eines Nestflüchters (*Acomys* (*cahirinus*) *minous* [Bate, 1906]) im Vergleich mit Nesthockern (Albinomaus, *Apodemus sylvaticus* [Linnaeus, 1758] und Albinoratte). *Morphol. Jahrb.* **112:**237–260.

Kruska, D. 1973. Cerebralisation, Hirnevolution und domestikationsbedingte Hirngrössenänderungen innerhalb der Ordnung Perissodactyla Owen, 1848 und ein Vergleich mit der Ordnung Artiodactyla Owen, 1848. *Z. Zool. Syst. Evolutionsforsch.* **11:**81–103.

Lande, R. 1979. Quantitative genetic analysis of multivariate evolution applied to brain:body size allometry. *Evolution* **33:**402–416.

Lapicque, L. 1907. Le poids encéphalique en fonction du poids corporel entre individus d'une même espèce. *Bull. Mem. Soc. Anthropol. Paris* **8:**313–345.

Mace, G. M., Harvey, P. H.,and Clutton-Brock, T. H. 1981. Brain size and ecology in small mammals. *J. Zool.* **193:**333–354.

Martin, R. D. 1980. Adaptation and body size in primates. *Z. Morphol. Anthropol.* **71:**115–124.

Martin, R. D. 1981. Relative brain size and basal metabolic rate in terrestrial vertebrates. *Nature* **293:**57–60.

Martin, R. D. 1983. *Human Brain Evolution in an Ecological Context* (52nd James Arthur Lecture on the Evolution of the Human Brain), American Museum of Natural History, New York.

Oboussier, H., and Schliemann, H. 1966. Hirn-Körpergewichtsbeziehungen bei Boviden. *Z. Saeugtierkd.* **31:**464–471.

Pakkenberg, H., and Voigt, J. 1964. Brain weight of the Danes. *Acta Anat.* **56:**297–307.

Portmann, A. 1962. Cerebralisation und Ontogenese. *Med. Grundlagenforsch.* **4:**1–62.

Riese, W., and Riese, H. 1952. Investigations on the brain weight of the baboon (*Papio papio* Desm.). *J. Comp. Neurol.* **96:**127–137.

Roderick, T. H., Wimer, R. E., and Wimer, C. C. 1976. Genetic manipulation of neuroanatomical traits, in: *Knowing, Thinking and Believing* (L. Petrinovich and J. L. McGaugh, eds.), pp. 143–178, Plenum Press, New York.

Röhrs, M. 1959. Neue Ergebnisse und Probleme der Allometrieforschung. *Z. wiss. Zool.* **162:**1–95.

Ronnefeld, U. 1970. Morphologische und quantitative Neocortexuntersuchungen bei Boviden, ein Beitrag zur Phylogenie dieser Familie. I. Formen mittlere Körpergrösse (25 kg bis 75 kg). *Morphol. Jahrb.* **115:**163–230.

Sacher, G. A. 1982. The role of brain maturation in the evolution of the primates, in: *Primate Brain Evolution: Methods and Concepts* (E. Armstrong and D. Falk, eds.), pp. 97–112, Plenum Press, New York.

Sacher, G. A., and Staffeldt, E. F. 1974. Relation of gestation time to brain weight for placental mammals: Implications for the theory of vertebrate growth. *Am. Nat.* **108:**593–616.

Schönheit, B. 1970. Die Väriabilität der Körpergrösse, des Hirngewichtes und der Form und Grösse des Bulbus olfactorius in der postnatalen Entwicklung der Albinomaus. *Anat. Anz.* **126:**363–390.

Shea, B. T. 1983. Phyletic size change in brain:body allometry: A consideration based on the African pongids and other primates. *Int. J. Primatol.* **4:**33–62.

Sholl, D. 1848. The quantitative investigation of the vertebrate brain and the applicability of allometric formulae to its study. *Proc. R. Soc. Lond. B* **135:**243–258.

Simpson, G. G. 1945. The principles of classification and a classification of mammals. *Bull. Am. Mus. Nat. Hist.* **85:**1–350.

Snedecor, G. W., and Cochran, W. G. 1967. *Statistical Methods,* Iowa State University Press, Ames, Iowa.

Snow, C. C., and Vice, T. 1965. Organ weight allometry and sexual dimorphism in the olive baboon *Papio anubis,* in: *The Baboon in Medical Research* (H. Vagtborg, ed.), pp. 151–163, University of Texas Press, Austin.

Sokal, R. R., and Rohlf, F. J. 1969. *Biometry,* Freeman, San Francisco.

Stephan, H. 1958. Vergleichend-anatomische Untersuchungen an Insektivorengehirnen. *Morphol. Jahrb.* **99:**853–880.

Stephan, H. 1972. Evolution of primate brains: A comparative anatomical investigation, in: *The Functional and Evolutionary Biology of Primates* (R. Tuttle, ed.), pp. 155–174, Aldine-Atherton, Chicago.

Stephan, H., Nelson, J. E., and Frahm, H. D. 1981. Brain size comparison in Chiroptera. *Z. Zool. Syst. Evolutionsforsch.* **19:**195–222.

Thonney, M. L., Touchberry, R. W., Goodrich, R. D., and Meiske, J. C. 1976. Intraspecies relationship between fasting heat production and body weight: A reevaluation of $W^{0.75}$. *J. Anim. Sci.* **43:**692–704.

Tyska, H. 1966. Das Grosshirnfurchenbild als Merkmal der Evolution: Untersuchungen an Boviden. *J. Zool. Mus. Inst.* (*Hamburg*) **63:**121–158.

Von Bertalanffy, L., and Pirozynski, W. J. 1952. Ontogenetic and evolutionary allometry. *Evolution* **6:**387–392.

Von Bonin, G. 1937. Brain-weight and body-weight of mammals. *J. Gen. Physiol.* **16:**379–389.

Wingert, F. 1969. Biometrische Analyse der Wachstumsfunktionen von Hirnteilen und Köpergewicht der Albinomaus. *J. Hirnforsch.* **11:**133–197.

Ontogenetic Allometry and Scaling 9

A Discussion Based on the Growth and Form of the Skull in African Apes

BRIAN T. SHEA

Introduction

The two great syntheses that established the study of allometry were D'Arcy Thompson's (1917, 1942, 1961) *On Growth and Form* and Julian Huxley's (1932) *Problems of Relative Growth.* The bulk of Thompson's volume dealt with the mechanical and physical factors underlying shape transformation. In his chapter entitled "The rate of growth," however, Thompson (1942, p. 79) stressed that "the *form* of an organism is determined by its rate of *growth* in various directions; hence rate of growth deserves to be studied as a necessary preliminary to the theoretical study of form." Huxley's (1932) work built on Thompson's foundation through extensive empirical and theoretical investigations of relative, or differential, growth in organisms. Huxley showed that a simple power function, $Y = bX^k$, could often describe correlated changes between growing parts (X and Y) of the body. He also analyzed ontogenetic allometry (relative growth, or heterogony, as Huxley labeled it) in terms of

BRIAN T. SHEA • Departments of Anthropology and Cell Biology and Anatomy, Northwestern University, Evanston, Illinois 60201.

growth gradients, taxonomy, genetics, and evolutionary transformations. Huxley's approach to the study of size and shape, therefore, was explicitly rooted in the analysis of growing form; we often term this *growth allometry,* in order to distinguish it from *size allometry,* which focuses on allometric changes in static adult series (Simpson, *et al.,* 1960).

Since the early work by Thompson and Huxley, a vast amount of research on allometry has been undertaken. Cock (1966) and Gould (1966) provide two of the most useful reviews of the subject and relevant literature. The current resurgence of interest in allometry is due in large measure to the work of Gould (e.g., 1966, 1968, 1969, 1971*a,b,* 1974, 1975*a,b,* 1977). The present volume attests to the continuing interest in problems of size and scaling.

Over the past several decades, the majority of allometric investigations have been concerned with identifying and explaining the biomechanical bases of broad patterns of interspecific allometry (again, the present volume illustrates this emphasis). The analysis of growth allometry has frequently been viewed as primarily of value only in elucidating the developmental pathway utilized by natural selection to produce adaptive changes in shape. In Mayr's (1961) terminology, growth becomes the proximate cause of change in form, while the ultimate causes relate to the external factors that, through natural selection, have produced evolutionary transformation. Recent work in a number of areas has suggested a more fundamental role for the comparative study of ontogenetic allometry, and development in general, however. In this chapter, I will briefly examine some of these issues in morphological analysis. Discussion will be based primarily on the growth and form of the skull in African apes. A more detailed analysis of the specific data base has been presented elsewhere (Shea, 1982, 1983*a–d,* 1984).

Ontogenetic Allometry

The various curves and mathematical functions used by biologists to study growth have been described by Medawar (1945), von Bertalanffy (1960), and Laird (1965), among others. Figure 1 illustrates four examples of growth curves for a given data set. Figure 1A depicts the curve of *growth,* or simply absolute weight plotted against time. The curve of *growth rate* (Fig. 1B) is the first derivative of the curve of growth, plotting the change in weight per time against time. Figure 1C is the curve of *specific growth,* given by plotting the *logarithm* of weight against time, thus providing a record of the *multiplication* (rather than the *addition*) of living substance. The *specific growth rate* (Fig. 1D) plots the change in log weight per time against time. The specific growth rate (for structure Y, for example) can also be described as the change in Y per unit amount of Y with respect to time; therefore it provides a quantitative indication of the intensity of underlying growth activity (Huxley, 1932;

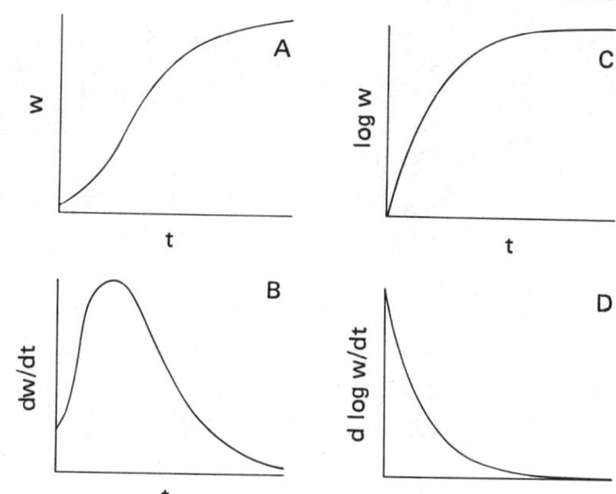

Fig. 1. Various growth curves for a given data set. Here w is weight, t is time. (A) Curve of growth, (B) curve of growth rate, (C) curve of specific growth, and (D) curve of specific growth rate. (Redrawn from Medawar, 1945.) See text for discussion.

Kowalski and Guire, 1974). The specific growth rate has sometimes been termed the "relative" or "percentage" growth rate, because it expresses growth in time relative to the amount of growth that has already occurred—this is a very different use of "relative growth" than the traditional usage of Huxley.

The primary distinction between the curve of growth and its derivatives, and the curve of specific growth and its derivatives, is that the former measures growth as *additive* increase whereas the latter gauges this change in *multiplicative* terms. Medawar (1945) also notes that use of specific growth rate erases the "prominent but perhaps not significant" feature of the curve of growth rate, i.e., the inflection point in Fig. 1B. Many workers (e.g., Medawar, 1945; Laird *et al.*, 1965) have favored a model of growth that denies any particular biological significance to this inflection point, for there seems to be little evidence that living substance is actually produced first with increasing and then decreasing energy. Rather, the curves of specific growth and specific growth rate suggest that living matter progressively loses the ability to multiply itself at the rate at which it was formed (Laird, 1965; Laird *et al.*, 1965).

It is the curve of specific growth and its derivatives that are of direct relevance to growth allometry. Huxley's (1932) formula of bivariate ontogenetic allometry, $Y = bX^k$, removes time as a variable, and relates growth of one structure, Y, to weight or a second structure, X. Thus the label "relative growth," for growth in one is measured *relative to* growth in the other. Huxley's use of the power function was originally explicitly based on his theoretical premise that growth is essentially a process of "self-multiplication." The relationship of the curve of specific growth to Huxley's allometry formula is made clearer by realizing that the equation $Y = bX^k$ is the solution of the differential equation

$$\frac{d}{dt}\log Y = k\frac{d}{dt}\log X$$

or

$$\frac{dY}{dt}\frac{1}{Y} = k\frac{dX}{dt}\frac{1}{X}$$

Since $dY/(Y\,dt)$ and $dX/(X\,dt)$ are the respective *specific growth rates* for structures Y and X, Huxley's k, or coefficient of ontogenetic allometry, actually equals the *ratio* of the specific growth rates of the structures involved (Kowalski and Guire, 1974). When Huxley's equation fits (i.e., when "simple allometry" holds), the specific growth rates maintain a constant ratio k throughout growth. Huxley (1932) analogized the specific growth rates to two capital investments that grow in the bank at different rates of interest, with the interest rates being a fixed multiple of each other, so that $dY/(Y\,dt) = k\,dX/(X\,dt)$. When simple allometry holds, a logarithmic plot of X and Y yields a straight line of slope k (log b represents the y intercept). When a logarithmic plot yields a curved line, the ratio of specific growth rates is changing throughout growth. This situation is in no sense less meaningful biologically, merely more difficult to assess statistically (Laird, 1965). Thus, Huxley's use of logarithmic axes was firmly based in a theoretical context of multiplicative growth. The pervasive concentration on size allometry, rather than growth allometry, in the past several decades has led to misunderstanding in this regard. For example, recent attempts to "rethink allometry" (Smith, 1980) appear to be based on the fallacious notion that log-transformation is undertaken primarily for empirical and statistical reasons. Nevertheless, not all growth is multiplicative, of course (e.g., Davenport, 1934), and Huxley ultimately argued that his original theoretical premises of ontogenetic allometry were untenable (Reeve and Huxley, 1945; Gould, 1966). Katz (1980) has recently presented an analysis of relative growth that would relate Huxley's k to the ratio of the frequencies of cell division between two growing parts, however, and thus Huxley's notions of multiplicative growth may be sound after all, at least in many cases.

Application of the principles of ontogenetic allometry to skull morphology permits analysis of proportion changes in terms of differential growth. Figure 2 illustrates the positively allometric growth of the muzzle (splanchnocranium) relative to the braincase (neurocranium) during ontogeny in the baboon. The shape of the skull changes dramatically during growth. Thus, studies of ontogenetic allometry can elucidate morphology by analyzing in a quantitative manner how changes in form are produced during development and evolution. These quantitative analyses must be combined with qualitative investigations, however, as Huxley and Reeve (1945) noted. This is particularly true in studies of skull growth and form, where, in addition to differential growth of the splanchnocranium and neurocranium, we also need to carefully examine reorientation of the face relative to the skullbase (e.g., Todd and Wharton, 1934). For example, in an interspecific comparison of baboons, Vogel (1968) distinguishes between the *size* of the muzzle and its *angulation* relative to the skullbase. In the case of the great apes, certain quantitative approaches reveal a strong similarity in the form of the face in *Pongo* and *Pan*/*Gorilla* (e.g., Moore and Lavelle, 1974; Moore, 1981), while a

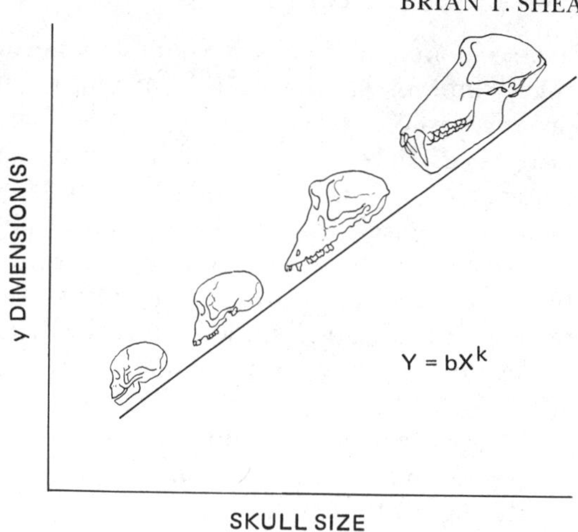

Fig. 2. An illustration of the dramatic changes in skull shape that occur during ontogeny in the baboon. Huxley's bivariate formula of relative growth is also given.

more complete examination indicates a significant change in position of the face relative to the skullbase in *Pongo,* however (Biegert, 1957, 1963).

Ontogenetic Scaling

The phenomenon of "ontogenetic scaling" occurs when an observed difference in body size (in experimental groups, subspecies, species, or a phyletic series) is produced or accompanied by an extension or truncation of common (or ancestral) growth allometries (Gould, 1966, 1975*a;* Shea, 1981). Many examples of ontogenetic scaling (although not always so labeled) have been given by Huxley (1932), Rensch (1959), and Gould (1966, 1977). Cock (1966) gives an especially good critical discussion of certain selected examples. In cases of ontogenetic scaling, size–shape relations between adults of the different populations or species are of the same nature as those distinguishing young from adult, smaller from larger, and female from male *within* each species or group. Evidence demonstrating that proportion differences between adults of sexually dimorphic species result from extension of a common growth allometry is given by Larson (1978), Wood (1976), Cheverud (1981), Dodson (1975*a,b*), and Shea (1981, 1982, 1983*b*), among others. Shea (1983*c,d*) has shown that even if ontogenetic scaling of proportions holds within and/or between species, the underlying shifts in rate of growth or time of growth may differ. It should be clear that ontogenetic scaling in evolution is related to the "hypermorphosis" of de Beer (1958) and Gould (1977), the "anaboly" of Rensch (1959) and earlier writers, and the "hypertely" of Delsol (1977) [see Shea (1983*d*) for additional discussion].

Ontogenetic scaling of limb proportions in chick embryos was demonstrated in an experimental situation by Lerner and Gunns (1938). They grew chick embryos at three different temperatures (98°, 101°, and 104°), and produced successively larger embryos (Fig. 3A). In spite of the overall size differences, however, a plot of tibiotarsus length against femur length reveals ontogenetic scaling of proportions. The ontogenetic trajectory of growth allometry is merely extended to larger ultimate sizes in the successive experimental subgroups. Although this is a phenotypic example in a single species, the experiment illustrates how patterns of growth in weight may be altered without affecting the underlying growth allometries. Interspecific cases of ontogenetic scaling that presumably reflect genotypic differences will be discussed below.

A good example of ontogenetic scaling in two subspecies of bottle gourd (*Lagenaria*) was given by Sinnott (1936). A logarithmic plot of length versus width (Fig. 4) in these gourds revealed that the two subspecies share a common growth trajectory, although they differ considerably in final adult size. Sinnott (1936, p. 251) noted:

> These bottle gourd types have the same slope and the same level [y intercept] for their length–width developmental line and differ only in their size at maturity, that is, the point along this line where growth stops. Genes which produce differences in size will therefore necessarily bring about differences in shape index whenever growth of the two dimensions is not equal. These two races of gourds may thus be regarded as differing primarily in size and only secondarily in shape.

This interesting conclusion could hardly have been reached without a careful analysis of the patterns of ontogenetic development in the respective forms.

Robb (1935) plotted facial length against total skull length in the ontogeny of the domestic horse (*Equus caballus*) and the phylogeny of fossil horses (*Hyracotherium* to *Equus*). Robb argued that the ontogenetic and phylogenetic gradients were identical. On this basis, Rensch (1959, p. 137) con-

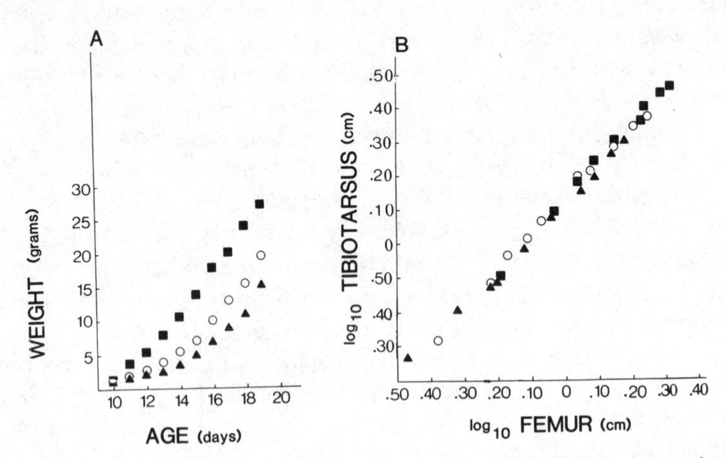

Fig. 3. (A) Growth in body weight and (B) ontogenetic scaling of limb proportions in chick embryos raised at different temperatures. Data from Lerner and Gunns (1938). (■) 104°F, (○) 101°F, (▲) 98°F.

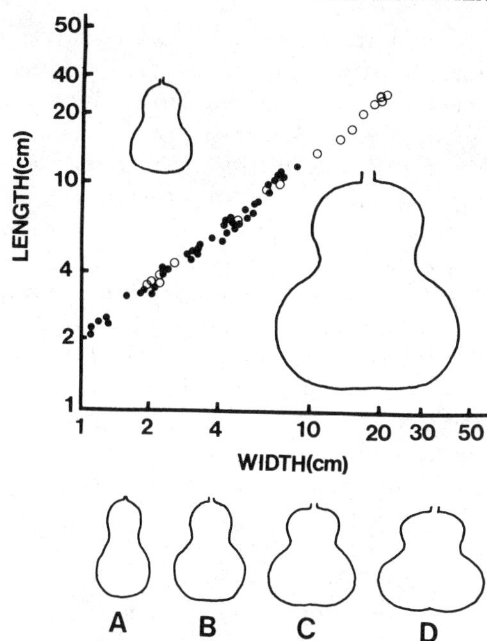

Fig. 4. Ontogenetic scaling of proportions in two subspecies of bottle gourd (*Lagenaria*). (A) Subadults of the smaller subspecies, (B) adults of this group, and (C) adults of the larger subspecies; (D) the shape expected if ontogenetic scaling were prolonged even farther. (Redrawn from Sinnott, 1936.)

cluded that "the phylogenetic transformation of the equine skull bones was predetermined by the growth gradient in the face bones of the ancestral type *Hyracotherium,* though this gradient had as yet no conspicuous effects in this relatively small animal. With increasing body size in the course of phylogeny, this growth gradient became more and more effective." Simpson (1944) made similar comments, although he later revised them (Simpson, 1953), based on the following information. Reeve and Murray (1942) demonstrated that certain of Robb's conclusions regarding the ontogeny and phylogeny of the horse skull were oversimplified, and that the actual situation is more complex. Cock (1966, p. 175) has summarized this argument. Nevertheless, Robb's study revealed a general similarity in patterns of size and shape change in the horse skull in ontogeny and phylogeny, and he established the importance of relative growth in the interpretation of these morphological changes (Simpson, 1953). Radinsky (1984) has recently reanalyzed this problem.

Reeve (1940) showed that the ontogenetic trajectories of nasal length versus cranium length in the anteaters *Tamandua* and *Cyclopes* coincided (and they both differed significantly from the pattern seen in the large *Myrmecophaga*). He concluded that the primary differences in skull shape between the former two genera were due to size differences, or in this case the differential extension of a shared growth trajectory. Lumer (1940) found several cases where the ontogenetic allometry of German Shepherd cranial dimensions coincided with the regression fit to mean adult values of his "Terrier tribe"; he also demonstrated examples where the relations differed. Cases of coincidence strongly suggest ontogenetic scaling in this example.

Alberch (1981; Alberch and Alberch, 1981) has analyzed a series of cranial and postcranial morphological differences between a small and derived lowland salamander (*Bolitoglossa occidentalis*) of Mexico and more generalized members of the genus. He finds that a general truncation of ontogenetic allometries is responsible for the primary size and shape differences between the forms. Again, this could not have been determined in the absence of detailed information on growth.

Dodson (1975*a–c*, 1976) has completed some of the best and most detailed work on growth allometry in recent years. He (Dodson, 1975*b*) took cranial and postcranial skeletal measurements on a growth series of two sympatric species of lizards, *Sceloporus undulatus* and *S. olivaceous;* maximum snout-to-sacrum length differs by a factor of almost two in these forms. Visual examination of 28 bivariate plots indicated obviously divergent growth allometries in only one case (although several slope comparisons yielded statistically significant differences). Figure 5 illustrates one such example of ontogenetic scaling of cranial proportions in the *Sceloporus* species. Most of the shape differences between these species can thus be accounted for by the differential extension of a common growth trajectory. In a second study, Dodson (1975*c*) analyzed growth allometry in lambeosaurine hadrosaurs (duck-billed dinosaurs) in order to determine whether these forms had been taxonomically "oversplit" because of failure to recognize juveniles (three genera and 12 species were recognized). Of 48 cranial dimensions measured, only five or six, confined to the skull crest of these forms, had any discriminatory value (i.e., exhibited divergent growth patterns). Facial dimensions of these hadrosaurs were strongly ontogenetically scaled. Studies similar to Dodson's have been carried out on *Diademodon* and *Aulacephalodon* by Grine *et al.* (1978) and Tollman *et al.* (1980), respectively.

Turning our attention to cranial form in primates, Freedman (1962) demonstrated that different species of baboons exhibited ontogenetic scaling in a comparison of facial versus calvarial length. Interspecifically as well as ontogenetically, a single trajectory of size/shape change proved to be the basis of the gross differences in proportions.

The work of the greatest direct relevance to that discussed throughout the remainder of this chapter is Giles' (1956) early study of cranial allometry

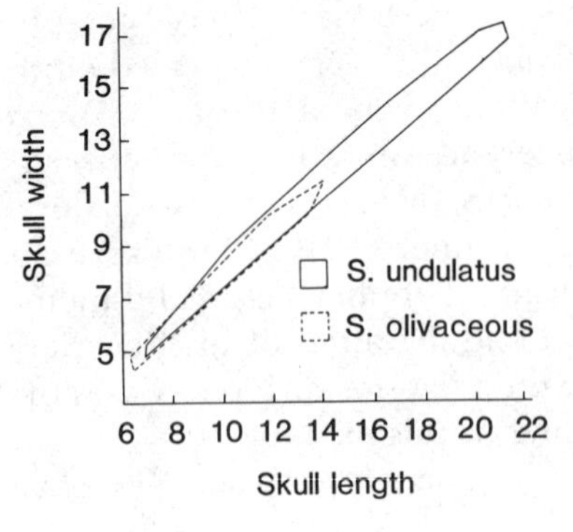

Fig. 5. Ontogenetic scaling of skull proportions in *Sceloporus* lizards. Skull length equals the total length of the skull roof; skull width equals the width across the posterior ends of the maxillae. The larger species has the proportions expected given an extension of the growth trajectory in the smaller species. (Redrawn from Lundelius, 1957.)

in the great apes. Giles examined several proportion comparisons among a series of adult and subadult *Gorilla, Pan,* and *Pongo.* Although sample sizes of subadults were inadequate for a detailed growth analysis, Giles found a close similarity between the coefficient of growth allometry within *Pan* and the interspecific coefficient of size allometry connecting *Pan* and *Gorilla* adult values for several dimensions [note that because adult *Pan* is a point on both regressions, no separate test for position differences is required, *contra* Gould (1975*a*)]. Interestingly, Giles (1956, pp. 56–57) concluded that "the apparent differences in these categories between the chimpanzee and the gorilla are due to similar ontogenetic growth patterns having been at some stage of evolution (so to speak) allowed to manifest different terminal overall morphological configurations through the mechanism of general body volume increase." Furthermore, he hypothesized that "the allometric growth patterns in the chimpanzee would, given an overall size increase, produce results quite similar to the exaggerated osteological morphology of the gorilla." Gould (1975*a*, p. 251) suggested that one way to more fully test the Giles hypothesis was to "construct an ontogenetic curve for chimpanzees and extrapolate it to gorilla sizes." This is essentially a test for ontogenetic scaling; the hypothesis is illustrated in Fig. 6.

Part of my own research on the great apes has involved a detailed test of the Giles (1956) hypothesis, based on a comparative ontogenetic study of cranial and postcranial proportions in *Pan paniscus, Pan troglodytes,* and *Gorilla gorilla.* Methodology and results are discussed in detail elsewhere (Shea, 1981, 1982, 1983*a–d,* 1984).

An Ontogenetic Criterion of Subtraction

Allometric analyses are frequently undertaken in order to establish a "criterion of subtraction," i.e., we control for size-related shape differences in

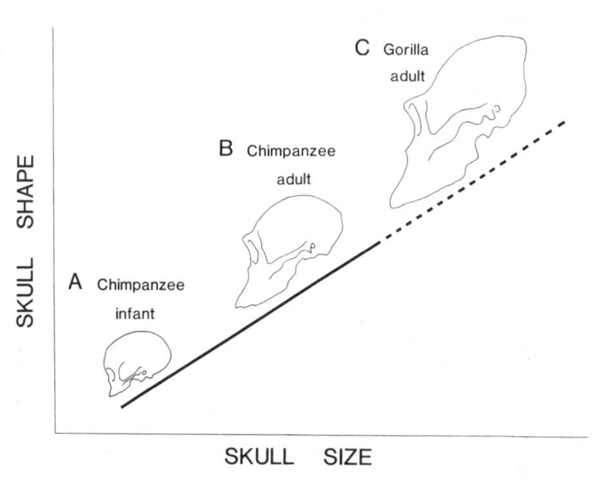

Fig. 6. A schematic illustration of Giles' (1956) hypothesis concerning skull form in the African apes. (A) Subadult and (B) adult skulls of *Pan troglodytes;* (C) adult male gorilla. The solid line depicts the ontogenetic trajectory of chimpanzees, and the dashed line illustrates its extension to gorilla size ranges.

order to more clearly assess the nonallometric adaptive changes (Gould, 1975*a*). In recent years, a number of approaches have been utilized in the attempt to control for size differences in studies of systematics, phylogeny, and functional anatomy (e.g., Corruccini, 1972, 1978; Susman and Creel, 1979; Gelvin, 1981; Albrecht, 1981). I would argue that many previous size correction techniques are statistical exercises that may have little biological rationale. Analysis of deviations from a regression line are only useful if we clearly understand the meaning of the underlying trend in a functional and biological sense. Thus, although Jerison's (1973) "EQ" correction for relative brain size has certainly been a significant advance over most previous attempts (as have other related regression approaches), the size-corrected numerical value, or residual, would take on much greater significance if we could unravel the basis of the broad scaling patterns in functional terms; in other words, if we knew such factors as the metabolic or surface area inputs into determining brain size. For reasons detailed by Cock (1966) and Shea (1983*a*), I would argue that no statistical technique of "size correction" of static adult intraspecific data can provide information that has much biological meaning (save in terms of the relative sizes of the parts being compared).

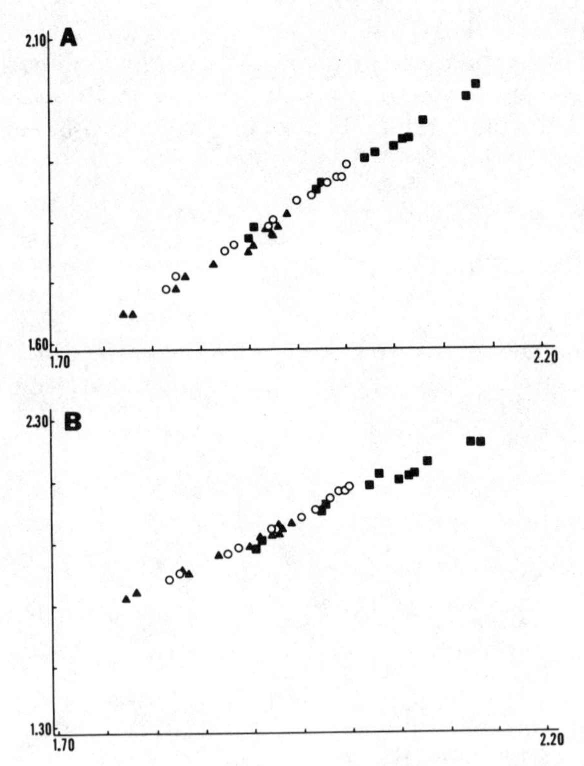

Fig. 7. Ontogenetic scaling of the facial region in the African apes, *Pan paniscus* (▲), *Pan troglodytes* (○), and *Gorilla gorilla* (■). The *x* axis in all cases represents basicranial skull length (basion–nasion). The *y* axis represents (A) basilar suture to nasion, (B) basilar suture to prosthion, (C) nasion to prosthion, (D) bimaxillary width. These are means of cross-sectional data aged via

Ontogenetic allometry provides an optimal approach to the problem of size correction. Here we clearly understand the biological meaning of the allometric trend: it is part of an integrated pattern of growth. And since phylogeny results from ontogenetic modifications, we can "read" the actions of natural selection by comparing growth allometries in closely related species. Divergent growth patterns are indicative of selection for altered proportions. For example, shifts in growth allometries of the limbs have been utilized to clarify particular locomotor adaptations in salamanders (Alberch, 1981; Alberch and Alberch, 1981), *Cebus* monkeys (Jungers and Fleagle, 1980), and African apes (Shea, 1981). I have used ontogenetic allometry and scaling as a criterion of subtraction in a study of dietary adaptations and the craniodental complex of the African apes (Shea, 1983*b*). This is an ideal case, because we are dealing with very closely related forms, which differ significantly in craniodental form, body size, and diet. Given the different dietary specializations of chimpanzees (frugivory) and gorillas (folivory), one might expect divergent growth adaptations for dealing with the different food types.

Results indicate a somewhat different situation. Although significant nonallometric differences in dental morphology clearly relate to the dietary

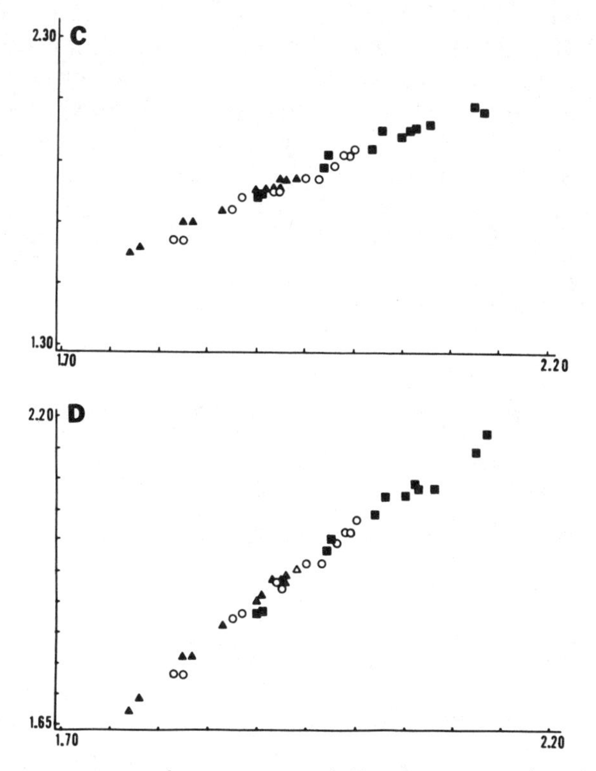

the dentition [six age groups and two sexes equal 12 points per species (Shea, 1982)]. The coincidence of the ontogenetic allometries in the three species for these dimensions yields a "three-dimensional" (length, height, width) case of ontogenetic scaling of gross facial proportions in these apes. See text for additional discussion.

divergence, no similar reorganization of gross facial proportions was indicated by the comparative growth study (Shea, 1983*b*). A plot of facial height, width, and length (Fig. 7) reveals that the three species share a common ontogenetic trajectory, differing only, in Sinnott's (1936, p. 251) words, in "the point along this line where growth stops." Overall, gorillas have the facial proportions expected in chimpanzees ontogenetically scaled to their skull sizes. Thus, Giles' (1956) hypothesis concerning the gorilla and the chimpanzee is largely confirmed. The implications of these results for our understanding of facial form and diet in the apes are discussed elsewhere (Shea, 1983*b*).

Adaptive Growth

What does a finding of such pervasive ontogenetic scaling mean for our interpretation of form, function, and adaptation? Huxley (1932, p. 215) spoke clearly on this issue, arguing that shape in such cases "is automatically determined as a secondary result of a single common growth-mechanism, and *therefore is not of adaptive significance*" (italics in original). Simpson (1953, p. 167) discussed the difficulties inherent in identifying and defining "adaptation" (see also Williams, 1966; Bock, 1980; Clutton-Brock and Harvey, 1979*a;* Gould and Lewontin, 1979; Lewontin, 1978; Gould and Vrba, 1982). I hope to bypass the traditional definitional and terminological mess in order to make a few simple points here. Simpson (1953, p. 275) did clearly attribute nonadaptive status to aspects of form resulting from allometric growth, however: "The genetic mechanism for correlated change and to a lesser extent the ontogenetic mechanisms have the peculiar feature that it is possible for them to produce nonadaptive or inadaptive changes under the influence of selection for correlated adaptive change."

Gould and Lewontin (1979, p. 152) offer the following example of a nonadaptive change in morphology.

> At least three times in the evolution of arthropods (mites, flies, beetles), the same complex adaptation has evolved, apparently for rapid turnover of generations in strongly *r*-selected feeders on superabundant but ephemeral fungal resources: females reproduce as larvae and grow the next generation within their bodies. Offspring eat their mother from inside and emerge from her hollow shell, only to be devoured a few days later by their own progeny. It would be foolish to seek adaptive significance in paedomorphic morphology *per se;* it is primarily a by-product of selection for rapid cycling of generations.

[This argument could be extended to the myriad cases of progenesis and hypermorphosis reviewed by Gould (1977).]

I think the above conclusion might be altered to read: "It would be foolish to view the paedomorphic morphology as the primary *focus* of natural selection *in this case;* it is primarily a by-product of selection for rapid cycling of generations." This seems necessary because the paedomorphic morphology, indeed the entire ancestral growth pattern, was presumably historically developed under the scrutiny of natural selection, which is Lewontin's

(1978) criterion for the label "adaptive." This is also why the term "exaptation" (Gould and Vrba, 1982) would probably not be applicable in this case.

Frazzetta (1975) has taken a slightly different approach to the interpretation of change in shape produced by paedomorphic truncation or peramorphic extension of growth patterns in response to selection for changes in body size or length of ontogeny. In an analysis of salamander neoteny, he stresses (Frazzetta, 1975, p. 226):

> Common examples include salamander populations whose members, instead of transforming to the adult stage of ontogeny, retain the juvenile adaptive form though they attain adult size and sexual maturity. The juvenile adaptive plan is obviously biologically "coherent" and offers an adaptive alternative in evolution.

In a chapter entitled "evolutionary anticipation from adaptability," Frazzetta (1975, p. 225) further probed the "flexibility of integrated systems to provide an array of varied adaptive expressions." I believe this notion of the "power of extrapolation" of such functionally integrated systems deserves more attention and application; it provides a much needed link between investigations of allometry (and development in general) and functional anatomy. Frazzetta's point is simply that functionally integrated systems can often be extrapolated to suit future functional demands under new circumstances. Thus, they are preadaptive—this is particularly central to the evolution of size change in closely related forms. This interpretation would only pose difficulties if the interaction of body parts during ontogeny was random or uncorrelated in a functional sense. It is unlikely in the extreme that this will be the case. The following passage from Goss' (1964) book, appropriately titled *Adaptive Growth,* makes a point similar to Frazzetta's notion of the preadaptive power of extrapolation of functionally integrated systems:

> Although it is not legitimate, *sensu strictu,* to attempt to equate comparisons between young vs. mature animals with small vs. large species, the close parallels between such systems may reflect the subordination of organ size and structure to considerations of physiological demands in both instances. The latter have unquestionably affected the course of evolution by providing the alternatives upon which natural selection operates. In a sense, therefore, evolution may be regarded as a kind of long range compensation for the physiological inadequacies engendered by changing environmental conditions and the necessity of organisms to adapt. Thus, organisms come equipped with genetic constitutions which tend to minimize, but never to abolish altogether, the necessity for the individual to undergo physiological adaptations to the demands of environmental changes.

My ontogenetic analysis of skull growth in the African apes provides an excellent example of Frazzetta's preadaptive power of extrapolation of functionally integrated systems. The broad pattern of size/shape change in the facial region not only "worked" in the smaller (and presumably ancestral) size ranges, but it was also "extrapolatable" to larger size ranges without major changes. This is particularly significant given the dietary shift from frugivory to folivory, although, as I noted above, we do find clear nonallometric changes in dental size and shape. I must argue, after careful allometric control, that the primary shape differences between skulls of adult chimpanzees and gorillas are related to differences in overall body size, and not diet (Shea, 1983*b*). Given the above arguments, however, this does not make the shape differences "nonadaptive." The allometric relationships observed during ontogeny reflect specific and important interrelationships of function and timing among the various functional components (Moss, 1973) of the skull (see below, section on Functional Cranial Analysis). I think this is an important point that should not be lost in attempts (admittedly justified in large measure) to criticize adaptationist arguments (e.g., Gould and Lewontin, 1979).

This notion of extrapolation applies to many closely related, size-differentiated, interspecific or phyletic series. Careful comparative study of ontogenetic allometries and testing for ontogenetic scaling should prove fruitful in such cases. As Frazzetta (1975, p. 241) notes, "the evidence that extrapolatable or anticipatory systems play a great role in evolutionary radiations is becoming harder to ignore."

Historical Factors

Lauder (1981) has recently presented a theoretical analysis of "historical factors" (Raup, 1972) in evolutionary transformations. He notes three basic factors that must be understood in an attempt to explain form—fabricational, functional, and historical. Historical factors may be elucidated by two types of analyses concerned with structural systems: (1) *equilibrium* anaylses, which focus on interactions between the organism and the environment through time, and (2) *transformational* analyses, which focus on structural and functional interactions and interrelationships within the organism. In the first case, explanations of structural change focus on *external events* (e.g., ecological factors), and thus are *extrinsic,* while in the second case, "historical patterns of structural changes are analyzed as a consequence of *intrinsic* organizational properties of structural systems" (Lauder, 1981, p. 431).

Although Lauder (1981) primarily focuses on emergent organizational properties of structural systems (see his example of the transformational consequences of the decoupling of the suspensorium from the maxilla in cichlid fishes), his discussion of historical factors is relevant to the present analysis of the morphology of the African ape skull. The significant component of ontogenetic scaling of cranial proportions in this series (here assumed for heuristic purposes to represent a phyletic series of increasing size) suggests that the primary morphological differences among the species are explicable given the *intrinsic* growth patterns interacting with overall size increase. Lauder (1981, p. 434) notes that:

> In a highly constrained system in which the structural elements are tightly coupled by both functional and morphological interactions, certain types of structural modification might be predicted to occur given a certain initial morphology, regardless of the nature of the extrinsic factors. These patterns of change do not require deter-

ministic explanations and are largely due to the network of constraints within the initial system.

Coupled with the arguments in the preceding section on the extrapolation of functionally integrated systems, I believe the notion of historical factors and transformational analyses lends insight into the examination of patterns of ontogenetic scaling in general, and the case of the African apes in particular. Ontogenetic allometry is merely a most simplified, quantitative way of probing the inherited pattern of structure–function networks. Of course, we may (and should!) seek to uncover potential external factors related to the shifts in body size through time, but, as noted above, it would be inappropriate to directly link, for example, the relatively enlarged facial mass of the gorilla to a change in diet. The ecological factors related to body size shifts are diverse (e.g., Bonner, 1968; Stanley, 1973; Gould, 1977), yet the morphological transition in such cases (apart from other influences) would be similar. The combination of both types of historical analyses (equilibrium and transformational) with analysis of fabricational and functional factors is required for a complete study of form, of course (Lauder, 1981).

Functional Cranial Analysis

Building on arguments advanced by Van der Klauuw (1945, 1948–52), Moss and colleagues have developed the approach to craniology known as "functional cranial analysis" (Moss, 1973, and references therein). This approach stresses that different functions (e.g., respiration, mastication) are carried out within the skull and associated elements, and that the form of a given skeletal unit is largely determined by the entire functional matrix, which loosely represents the soft tissues and functioning spaces required to complete a given function. A central tenet of functional cranial analysis is that an examination of skeletal form must consider the potential independence (in development and evolution) of the various functional cranial components. A second school based on Van der Klauuw's work—that of Dullemeijer and colleagues (see Dullemeijer and Barel, 1977, and references therein)—stresses the mutual structural interdependence of functional cranial components.

Ontogenetic allometry and scaling have some relevance and application to the study of functional cranial analysis, although this is limited by the fact that allometric studies (as the present one on the African apes) are usually carried out on skeletal parts, and thus we can at best gain a glimpse of the entire functional cranial component (Moyers and Bookstein, 1979). This criticism is ameliorated by the footnote that paleontologists have no choice in these matters. Nevertheless, given proper selection of variables, allometric investigations of skull form can elucidate functional relationships. Qualitative (e.g., Biegert, 1963) and quantitative (e.g., Freedman, 1962) allometric approaches have provided additional verification of Van der Klauuw's claim of potential independence of neurocranial and splanchnocranial components. The actual numerical values of the coefficients of growth and size allometry can also reveal information of functional import; relative growth and size increase of the brain provide an excellent example (Gould, 1975*a*).

Shifts or marked slope differences in growth regressions (i.e., clear *departures* from ontogenetic scaling) also suggest and require functional interpretations. The ontogenetic criterion of subtraction discussed above (p. 185) particularly can be used to gain a quantitative demonstration that a functional shift has required changes in a functional matrix and its associated "skeletal unit" (Moss, 1973). This is determined by changes in the growth allometries. In the case of skull form in the African apes, I (Shea, 1983*b*) demonstrated that successive upward shifts or transpositions in growth allometries (see Fig. 8) characterized those structures that grow predominantly in early ontogeny (such as brain and eyes). Even if the resulting interspecific trend is strongly negatively allometric, any *absolute* increase in the size of such structures must be accomplished during early growth. These timing differences in the growth of the various functional regions of the skull provide a simple but direct basis for rejecting any expectation of total and rigid ontogenetic scaling, or recapitulation, of skull form in even the most closely related species (Rensch, 1948; von Bertalanffy and Pirozynski, 1952; Shea, 1982).

A final example of functional inferences based on quantitative allometric analyses of skeletal form involves the pervasive differences in skull shape between *Pongo* and *Pan*/*Gorilla*. Allometric patterns seem to exhibit some clear divergences here (Krogman, 1931*a–c;* Giles, 1956; Shea, 1982), and these presumably relate to important, though at present not well understood, differences in the respective functional matrices of various regions. Biegert's (1957, 1963) suggestion that specialization of the laryngeal apparatus in orangutans has resulted in numerous correlated changes in the bony elements of the skull (such as palate, orbits, basicranium) requires additional investigation. Comparative examination of skeletal growth allometry may contribute to such questions, but qualitative and experimental techniques are obviously also required.

The central lesson of functional cranial analysis for studies of ontogenetic allometry and scaling is that data on skeletal form must be tanslated into more meaningful information concerning the entire relevant functional matrix.

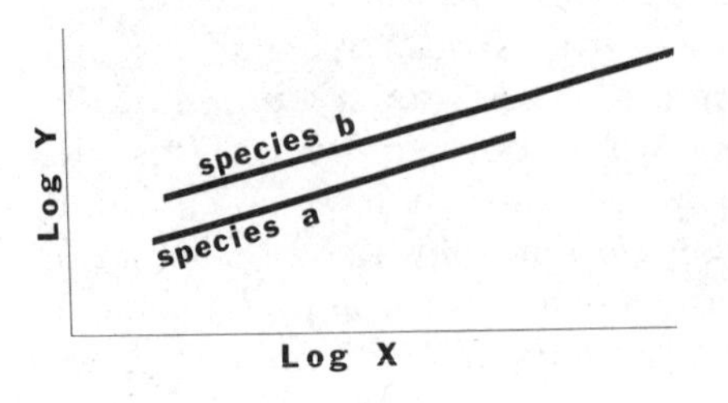

Fig. 8. An illustration of a bivariate shift or transposition between the growth allometries of two species, a and b. This pattern characterizes growth comparisons of the cranial vault and orbit in African apes, reflecting the early ontogenetic growth of the brain and eyes. See text for additional discussion.

Heterochrony

Modern usage of the term heterochrony refers to the displacement in time in descendants of the development of a given feature seen in ancestors (de Beer, 1958; Gould, 1977). Gould's (1977) recent synthesis, *Ontogeny and Phylogeny,* dealt with the analysis of morphology in terms of heterochrony, and related this theme to current developments in evolutionary biology and ecology. Alberch *et al.* (1979) have refined Gould's terminology (based on DeBeer and many earlier workers), and attempted to describe morphological transformations in terms of about a dozen basic parameters controlling ontogenetic development. Detailed application to the salamander genus *Bolitoglossa* can be found in Alberch (1980, 1981) and Alberch and Alberch (1981). General discussion of these and related issues is presented in Bonner (1982). I have analyzed the cranial and postcranial morphological differences among the African apes in terms of allometry and heterochrony in detail (Shea, 1982, 1983*c,d,* 1984). The present discussion is primarily based on this work.

An analysis of morphology in terms of heterochrony requires data on size, shape, and age (Gould, 1977). Ontogenetic allometries provide the requisite information on the first two categories, while reliable data on length of ontogeny in time, time of maturation, and so forth, must be obtained from other sources. Table 1 outlines the major heterochronic processes and their morphological results in terms of shifts in the controlling parameters of development (Alberch *et al.,* 1979). An important distinction between hypermorphosis and hypomorphosis on the one hand, and neoteny and acceleration on the other, relates to the difference between ontogenetic allometry and ontogenetic scaling. That is, although we use the data of ontogenetic allometry to demonstrate and test for the various processes and their morphological results of paedomorphosis and peramorphosis, it is only in the case of (rate or

Table 1. A Categorization of Heterochronic Processes in Relation to the Morphological Results Observed, the Shifts in Size, Shape, and Age Involved, and the Presence of Ontogenetic Scaling[a]

Heterochronic process	Morphological result	Size, shape, and age change	Ontogenetic scaling or dissociation
Time hypermorphosis	Peramorphosis	Age and size increase	Ontogenetic scaling
Time hypomorphosis	Paedomorphosis	Age and size decrease	Ontogenetic scaling
Rate hypermorphosis	Peramorphosis	Size/shape increase	Ontogenetic scaling
Rate hypomorphosis	Paedomorphosis	Size/shape decrease	Ontogenetic scaling
Acceleration	Peramorphosis	Shape increase	Dissociation
Neoteny	Paedomorphosis	Shape decrease	Dissociation

[a]Modified from Alberch *et al.* (1979). Dissociation or "uncoupling" of ancestral ontogenetic allometries occurs with acceleration and neoteny, even though morphological results may be quite similar to cases where ontogenetic scaling holds. The notation "size/shape" is here taken to mean an allometric (i.e., nonisometric) pattern; "shape" refers to nonallometric change in shape with no concomitant size change.

time) hypomorphosis and hypermorphosis that ontogenetic *scaling* of proportions is observed. With neoteny and acceleration, significant departures from ontogenetic scaling result, because ancestral patterns of size and shape change (i.e., ontogenetic allometries) have been dissociated, or uncoupled (Gould, 1977; Alberch *et al.*, 1979; Alberch and Alberch, 1981; Shea, 1983*c,d*). Nevertheless, the relevance of ontogenetic scaling to studies of heterochrony is clear. It provides a quantitative test of the criterion of fit for a given prediction of hypomorphosis or hypermorphosis, and it yields a baseline against which to gauge dissociations and deviations from ontogenetic scaling (again, in quantitative terms) in an assessment of neoteny and acceleration. A finding of paedomorphosis or peramorphosis via neoteny or acceleration implies specific selection for a given morphological configuration, whereas similar morphological results produced via rate or time hypomorphosis and hypermorphosis may simply be the correlated result (via ontogenetic scaling) of selection for either shifts in the length of ontogeny in time or the rate of body weight growth.

Results of my study of morphology and heterochrony in the African apes may be briefly summarized as follows (from Shea, 1983*c,d,* 1984). Data on developmental timing suggest no firmly established difference in length of the growth period in time among *Pan paniscus, Pan troglodytes,* and *Gorilla gorilla.* The significant component of ontogenetic scaling in the cranium and postcranium of these apes indicates that *rate* hypomorphosis or hypermorphosis, depending on whether small size is considered derived or primitive (with three taxa, the trend need not be unidirectional, of course), is responsible for many of the morphological differences among the species. Paedomorphosis or peramorphosis result in such cases. Figure 9 summarizes the morphological differences among the African great apes in terms of size, shape, and age. Of course, some departures from ontogenetic scaling are observed in the comparison of growth patterns in these pongids (Shea, 1981, 1982, 1983*b*). These are suggestive of selection for specific proportion altera-

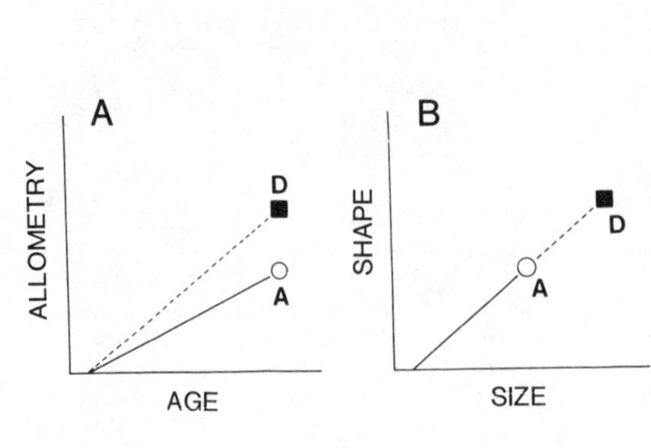

Fig. 9. Schematic representation of ontogenetic scaling in ancestral (A) and descendant (D) forms in the analysis of heterochrony. In part A the x axis represents age and the y axis represents differential travel along a given ontogenetic trajectory for two (or more) variables (i.e., ontogenetic scaling). In part B age (time) is dropped from the comparison and the axes represent different dimensions, as in the usual allometric plots. In this example, the descendant form is produced by peramorphosis via rate hypermorphosis. The reverse case (D to A) would be paedomorphosis via rate hypomorphosis.

tions, perhaps related to the size change in a biomechanical sense (Gould, 1966; Shea, 1981), or perhaps in relation to other aspects of an animal's adaptations. The concordance in many cases with the predictions of ontogenetic scaling implies that the preponderant amount of morphological divergence among these apes can be explained as the correlated result of a shift in one of the underlying parameters controlling development, however. In other words, the shifts in overall rates of body weight growth (and ultimate adult size attained) result in differential "travel" along the shared trajectory of size/shape change; to paraphrase Sinnott (1936) once more, the adults differ primarily in their size at maturity and in the end point along the line at which their growth stops. Recent work in progress on the question of ontogenetic scaling in the African apes has suggested the possibility that different regions of the body, while often demonstrating ontogenetic scaling within a given region (e.g., head, trunk, hindlimbs), may not actually be ontogenetically scaled relative to one another (Shea, 1983*c*).

The ecological bases for the shifts in growth rate patterns are unknown in detail, of course. It seems likely, however, that selection for size differences related to niche divergence may have been partially responsible (Shea, 1983*b*). Clutton-Brock and Harvey (1979*b*) suggest that the larger size in gorillas evolved in response to a specialization for terrestrial folivory.

Gould (1977) and Zuckerkandl (1976) have suggested that heterochronic analyses might elucidate the link between morphological and genetic changes in terms of basic regulatory phenomena. The increase in rates of body weight growth seen in the African apes may be the result of such a simple regulatory shift at the genetic level. The regularity of the shape transformation going from *P. paniscus* to *P. troglodytes* to *G. gorilla* illustrates the diversity of form that can be produced via such regulatory phenomena. The outline of Alberch *et al.* (1979) provides a rigorous basis for the study of morphology in terms of these regulatory shifts. Additional emphasis in genetic studies of hominoid primates should be directed toward examining the genetic distances between species that differ primarily in allometric relations and size, and which are seemingly very closely related to one another (Delsol, 1977), such as the chimpanzees and gorillas.

Genetics

A number of investigators have recently criticized aspects of morphometric and allometric research, arguing that "hypotheses about correlated response to selection or correlated patterns in morphogenesis imply specific relationships about genetic parameters which may not be accurately described by phenotypic statistics alone" (Atchley *et al.*, 1981, p. 1038). Details of this argument are presented in various papers (Lande, 1979; Atchley and Rutledge, 1980; Atchley *et al.*, 1981; Cheverud, 1982), but basically center about the claim that patterns of phenotypic change may not be a reliable guide to underlying patterns of genetic relationships (Lande, 1979, and this volume), which are what is relevant if we are attempting to reconstruct selective scenarios (e.g., selection for increased body size and extension of ancestral growth allometries). In the case of allometric studies, Lande (1979) and Cheverud (1982) have noted that the bivariate slope k will only provide a reliable estimate of shape changes produced as a correlated response to selection for size increase if patterns of genetic and phenotypic covariance are similar (i.e., if phenotypic and evolutionary bivariate allometry slopes resemble one another).

These studies and arguments are obviously of direct relevance to the present analysis of ontogenetic scaling in general, and the observed patterns of phenotypic change in the African apes in particular. It is worth pointing out, however, that these analyses of our ability to reconstruct or decipher selective events based on phenotypic data alone apply to virtually the entire field of morphometrics, save the very few cases where the appropriate genetic information is available (Atchley *et al.*, 1981). Furthermore, Lande (1979, p. 405) has argued that phenotypic, genetic, and environmental covariance patterns are usually *markedly similar,* especially for character sets such as a group of linear skeletal measurements [see references cited in Lande (1979)]. In addition, the empirical arguments of Lande (1979) and Cheverud (1982) apply to *static adult* patterns and not true ontogenetic trends, as in the present case. Assumptions concerning phenotypic and genetic relationships may be the same in both cases, but in the ontogenetic example we are dealing with phenotypic change directly interpretable in terms of true growth (Cock, 1966). Extrapolation of static adult scatters of variation makes little biological sense (Shea, 1983*a*). Finally, in the present case we are not trying to *infer* shapes that would be observed if selection acted on an ancestral trajectory to increase overall size [although we might in fact do this; see Atchley *et al.* (1981)]. Rather, we are comparing observed growth patterns of species that are known to be very closely related and differentiated in size. Interpretation of the basis of between-species differences simply makes recourse to the observed within-species ontogenetic trends. I would simply argue that, whatever the genetic and environmental factors producing the observed (primitive) pattern of phenotypic growth, these have in large part merely been extended to greater sizes in the bigger and different species. Nevertheless, the genetic arguments certainly caution against an oversimplified reconstruction based on specific selection for increased body size, with given shape changes merely being correlated consequences. As Simpson (1953, p. 277) has stressed (and as I elaborated in the sections on Adaptive Growth and on Historical Factors), it is the entire growth pattern, with its myriad results and interrelationships, that is presumably the target of selection, and not simply final adult body size.

Cheverud (1982) notes that Gould's (1977) central contention that shifts in the length of ontogeny in time, or in the rate of body weight growth per time, may occur without changes in underlying allometric relationships *as-*

sumes no genetic correlation between length and slope of ontogenetic vectors. Cheverud offers some evidence to suggest such a correlation, and he argues:

> Whether increases in ontogenetic vector length can occur without changes in slope by selection on adult size is highly problematical. It would require rather specific patterns of genetic variance and covariance between adult size and size at earlier ages and between adult size and age-specific trait values of the other character under consideration.

I (Shea, 1982) noted many cases of very slight (but often statistically significant) differences among bivariate slopes of *P. paniscus, P. troglodytes,* and *G. gorilla,* even when the species clearly seemed to closely follow a common ontogenetic trend. Perhaps the genetic correlations noted by Cheverud play a role here, although it would appear to be impossible in this case to distinguish such effects from (inferred) selection for changes in phenotypic slopes.

In conclusion, investigations of the genetic components of size and shape change suggest that greater caution needs to be exercized (and assumptions clearly stated) in the interpretation of phenotypic patterns of ontogenetic scaling in terms of underlying genetic relationships and the reconstruction of evolutionary events and selective scenarios. Additional studies of quantitative genetics will further enhance our understanding of the evolution of patterns of size and shape change. Nevertheless, the similarity in patterns of phenotypic growth seen in the cranium and postcranium of the African apes, and their extension to larger size ranges, most probably reflects underlying patterns of genetic allometry, and thus selection for increased rates of body weight growth, larger overall size, and extension of many of the ancestral ontogenetic allometries.

Taxonomy

Shape differences are usually considered of greater importance than are size differences in taxonomic studies (Giles, 1956; Corruccini, 1972). Allometry elucidates the size-related or size-required aspects of many shape differences, however, as Gould (1966, p. 609) stressed:

> New shapes produced by continuation of an allometric relationship into new size ranges are not independent taxonomic criteria. Size increase itself may be a mark of taxonomic distinction, but failure to recognize the allometric consequences interdependent with such increase has often led to unwarranted taxon distinctions between organisms differing in only one characteristic whose complex effects were not appreciated.

Gould (1966) also cited several cases where a detailed allometric investigation resulted in taxonomic revision. For example, Reeve (1941) demonstrated that three subspecies of anteaters (*Tamandua tetradactyla*), which were originally defined on the basis of shape differences, actually lie on the same allometric regression (i.e., they are ontogenetically scaled). Dodson (1975*a*, p. 421) noted that his studies on living and extinct reptiles were designed "to test whether application of principles of ontogenetic allometry can demonstrate that 12 species of lambeosaurine hadrosaurs (duck-billed dinosaurs) of variable size and shape from a single formation may actually constitute a single ontogenetic series." Dodson's analysis utilized ontogenetic allometry and scaling to reduce the previous classification (three genera, 12 species) to but two genera and three species. In addition, *departures* from ontogenetic scaling suggest divergent growth patterns, and thus can be of value in distinguishing between closely related species. For instance, Dodson's (1975*b*) application also clearly revealed several different species of duck-billed dinosaurs (rather than a single sexually dimorphic one) in the Campanian deposits of Alberta. A similar approach to morphological diversity and taxonomy in fossil reptiles has recently been utilized in studies of *Diademodon* (Grine *et al.*, 1978) and *Aulacephalodon* (Tollman *et al.*, 1980). In a neontological example, I (Shea, 1984) have utilized ontogenetic allometries to demonstrate that the pygmy chimpanzee (*Pan paniscus*) is indeed a species distinct from *Pan troglodytes.*

Most recent authorities have placed the gorilla in a separate genus, distinct from the chimpanzees. Notable exceptions include Mayr (1950, 1963), Simpson (1961), Tuttle (1969), and Szalay and Delson (1979). There is no question that my morphological studies, by revealing the similarity of growth patterns and the importance of size differences among these pongids, have strengthened the position of those who wish to argue for the inclusion of the gorillas and chimpanzees in a single genus. But, unfortunately, there can be no simple taxonomic solution given a finding of general ontogenetic scaling. First, there will always be *some* departures from ontogenetic scaling (whether we have measured these or not), and there is of course no consistent, quantitative measure matching taxon level to degree of similarity. The enumeration of features that are ontogenetically scaled, those that are biomechanically scaled, and those that are unrelated to the size differences *per se* poses difficulties for as clean a solution as we would often prefer (Shea, 1981, p. 196). Also, we must take into consideration information other than morphological data, and diet, ecology, and social structure of chimpanzees and gorillas differ substantially, and thus many feel that these differences warrant generic distinction (e.g., Dixson, 1981).

In this and previous work, I have followed the taxonomic suggestion of Oxnard (1978) regarding the African apes, while simultaneously probing the morphological bases for this classification. He writes (Oxnard, 1978, p. 209):

> An example of controversy in primate systematics is . . . the difficulties surrounding the generic names *Gorilla* and *Pan.* Here also there is probably no doubt in the minds of most workers that the gorilla and chimpanzee are far closer than is implied by the use of separate generic designations; and a number of workers have, on reasonable evidence, suggested that they be grouped as a single genus *Pan.* This usage is starting to catch on. But the consensus about the basic information is probably good enough that many believe there is no real need to make the nomenclatorial change; most workers will undoubtedly try to reduce confusion in liter-

ature comparisons by retaining the older terms while still accepting the newer relationship.

My primary interests are in elucidating these relationships in terms of evolutionary morphology, and the similarities in growth allometries among the African apes provide important new information, which probably suggests congeneric status. The key here is how much emphasis we wish to place on the *differences* in pattern (such as departures from ontogenetic scaling), however. Thus, I reserve final judgement until a detailed comparison with the growth allometries of the orangutan (*Pongo pygmaeus*) is completed. The central point of relevance to this discussion, however, is that the use of ontogenetic allometry and scaling in morphological investigations produces important information that should be carefully examined by the taxonomist.

A more general point about the usefulness of developmental information in systematic studies has been made by Alberch and Alberch (1981). They argue that systematic studies must carefully examine ontogenetic information in addition to static adult data. Several workers have commented on the potential benefits of basing systematic analyses on the distribution of symplesiomorphic and synapomorphic *patterns* of ontogenetic development (Rensberger, 1981). Alberch and Alberch (1981) note that this approach will help control for placing too much emphasis on the correlated consequences of a single developmental alteration (such as size increase acting to extend ancestral allometries); they also point out that dissociations, while perhaps not yielding great morphological differences in many cases, may require significant genetic change. The argument, in essence, is for classifications built on dynamic processes rather than static results (Sinnott, 1937; Gould, 1977; Alberch and Alberch, 1981; Bonner, 1968, 1982). Only detailed studies of ontogeny can provide the necessary data.

Ontogenetic Allometry and Australopith Morphology

The previous discussions of ontogenetic allometry and scaling of the cranium may be extended to the much-discussed differences in skull form between gracile and robust australopiths. Grine (1981) has presented a detailed review of the relevant literature on australopith studies in general and allometric arguments in particular. Almost all quantitative investigation of Pilbeam and Gould's (1974) contention that the primary morphological differences between gracile and robust australopiths are allometric has been focused on the teeth. Pilbeam and Gould (1974) themselves discussed cranial anatomy in only the most general and qualitative fashion, in terms of, e.g., cresting or robusticity. No direct attempt has yet been made to apply tests based on ontogenetic allometry or scaling to the issue of variation in skull morphology in early hominids. Here I present results of a very brief and exploratory application based, like earlier allometric analyses of the australopiths, on a comparison and analogy with the African pongids.

I have plotted a series of measures of facial dimensions taken on the Taung infant, "Mrs. Ples" (STS 5), and "Zinj" (OH 5) in Fig. 10. These measurements were made on casts at the American Museum of Natural History in New York, or taken from Tobias (1967). The trajectory of Taung to STS 5 is considered to roughly approximate the ontogenetic vector in gracile australopiths; a hypothesis of ontogenetic scaling suggests that OH 5 should lie on a linear extension of this vector. Since the adult STS 5 lies on both lines, no separate test for position differences is required. Tobias' (1973) suggestion

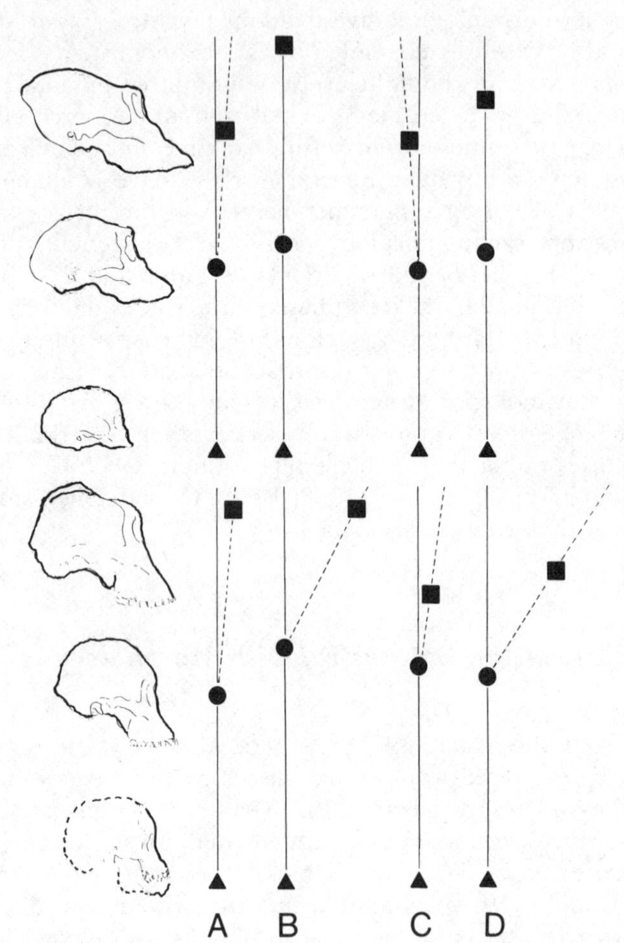

Fig. 10. A parallel test of ontogenetic scaling in African apes and early hominids. Apes: (▲) infant chimpanzees, (●) adult chimpanzees, and (■) adult gorillas; hominids: (▲) infant *A. africanus* (Taung), (●) adult *A. africanus* (STS 5), (■) and adult *A. boisei* (OH 5). Bivariate comparisons: (A) bimaxillary breadth/superior facial height; (B) superior facial height/palate length; (C) palate breadth palate length; (D) bimaxillary breadth/palate length. The strong divergence of OH 5 from the expectations of ontogenetic scaling in parts B and D reflects a reorganization of facial proportions. See text for discussion.

that the Taung subadult is actually a robust australopith would slightly alter the procedure, but a hypothesis of ontogenetic scaling would still require a straight-line fit to be confirmed. I am aware of the many potential difficulties inherent in this analysis; indeed, the potential errors introduced by using casts of possibly distorted material (e.g., STS 5) are dwarfed by the problems of reconstructing a growth pattern based on two individual points. Nonetheless, ontogenetic investigations of fossil material will never be very complete, and the present endeavor is viewed as a brief methodological outline of the procedures to be utilized if we ever are blessed with more complete subadult and adult crania. Moreover, I wish to make the simple point, in continuing to emphasize the value of ontogenetic data in this chapter, that one of the most basic and straightforward allometric tests has been ignored in the long debate over australopith morphology.

For comparison with the hominid faces, I have used identical dimensions for gorillas and chimpanzees from Shea (1982). The three ape points are mean values for *Pan troglodytes* subadults (deciduous dentition plus M^1), *Pan troglodytes* adults, and *Gorilla gorilla* adults (sexes lumped in all cases). This provides a three-point ontogenetic and interspecific vector analogous to that composed of Taung–STS 5–OH 5. In Fig. 10, the allometric trends of African apes and australopiths are compared. The original axes have been removed and the trends are placed side by side in vertical position for ease of comparison. A great divergence of the dashed line from the extended solid line is strongly suggestive of different growth trajectories and, presumably, selection for facial proportions altered from the expectations of size increase acting via simple extension of ancestral growth patterns in the gracile australopiths.

Results clearly suggest departures from ontogenetic scaling in the hominids, but not in the African apes (Fig. 10). Robust australopiths have higher ratios of facial height or width to palate length than expected in gracile forms ontogenetically scaled to their sizes. This provides a quantitative confirmation that, having controlled for size differences, robust australopiths have relatively short (anteriorly to posteriorly), deep faces. This is contrary to the interspecific trend observed in a chimpanzee–gorilla comparison. Implications of such allometric analyses for reconstruction of changes in diet and skull morphology are explored more fully elsewhere (Shea, 1983*b*), but this brief comparison reveals a reorientation of facial structure in the robust australopiths that is suggestive of a dietary concentration on hard food items. This conclusion accords with the suggestions of many previous workers, of course (e.g., Robinson, 1972; DuBrul, 1977; Hylander, 1979; Johanson and White, 1979; Grine, 1981). It does not support Pilbeam and Gould's (1974) hypothesis that the morphological distinctions between gracile and robust australopiths are simple allometric consequences that provide no independent evidence of dietary divergence. Grine (1981, p. 221), citing White's (1977) morphological studies, concludes that "most, if not all, of the cranial and mandibular features characteristic of the 'robust' australopithecines [are] related to a craniofacial adaptation that maximizes the generation and distribution of vertical occlusal force." Analysis of gnathic morphology in subadult and adult australopiths reveals differences among subadults of the two species that parallel those observed between full adults (Grine, 1981). Furthermore, Grine's (1981) own studies of dental enamel suggest that the robust forms habitually or seasonally masticated harder food items than the gracile australopiths. DuBrul's (1977) qualitative analysis of the relevant morphology yields a similar conclusion, as does the consensus of the literature on dental allometry (Grine, 1981). This cursory look at facial morphology supports these conclusions and illustrates the value of ontogenetic allometry and scaling as a criterion of subtraction to provide insight into problems in functional and evolutionary morphology of the early hominids. Thus, the analogy between the australopiths and the African apes (as closely related species differing in body size) is enlightening after all, but not in the fashion proposed by Pilbeam and Gould (1974).

Conclusions

In this chapter, I have analyzed some ways in which ontogenetic allometry and scaling may elucidate the growth and form of the skull in African apes and other animals. I suggest that use of ontogenetic allometry and scaling within closely related groups provides the "criterion of subtraction" that is biologically most appropriate. Application of this approach to the African apes indicates that the primary differences in skull form among pygmy chimpanzees, common chimpanzees, and gorillas relate to variations in ultimate body size attained and differential extrapolation of a common growth allometry. A clear shape factor related to diet, as opposed to growth or size, does emerge in the study of the teeth, however. A brief consideration of australopith facial morphology using ontogenetic scaling suggests a situation quite different from that observed in the African great apes. In the early hominids, shape changes of the skull between gracile and robust forms clearly indicate significant dietary divergence. The robust australopiths have relatively short, deep faces compared to the expectations of ontogenetic scaling.

I have also tried to stress other benefits of an ontogenetic approach. Studies of growth need not be merely descriptivist expositions of the pathways utilized by natural selection to produce interspecific differences in form, as noted by Thompson (1917) and Huxley (1932) many years ago. Indeed, such studies can increase our understanding of adaptation and evolutionary transformations in their own right. Investigations of ontogeny can elucidate historical factors in evolutionary transformations and provide valuable information on the functional integration of body parts during development and evolution. Frazzetta's (1975) notion of the preadaptive potential for the extrapolation of functionally integrated systems is especially appropriate to studies of ontogenetic scaling and size change in evolution. This provides a vital link between studies of allometry (including many of the old orthogenesis arguments), adaptive growth, and functional anatomy.

Studies of ontogeny and development promise to play an important role in some central issues in evolutionary biology in the near future. These include the relationship between morphological transformations and genetic regulatory mechanisms (Gould, 1977; Zuckerkandl, 1976), as well as the study of "constraints" in evolutionary change (Alberch, 1980; Gould, 1980; Lauder, 1981). In the edited volume from the Dahlem Conference on evolution and development, Bonner (1982) notes that renewed interest in development and ontogeny may produce a more complete integration of this area with the traditional evolutionary synthesis of population genetics and Darwinian biology. Gould's (1977) book *Ontogeny and Phylogeny* has made a major contribution in this regard. Further comparative analysis of patterns of ontogenetic development should provide new insights into morphological transformations and the evolution of size, shape, and age.

Acknowledgments

I would like to thank Drs. R. F. Kay, M. Cartmill, W. L. Hylander, E. Simons, and I. M. Tattersall for their support and helpful comments on aspects of this work. I also wish to acknowledge my indebtedness to S. J. Gould, who, through his conversations and writing, has clarified many important concepts in allometry for me. All ideas expressed here are my own responsibility, however.

The assistance of Lester Barton at the Powell-Cotton Museum in Birchington, England, and Dr. Thys van den Audenaerde at the Central African Museum in Tervuren, Belgium, greatly facilitated my research. Financial support was provided by the National Science Foundation, Duke University Graduate School, Sigma Xi, and the Southern Regional Education Board.

I would also like to express my gratitude to W. L. Jungers, for inviting me to participate in the Atlanta IPS symposium and to contribute to this volume. Thanks are also due to Nicholas Amorosi for help with the figures and Pat Bramwell for typing the manuscript. This chapter is publication #6 from the Richard Lounsbery Laboratory of Physical Anthropology, American Museum of Natural History, New York. The financial support of a Lounsbery Postdoctoral Research Fellowship is gratefully acknowledged.

References

Alberch, P. 1980. Ontogenesis and morphological diversification. *Am. Zool.* **20**:653–667.

Alberch, P. 1981. Convergence and parallelism in foot morphology in the neotropical salamander genus *Bolitoglossa*. I. Function. *Evolution* **35**:84–100.

Alberch, P., and Alberch, J. 1981. Heterochronic mechanisms of morphological diversification and evolutionary change in the neotropical salamander, *Bolitoglossa occidentalis*. *J. Morphol.* **161**:249–264.

Alberch, P., Gould, S. J., Oster, G. F., and Wake, D. B. 1979. Size and shape in ontogeny and phylogeny. *Paleobiology* **5**:296–317.

Albrecht, G. H. 1981. Double regression adjustment for within-group versus between-group allometric scaling in morphometrics. *Am. J. Phys. Anthropol.* **54**:196 (abstract).

Atchley, W., and Rutledge, J. 1980. Genetic components of size and shape. I. Dynamics of components of phenotypic variability and covariability during ontogeny in the laboratory rat. *Evolution* **34**:1161–1173.

Atchley, W., Rutledge, J., and Cowley, D. 1981. Genetic components of size and shape. II. Multivariate covariance patterns in the rat and mouse skull. *Evolution* **35**:1037–1055.

Biegert, J. 1957. Der Formwandel des Primatenschadels und seine Beziehungen zur ontogenetischen Enwicklung und den phylogenetischen Spezialisation der Kopforgane. *Morphol. Jahrb.* **98**:77–199.

Biegert, J. 1963. The evaluation of characteristics of the skull, hands and feet for primate taxonomy, in: *Classification and Human Evolution* (S. L. Washburn, ed.), pp. 116–145, Aldine, Chicago.

Bock, W. J. 1980. The definition and recognition of biological adaptation. *Am. Zool.* **20**:218–227.

Bonner, J. T. 1968. Size change in development and evolution. *J. Paleontol.* **42**(5):1–15.

Bonner, J. T. (ed.). 1982. *Evolution and Development.* Springer-Verlag, New York.

Cheverud, J. M. 1981. Epiphyseal union and dental eruption in *Macaca mulatta*. *Am. J. Phys. Anthropol.* **56**:157–168.

Cheverud, J. M. 1982. Relationships among ontogenetic, static, and evolutionary allometry. *Am. J. Phys. Anthropol.* **58**:1–11.

Clutton-Brock, T. H., and Harvey, P. H. 1979*a*. Comparison and adaptation. *Proc. R. Soc. Lond. B* **205**:547–565.

Clutton-Brock, T. H., and Harvey, P. H. 1979*b*. Home range size, population density, and phylogeny in primates, in: *Primate Ecology and Human Origins* (I. S. Bernstein and E. O. Smith, eds.), pp. 201–214, Garland Press, New York.

Cock, A. G. 1966. Genetical aspects of metrical growth and form in animals. *Q. Rev. Biol.* **41**:131–190

Corruccini, R. S. 1972. Allometry correction in taximetrics. *Syst. Zool.* **21**:375–383.

Corruccini, R. S. 1978. Morphometric analyses: Uses and abuses. *Yearb. Phys. Anthropol.* **21**:134–150.

Davenport, C. B. 1934. Critiques of curves of growth and relative growth. *Cold Spring Harbor Symp. Quant. Biol.* **2**:203–208.

De Beer, G. R. 1958. *Embryos and Ancestors,* Clarendon Press, Oxford.

Delsol, M. 1977. Embryogenesis, morphogenesis, genetics and evolution, in: *Major Patterns of Vertebrate Evolution* (M. K. Hecht, P. C. Goody, and B. M. Hecht, eds.), pp. 119–138, Plenum Press, New York.

Dixson, A. F. 1981. *The Natural History of the Gorilla,* Columbia University Press, New York.

Dodson, P. 1975*a*. Functional and ecological significance of relative growth in *Alligator*. *J. Zool. Lond.* **175**:315–355.

Dodson, P. 1975*b*. Relative growth in two sympatric species of *Sceloporus*. *Am. Midl. Nat.* **94**(2):421–450.

Dodson, P. 1975*c*. Taxonomic implications of relative growth in lambeosaurine hadrosaurs. *Syst. Zool.* **24**:37–54.

Dodson, P. 1976. Quantitative aspects of relative growth and sexual dimorphism in *Protoceratops*. *J. Paleontol.* **50**:929–940.

DuBrul, E. L. 1977. Early hominid feeding mechanisms. *Am. J. Phys. Anthropol.* **47**:305–320.

Dullemeijer, P., and Barel, C. D. N. 1977. Functional morphology and evolution, in: *Major Patterns of Vertebrate Evoluation* (M. K. Hecht, P. C. Goody, and B. M. Hecht, eds.), pp. 83–117, Plenum Press, New York.

Frazzetta, T. H. 1975. *Complex Adaptations in Evolving Populations,* Sinauer, Sunderland, Massachusetts.

Freedman, L. 1962. Growth of muzzle length relative to calvaria length in *Papio. Growth* **26:**117–128.

Gelvin, B. R. 1981. Size correcting properties of some anthopometric data transformation techniques. *Am. J. Phys. Anthropol.* **54:**224 (abstract).

Giles, E. 1956. Cranial allometry in the great apes. *Hum. Biol.* **28:**43–58.

Gould, S. J. 1966. Allometry and size in ontogeny and phylogeny. *Biol. Rev.* **41:**587–640.

Gould, S. J. 1968. Ontogeny and the explanation of form: An allometric analysis, in: *Paleobiological Aspects of Growth and Development, A Symposium* (D. B. Macurda, ed.). *Paleontol. Soc. Mem.* **2:**81–98.

Gould, S. J. 1969. An evolutionary microcosm: Pleistocene and Recent history of the land snail *P.* (*Poecilozonites*) in Bermuda. *Bull. Mus. Comp. Zool.* **138:**407–532.

Gould, S. J. 1971*a*. D'Arcy Thompson and the science of form. *New Lit. Hist.* **2:**229–258.

Gould, S. J. 1971*b*. Geometric scaling in allometric growth: A contribution to the problem of scaling in the evolution of size. *Am. Nat.* **105:**113–136.

Gould, S. J. 1974. The evolutionary significance of "bizarre" structures: Antler size and skull size in the "Irish Elk," *Megaloceros giganteus. Evolution* **28:**191–220.

Gould, S. J. 1975*a*. Allometry in primates, with emphasis on scaling and the evolution of the brain, in: *Approaches to Primate Paleobiology* (Contrib. Primatol., Vol. 5, F. Szalay, ed.), pp. 244–292, S. Karger, Basel.

Gould, S. J. 1975*b*. On the scaling of tooth size in mammals. *Am. Zool.* **15:**351–362.

Gould, S. J. 1977. *Ontogeny and Phylogeny,* Harvard University Press, Cambridge.

Gould, S. J. 1980. The evolutionary biology of constraint. *Daedalus* **109**(2)**:**39–52.

Gould, S. J., and Lewontin, R. C. 1979. The spandrels of San Marco and the Panglossian paradigm: A critique of the adaptationist programme. *Proc. Roy. Soc. Lond. B* **205:**581–598.

Gould, S. J., and Vrba, E. S. 1982. Exaptation—A missing term in the science of form. *Paleobiology* **8**(1)**:**4–15.

Goss, R. J. 1964. *Adaptive Growth,* Academic Press, New York.

Grine, F. E. 1981. Trophic differences between 'gracile' and 'robust' australopithecines: A scanning electron microscope analysis of occlusal events. *S. Afr. J. Sci.* **77:**203–230.

Grine, F. E., Hahn, B. D., and Gow, C. E. 1978. Aspects of relative growth and variability in *Diademodon* (Reptilia: Therapsida). *S. Afr. J. Sci.* **74:**50–58.

Huxley, J. S. 1932. *Problems of Relative Growth,* Methuen, London.

Hylander, W. L. 1979. The functional significance of primate mandibular form. *J. Morphol.* **160:**223–240.

Jerison, H. J. 1973. *Evolution of the Brain and Intelligence,* Academic Press, New York.

Johanson, D. C., and White, T. D. 1979. A systematic assessment of early African hominids. *Science* **202:**321–330.

Jungers, W. L., and Fleagle, J. G. 1980. Postnatal growth allometry of the extremities in *Cebus albifrons* and *Cebus apella:* A longitudinal and comparative study. *Am. J. Phys. Anthropol.* **53:**471–478.

Katz, M. J. 1980. Allometry formula: A cellular model. *Growth* **44:**89–96.

Kowalski, C. J., and Guire, K. E. 1974. Longitudinal data analysis. *Growth* **38:**131–169.

Krogman, W. M. 1931*a*. Studies in growth changes in the skull and face of anthropoids. III. Growth changes in the skull and face of the gorilla. *Am. J. Anat.* **47:**89–115.

Krogman, W. M. 1931*b*. Studies in growth changes in the skull and face of anthropoids. IV. Growth changes in the skull and face of the chimpanzee. *Am. J. Anat.* **47:**325–342.

Krogman, W. M. 1931*c*. Studies in growth changes in the skull and face of anthropoids. V. Growth changes in the skull and face of orang-utan. *Am. J. Anat.* **47:**343–365.

Laird, A. K. 1965. Dynamics of relative growth. *Growth* **29:**249–263.

Laird, A. K., Tyler, S. A., and Barton, A. D. 1965. Dynamics of normal growth. *Growth* **29:**233–248.

Lande, R. 1979. Quantitative genetic analysis of multivariate evolution, applied to brain:body allometry. *Evolution* **33:**402–416.

Larson, S. G. 1978. Scaling of organ weights in *Macaca arctoides. Am. J. Phys. Anthropol.* **49:**95–102.

Lauder, G. V. 1981. Form and function: Structural analysis in evolutionary morphology. *Paleobiology* **7**(4)**:**430–442.

Lerner, I. M., and Gunns, C. A. 1938. Temperature and relative growth of chick embryo leg bones. *Growth* **2:**261–266.

Lewontin, R. C. 1978. Adaptation. *Sci. Am.* **239**(3)**:**156–169.

Lundelius, E. L., Jr. 1957. Skeletal adaptations in two species of *Sceleporus. Evolution* **11:**65–83.

Lumer, H. 1940. Evolutionary allometry in the skeleton of the domesticated dog. *Am. Nat.* **74:**439–467.

Mayr, E. 1950. Taxonomic categories in fossil hominids. *Cold Spring Harbor Symp. Quant. Biol.* **15:**109–118.

Mayr, E. 1961. Cause and effect in biology. *Science* **134:**1501–1506.

Mayr, E. 1963. *Animal Species and Evolution,* Harvard University Press, Cambridge.

Medawar, P. B. 1945. Size, shape, and age, in: *Essays on "Growth and Form" Presented to D'Arcy Wentworth Thompson* (W. E. le Gros Clark and P. B. Medawar, eds.), pp. 157–187, Clarendon Press, Oxford.

Moore, W. J. 1981. Facial growth in primates with special reference to the Hominoidea. *Symp. Zool. Soc. Lond.* **46:**37–62.

Moore, W. J., and Lavelle, C. L. B. 1974. *Growth of the Facial Skeleton in the Hominoidea,* Academic Press, London.

Moss, M. L. 1973. A functional cranial analysis of primate craniofacial growth, in: *Symposium 4th International Congress Primatol.,* pp. 191–208. Vol. 3: Craniofacial Biology of Primates, (F. S. Szalay, ed.), Karger, Basel.

Moyers, R. E., and Bookstein, F. L. 1979. The inappropriateness of conventional cephalometrics. *Am. J. Orthodontics* **75**(6)**:**599–617.

Oxnard, C. E. 1978. One biologist's view of morphometrics. *Annu. Rev. Ecol. Syst.* **9:**219–241.

Pilbeam, D. R., and Gould, S. J. 1974. Size and scaling in human evolution. *Science* **186:**892–901.

Radinsky, L. 1984. Ontogeny and phylogeny in horse skull evolution. *Evolution* **38:**1–15.

Raup, D. M. 1972. Approaches to morphologic analysis, in: *Models in Paleobiology* (T. J. M. Schopf, ed.), pp. 28–44, Freeman, San Francisco.

Reeve, E. C. R. 1940. Relative growth in the snout of anteaters. *Proc. Zool. Soc. Lond* **110:**47–80.

Reeve, E. C. R. 1941. A statistical analysis of taxonomic differences within the genus *Tamandua* Gray (Xenarthra). *Proc. Zool. Soc. Lond.* **111:**279–302.

Reeve, E. C. R., and Huxley, J. S. 1945. Some problems in the study of allometric growth, in: *Essays on "Growth and Form" Presented to D'Arcy Wentworth Thompson* (W. E. le Gros Clark and P. B. Medawar, eds.), pp. 121–156, Clarendon Press, Oxford.

Reeve, E. C. R., and Murray, P. D. F. 1942. Evolution in the horse's skull. *Nature* **150:**402–403.

Rensberger, B. 1981. The evolution of evolution. *Mosaic* **12**(5)**:**14–23.

Rensch, B. 1948. Histological changes correlated with evolutionary changes in body size. *Evolution* **2:**218–230.

Rensch, B. 1959. *Evolution above the Species Level,* Columbia University Press, New York.

Robb, R. C. 1935. A study of mutation in evolution. Part 2: Ontogeny in the equine skull. *J. Genet.* **31:**47–52.

Robinson, J. T. 1972. *Early Hominid Posture and Locomotion,* University of Chicago Press, Chicago.

Shea, B. T. 1981. Relative growth of the limbs and trunk in the African apes. *Am. J. Phys. Anthropol.* **56:**179–202.

Shea, B. T. 1982. Growth and Size Allometry in the African Pongidae: Cranial and Postcranial Analyses, Ph.D. Thesis, Duke University, Durham, North Carolina.

Shea, B. T. 1983*a*. Phyletic size change and brain/body scaling: A consideration based on the African pongids and other primates. *Int. J. Primatol.* **4:**33–62.

Shea, B. T. 1983*b*. Size and diet in the evolution of African ape craniodental form. *Folia Primatol.* **40:**32–68.

Shea, B. T. 1983*c*. Paedomorphosis and neoteny in the pygmy chimpanzee. *Science* **222:**521–522.

Shea, B. T. 1983*d*. Allometry and heterochrony in the African apes. *Am. J. Phys. Anthropol.* **62:**275–289.

Shea, B. T. 1984. An allometric perspective on the morphological and evolutionary relationships between pygmy (*Pan paniscus*) and common (*Pan troglodytes*) chimpanzees, in: *The Pygmy Chimpanzee: Evolutionary Biology and Behavior* (R. L. Susman, ed.), pp. 89–130, Plenum Press, New York.

Simpson, G. G. 1944. *Tempo and Mode in Evolution,* Columbis University Press, New York.

Simpson, G. G. 1953. *The Major Features of Evolution,* Columbia University Press, New York.

Simpson, G. G. 1961. *Principles of Animal Taxonomy,* Columbia University Press, New York.

Simpson, G. G., Roe, A., and Lewontin, R. C. 1960. *Quantitative Zoology,* 2nd ed., Harcourt, Brace and World, New York.

Sinnott, E. 1936. A developmental analysis of inherited shape differences in cucurbit furits. *Am. Nat.* **70:**245–254.

Sinnott, E. W. 1937. Morphology as a dynamic science. *Science* **85:**61–65.

Smith, R. J. 1980. Rethinking allometry. *J. Theor. Biol.* **87:**97–111.

Stanley, S. M. 1973. An explanation for Cope's rule. *Evolution* **27:**1–35.

Susman, R. L., and Creel, N. 1979. Functional and morphological affinities of the subadult hand (OH 7) from Olduvai Gorge. *Am. J. Phys. Anthropol.* **51:** 311–332.

Szalay, F. S., and Delson, E. 1979. *Evolutionary History of the Primates,* Academic Press, New York.

Thompson, D'Arcy W. 1917. *On Growth and Form,* Cambridge University Press, Cambridge.

Thompson, D'Arcy W. 1942. *On Growth and Form,* 2nd ed., Cambridge University Press, Cambridge.

Thompson, D'Arcy W. 1961. *On Growth and Form,* abridged ed. (J. T. Bonner, ed.), Cambridge University Press, Cambridge.

Tobias, P. V. 1967. The craniun and maxillary dentition of *Australopithecus* (*Zinjanthropus*) *boisei,* in: *Olduvai Gorge* (L. S. B. Leakey, ed.), Vol. 2, pp. 1–264, Cambridge University Press, Cambridge.

Tobias, P. V. 1973. The Taung skull revisited. *Nat. Hist.* **83:**38–43.

Todd, T. W., and Wharton, R. E. 1934. Later postnatal skull growth in the sheep. *Am. J. Anat.* **55:**79–95.

Tollman, S. M., Grine, F. E., and Hahn, B. D. 1980. Ontogeny and sexual dimorphism in *Aulacephalodon* (Reptilia, Anomodontia). *Ann. S. Afr. Mus.* **81**(4)**:**159–186.

Tuttle, R. H. 1969. Knuckle walking and the problem of human origins. *Science* **166:**953–961.

Van der Klauuw, C. J. 1945. Cerebral skull and facial skull. *Arch. Neer. Zool.* **7:**16–37.

Van der Klauuw, C. J. 1948–1952. Size and position of the functional components of the skull. *Arch. Neer. Zool.* **8:**1–599.

Vogel, C. 1968. The phylogenetical evaluation of some characters and some morphological trends in the evolution of the skull in catarrhine primates, in: *Taxonomy and Phylogeny of Old World Primates With Special Reference to the Origin of Man* (B. Chiarelli, ed.), pp. 21–55, Rosenberg and Sellier, Torino.

von Bertalanffy, L. 1960. Principles and theory of growth, in: *Fundamental Aspects of Normal and Malignant Growth* (W. W. Nowinski, ed.), pp. 137–259, Elsevier, Amsterdam.

von Bertalanffy, L., and Pirozynski, W. J. 1952. Ontogenetic and evolutionary allometry. *Evolution* **6:**387–392.

White. T. D. 1977. The Anterior Corpus of Early African Hominidae: Functional Significance of Size and Shape, Ph.D. Thesis, University of Michigan, Ann Arbor, Michigan.

Williams, G. C. 1966. *Adaptation and Natural Selection,* Princeton University Press, Princeton.

Wood, B. A. 1976. The nature and basis of sexual dimorphism in the primate skeleton. *J. Zool. Lond.* **180:**15–34.

Zuckerkandl, E. 1976. Programs of gene action and progressive evolution, in: *Molecular Anthropology* (M. Goodman and R. E. Tashian, eds.), pp. 387–447, Plenum Press, New York.

Modeling Differences in Cranial Form, with Examples from Primates

10

FRED L. BOOKSTEIN

Introduction

The cranium of living primates fascinates scientists of many different persuasions. As an experimental system, it serves as test-bed for craniofacial medical procedures; as an evolutionary sketchbook, it aids in solving certain questions of comparative functional anatomy and posing others.

For most of this century there has been an analytic morphological technique generally suited to the scientific contexts in which the primate cranium is studied. This is the method of radiographic cephalometrics—serial metric study of the living skull. A routine cephalogram in standardized position gathers information useful for a variety of applications. We are fortunate that cranial form is so expressive of the function of its contents, for the other tissues of the head are more difficult to measure.

However, the numerical analysis of this information has not changed its complexion since the beginnings of the method. One still sees contrasts and trends of biological interest characterized by their mere reflections in arbitrary suites of cephalometric measures: distances, angles, and ratios chosen *a priori*. Even when the measures are statistically rebuilt into linear discriminators, principal or canonical components, and factors, the underlying arbitrariness remains unchanged. Our ability to gather cephalometric data has

FRED L. BOOKSTEIN • Center for Human Growth and Development and Department of Radiology, University of Michigan, Ann Arbor, Michigan 48109.

outstripped our customary analyses of them for whatever scientific purpose. Although measurement of human and other primate crania is important for work at a variety of levels of sophistication, there is little guidance to be had about what to measure or why. Cephalometrics is anarchic: every investigator seems to proceed as he or she pleases.

In recent years I have attempted to build a *foundation* for cephalometrics, so that the information gleaned by the method can be matched to the application—the discrimination of populations, the canonical description of growth, or the detection of effects that experimental alterations or natural variation may have upon growth. In all these contexts cephalometrics manifests a common theme: it is a technique of form *comparison,* not primarily a technique of form recording at all (Bookstein, 1982*b*).

Examination of these matters is crucially bound up with one fundamental biomathematical model, D'Arcy Thompson's *Cartesian transformation.* It was Thompson who realized that shape change was not to be measured by numerical differences among measures of shapes separately. Rather, shape change is a geometric object in its own right, the deformation taking one form into another in accord with biological homology. Thompson suggested this object be depicted by its effect on a grid laid over one form; he always began with a Cartesian (square) grid. As it happens, that particular grid is usually misleading; but the idea of transformation as measurement is seminal.

Cephalometrics can be rebuilt as the general analytic elaboration of this single fruitful model. In this chapter I explain a simple statistical realization of Thompson's idea, which allows the computation of mean transformations and comparisons among these means. The method has an unusual rhythm: one first observes quantitative effects upon form and only later determines conventional characters—distances, angles, ratios—that measure those effects. By measuring forms two at a time one eventually arrives at efficient ways to characterize them one at a time.

Data: Landmarks and the Homology Map

Morphometrics deals with two kinds of information: geometric form and biological homology. The *geometric form* is the collection of landmark points and curving arcs between them that constitute a sketch of the organ or organism under discussion. A *biological homology* is a spatial or ontogenetic correspondence among definable structures or "parts"—bones, nerves, muscles, other tissues. In the model it becomes a homology *function,* a deformation (smooth geometric mapping) not of parts to parts but of points to points. For any choice of point or curve upon or inside any particular form, the homology function associates well-defined and biologically acceptable counterparts, the *homologues* of the point or curve, on all the other geometric forms in the data set. In this manner the mathematical model permits one to refer to points as homologues of other points, even though the biological information permitted only the assertion of homologies among parts.

Computation of a homology function from geometric data begins with a sample of the correspondence: *landmark* sets, which must be located reliably in all the forms of a series. The computed homologues for all the other points—the nonlandmarks—are interpolated among the locations of the homologous landmarks. Operational criteria for landmarks are affected by the context of the deformation model. The precise biological or histological nature of their correspondence, which will vary from case to case, should not be judged dogmatically. Rather, the test of landmark quality is the statistical strength of the contrasts or trends that emerge at the end of the analysis. But some general rules are worth considering. Usually it is inappropriate to choose as landmarks points having extreme values relative to a coordinate system. The designation of "lowest" or "most anterior superior" points of a form is arbitrarily dependent on orientation of the coordinate axes upon the form, and thus is measuring not only the landmark but also the points in terms of which the axes were set down. Points constructed as geometric combinations of landmarks at a distance are usually unhelpful, as are intersections of lines or "axes," points at extreme distance from landmarks defined earlier, points of contact of lines tangent along two separate arcs, and intersections of shadows. All these proscribed categories ought to be supplanted by new versions tied to local features. Although landmarks will be assigned coordinate values in the course of digitizing, a sound landmark needs to have a coordinate-free definition in terms of the anatomy or boundary curvature in its vicinity. We have dealt with these matters elsewhere (Moyers and Bookstein, 1979; Bookstein *et al.*, 1984).

The metallic implants so useful in studies of rigid motion, such as jaw kinematics, are unacceptable as landmarks because they are not homologous from form to form. Of course they represent perfectly valid data in the context of another, simpler model. Also excluded from this discussion is the information about curving form between landmarks available from tracings of continuous biological outlines. There is an extension of the tensor method taking these curves into account (Bookstein, 1978); or one may study curving form in a wholly landmark-free manner by the method of medial axes (Bookstein, 1981).

Description of the Tensor Method

Summary. Biological shape changes may be modeled as uniform over small regions. They are described by the rate of change of length, or dilatation, as a function of direction. Dilatation is measured as a ratio, not a difference, of homologous lengths. It is dimensionless. The core of the tensor method is the summary of the dilatations by the largest and smallest at any point. These are called the principal dilatations. The directions along which they lie, the principal axes or biorthogonal directions, are at exactly 90° both before and after transformation. (Bookstein, 1983*a*)

For expository purposes we may take the basic unit of analysis to be a homologous pair of triangles of landmarks, as in Fig. 1. These triangles are not to be thought of as biologically real. Their sides may pass through various tissues or through air. They merely provide a convenient graphic formalism for describing changes in the configuration of their vertices, the landmarks whose locations are the real biometric data of these examples. Of course, the method is not limited to landmarks in sets of three. Extended polygons can be studied by use of alternate triangulations or by a tensor field that is not uniform, as represented in the curving biorthogonal grid (Bookstein, 1978, Bookstein *et al.*, 1984). In the absence of other information we may take the transformation sampled by these limited data to be uniform between each homologous pair of edges and throughout the interiors of the triangles. The homogeneity of the transformation is indicated clearly in the transformation grid after the style of D'Arcy Thompson (Fig. 1a).

But we may draw the transformation just as clearly in terms of the collection of lines in all directions (Fig. 1b). The deformation we are observing, driven by the displacements of those landmarks at the corners, will deform these lines into others that divide the edges in the same fractions. That is, the deformation takes edges to edges, median lines (dividing the opposite sides in the ratio 50 : 50) to medians, and so on.

We are interested in the ratios of lengths of corresponding lines in the two triangles, the dilatations. We could compute them by taking quotients of

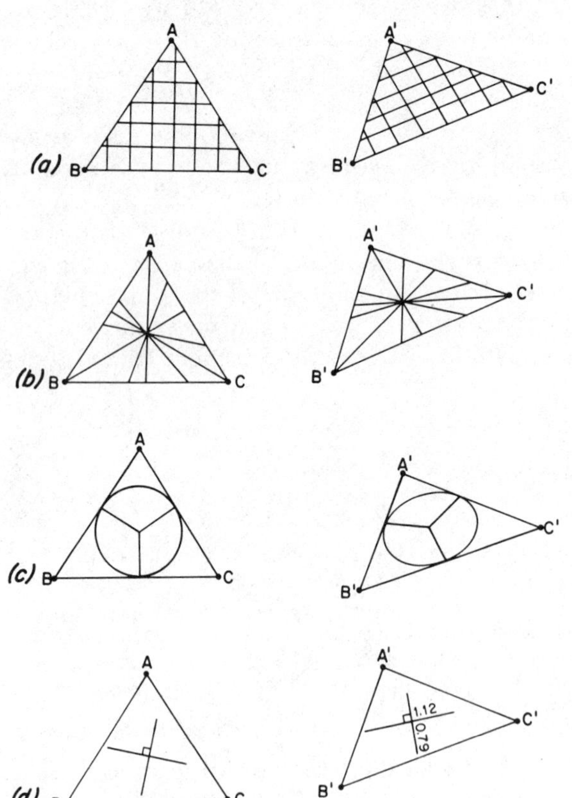

Fig. 1. The method of biorthogonal directions for two homologous triangles. (a) The uniform transformation of interiors they suggest. (b) Dilatations of lengths in various directions. (c) Dilatations may be represented by the radii of the ellipse into which a circle is deformed. (d) The principal directions are axes of this ellipse, and the principal dilatations are proportional to their lengths.

corresponding lengths, direction by direction. But instead we may observe the dilatations explicitly as the lengths of the deformations of lines of constant length, that is, of radii of a circle. We can actually *draw* a circle (Fig. 1c) whose deformation we wish to observe, and the oval into which the uniform shear takes it. The dilatations of line elements are proportional to the radii of this oval.

The reader will be convinced by an accurate drawing that this oval is remarkably like an ellipse. In fact, it is an ellipse, exactly. The image of the circle under growth has two axes of symmetry, which lie at 90°. One is the largest diameter of the ellipse, one the smallest. The diameters of the circle that transform into them are likewise at 90°.

Recall that the lengths of the radii embody the dilatations as a function of direction. Then the principal axes of the ellipse into which a circle is taken *are* the principal directions of the deformation as they lie upon the right-hand form. The diameters that were mapped into them (the same tissue at the earlier time, according to the model) are determined by corresponding fractions of intersection along edges of the triangles. Figure 1d presents them without their ovals; they are the *principal axes* of the deformation. The dilatations indicated on the figure were computed by division of lengths of homologous segments. Dilatations in any other direction may be computed from these two according to the formula

$$d^2(\theta) = d_1{}^2 \cos^2\theta + d_2{}^2 \sin^2\theta$$

where θ is the angle (in the left-hand form) between the axis of dilatation d_1 and the direction considered.

The analysis separates the observed change into one component for size change and a second for shape change. The product of the dilatations, $1.12 \cdot 0.79 = 0.885$, is the ratio by which the area of the triangle has increased; their quotient, $1.12/0.79 = 1.42$, is a measure of the *anisotropy*, or directionality, of this size change. One may think of any distortion as the composition of a pure size change, altering nothing but scale, and a pure shape change, leaving area alone. Even though form cannot be decomposed as "shape plus size," form change can be decomposed so, triangle by triangle.

To this point the method is equivalent to the simplest form of *finite-element analysis* as invoked in biomechanics and elsewhere (Skalak *et al.*, 1982).

From Tensors to Variables

The simple geometry of this construction directs our attention to a particularly useful selection of ordinary measures of single forms: distances, angles, and ratios that either optimally reflect the deformation or else are left precisely unchanged by it. For additional exploration of these configurations, see Bookstein (1983*b*).

Angle. The principal cross itself is unchanging at 90°. Other angles remain constant for which, in the notation of Fig. 2a, $\tan\theta_1 \tan\theta_2 = d_2/d_1$. For small changes, where d_2/d_1 is very close to unity, these correspond to angles with $\theta_1 + \theta_2 = 90°$—lines symmetrically placed about the bisectors of the principal axes. Whenever a bisector of the principal cross is aligned with the bisector of one of the triangle's angles (Fig. 2b), that angle will remain approximately constant over the transformation.

Proportional Division. If one principal axis is parallel to a side of the original triangle, than whatever the dilatations, the proportion in which the foot *X* of the perpendicular from the third vertex divides that edge is unchanged over the deformation. (See Fig. 2c.) Only in this case can we say unambiguously that landmark *C* is "moving away from" or "moving toward" edge *AB*, for in general, up to a general change of scale, every landmark is moving away from one point of the opposite edge *and* toward another point of that edge.

Ratio of Perpendiculars. If each of the principal axes makes an angle of 45° with one side of the starting triangle, then both that side and the altitude to it

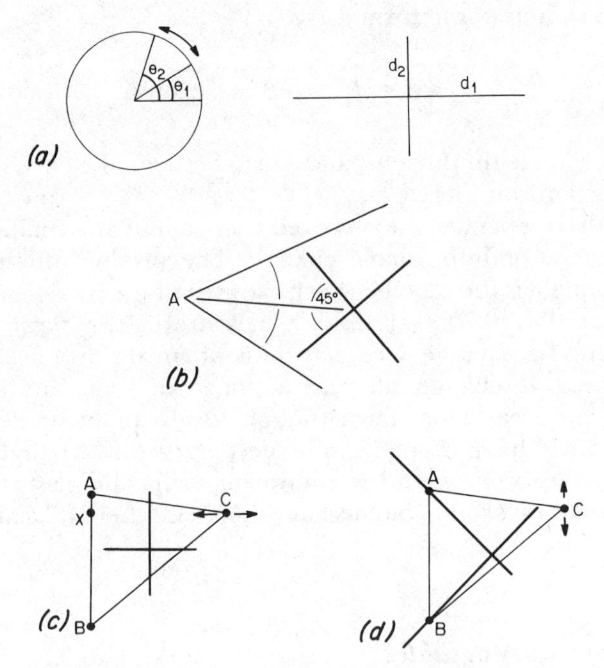

Fig. 2. Some invariants inferred from the tensor representation of a deformation. (a) Those angles are constant for which $\tan\theta_1 \tan\theta_2 = d_2/d_1$. (b) For small changes of shape these are angles bisected by the bisectors of the principal axes. (c) When one principal axis is parallel to side *AB* of the triangle, the ratio *AX*:*XB* is unchanged over the deformation, where *X* is the foot of the perpendicular to *AB* from the third vertex *C*. (d) When the principal axes are at 45° to one side of the triangle, then that side and its perpendicular have the same dilatation, and the third vertex *C* may be said to be moving *along* the opposite edge.

from the third vertex are dilatated at the same rate (as their directions are symmetric with respect to the axes). Hence, as in Fig. 2d, the aspect ratio of this triangle, its height divided by its base, will be approximately invariant over the deformation.

Generality of These Constructs. It is not necessary that any of the sides or angle bisectors of the triangle of landmarks lie along the principal axes or their bisectors for there to be invariants. We may define homologous segments through any points of the triangles, not only those that happen to be named landmarks. In Fig. 3a, for instance, the lines at 45° to the principal axes are parallel to transects connecting vertex A to points 0.2 and 0.8 of the distance from vertex B to vertex C, segments whose ratio is therefore unchanged by the deformation. In Fig. 3b, the principal axes lie along lines from A to a point 0.7 of the way from B to C and from B to a point 0.8 of the way from A to C; so the angle between these lines is exactly unchanged at 90° over the deformation.

Duality of Invariants and Covariants. A slowly changing length ratio corresponds to a rapidly changing angle, and conversely. In Fig. 3a, the dashed lines, at 45° to the principal axes, are dilatated at the same rate, whereas the angle between them has altered a great deal. In Fig. 3b, because the dashed lines are at an invariant angle of 90°, the proportion between them has altered more than any other shape measure. It thereby becomes a best scalar for such operations as discriminating before from after, diagnosing anomalies, and characterizing experimental effects.

Populations of Shape Comparisons

The extension of this method to whole populations of shape changes or shape comparisons proceeds in four steps (Bookstein, 1982*a*, 1983*b*).

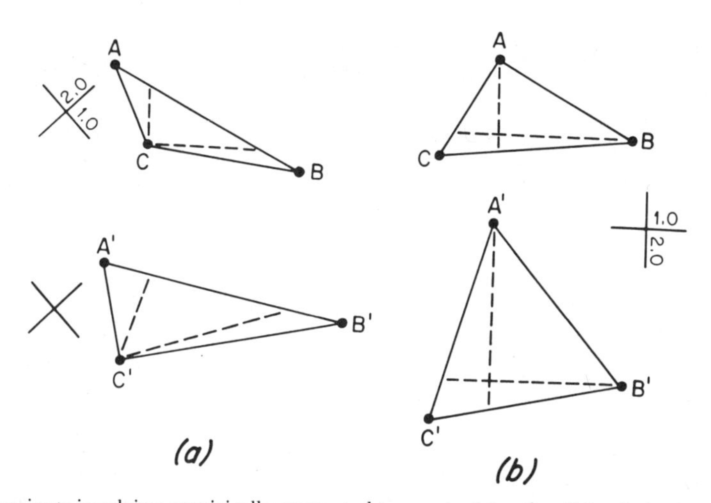

Fig. 3. Invariants involving empirically computed transects: (a) ratio of the dashed segments; (b) angle between them.

1. Triangle by triangle, we average dilatations in directions connecting any vertex to a point dividing the opposite edge in a fixed fraction. For instance, an average dilatation is computed separately for each of the three edges of each triangle of landmarks; for each of the three median lines, which divide the sides opposite in the ratio 50 : 50; and so on. This accords with the homology function induced by the uniform transformation of Fig. 1 for deformations between triangles of a series.

2. Triangle by triangle, we determine for each population group the greatest and least mean dilatations over all directions within the triangle. Ordinarily the directions along which they lie will be nearly at 90° in the forms of the series. These may be taken as the appropriate mean biorthogonal analysis (Fig. 4a).

If all the crosses for all the shape changes of a population are aligned in the same directions (as determined by our homology convention), the maximum of the mean dilatations would be the mean of the separate maxima, and likewise the minimum. The anisotropy of the mean change would then be just the mean of the anisotropies of the separate changes being averaged (Fig. 4b). If the directions of extreme dilatation for the individual deformations wander

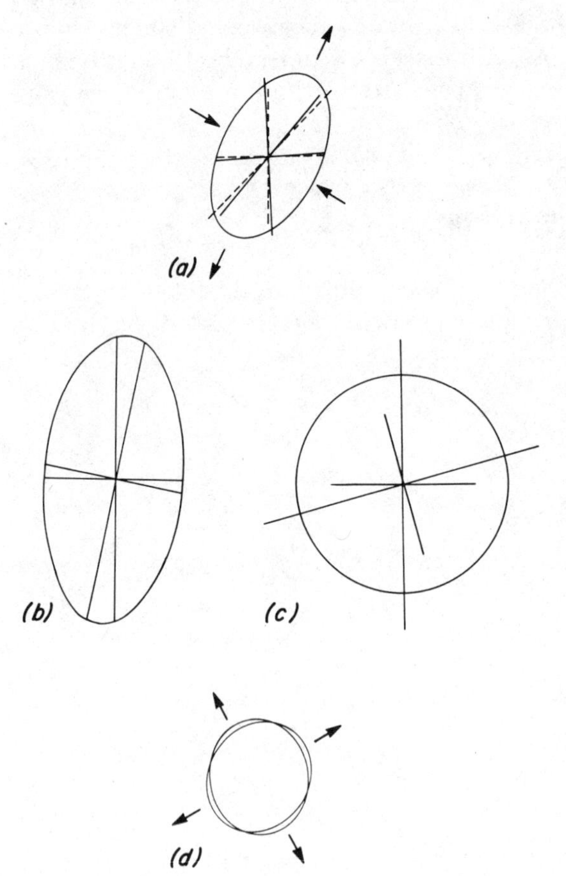

Fig. 4. (a) Computing a summary strain oval, nearly an ellipse, by taking means of dilatations direction by direction over a population of strain ellipses. (b) A population of strains with a high fraction of variance explained by shape change: principal dilatations tightly aligned. (c) A population of strains with no variance explained by shape change: principal dilatations wander widely. (d) Computing the relative dilatations from two strain ellipses: directions of algebraically greatest and least difference between the dilatations.

far from mutual alignment, then the averages will mix extremal dilatations aong principal directions of certain shape changes with middling dilatations along others, attenutating the extrema of those averages (Fig. 4c). The ratio of anisotropy of the means to mean anisotropy is a useful statistic, a sort of fraction of variance explained by shape change. We will refer to it in the course of Examples 1 and 3 below.

3. Group by group in pairs, for each triangle of landmarks we determine and depict as well the algebraically greatest and least *differences* in dilatation (Fig. 4d). These lie in the directions along which rate of change of distance in one group most exceeds, or most falls short of, the homologous rate in the other group. The dilatations along these (homologously defined) directions of algebraically greatest and least difference in dilatation between the groups constitute the quantification of group differences in deformation, such as the "effects" of age range or sex on growth (see Example 3).

4. Finally, the summaries of deformation by these crosses at nearly 90° are reduced to ordinary measures of the forms separately according to the heuristic of invariants and covariants explained above. (These are now approximate, as the mean crosses are not precisely at 90°.) The ratio between lengths measured in the two principal directions is changing faster, on the average, than any other proportion homologously measured, and serves as the optimal dimensionless discriminator; the ratio between lengths measured at 45° to those axes is changing, on the average, hardly at all, but their angle is changing nearly fastest of all angles homologously measured; and so on. Likewise, by multiplying or dividing the mean dilatations, we may ascertain how much of a typical transformation or contrast between deformations involves size change and how much involves shape change. Note that we do not measure the forms separately until just before publication. We have only archived the landmark locations, coordinate by coordinate, then very efficiently restricted our measurement to the subject matter of form change, directly characterized after the style of D'Arcy Thompson.

Types of Studies

This general method applies to three different study designs for contrasts or trends of prior biological interest.

Mode I. Comparison of Two Populations of Single Forms. Using the two mean deformations from any convenient reference form to each member of two populations, we may compute the population comparison as their relative difference tensor. The findings constitute the distances, angles, and ratios that most differ between the typical members of the populations being compared, as in Example 2 below.

Mode II. Populations of Specific Comparisons. This includes the study of typical growth patterns in a single group; here the triangles being compared have for their vertices the same landmarks at two ages. This design also includes other conventional "matched" designs, for instance the study of asymmetry, or the characterization of syndromes in mixed samples (Example 1 below), for which the deformation of each specimen is computed with reference to an age- and sex-specific normative mean. The findings include distances, angles, and ratios that most differ, or (for data over time) are changing most rapidly, across the comparison. In addition to the scalar quantities that best discriminate between the poles of the comparison, there are also produced the invariants that have changed little on the average.

Mode III. Comparisons between Populations of Contrasts. This involves series of analyses as in mode II, followed by extraction of the distances, angles, and ratios that best discriminate between the systematic changes, group versus group, as in step 3 of the general population protocol. (See Example 3 below). Group differences in growth, such as experimental effects, are also computed in this way. The measures optimally describing differences in growth are not equivalent either to the measures of differences in starting form or the measures of growth; they explicate a four-form comparison, growth versus growth or form A/norm A versus form B/norm B.

Statistical Method

Statistical significance tests for these means and differences have been published in Bookstein (1984). The test of any mean tensor proceeds in two phases: First, a test for significance of the shape change or shape difference component is conducted; if the null hypothesis of no shape component cannot be rejected, then (and only then) is a test for significance of the size change or size difference component conducted. All significance levels are computed with respect to a model involving identical circular noise at each vertex of a mean triangle.

In study design II, the matched comparison, $\sqrt{(2N/\pi)}$ times the fraction-of-variance explained is distributed approximately as a χ_2 variate; as a rule of thumb, the 5% level of significance of that fraction is $2/\sqrt{N}$. If this test proves insignificant, one tests for change in size by an ordinary matched t-test on the mean change of area of the triangle.

In study designs I and III, the difference between the principal relative dilatations, divided by the sum of their ordinary sampling standard errors, is distributed approximately as $\sqrt{(\pi/4)}$ times a t-ratio. (More accurately, the quotient is the square root of an F-ratio with two degrees of freedom in the numerator.) Its 5% level of significance is approximately 1.77. If this level is not attained, one may test the same relative tensor for size change by the ratio of the sum of the principal relative dilatations to the same sum of their standard errors of sampling. This quotient goes as $1/\sqrt{2}$ times an ordinary t-ratio, and its 5% significance level is about 1.41.

Examples

Example 1. The Cranial Base in Apert's Syndrome

Craniofacial synostosis is a class of human craniofacial deformity generally manifesting premature closure (synostosis) of the intracranial bony sutures about the upper jaw (maxilla) and frontal bone. Dr. Joseph McCarthy and the staff of the Institute for Reconstructive Plastic Surgery at New York University supplied me data regarding 11 patients who bear one such deformity, acrocephalosyndactyly or Apert's syndrome. As the Greek implies, this condition is characterized by deformities of the extremities as well as of the craniofacial skeleton. Facially, the deformity typically includes a high, bulging forehead and a short maxilla positioned farther back than normal. As a first example of the tensor analysis and the manner of its reporting, I report here on the cranial base deformity, which is thought primary. This is part of a larger project to be reported elsewhere (Grayson *et al.,* 1984).

Landmark Selection

The data for this exercise consist of coordinates of five landmarks along the cranial base: basion (Bas), sella (Sel), sphenoethmoid intersect (SEI), frontomaxillonasal suture (FMN), and nasion (Nas). Figure 5 shows them *in situ.* Of these, nasion is a fine anatomical landmark, an intersection of three sutures (under the bridge of the nose) easily detectable on a solid specimen. Basion, too (midsagittal extremum of the cranial base anterior to the foramen magnum), is plain. Sella is the "center" of the pituitary fossa; it can be reproducibly located by a trained digitizing technician even though, technically, there is nothing there to locate. FMN is the overlapping image of a laterally symmetric pair of landmarks, each the intersection of three sutures. But SEI is not a landmark at all: it is the intersection of two shadows, one representing the lower margin of the sphenoid bone in the midsagittal plane, the other the (averaged) anterior margin of the greater lateral wings. The location of this "point" expresses the position of two structures, not one. We will confront this difficulty after reviewing the findings of the tensor analysis.

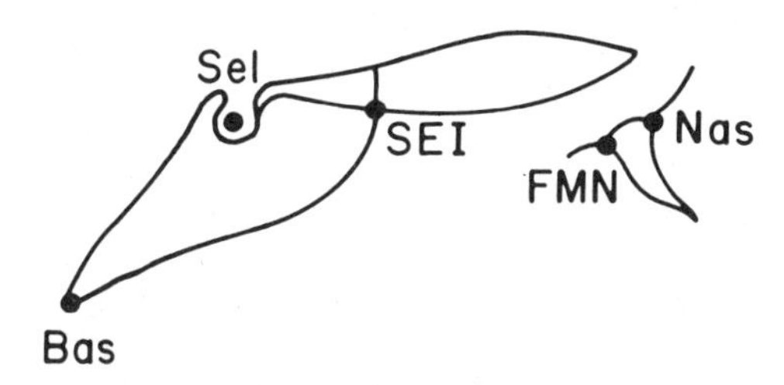

Fig. 5. Lateral schematic of the human cranial base, with five landmarks indicated.

Data

In arbitrary Cartesian coordinate systems we recorded the locations of the five landmarks for the 11 patients. From archives at the Center for Human Growth and Development we drew the same locations for the 83 subjects of the University of Michigan University School Study (Riolo *et al.*, 1974), which is a "normative" study of an upper middle class white Ann Arbor sample of growing children. From those data we extracted normal mean forms of the cranial base, age- and sex-specific, merely by averaging coordinates of each landmark in a consistent coordinate system.

The Measurement Design: Deformation of Normal Mean into Apert's

Each Apert's cranial base may be construed as a deformation of the appropriate normal mean cranial base: this is mode II of the three designs reviewed above. To characterize the Apert's cranial base most efficiently, we compute the mean tensor expressing the average of these deformations, normal to Apert's, over the population of 11 Apert's cases, each matched to its appropriate normal mean.

From the five landmarks one may assemble a total of 10 triangles for analysis. Of the 10, six are too narrow for the deformation model to be usefully applied (Bookstein, 1982*a,* p. 186). The four relatively useful triangles are diagrammed in Fig. 6, in which their vertices are positioned according to a normal mean form. Drawn over each triangle are the two principal

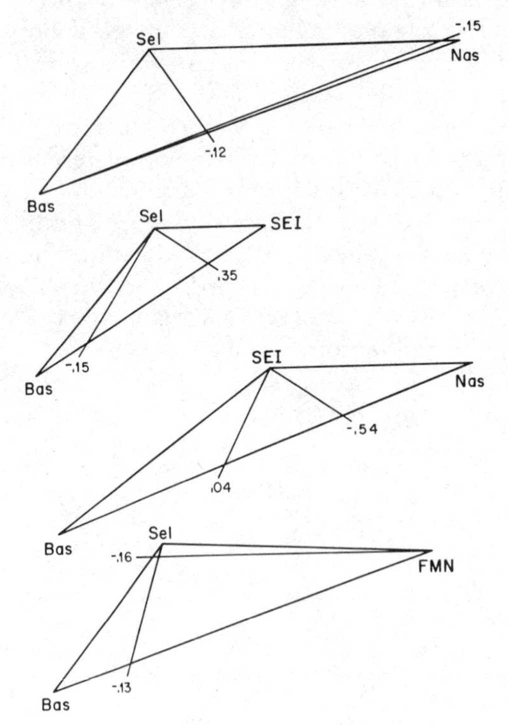

Fig. 6. Cranial base deformity in Apert's syndrome: mean tensor analysis of the deformation of 11 Apert's cranial bases from age- and sex-matched normals.

axes of its deformity and the dilatations along each axis. The dilatations are printed as the fractions by which lengths in the typical Apert's case fall short of, or exceed, the corresponding lengths in the matched normal mean forms. Some helpful statistics associated with these mean tensors are collected in Table 1. In a sample of $N = 11$, shape fractions greater than 0.60 are significant at the 5% level according to the approximate χ_2 test described in the text.

Findings

Consider first the large triangle Bas–Sel–Nas (Fig. 6, top). The diagram shows two principal dilatations: a largest relative dilatation of −0.12 (a compression by 12%) along the direction from sella to a point some 35% of the way from basion to nasion; and a smallest relative dilatation of −0.15 (a 15% compression) aligned very nearly with the segment basion–nasion itself. These two dilatations are not very different from each other. Most of the systematic difference between the two versions of this triangle, the normal and the deformed, is mere size difference, shrinkage by 12–15% in every direction. The shape component is far from significant; the *t*-ratio for change of area is 4.4, highly significant. The angle usually taken to represent the flexion of the cranial base, the angle between the segments sella–basion and sella–nasion at sella, is nearly the best discriminator of this deformity in regard to this triangle, as one of the principal axes lies almost along the bisector of that angle; but the net mean difference is still very little, a mere 3% or so (15% − 12%) in the tangent of the angle.

The futility of diagnosing the syndrome from the shape of this particular triangle is indicated as well in Table 1. The average Apert's case differs from the age-matched normal by about 18% in *some* proportion involving this triangle (entry in the column labeled "mean anisotropy"). But only 3% out of this 18% is systematic; the rest is variability within the syndrome.

We gain rather more helpful information from the second triangle in Fig. 6, Bas–Sel–SEI. This triangle shows a mean relative dilatation of +0.35 in

Table 1. Statistics of Population Variation (N = 11) in Apert's Deformity

	Principal dilatations					
	Maximum		Minimum			
Triangle	Mean (a)	SD	Mean (b)	SD	Mean anisotropy (c)	Shape fraction $(a - b)/c$
Bas–Sel–Nas	−0.115	0.159	−0.147	0.037	0.181	0.177
Bas–Sel–SEI	0.351	0.199	−0.146	0.101	0.537	0.925
Bas–SEI–Nas	0.043	0.113	−0.538	0.223	0.641	0.906
Bas–Sel–FMN	−0.130	0.127	−0.162	0.038	0.178	0.180

one direction, with a standard deviation of 0.199 (that is, a standard error of 0.062), and a mean dilatation of −0.15 in the perpendicular direction, with a standard error of 0.032. The net deformity is thus a disproportion (anisotropy) of 1.35–0.85 = 50%. Table 1 indicates that this anisotropy of the mean taps 92.5% of the mean anisotropy over the patients separately, which is 0.537; it is, in other words, very strongly typical of the syndrome, and is statistically significant well past the 1% level. The principal axes of this deformation (Fig. 6) represent it as an extension of the distance from sella to a point 80% of the way from basion to SEI and a compression along the direction from sella to a point 10% of the way from basion to SEI. The net mean deformation may thus be drawn as the Cartesian grid of Fig. 7a. Note that its alignment on the triangle does not correspond to any familiar anatomical orientation: the coordinate grid is to be aligned with the shape change, not the shape.

Consider now the third triangle in Fig. 6, the triangle Bas–SEI–Nas. It, too, is highly misshapen. In comparison to the matched normals, SEI is typically more than 50% too close to a point 80% of the way from basion to nasion, although it is at just about the right distance from a point 40% of the way from basion to nasion. We saw in Fig. 6 that the segment basion–nasion

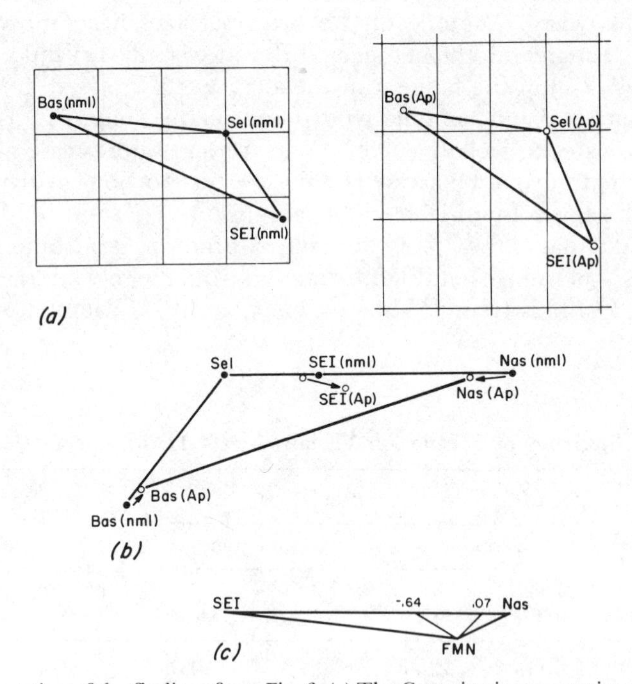

Fig. 7. Interpretation of the findings from Fig. 6. (a) The Cartesian interpretation of the displacement of SEI from posterior structures. (b) Another view: size change in the large together with imputed displacement of SEI forward and a little downward. (c) Discrepancy between FMN and nasion in expressing the anomaly.

contracted by 15%, along with basion–sella and sella–nasion at 12% each. I believe it is most useful to view SEI as further displaced (from the position imputed to it in that general contraction) by a translation forward and a little downward (Fig. 7b). This deficit of 54% (Table 1) is the most diagnostic single measure for Apert's in this little data set, some seven times the standard error imputed to it by variation among the patients. (There is an additional component to that standard error, the sampling variation of the normal Ann Arbor student means, which is not invoked here.)

Because landmark FMN is very near to nasion, the triangle Bas–Sel–FMN (Fig. 6, bottom) shows the same size change without shape change we noted for Bas–Sel–Nas. But the dilatations are slightly different. There seems to be some discrepancy between the relative displacements of FMN and nasion with respect to posterior structures. The best triangle for visualizing this discrepancy is SEI–FMN–Nas, drawn in Fig. 7c. Even correcting for the compression of 16% in the segment SEI–nasion, FMN is displaced almost another half of the normal distance (which is quite small) from about the midpoint of that segment. We cannot tell if this represents displacement of the Apert's nasion "downward" or of his FMN "upward" with respect to the normal position, as there is no other information in the vicinity.

Although the major cranial base anomaly of these Apert's syndrome cases is in the position of SEI, we must recall that that point fails to qualify as an anatomical landmark. It is only an intersection of shadows generated by two separate parts of the sphenoid bone. In the anatomy of the normal cranial base (Fig. 5) the border of the body of the sphenoid runs more or less horizontally and the ala vertically, so that the observed relative displacement of SEI would be compounded of a dropping of the body together with a forward repositioning of the ala. However, should the orientations of those curves itself be abnormal in the syndrome, the interpretation must be suitably adjusted. This matter can be resolved by the continuous tracings of cephalometric outlines between the landmarks we sampled; but that resolution is outside the scope of this exercise.

Example 2. Sex Differences in Macaque Skulls

Prof. David Carlson of the University of Michigan Department of Anatomy heads a group managing a colony of macaques used as subjects in craniofacial experiments. The birthdates of these monkeys are unknown, but thier ages can be grouped by dental stage into five categories: infant, juvenile, adolescent, young adult, and adult. As part of a larger study, we are interested in sexual dimorphism among macaques of comparable age.

Data

Lateral cephalograms were digitized for those monkeys not yet subjected to any surgical procedure. Of these films, nine were from juvenile males, 14 from juvenile females, eight from adult males, and 32 from adult females. We may therefore study dimorphism at each of two ages. To help in the interpretation we will also note the mean deformation between the male age classes, that is, the male *growth deformation*. (This is an inefficient way to study that phenomenon when true longitudinal records are available—see the next example.) From the digitizing scheme we extracted a subset of nine landmarks sampling some of the osteological components in which we are interested, and assembled them into the triangles shown in Fig. 8. Of the landmarks, eight (whose operational definitions are not given here) are conventional: condylion, gonion, menton, supradentale (SPD), sella, basion, and the tips I^1 and I_1 of the upper and lower central incisors. In addition we use, in place of nasion (which cannot be seen in the macaque), the point labeled "N" at the deepest concavity of the arc from glabella to the tip of the nasal bones. The diagrams draw these triangles using the landmark locations of a typical adult female. Figure 8a keys the triangles.

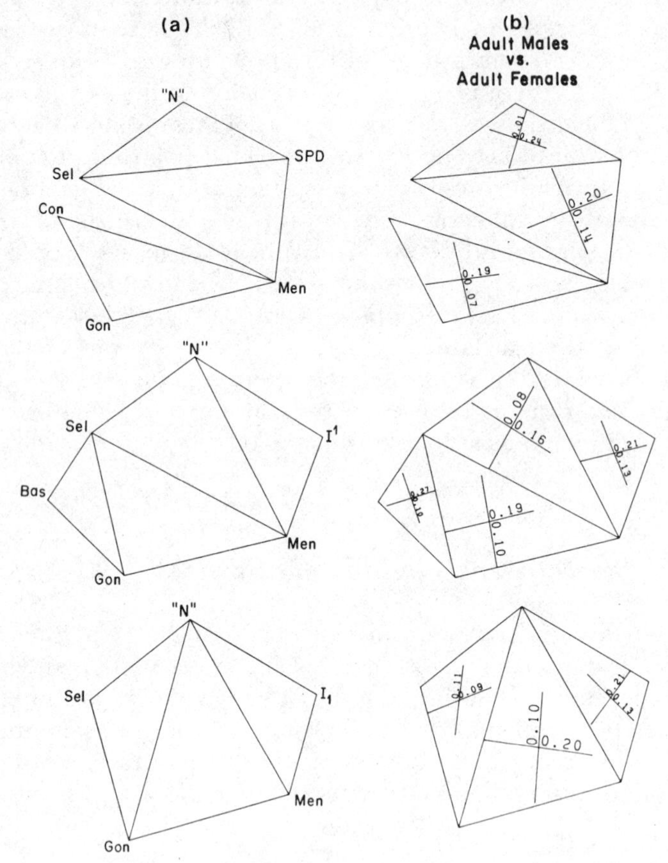

Fig. 8. Tensor analysis of mean deformations relating three populations of the rhesus monkey. (a) Key to landmarks and triangles. (b) Deformation from typical adult female to typical adult

These triangles were compared in three separate mode I studies. Each computed an average deformation: from juvenile females to juvenile males, from adult females to adult males, or from juvenile males to adult males. In this context, the principal dilatations indicated along the diagrams of Figs. 8b–8d are along the lengths whose ratios are, on the average, greatest and least between the poles of the comparisons. These dilatations are printed as fractions of mean increase. As there is considerable size variation within each of our age classes, our primary concern is with the principal shape discriminator, the ratio between lengths measured along the arms of the principal cross.

Findings

In either age group, the males are larger than the females in every homologous linear measure upon these triangles. (Of course, there is no way to

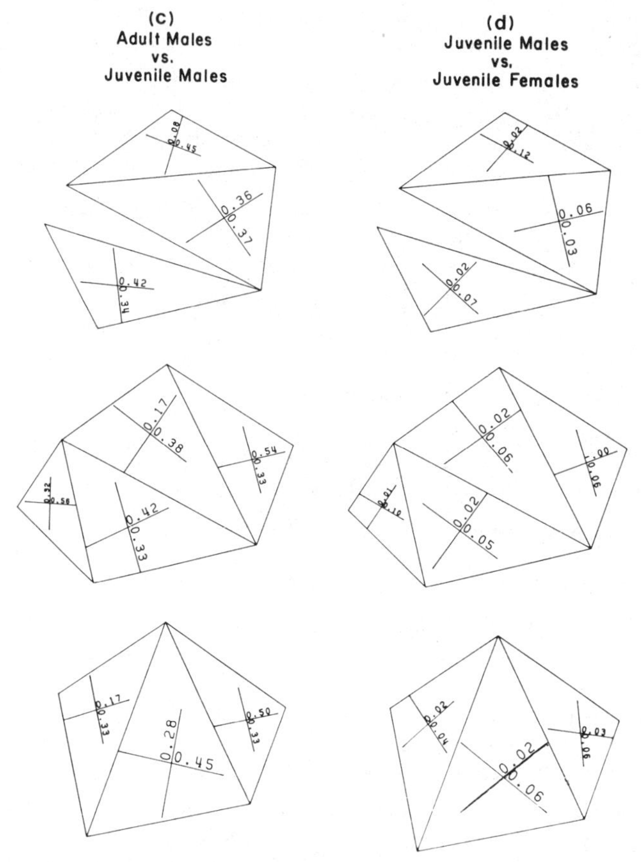

male. (c) Deformation from typical juvenile male to typical adult male. (d) Deformation from typical juvenile female to typical juvenile male.

verify that the two juvenile samples have the same average chronological age.) None of the reported minimum dilatations are negative. For the adult comparison, the lowest dilatation is 0.01 (Fig. 8b, top), along a line from "nasion" toward a point 35% of the way from sella to SPD. This distance is the same in the two adult samples. Perpendicular to it, from SPD to the point 75% of the way from sella to nasion, the males considerably exceed the females in length, by 24% of the mean female length. This segment is nearly aligned with the segment "nasion"–SPD, the upper facial line. That "maxillary triangle" shows the greatest anisotropy of any triangle analyzed here: a 23% disproportion between the adult males and the females. The same shape variable, ratio of these measured lengths, also best discriminates the adult male from the juvenile (Fig. 8c, top), with an anisotropy of 37%, and furthermore distinguishes the sexes among the juveniles, with an anisotropy of 10% (Fig. 8d, top). The angle most sensitive to this discrimination is the angle at SPD between the segment toward sella and the segment toward nasion.

The triangle Bas–Sel–Gon (Figs. 8b–8d, middle) has a disproportion of 17% between typical adult forms of opposite sex. The male is 10% longer from sella to gonion but 27% longer in the distance of basion *to* that line. This latter distance shows the greatest rate of increase from juvenile to adult males of any segment in the data set—55%; the next highest allometric coefficient in this comparison is of the displacement of the upper central incisor tip from the line "nasion"–menton, at the right in the same panel of Fig. 8.

Other triangles show very little shape dimorphism in certain comparisons. For instance, between the two adult populations, the triangle Sel–Gon–"N" (Fig. 8b, bottom) bears a dimorphism of 9.8% ± 1.1% in every direction. This is the smallest size dimorphism of all the triangles in the adult–adult comparison. The triangle Sel–SPD–Men, a "bimaxillary" triangle, shows very little shape dimorphism in *any* comparison. In particular, in the mean deformation from juvenile males to adults its dilatations range only between 36% and 37% as a function of direction (Fig. 8c, top)—a large size change with no associated shape change at all.

The organization of these tensors in the large corroborates the conventional understanding of primate splanchnocranial allometry. Consider the region between anterior cranial base and mandibular plane, the quadrilateral sella-"nasion"-menton-gonion. In the comparisons involving adult males (Figs. 8b and 8c, middle), the larger principal dilatations in triangles involving menton point away from that landmark toward gonion or toward sella. But of these triangles only one, the maxillary triangle, shows great anisotropy. The triangles below that one, such as Sel–SPD–Men, manifest rather more size change than shape change. This confirms the familiar notion of bimaxillary prognathism as the main size allometry discriminating adult males from females or juveniles. The deformation mainly *enlarges* the jaws. They come to appear prognathic partly because the anterior–posterior dilatations are slightly greater than those running craniocaudally, but mainly because of the increase in general size. In the growth of the adult male, for instance, the chin moves as far "downward" as "forward" with respect to the juvenile maxilla.

The global comparison of sexes among the juveniles is less marked. Just as for adults, in no triangle is any mean distance less in the males than in the females. All triangles except the maxillary triangle and the triangle Sel–Bas–"N" show a sexual dimorphism of about 4% ± 2% in all directions. This represents a somewhat uniform disproportion of about 4%, much less than between the adult forms, with principal directions of male excess predominantly oriented toward menton. In general, juvenile females are most nearly equal in size to males along arcs of circles about menton.

Example 3. Mean Growth Deformations

Prof. Robert E. Moyers and I are studying growth in the normal human cranium by way of mode III investigations into mean growth deformations of Ann Arbor normal children over various age ranges.

Data

In the large University of Michigan University School Study data base (recall Example 1) we identified 42 boys and 29 girls having cephalograms on file in each of the age ranges 6–8, 10–12, and 14–16 years. When several films were available in an interval, we chose the earliest. Six interesting landmarks [operationally defined in Riolo *et al.* (1974)] and three of the triangles that can be formed from them are shown in Fig. 9, with a splanchnocranial summary triangle Sel–Nas–Men in Fig. 10. The key to the triangles is at the lower left. In the top row of panels are the annualized mean increases of the two principal lengths, separately by sex, over the two age ranges. These are printed in units of percent per year: thus, the decimal "2.93" at upper left in Fig. 9 represents length increase by a factor of 1.0293 per year or 11.7% over a 4-year interval. Table 2 presents the standard errors of these principal dilatations and the fraction of mean anisotropy captured cross by cross.

The bottom row of triangles presents three relative tensors comparing pairs of mean growth deformations from the top row. Those in the second and third columns relate their mean growth tensors to that observed in the growth of males from 6 to 10, column 1; that in the lower right corner compares the mean growth of females to that of males over the age range from 10 to 14. The triangles are drawn using the means of landmark coordinates, in a sella–nasion registration, for the 6-year-old males.

Findings

These analyses are all consistent with the familiar notion of a *growth axis* (Bookstein, 1983*a*,*b*) running from menton through a point some one-third to one-half of the way from sella to nasion. Growth along this axis proceeds at

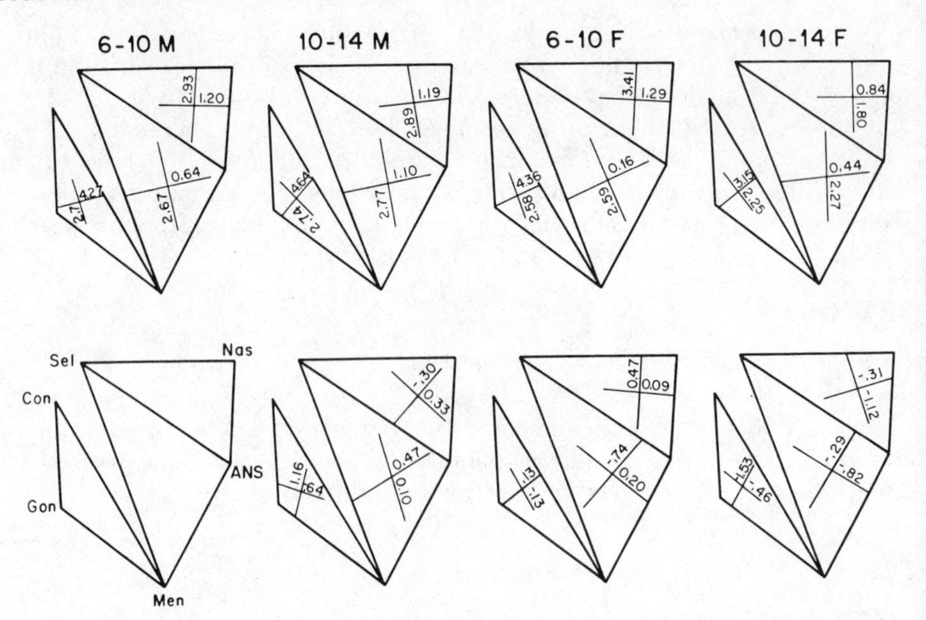

Fig. 9. Tensor analysis of "normal" human growth between the ages of 6 and 14 years. Within triangles in the upper row are mean growth deformations by sex and period; in the triangles in the lower row are drawn comparisons of pairs of means from the top row. For each age range, mean growth of the females is compared to mean growth in the males; and the mean growth of males aged 10–14 is compared to that of males aged 6–10.

somewhat more than 2%/year, and growth perpendicular to it at not much more than 1%/year, for a net change in proportion of some 1.5%/year in all four mean comparisons. The triangle Sel–Nas–ANS, a "maxillary triangle," shows the stablest growth direction of any region of the cephalogram: the angle Sel–Nas–ANS is very nearly unchanging, in mean, across *all* the comparisons. (Orthodontists noticed this many years ago.) The change in shape of the "mandibular triangle" Con–Gon–Men is also surprisingly stable. There is more change of size than of shape in all four populations of mandibular

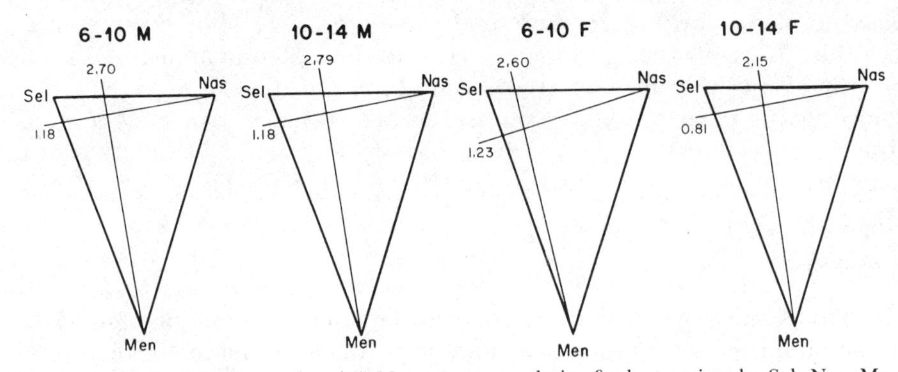

Fig. 10. Analysis of the "growth axis." Mean tensor analysis of a large triangle, Sel–Nas–Men, summarizing the two facial triangles in Fig. 9.

Table 2. Statistics of "Normal" Human Growth over Four Years, by Sex

Triangle, age range, sex	Principal dilatations, %/year: Maximum Mean	Maximum SD	Minimum Mean	Minimum SD	Mean anisotropy, %/year	Shape fraction, %
Sel–Nas–ANS						
6–10 ♂	2.93	1.01	1.20	0.38	2.13	81
10–14 ♂	2.90	0.95	1.19	0.42	1.93	88
6–10 ♀	3.41	0.95	1.29	0.38	2.26	94
10–14 ♀	1.80	0.84	0.84	0.39	1.50	64
Sel–ANS–Men						
6–10 ♂	2.67	0.69	0.64	1.07	2.48	82
10–14 ♂	2.77	0.73	1.10	0.85	2.00	83
6–10 ♀	2.59	0.72	0.16	1.18	2.79	87
10–14 ♀	2.27	0.62	0.44	1.33	2.20	83
Con–Gon–Men						
6–10 ♂	4.27	2.17	2.70	0.99	3.05	52
10–14 ♂	4.64	1.58	2.74	0.95	2.96	64
6–10 ♀	4.36	1.55	2.58	1.16	2.93	61
10–14 ♀	3.15	1.71	2.25	0.76	2.21	40
Sel–Nas–Men						
6–10 ♂	2.70	0.64	1.18	0.38	1.74	87
10–14 ♂	2.79	0.68	1.18	0.40	1.77	91
6–10 ♀	2.60	0.73	1.23	0.35	1.71	80
10–14 ♀	2.15	0.55	0.81	0.43	1.59	84

growth deformations. Such shape change as there is is adequately quantified by changes in the so-called gonial angle, the angle between the segments Gon–Con and Gon–Men.

In its principal directions and also in the ratio between its dilatations, the mean growth of the summary triangle Sel–Nas–Men for males (column 2 of Fig. 10) corresponds to the growth deformation for the triangle Sel–"Nas"–SPD (Fig. 8c, middle) in the macaque males. (Note also that the point ANS deforms homogeneously with this transformation; in column 2 of Fig. 9, the mean tensors for Sel–Nas–ANS and Sel–ANS–Men are nearly identical.) In both species the angle changing most rapidly is that measured at menton between the segments Men–Sel and Men–Nas. Likewise, the growth deformation of the macaque triangle Sel–"Nas"–SPD is analogous to that of the human maxillary triangle Sel–Nas–ANS: each shows a dominant dilatation along the facial plane. Other changes are inconsistent between the two species—for instance, growth changes in the mandibular triangle.

In the lower row of Fig. 9 we see the direct visualization of *differences* (dimorphisms or changes over time) in growth deformation. For instance, in the second column is the comparison of annual growth deformation in 10- to 14-year-old males to the same in 6- to 10-year-old males. In the lower face the growth directions are nearly the same; but there is one direction of acceleration (probably expressing the circumpubertal growth spurt). In the maxillary triangle, however, the growth directions have altered. The relative dilatations in the bottom diagram indicate that the later growth pattern differs from the earlier by an additional shear of the sort indicated in Fig. 2d. While sella alters its distance from the facial plane at the same rate in the two growth deformations, in the later period ANS may be modeled as additionally moving forward relative to the segment sella–nasion.

Sexual dimorphism in growth is considered in the lower panels of columns three and four in Fig. 9. Between the ages of 10 and 14 the females are growing more slowly than the males in every direction of every triangle considered here. Notice, however, that the relative tensor for the triangle Sel–ANS–Men has changed its polarity between the age ranges. Over the age range 6–10, the female growth rate actually exceeds the male's along sella–ANS, and falls most short of the male's along ANS–menton; but over the age range 10–14, the growth rate along sella–ANS shows the greatest deficit of all directions, with the least deficit along ANS–menton. Neither of these directions is aligned with the principal growth directions for either sex separately. To a different subject of study—dimorphism in growth, not growth *per se*—corresponds a different set of optimal descriptors.

Concluding Remarks

These examples reviewed the modes of mean shape change analysis to indicate how the tensor diagrams may be conveniently interpreted in typical biological contexts. A great deal of work remains to be done in many aspects of this general method.

1. We need computational techniques for averages of the nonuniform tensor fields that describe changes in configurations of four or more landmarks.
2. The statistical method must be extended to study covariates of deformation other than group membership: starting form, perhaps, measures of experimental insult, or changes in other regions.
3. The model needs extension to include types of biological shape change other than deformation, such as creation of new regions unrepresented in comparison forms. (See Skalak *et al.*, 1982.)

But these unsolved problems should not obscure the reader's sense of how far we have come. The direct visualization of biological shape change, D'Arcy Thompson's great contribution, is here rigorously realized as a statistical technique suitable for routine comparative morphology—contrasts, growth trends, and contrasts of growth trends within and between groups or conditions. Out of a data base combining geometric location and biological homology—landmarks conscientiously defined and digitized—the tensor technique automatically generates the distances, angles, and ratios that best characterize the effects under study. In this way cephalometrics at last becomes subordinated to biological process: it becomes the study not of static forms but of phenomena.

ACKNOWLEDGMENTS

I am grateful for the support of NIDR grants DE-03610 to Robert E. Moyers, DE-03568 to Joseph G. McCarthy, and DE-05410 to myself in the course of the studies reported here. The discussion of homology owes much to conversations with the Morphometrics Study Group at the University of Michigan Museums: Barry Chernoff, Ruth Elder, Julian Humphries, Gerry Smith, and Richard Strauss.

References

Bookstein, F. L., 1978. *The Measurement of Biological Shape and Shape Change* (Lecture Notes in Biomathematics, Vol. 24), Springer-Verlag.

Bookstein, F. L., 1981. Looking at mandibular growth: Some new geometrical methods, in: *Craniofacial Biology* (D. S. Carlson, ed.), pp. 83–103, Center for Human Growth and Development, University of Michigan, Ann Arbor.

Bookstein, F. L., 1982*a*. On the cephalometrics of skeletal change. *Am. J. Orthodontics* **82:**177–198.

Bookstein, F. L., 1982*b*. Foundations of morphometrics. *Ann. Rev. Ecol. Syst.* **13:**451–470.

Bookstein, F. L., 1983*a*. Measuring treatment effects on craniofacial growth, in: *Clinical Alteration of the Growing Face* (D. S. Carlson, ed.), pp. 65–80, Center for Human Growth, University of Michigan, Ann Arbor.

Bookstein, F. L., 1983*b*. The geometry of craniofacial growth invariants. *Am. J. Orthodontics* **83:**221–234.

Bookstein, F. L., 1984. A statistical method for biological shape comparisons. *J. Theor. Biol.* **107:**475–520.

Bookstein, F. L., Chernoff, B., Elder, R., Humphries, J., Smith, G., and Strauss, R. 1984. *Shape Comparisons in Fishes: An Introduction to Morphometrics.* University of Michigan, Ann Arbor (in press).

Grayson, B., Weintraub, N., Bookstein, F. L., and McCarthy, J. 1984. A comparative cephalometric study of the cranial base in craniofacial syndromes. *Cleft Palate J.* (submitted).

Moyers, R. E., and Bookstein, F. L. 1979. The inappropriateness of conventional cephalometrics. *Am. J. Orthodontics* **75:**599–617.

Riolo, M. L., Moyers, R. E., McNamara, J. A., and Hunter, W. S. 1974. *An Atlas of Craniofacial Growth,* Center for Human Growth and Development, University of Michigan, Ann Arbor.

Skalak, R., Dasgupta, G., Moss, M. L., Otten, E., Dullemeijer, P., and Vilmann, H. 1982. A conceptual framework for the analytical description of growth. *J. Theor. Biol.* **94:**555–577.

Ontogenetic Allometry of the Skull and Dentition of the Rhesus Monkey (*Macaca mulatta*) 11

LARRY R. COCHARD

Introduction

The study of cranial variation has played an important role in primate systematics. Many different research strategies have been employed to investigate functional anatomy of the mammalian skull; but to fully understand the adaptive significance of cranial structure, it is necessary to consider cranial allometry, the relationship between the size and shape of the skull and body size. Over 20 years ago, le Gros Clark (1963, pp. 156–157) noted that

> It is of the utmost importance to understand the implications of allometry in evolutionary development as well as in the growth of the individual . . . As the result of allometric growth, the general shape and proportions of the skull (and also of other parts of the skeleton) may be very different in two quite closely related types. Such differences clearly have no great taxonomic value, since they may be related only to one major factor, i.e. body size.

Allometric analysis of the skull can have implications for systematics in two different ways. First, the *extent* to which the skull participates in a general growth phenomenon that affects the body as a whole can be identified. Sec-

LARRY R. COCHARD • Department of Cell Biology and Anatomy, Northwestern University Medical School, Chicago, Illinois 60611.

ond, any deviation of two parts of the skull in the same direction from a common allometric pattern in species or sexes of different size may indicate correlation in function. The investigation of cranial allometry, then, can direct attention to areas of the skull that may have functional and taxonomic significance and provide clues for functional interrelationships.

There have been few studies of cranial allometry despite continued emphasis on their importance. Giles (1956) was one of the first to study cranial allometry in primates. He regressed four pairs of cranial dimensions in the great apes and concluded that the chimp and gorilla are closely related through common allometric growth patterns that are distinct from those of the orang. Fooden (1969) conducted a similar analysis of the skulls of closely related species of Southeast Asian macaques. Shea (1983) and Albrecht (1978) followed Giles (1956) and Fooden (1969) with more comprehensive analyses of cranial allometry in great apes and macaques, respectively. Other studies of cranial allometry in primates have focused on baboons—on the growth of the snout in relation to growth of the neurocranium (Freedman, 1962) and the bending of the face on the braincase (Dechow, 1980). Most of the more recent interest in cranial allometry has been generated by the problem of cranial adaptations of the Plio-Pleistocene hominids. Many investigators (Tobias, 1967; Brace, 1973; Pilbeam and Gould, 1974; Corruccini and Ciochon, 1979) have argued that the difference in cranial anatomy between gracile and robust species is due to differences in body size and not to any significant dietary differentiation.

The allometry of sexual dimorphism in the skull has received even less attention. There have been many descriptive studies of sexual dimorphism in the primate skull that have identified the cranial dimensions that best discriminate between the sexes in different species (Cave and Steel, 1964; Napier and Napier, 1967; Fenart and Deblock, 1974; Alexander, 1981). The diagnostic value of this type of information is limited, however, when the significance of the dimorphism is unknown. The skulls of males and females may differ in size because the males are larger animals than females or because the male skull, in part, differs in function. This can only be determined through allometric analysis. If it is not known *why* male and female skulls are dimorphic, the results of discriminant studies on contemporary primates cannot be applied with confidence to the crania of extinct species.

Despite the investigation of specific allometric problems, our understanding of the relative growth of the skull is incomplete. Most studies focus on allometry of the brain and teeth or on growth of one part of the skull in relation to another with no external measure of body size. The scaling of the viscerocranium with body size is usually addressed by simply describing in a general way the positive allometry of the face during ontogeny and between species.

The purpose of this chapter is to test the hypothesis that the difference in cranial size and shape between male and female rhesus monkeys is a result of their difference in body size, measured as weight. The investigation of ontogenic allometry in this study will offer the following advantages. (1) Dental, cranial, and body size data were collected from the same animals; all three types of data have not been used in any study of cranial allometry (to my knowledge). (2) The relationship between dental and cranial variables and body size throughout growth is virtually unexplored. Tooth size is treated as a continuous variable in this cross-sectional study so that ontogenetic slopes for dental and cranial variables can be compared. (3) There is significant sexual dimorphism in tooth size, cranial size (Fig. 1), and body size in *Macaca mulatta*, so comparison of the three from an allometric perspective can offer some insight into the significance of cranial variation in general.

The study can also contribute to our understanding of dental function. The relative size of the postcanine teeth has yet to be explained (Kay, 1978), since interspecific studies of postcanine allometry have yielded slopes lower than those for masticatory requirements (based on body weight or basal metabolism), and females have relatively larger postcanine teeth than males in many primate species. The comparison of ontogenetic with interspecific dental allometry will have implications for the significance of relative tooth size.

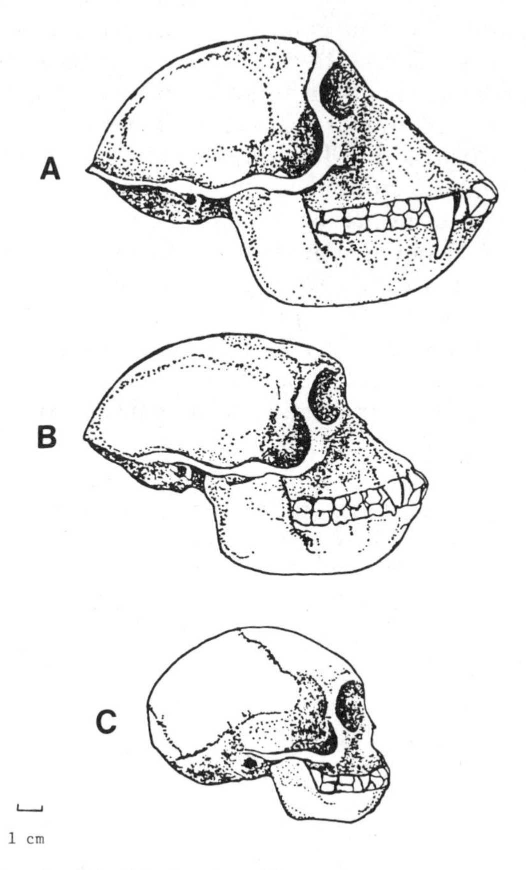

Fig. 1. (A) Adult male, (B) adult female, and (C) infant rhesus monkey skulls.

Materials and Methods

Body weights, lateral and superior cephalograms, and dental casts were taken on a total of 213 anesthetized monkeys (Table 1) at the Wisconsin Regional Primate Research Center. Monkeys younger than 3 years were housed with peers in large pens, while most older animals were maintained individually in 3 × 3 × 4 ft wire mesh cages. They were fed a standard diet of Purina Monkey Chow supplemented with fruit, similac, and molasses. Attempts were made to select healthy animals not subject to experimentation that might disrupt growth patterns. Monkeys specifically eliminated from data collection were hermaphrodites, castrates, monkeys injected with hormones, diseased animals, and emaciated older monkeys with tooth loss.

Mesiodistal and buccolingual diameters of all left teeth were measured on the dental casts. Mesiodistal length was measured parallel to the long axis of each tooth, and buccolingual diameter was taken perpendicular to length. For the incisors, mesiodistal length was measured along the cutting edge of the tooth, breadth at the base of the crown. On 50 casts, the mesiodistal length of the first molar, canine, and central incisor were also measured on the right side. There were no significant differences ($p < 0.05$) between the size of left and right teeth, so diameters of the antimeres of seven missing or damaged left teeth were substituted. Since tooth size was treated as a continuous variable in this study, partially erupted crowns were also measured. All teeth were measured twice, with mean crown diameters used in the analysis. The repeated measurements resulted in an average error just below the 0.2 mm reported for many studies of variation in tooth size (Schuman and Brace, 1955; Leutenegger, 1971; Swindler, 1976). The range was from 0.10 mm for breadth of the first permanent molar to 0.23 mm for mesiodistal length of the maxillary lateral incisor. The lowest and highest correlations between duplicate measurements were 0.969 and 0.989, respectively.

The cephalograms were traced on acetate paper, measured with a digi-

Tabe 1. Sample Size of Monkeys by Age and Sex

Age, years	Males	Females
0–1	2	7
1–2	16	7
2–3	10	20
3–4	11	10
4–5	14	20
5–6	10	7
6–7	8	2
>7	42	27
Total	113	100

tizer, then corrected for enlargement. Standard measuring points were used, with the following exceptions: (1) points for neurocranial height and breadth were located on the internal contours of the vault to more closely approximate size of the brain and associated tissues; (2) superior orbital point (Sirianni and Van Ness, 1978; Sirianni and Swindler, 1979), the intersection of the superior margin of the orbit with the contour of the anterior cranial fossa, was substituted for nasion, not only because it is close to the anteriormost extent of the anterior cranial fossa, but because the latter was difficult to locate on X rays of older monkeys; and (3) the occlusal plane (Carlson, 1976; McNamara, 1980), defined here as the line of best fit through the tips of the cusps of the maxillary postcanine teeth, was used as a reference for measuring mandibular length and facial prognathism instead of the traditional Frankfort Horizontal Plane. Body weight posed a problem in that many of the adults were visibly obese due to the lack of exercise in the restricted laboratory environment. The problem was minimized by using longitudinal weight data from the Primate Center records. Weights are taken on every monkey in the colony at least monthly, so for monkeys considered adult in this study (males older than 7 years, females older than 5 years), a mean of 5–10 weights

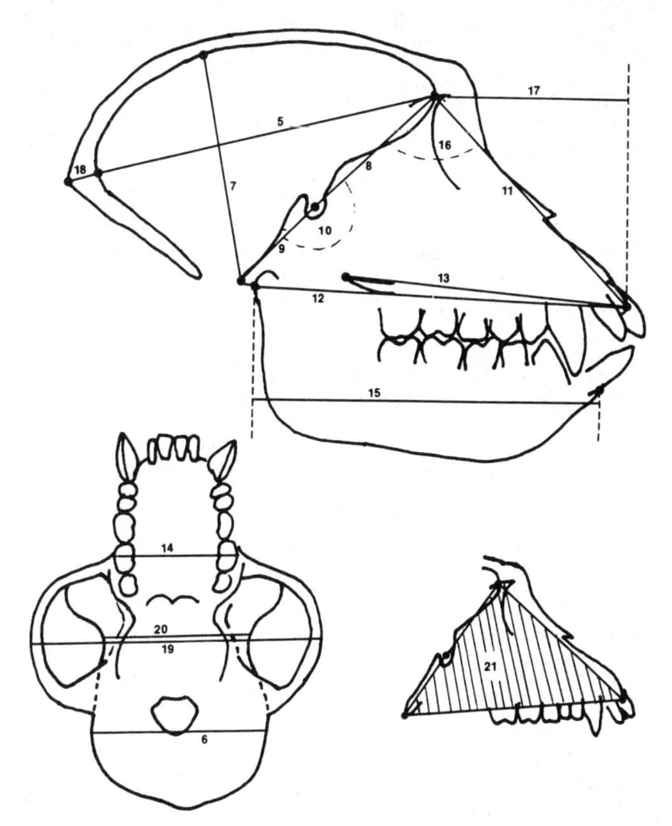

Fig. 2. Cranial measurements included in the analysis.

recorded when each animal was a young adult was used to represent adult body weight. Growing monkeys did not appear to be obese. This was substantiated by lower mean subscapular skinfold measurements in every subadult age class for males and females compared with the adult mean values (Cochard, 1981). The single weight taken during data collection was used for subadults.

Variables are defined as follows (Fig. 2):

1. Body size: cube root of body weight.
2. Postcanine tooth size: square root of the sum of mesiodistal length times buccolingual breadth for each maxillary postcanine tooth.
3. Canine size: mesiodistal length of the maxillary canine.
4. Incisor size: the sum of the mesiodistal crown lengths of the maxillary central and lateral incisors.
5. Neurocranial length: superior orbital point to a point on the internal table of bone at the level of inion.
6. Neurocranial breadth: euryon to euryon.
7. Neurocranial height: basion to a point on the internal table of bone at the top of the cranium chosen where height is perpendicular to length (visually determined).
8. Anterior limb of the cranial base: superior orbital point to sella turcica.
9. Posterior limb of the cranial base: sella turcica to basion.
10. Flexion of the cranial base: angle formed by superior orbital point to sella turcica to basion.
11. Upper facial height: superior orbital point to prosthion.
12. Basicranial length: basion to prosthion.
13. Palate length: prosthion to posterior nasal spine.
14. Dental arcade breadth: maximum distance across the maxillary postcanine tooth rows measured from the buccal surfaces of the tooth crowns.
15. Mandibular length: posterior margin of the mandibular condyle to interdentale inferior, measured parallel to the occlusal plane.
16. Flexion of the face on the neurocranium (facial angle): angle formed by sella turcica to superior orbital point to prosthion.
17. Facial prognathism: distance from superior orbital point to prosthion measured parallel to the occlusal plane.
18. Occipital bone thickness (nuchal crest thickness): maximum thickness at inion.
19. Bizygomatic diameter: distance between the most lateral points on the zygomatic arches.
20. Minimum frontal diameter: minimum distance across the frontal bone behind the orbits.
21. Facial area: the square root of the area of the trapezoid constructed from prosthion, superior orbital point, sella turcica, and basion.

Fifty films selected at random were measured on a second occasion. None of the means of the repeated measurements were significantly different ($p < 0.05$) from the originals. Correlation coefficients for the duplicate measurements ranged from 0.978 for occipital bone thickness to 0.993 for cranial breadth.

All the variables were converted to base-10 logarithms and analyzed on a Univac 1100 computer using programs from the Statjob series of the Madison Academic Computing Center, University of Wisconsin at Madison. Male and female ontogenetic regressions for each variable against body size were computed using principal components (Jolicoeur, 1963*a,b*).

Results and Discussion

Sexual Dimorphism and Cranial Growth

The allometric problem can be more specifically described by presenting an overview of growth changes in the rhesus skull as a function of age and by describing how sexual dimorphism in cranial size is related to growth and development. As in most primates, the skull of the infant rhesus monkey (Fig. 1) is dominated by the relatively large braincase. The face is very small in comparison: the zygomatic arches do not even project beyond the maximum width of the neurocranium. There is also very little sexual dimorphism in cranial and dental size. Table 2 gives the size difference between the male and female variables as a percentage of size in the females for monkeys between 1 and 2 years and for adults. Negative values indicate that the female means are larger than the male. For 17 of the 21 variables, young males are less than 4% larger than females, and of the 19 linear dimensions, females are larger than males on the average in eight cases, including all three dental variables. The only differences that are significant at the 5% level are maximum occipital thickness, incisor size (with females larger than males for both traits), upper facial height, and facial prognathism.

The major trend throughout ontogeny in primates is the rapid growth of the viscerocranium in relation to the neurocranium (Gould, 1977). This can be characterized in the rhesus sample by plotting the cross-sectional male and female means for bizygomatic diameter and cranial module (the average of neurocranial length, height, and breadth) by year (Fig. 3). There is little sexual dimorphism in size of the neurocranium at any age. In fact, there is little postnatal increase at all in the size of the braincase. After the rapid prenatal growth of the brain, its size and the size of the cranial vault are essentially a constant in males and females. Because of this lack of sex- and age-related variation in the size of the neurocranium, the growth of the viscerocranium in relation to the growth of the body will be the focus of this study.

The developmental pattern of growth in bizygomatic diameter in males

Table 2. Sexual Dimorphism[a] in the Cranial Traits of Infant and Adult Rhesus Monkeys

Trait	Age 1–2	Adult
1. Canine size	−1.76%	56.0%
2. Facial prognathism	15.20	32.40
3. Dental arcade breadth	−0.89	28.30
4. Occipital bone thickness	−11.70	25.30
5. Upper facial height	6.68	22.30
6. Palate length	3.86	20.90
7. Mandibular length	−0.11	20.20
8. Bizygomatic breadth	2.96	18.90
9. Basicranial length	−0.07	18.70
10. Cube root of body weight	1.38	17.60
11. Square root of facial area	2.81	13.90
12. Posterior cranial base length	2.08	11.90
13. Postcanine surface area	−1.90	6.92
14. Neurocranial height	1.89	5.67
15. Anterior cranial base length	2.42	5.51
16. Neurocranial breadth	3.82	4.83
17. Neurocranial length	1.33	3.12
18. Facial angle	−3.65	2.42
19. Incisor size	−6.67	0.66
20. Minimum frontal diameter	−2.26	0.63
21. Cranial base angle	−0.39	−2.49

[a][(Male − female)/female] × 100.

and females is typical for many of the dimensions of the face. Unlike growth of the cranial vault, males and females are similar in bizygomatic size at an early age, but their means begin to diverge around age 4. The allometric significance of the male and female ontogenetic trajectories will depend on the rate at which males and females diverge in body size. In an allometric analysis, time is removed as a variable so that increase in one dimension is considered only in relation to the change in size of another (Gould, 1966). If the rate of increase in bizygomatic diameter (for example) is equal to the rate of increase in body weight in both sexes, an ontogenetic plot of bizygomatic breadth against body weight would be reduced to a single curve, with males simply traveling a further distance along it. The cranial difference between the sexes would be an allometric phenomenon, related only to a difference in body size and implying no difference in function between males and females for the dimension.

While there is little size difference in the skulls of males and females at an early age, among the adults, sexual dimorphism ranges from 0.63% for minimum frontal diameter to 56% for canine size (Table 2). At 17.6% dimorphism, the cube root of body weight falls right in the middle of the variables ranked in order for the amount of sexual dimorphism. Not unexpectedly, the measurements relating to size of the neurocranium rank below body weight

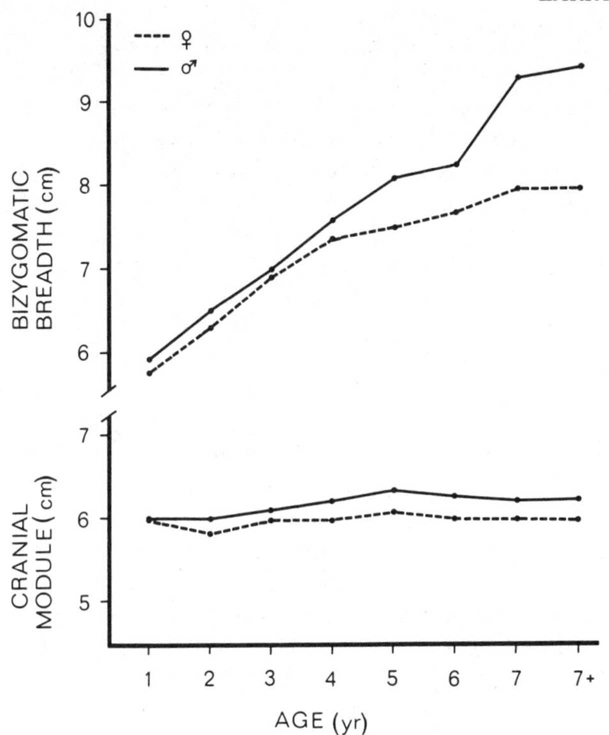

Fig. 3. Cranial module and bizygomatic diameter as a function of age in the rhesus monkey.

dimorphism, while most viscerocranial variables are above. Most of the traits are not very close to the cube root of body weight in percentage dimorphism. This does not necessarily mean that the difference between males and females for those traits is not closely related to body size; it only suggests that they may not scale isometrically with the cube root of body weight. Those characters that are closest to weight dimorphism are variables that relate to overall size of the face: basicranial length (basion to prosthion) and the square root of the cross-sectional area of the face in the midsagittal plane. Of particular interest is the amount of sexual dimorphism for the dental variables. While mesiodistal diameter of the canine ranks at the top of the list for dimorphism, there is virtually no difference between the size of male and female permanent incisors. And while males are 17.6% larger than females in body size (cube root of weight), they are only 6.9% larger in postcanine occlusal surface area.

Allometry Coefficients

The eigenvectors of the first principal components for all males and all females are given in Table 3. Each value is the cosine of the angle between the axis for the scatter of points for a given variable and the "major" axis of the point swarm for the entire data set in multidimensional space. Most of the values are positive, indicating a simultaneous increase in the variables. For example, an increase of 100 units along the major axis of the first component for all males corresponds to an increase of approximately 26 units in mandibular length and 18 units in bizygomatic diameter. Because the variables increase simultaneously in the first principal component, it is interpreted as a growth or size component (Jolicoeur, 1963*b*). The ratio of any two cosines measures the rate of increase in one variable relative to the other and is proportional to the bivariate allometry coefficient (Jolicoeur, 1963*a,b;* Jungers and German, 1981; Hills, 1982). Two loadings on the ontogenetic component for males are negative, which is not usually expected on the first principal component since most characters increase in size during ontogeny. One is the angle of the cranial base, which also has a negative loading for females. This angle decreases in both sexes as the cranial base becomes more flexed during growth. The other negative value is for minimum frontal diameter, which is closely related to brain size. While negative, it is close to zero, as are all the cosines for the variables that relate to brain size, which show little postnatal increase in the rhesus monkey (Fig. 3). The negative sign reflects a smaller mean value in adult males compared to infants and can be attributed to sampling error in the cross-sectional data.

Table 3. The First Principal Components for Cranial Traits and Body Weight in the Rhesus Monkey

Trait	All males	All females
1. Occipital bone thickness	0.369	0.335
2. Neurocranial length	0.003	0.001
3. Neurocranial height	0.040	0.030
4. Neurocranial breadth	0.023	0.014
5. Minimum frontal diameter	−0.006	0.030
6. Anterior cranial base length	0.047	0.061
7. Posterior cranial base length	0.152	0.133
8. Cranial base angle	−0.059	−0.060
9. Facial angle	0.054	0.052
10. Basicranial length	0.250	0.255
11. Palate length	0.285	0.321
12. Dental arcade breadth	0.169	0.203
13. Upper facial height	0.272	0.298
14. Bizygomatic diameter	0.184	0.137
15. Mandibular length	0.263	0.269
16. Square root of facial area	0.162	0.177
17. Facial prognathism	0.425	0.485
18. Postcanine surface area	0.251	0.325
19. Incisor size	0.091	0.089
20. Canine size	0.398	0.205
21. Cube root body weight	0.209	0.218
Percent total variance	87.0	73.9

Each variable (log-transformed) from the viscerocranium was regressed against the log of the cube root of body weight in males and females. The multivariate allometry coefficients, i.e., regression slopes, were computed by dividing the appropriate directional cosines from the first principal component. The multivariate slopes are compared with the bivariate reduced major axis slopes in Table 4. The Spearman rank order correlation coefficient between the multivariate and bivariate slopes is 0.99 for males and 0.98 for females. Plots of the regressions (Figs. 4–7) do not include the scatter of points for ease of comparison of the male and female curves. However, linearity of the regressions is suggested by the high correlation coefficients (Table 4). Of particular note are the correlations above 0.9 for postcanine occlusal surface area against weight in males and females. Linearity for these regressions justifies the treatment of postcanine tooth size as a continuous variable during ontogeny. The only distributions of points that did not appear linear and have the lowest correlations were canine and incisor size. This is due to the replacement of only one or two deciduous teeth by their permanent successors, so that measurement of partially erupted permanent teeth resulted in a downward bend in the middle of the distribution. For the purpose of this study, though, straight lines were forced through the point scatter only to demonstrate the general pattern of scaling in the anterior teeth. For the rest of the variables, the lower correlations are for the angles, measurements of the anterior and pos-

Table 4. Correlations (r) and Ontogenetic Allometry Coefficients[a] for Each Cranial Variable Regressed against the Cube Root of Body Weight in Male and Female Rhesus Monkeys

Trait[b]	Male			Female		
	PC	RMA	r	PC	RMA	r
1. Facial prognathism	2.03	2.03	0.93	2.22	2.12	0.88
2. Canine size	1.91	1.92	0.83	0.94	0.99	0.64
3. Occipital bone thickness	1.60	1.93	0.83	1.13	1.90	0.60
4. Palate length	1.36	1.35	0.95	1.47	1.37	0.91
5. Upper facial height	1.30	1.29	0.94	1.37	1.34	0.87
6. Mandibular length	1.26	1.24	0.96	1.23	1.15	0.94
7. Basicranial length	1.20	1.18	0.96	1.17	1.10	0.94
8. Square root of postcanine surface area	1.20	1.23	0.95	1.49	1.45	0.91
9. Bizygomatic breadth	0.88	0.92	0.96	0.63	0.76	0.94
10. Dental arcade breadth	0.81	0.83	0.94	0.93	0.91	0.92
11. Square root facial area	0.77	0.77	0.96	0.81	0.76	0.93
12. Posterior cranial base length	0.73	0.81	0.87	0.61	0.69	0.85
13. Incisor size	0.43	0.66	0.68	0.41	0.59	0.60
14. Facial angle	0.26	0.36	0.69	0.24	0.41	0.58
15. Anterior cranial base length	0.22	0.37	0.64	0.29	0.38	0.70
16. Cranial base angle	−0.28	−0.30	−0.87	−0.28	−0.29	−0.81

[a]PC, allometry coefficients computed from principal components; RMA, Reduced major axis allometry coefficients.

[b]All variables log-transformed.

terior limbs of the cranial base, and maximum occipital bone thickness. The very high correlations for the remaining variables imply linearity and are evidence that the body weights for the adults are estimates of their "normal" body sizes. If the adult weights were obese weights, "normal" body size would be overestimated compared with subadults, and a downward bend at the upper end of the regression curves and lower correlations would result.

Dental Allometry

The male and female regressions for the incisors, canines, and postcanine teeth are given in Fig. 4. The regression lines were fit through the means of the two variables, oriented with multivariate slopes, and were drawn to span the range in weight for males and females in the sample. The slopes are listed in Table 4.

The difference between the amount of occlusal surface area on the molars and premolars of adult males and females is not an allometric phenomenon. The ontogenetic slopes for each sex are both positively allometric, but the female coefficient is 24% higher than the male, 1.49 versus 1.20. The

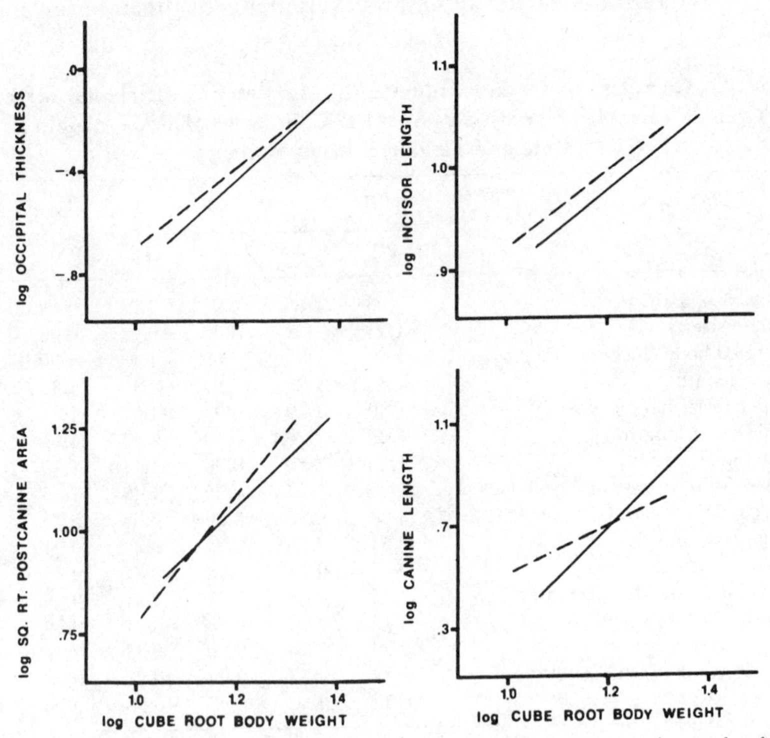

Fig. 4. Ontogenetic allometry of maximum occipital bone thickness, postcanine occlusal surface area, incisor length, and canine length in rhesus monkeys. The lines span the range in weight for males (solid line) and females (dashed line).

deciduous molars are of similar size in both sexes (Table 2), but adult females have much larger postcanine teeth for their body size than do males. Data from Albrecht (1978) (tooth size measured as the length of the maxillary postcanine tooth row and body size represented as mandibular length) show the same pattern for different species of Sulawesi macaques, and Kay (1978) has noted that females have relatively large second molars compared with males in many species of Old World monkeys. In the rhesus monkey at least, enlarged postcanine teeth in females appear to be functionally related to pregnant body size (Cochard, 1981). Near-term pregnant weights were available for 18 females in the sample. When pregnant weights are substituted for the nonpregnant weights of sexually mature females (older than 3 years), the ontogenetic male and female least-squares regressions are insignificantly different from each other ($p < 0.05$). Adult females require relatively large postcanine teeth because they must masticate more food to satisfy the elevated energetic costs of pregnancy and lactation that they repeatedly experience during their lifespan.

The interspecific scaling of postcanine tooth size in primates is close to isometry (Goldstein *et al.*, 1978) or even negatively allometric (Kay, 1975, 1978). The ontogenetic slopes in the rhesus monkey are higher than their interspecific analogs, and they are also unrelated to the scaling of basal metabolism during growth and development. Basal metabolic rate during ontogeny is related to body surface area (Portman, 1970), which would result in a lower slope than the positive allometry of postcanine tooth size in growing monkeys.

The relative size of the incisors during ontogeny is also unrelated to metabolism and interspecific allometry. Hylander (1975) plotted the combined width of the four permanent maxillary incisors against body mass for males of 57 species of anthropoids, most of which were cercopithecoid monkeys, and computed a near-isometric regression slope of 0.312 (a slope of 0.94 would result if the cube root of body weight were used). The point for *Macaca mulatta* falls slightly above the regression line. The incisors in the rhesus monkey, however, show strong negative allometry during ontogeny (Fig. 4, Table 4). The male and female slopes are 0.43 and 0.41, respectively. The deciduous incisors are nearly as large in absolute size as the permanent incisors. So for their body size, young males and females have much larger incisors than both adult rhesus monkeys and adults from "average" species similar in body size to the growing macaques. The cutting edge of the incisors during growth scales at a lower power than basal metabolism and at a much lower power than is found for interspecific allometry. Incisor allometry is similar to postcanine allometry not only in its independence in pattern, but also in a difference in scaling between the sexes. The male and female slopes are very close, but the regressions (Fig. 4) show that the y intercept is higher in females. Females have larger incisors for their body size at any age compared with males. It is not clear why female infants have relatively larger incisors than do males, but the metabolism/pregnancy argument to explain postcanine scaling can also be invoked to account for the difference in incisor size among adults. When pregnant weights are substituted for sexually mature females, the slope of the female regression is lowered so that the curves overlap at the adult end of the distributions.

Ontogenetic allometry of the canines is related to the function of the adult male canine. Primates are notable for the large male canine in many species (Napier and Napier, 1967; Swindler, 1976). Female canine size is conservative, perhaps having some minimal dietary function (Wolpoff, 1978; Smith, 1981). Male canines are enlarged in relation to the "baseline" size in females as an adaptation for predator defense (DeVore and Hall, 1965; Brace, 1973) and sexual selection, i.e., enlarged canines offer an advantage in aggressive encounters between males competing for females (Clutton-Brock and Harvey, 1976; Leutenegger and Kelly, 1977). Females have slightly larger deciduous canines than males, while the mesiodistal length of the adult male canine is 56% larger than the length of the female canine (Table 3). The ontogenetic scaling of canine size is a reflection of the difference in function of the permanent canines and accounts for the high degree of sexual dimorphism in adult canine size. Females are nearly isometric in relative canine size, with a slope of 0.94 (the reduced major axis slope is 0.99), while the male slope of 1.91 is the second highest of all the cranial variables. The difference in scaling between the sexes is simply a function of the enlarged permanent canine relative to body size in the male. The conservative nature of canine size in the female is reflected in the isometric scaling of the canine both ontogenetically and interspecifically. A variety of slopes distributed around 1.0 is reported in the literature, depending on taxa included and measurements representing canine size and body size. Wood (1979) suggests that female canine size is positively allometric, Smith (1981) and Wolpoff (1978) argue that female canines scale with moderate negative allometry, while Harvey *et al.* (1978) compute slopes showing perfect isometry.

Since the relationship between tooth size and body size during ontogeny has not been investigated previously (to my knowledge), the results are descriptive and can be used for generating hypotheses. With no prior information, the interspecific slopes can be considered null hypotheses for the ontogenetic dental allometry, but they are more useful simply as a basis for comparison and interpretation given the static versus dynamic nature of the different types of analyses. Three general points emerging from the results bear on the significance of ontogenetic dental allometry: (1) ontogenetic scaling is different from interspecific scaling; (2) relative incisor and postcanine tooth size during the growth period appears to be unrelated to masticatory requirements (based on basal metabolism); and (3) the patterns of ontogenetic scaling are related to function of the dentition in the adult. Based on these considerations, I suggest that allometry of the incisors, canines, and postcanine teeth in the growing rhesus monkey is primarily a function of the dynamic relationship between characters during ontogeny—the constraints of the process of growth and development are imposed upon functional considerations. Postcanine occlusal surface area scales with positive allometry not because adults require relatively larger teeth than infants, but because of size restrictions on the deciduous molars. The amount of space available in the mouth and jaws is limited, particularly in infants and during the prenatal period when the deciduous molars are developing. The relatively large amount of food that must be processed by relatively small postcanine teeth in juveniles is a transient situation that can be tolerated by simply suffering an increased amount of wear on the deciduous molars. But while the deciduous molars are relatively small, the deciduous incisors are nearly as large in absolute size as the permanent incisors. It makes no sense to suggest that this indicates that young monkeys incorporate a much higher percentage of fruit or other food items requiring extensive incisal preparation compared with adults. It is more reasonable to assume that deciduous incisors are relatively large only because the length of the jaws is limited more during ontogeny than is breadth (see section on cranial allometry). The width of the jaws is relatively unaffected by the transformation of a globular head in a neonate to a long-faced adult. All deciduous teeth will simply be as large as they can be within these size constraints.

That the morphology of a growing monkey is constrained by the process of growth and development does not suggest that function is not a consideration. Deciduous incisors, for example, are not required to be large only because there is space available. Since juveniles have the same general diet as adults (Clutton-Brock, 1977), there may be a minimum length for the combined cutting blade of the incisors to be effective in processing the diet of the species independent of size (age). It is just that deciduous molars are constrained more in size than incisors, and that at any given time, subadult morphology only has to approximate a functional optimum because functional requirements are transient until adulthood. But it is clear that (1) tooth size and body size are similar in infant males and females because their intrauterine development results in similar constraints in both sexes, (2) relative tooth size during growth is not necessarily related to function, and (3) relative tooth size in adults *is* related to function. Females have relatively larger incisors and postcanine teeth because of the elevated energetic costs of pregnancy and lactation, and males have relatively larger canines because of the social and defense functions of the canine. The results from dental allometry during ontogeny suggest that characters develop in the most expedient manner from the fetal state so that the morphology of the adult is adaptive. The masticatory apparatus in young monkeys is not a scaled-down version of the adult, but a compromise between the function required of the system and the mechanics and dynamics of the developmental state.

Regardless of the significance of dental allometry in the rhesus monkey, the ontogenetic results offer an ideal situation for the investigation of cranial allometry. The patterns of dental allometry are different in each part of the jaw and also different between the sexes. The incisors scale with marked negative allometry, with females having a larger y intercept. The canines and postcanine teeth are positively allometric (with the exception of the female canine), but males have a much larger slope for the former, females for the latter. It will be easy to test if cranial size in males and females is related to tooth size, body size, or neither.

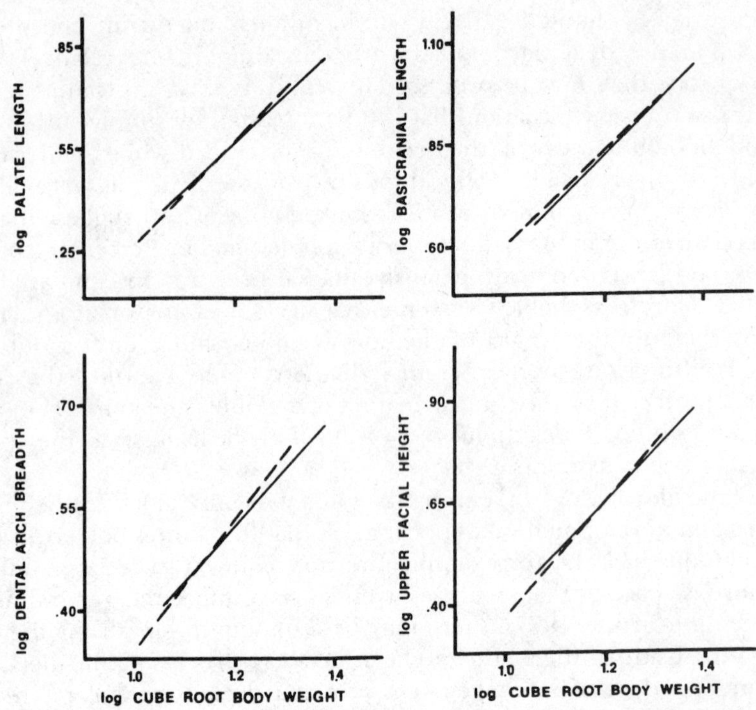

Fig. 5. Ontogenetic allometry of palate length, dental arcade breadth, basicranial length, and upper facial height in rhesus monkeys. The lines span the range in weight for males (solid line) and females (dashed line).

Cranial Allometry

The ontogenetic regressions for the male and female cranial variables are given in Figs. 4–7. Excluding the dental variables, which have already been discussed, there is a general pattern among the cranial regressions: the male and female curves are very close together. There are no statistics available to test for significant differences between any two slopes computed with principal components (though for crude estimates, the 95% confidence intervals for the equivalent least-squares regressions are on the order of 0.07–0.08), but the male–female allometry coefficients can be tested as a group with the Spearman rank correlation coefficient (Snedecor and Cochran, 1967). For 13 pairs of slopes (excluding the dental and brain size variables), the rank correlation between male and female values is 0.98, significantly different from zero ($p < 0.001$). Any particular pair of slopes for the two sexes may not be very close, but as a group they correspond very well.

The male and female regressions that are most similar are for the variables that measure the overall size and length of the viscerocranium: facial area and prognathism, upper facial height, palate length, and length of the posterior limb of the cranial base. The male and female curves for the angle

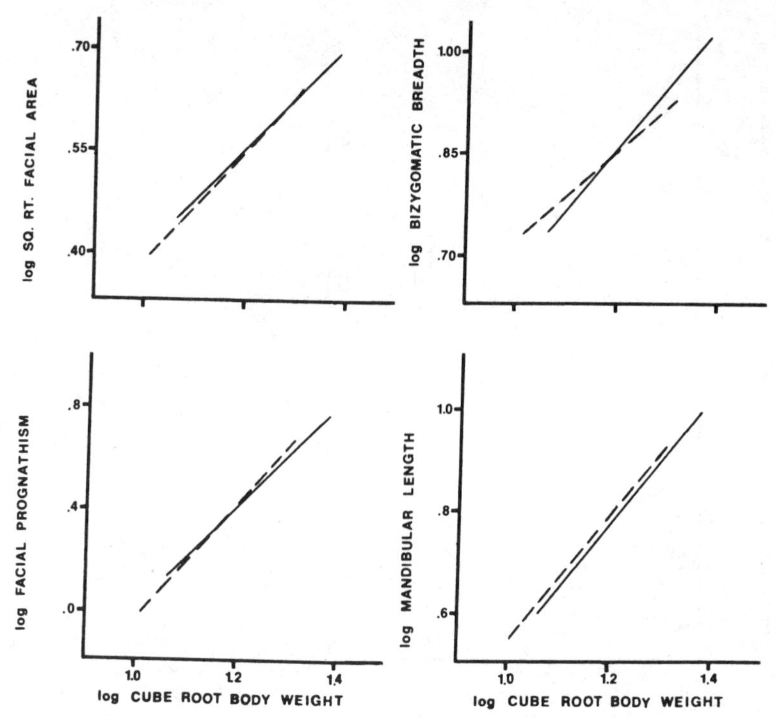

Fig. 6. Ontogenetic allometry of facial area, facial prognathism, bizygomatic breadth, and mandibular length in rhesus monkeys. The lines span the range in weight for males (solid line) and females (dashed line).

of the cranial base are also nearly superimposed upon each other. The facial skeleton of male and female rhesus monkeys throughout ontogeny is a single allometric function of body weight. The size and shape of the female viscerocranium and its position with respect to the neurocranium would be very similar to those in the male if females grew to the same body size as males.

The ontogenetic regressions also quantify the change in shape of the skull with increase in body size. The growth of the face anteriorly is positively allometric compared with body weight, while growth in breadth (and of the braincase) is negatively allometric. The five linear dimensions with an anteroposterior orientation—palate length, basicranial length, mandibular length, upper facial height, and facial prognathism—have a mean allometry coefficient of 1.46 for both sexes, with a range of 1.17–2.22. The mean slopes for breadth of the dental arcade and bizygomatic breadth are 0.87 and 0.76, respectively. The viscerocranium is relatively longer and narrower in bigger monkeys. Although positive allometry of the face is well documented for primates in general (Lavelle *et al.*, 1977), the results underscore the importance of using caution when interpreting the significance of sexual dimorphism in cranial characters. The variables in this sample are different from each other and from body weight in the amount of sexual dimorphism (Table 2). But most

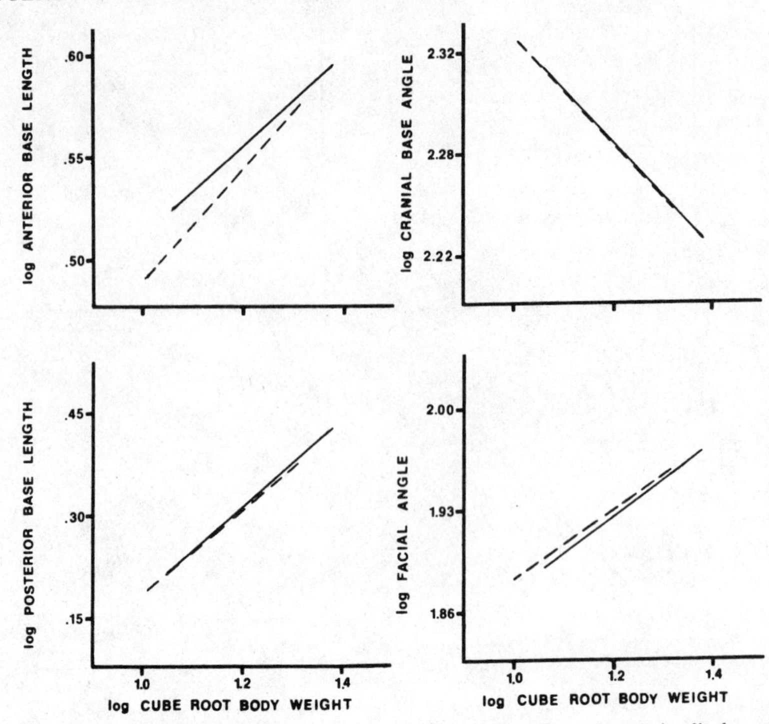

Fig. 7. Ontogenetic allometry of the anterior limb of the cranial base, posterior limb, cranial base angle, and facial angle in rhesus monkeys. The lines span the range in weight for males (solid line) and females (dashed line).

of the differences have no special biological significance, because proportions of the skull change with increase in size. If males and females follow the same ontogenetic trajectory, the steeper the regression slope, the greater the difference between male and female adult values; the lower the slope (i.e., the more negatively allometric), the closer together the adult values for males and females. In fact, none of the allometry coefficients in either sex is close to isometry. Because of the variance of the ontogenetic slopes, it is impossible to judge from adult means alone whether or not the amount of dimorphism for a particular character is due to allometry.

For all the cranial regressions, nonpregnant body weight was used for sexually mature females. The similarity between male and female curves suggests that none of the variates in the female scales specifically with the size of the postcanine teeth, which shows a close relationship with pregnant body size. Substituting pregnant weights would lower the slopes of the female regressions below those of the male curves. The evidence from cranial scaling suggests that increase in the average size of the viscerocranium in the female appears to be related to a general growth phenomenon influencing the body as a whole and not to any additional demands placed on the masticatory system.

The lack of correspondence between cranial and dental allometry is not surprising when the functions of the facial skeleton are considered. The bony snout in many large-faced monkeys has been described as a "dental muzzle" (Napier and Napier, 1967). While it is clear that larger mammals possess larger teeth (Creighton, 1980), many other factors are related to size of the viscerocranium. The skull is composed of a number of functional cranial components (Moss, 1973), which consist of the spaces, soft tissues, and skeletal units necessary to carry out the various functions of the head, such as digestion, respiration, or olfaction. These functions and their corresponding functional cranial components are interrelated to varying degrees. None of the measurements in this study is directly related to the length of the tooth row or size of the dentition. The closest, perhaps, is width of the dental arcade, and the ontogenetic slope is 0.12 higher in females than in males, which corresponds to the relatively larger incisor and postcanine size in females. But in the midsagittal plane, the cross-sectional area of the face includes the orbits, nasal and oral cavities, nasal and oral pharynx, and associated tissues. The face does not serve exclusively as an anchor for the dentition. There is a lot of room in the viscerocranium for tooth size to vary between individuals, sexes, or species and still have the size of the face relate to size of the body along a single allometric curve.

While the similar male and female regressions for cranial traits do not indicate any direct relation to dental allometry, the curves that differ between the sexes have a more direct bearing on the subject. Three variables show marked deviations in scaling between the sexes compared to the other variables: maximum occipital bone thickness, length of the anterior limb of the cranial base, and bizygomatic diameter. The latter has the greatest difference in slope between the sexes and is the most functionally important of the cranial variables in relation to mastication, since it is largely a measure of the size of the temporalis muscle. Bizygomatic breadth is negatively allometric in the male, with a slope of 0.88. It is even more negatively allometric in the female, scaling at the 0.63 power of the cube root of body weight. A large component of the dimension is related to brain size, which accounts for the negative allometry. And since brain size is nearly a constant in both sexes (Fig. 3), the steeper slope in the male indicates a relatively larger infratemporal fossa (and inferred size of the temporalis muscle) in males compared with females. The only other pattern that is similar to bizygomatic allometry is the scaling of canine size. Compared with males, females have relatively larger incisors and postcanine teeth, but males have much larger canines for their body size, implying that the temporalis muscle is important for function of the canine in the male. Studies of muscle size and function in the rhesus monkey support this contention. Grant (1973) investigated the size and biomechanics of the muscles of mastication in *Macaca mulatta* and found that the superficial masseter and the anterior and posterior temporalis muscles are hypertrophied in the male beyond what would be expected on the basis of differences in body size between the sexes. The physiological cross-sectional area of the posterior temporalis is 3.2 times greater in the male than in the female, and the anterior temporalis and superficial masseter are 2.2 and 2.6 times larger, respectively. Electromyographic studies of jaw function indicate that the superficial masseter is the dominant elevator of the jaw during mastication in the adult rhesus monkey (McNamara, 1974; Byrd and Garthwaite, 1981). Even though the anterior temporalis is in a mechanical position for postcanine mastication (Grant, 1973), it is apparently recruited only when extreme force is required. The primary active function of the temporalis is with use of the anterior teeth. If size of the temporalis muscle is not related to canine function, then aspects of theories on form and function will have to be reconsidered. An enlarged temporalis muscle in males is inconsistent with the function of relatively small molars and premolars, and, to a lesser degree, incisors; and the metabolic cost of the hypertrophied muscles in the male would seem to be too high to be simply a passive result of the male hormonal environment.

The other two cranial variables that scale differently between the sexes—maximum occipital bone thickness and anterior limb of the cranial base—have no clear explanation. They may be related to endocrine or other intrinsic sexual factors related to the growth process. This is particularly significant for maximum occipital thickness (Fig. 8). Females have a relatively larger nu-

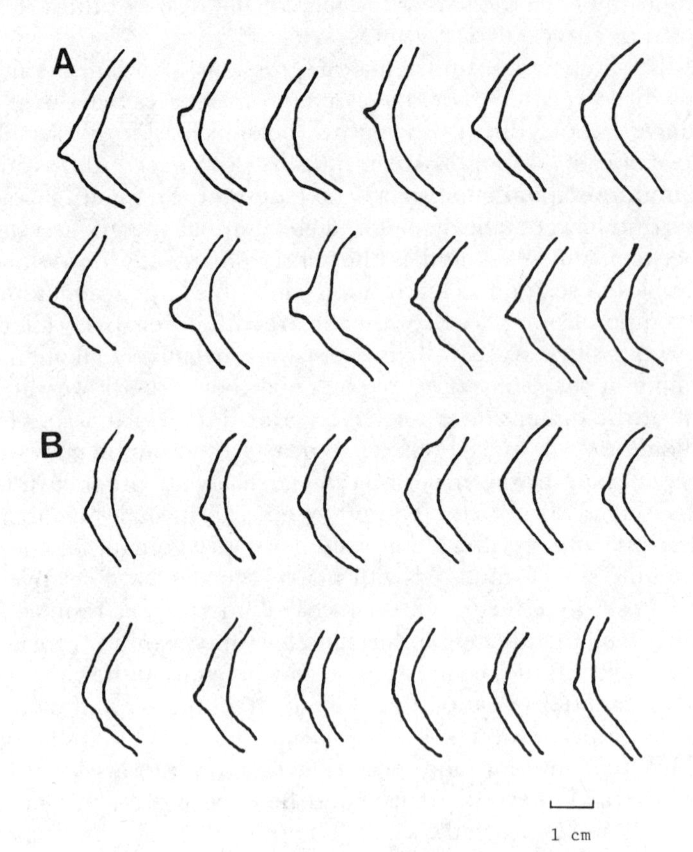

Fig. 8. Variation in the contour of the nuchal region in (A) male and (B) female rhesus monkeys.

chal crest than do males at every age. Cranial cresting has been traditionally related to the relative size of the musculature, with the nuchal crest acting as an anchor for the neck muscles, which must balance the weight of the viscerocranium on the vertebral column (Lavelle *et al.*, 1977). If this is true, cresting should be related to muscle attachment and function independent of age, sex, and species. But while female macaques have relatively thicker occipital bones than do males, both sexes have facial skeletons of equal proportion compared to body size. This indicates that the degree of cresting should not be strictly equated with size. There may be a sex- or species-specific component of variation in crest development that is unrelated to muscle size and function.

Problems for Future Research

Is there a common pattern of relative growth of the skull in primates? The size of the viscerocranium in male and female rhesus monkeys can be explained in terms of body size, but is cranial variation in higher taxonomic categories the result of extrapolation along the same allometric trajectory during ontogeny? The studies of Albrecht (1978) and Shea (1983), based on regressions among cranial variables in closely related primate species, are suggestive of common allometric trends, but it would be informative to extend allometric analyses to more diverse taxa and include a measure of body size external to the skull. Studies of the absolute growth of the primate skull are also suggestive of common relative growth patterns. Swindler *et al.* (1973) found that while the rate of growth of the skull of *Macaca nemestrina* and *Papio cynocephalus* from 3 months to 3 years is similar, the baboon had a larger skull than the macaque at all ages during the time interval. But they also noted that the baboon was a larger animal at birth—if body size had been taken into account, both the absolute *and* relative growth rates of the skull in the two genera may have been similar.

It is important to identify the mechanisms responsible for change in cranial proportions. Is a large skull compared to body size the result of extrapolation along a positively allometric curve, or an "adaptive shift" related to proportional changes occurring in early embryogeny, or a result of heterochrony (Gould, 1977), in this case an upward bend in the allometric regression due to increased cranial growth long after the body ceases growth? Considerations such as these would improve our understanding, for example, of species close to the early hominid radiation (e.g., *Sivapithecus* and *Gigantopithecus*), which apparently had large skulls and teeth for their body size (Pilbeam *et al.*, 1977).

The study of skull size from a developmental and allometric perspective also sheds some light on the significance of sexual dimorphism by eliminating one possible mechanism for producing the dimorphism. The larger primates tend to be the most dimorphic in size (Post *et al.*, 1978; Leutenegger and

Cheverud, this volume). If the allometric curves for the skull during ontogeny have steeper slopes in males than in females, extrapolation to larger body sizes would result in more dimorphism. But the results of this study indicate that sexual dimorphism in skull size is determined by the distance that males and females travel along a single ontogenetic trajectory and not a result of different trajectories. More cranial dimorphism in larger primates can still be an "artifact" or correlate of size [see the genetic variance model of Leutenegger and Cheverud (Chapter 3)], but it is not a result of sexual differences in ontogenetic allometry. More research needs to be directed to other species to see if the results of this study apply to primates with different degrees of sexual dimorphism.

An important remaining problem is the allometry of the muscles of mastication. The size of the muscles of mastication has a major impact on cranial structure (Tattersall, 1973), yet it is difficult to reconcile the results of allometric studies that bear on the subject. In the present study, the scaling of the inferred size of the temporalis muscle differs between the sexes and is apparently related to male canine function, even though the female has relatively greater masticatory requirements. Shea (1983) showed that the skull of the gorilla is a scaled-up version of the chimpanzee skull, including bizygomatic diameter, even though the species differ in diet and the amount of dimorphism in canine size. And Stockmann (1979) has demonstrated that the inferred size of the masseter muscle in 51 species of African bovids is related to diet independent of tooth size and inferred body size. Considering the importance of the muscles of mastication in dental function and in relation to cranial structure, it is surprising so little attention has been devoted to the allometry of muscle size.

Summary

The hypothesis that cranial size and shape in the rhesus monkey is an allometric function of body size independent of age and sex is supported for the facial skeleton and rejected for nuchal crest thickness and bizygomatic breadth (and inferred size of the temporalis muscle). The size of the viscerocranium is closely related to body size, but is independent of the size of the dentition. Ontogenetic dental allometry differs between the sexes and between each class of teeth, while a single curve describes the relative size of the bony face in the entire sample. Males have larger skulls than do females because they are larger animals, not because they have larger teeth.

The ontogenetic scaling of the dentition is related to function in the adult. Females have steeper ontogenetic slopes for the postcanine teeth and incisors because of the extra nutritional requirements of pregnancy and lactation. Males have a steeper slope for canine size because of the social and defense functions of their permanent canines. Ontogenetic slopes differ from interspecific ones because tooth size in young monkeys may be related more to constraints imposed by the dynamic process of ontogeny than to functional considerations.

More research is necessary to determine whether relative growth of the skull in higher taxonomic categories follows a single ontogenetic trajectory and to identify the mechanisms responsible for producing cranial proportions in species that deviate from a common allometric curve.

ACKNOWLEDGMENTS

A version of this chapter was presented in April 1983 at the meetings of the American Association of Physical Anthropologists in Indianapolis. I thank Dr. Walter Leutenegger, Dr. Karen Steudel, Dr. Kenneth Bennett, Dr. Jack Rutledge, and Dr. James Petterson for their help and guidance with the research on which this chapter is based. I am grateful to Dr. James Cheverud and Dr. Susan Larson for help with data collection and for many enjoyable discussions of allometry. Susan Finner kindly assisted in all phases of the research. This work was supported by a computer grant from the University of Wisconsin Graduate School, the Wisconsin Regional Primate Research Center, and National Science Foundation grant BNS-7910180.

References

Albrecht, G. A. 1978, *The Craniofacial Morphology of the Sulawesi Macaques,* Contr. Primat., Vol. 13 (F. Szalay, ed.), pp. 1–151, S. Karger, Basel.

Alexander, R. W. 1981. Sexual dimorphism in the cranium of *Cercopithecus aethiops*. *J. Hum. Evol.* **10:**311–317.

Brace, C. L. 1973. Sexual dimorphism in human evolution. *Yearb. Phys. Anthropol.* **16:**31–49.

Byrd, K., and Garthwaite, C. 1981. Contour analysis of masticatory jaw movements and muscle activity in the rhesus monkey (*Macaca mulata*). *Am. J. Phys. Anthropol.* **54:**391–400.

Carlson, D. S. 1976. Patterns of morphological variation in the human midface and upper face, in: *Factors Affecting Growth of the Midface* (J. McNamara, ed.), pp. 277–299, Center for Human Growth and Development, University of Michigan, Ann Arbor.

Cave, A. J., and Steel, F. L. 1964. Craniometric sex determination in the *Colobus* skull. *Proc. Zool. Soc. Lond.* **143:**503–510.

Clutton-Brock, T. H. 1977. Some aspects of intraspecific variation in feeding and ranging behavior in primates, in: *Primate Ecology: Studies of Feeding and Ranging Behaviour in Lemurs, Monkeys, and Apes* (T. H. Clutton-Brock, ed.), pp. 539–556, Academic Press, New York.

Clutton-Brock, T., and Harvey, P. 1976. Evolutionary rules and primate societies, in: *Growing Points in Ethology* (P. Bateson and R. Hinde, eds.), pp. 195–237, Cambridge University Press, Cambridge.

Cochard, L. R. 1981. Allometry of the Skull and Dentition in the Rhesus Monkey (*Macaca mulatta*), Ph.D. Diss., University of Wisconsin at Madison.

Corruccini, R., and Ciochon, R. 1979. Primate facial allometry and interpretations of australopithecine variation. *Nature* **281:**62–64.

Creighton, G. K. 1980. Static allometry of mammalian teeth and the correlation of tooth size and body size in contemporary mammals. *J. Zool. Lond.* **191:**435–443.

Dechow, P. 1980. Masticatory muscles and neurocranial–facial relationships in baboons. *Am. J. Phys. Anthropol.* **52:**219–220.

Devore, I., and Hall, K. 1965. Baboon ecology, in: *Primate Behavior* (I. Devore, ed.), pp. 20–52, Holt, Rhinehart, and Winston, New York.

Fenart, R., and Deblock, R. 1974. Sexual differences in adult skulls of *Pan troglodytes, J. Hum. Evol.* **3:**123–133.

Fooden, J. 1969. *Taxonomy and Evolution of the Monkeys of Celebes* (*Primates: Cercopithecidae*) Biblioteca Primatol., No. 10), S. Karger, New York.

Freedman, L. 1962. Growth of muzzle length relative to calvaria length in *Papio. Growth* **26:**117–128.

Giles, E. 1956. Cranial allometry in the great apes, *Hum. Biol.* **28:**43–58.

Goldstein, S., Post, D., and Melnick, D. 1978. An analysis of cercopithecoid odontometrics. I. The scaling of the maxillary dentition, *Am. J. Phys. Anthropol.* **49:**517–532.

Gould, S. J. 1966. Allometry and size in ontogeny and phylogeny. *Biol. Rev.* **41:**587–640.

Gould, S. J. 1977. *Ontogeny and Phylogeny,* Harvard University Press, Cambridge.

Grant, P. 1973. Biomechanical Analysis of the Masticatory Apparatus in the Rhesus Macaque (*Primates, Macaca mulatta*), Ph.D. Diss., University of California, Berkeley.

Harvey, P. H., Kavanagh, M., and Clutton-Brock, T. 1978. Sexual dimorphism in primate teeth. *J. Zool. Lond.* **186:**475–485.

Hills, M. 1982. Bivariate vs. multivariate allometry: A note on a paper by Jungers and German, *Am. J. Phys. Anthropol.* **59:**321–322.

Hylander, W. L. 1975. Incisor size and diet in anthropoids with special reference to the cercopithecidae. *Science* **189:**1095–1098.

Jolicoeur, P. 1963*a*. The multivariate generalization of the allometry equation. *Biometrics* **19:**497–499.

Jolicoeur, P. 1963*b*. The degree of generality of robustness in *Martes americana. Growth* **27:**1–27.

Jungers, W., and German, R. 1981. Ontogenetic and interspecific skeletal allometry in nonhuman primates: Bivariate and multivariate analysis. *Am. J. Phys. Anthropol.* **55:**195–202.

Kay, R. 1975. Allometry and early hominids. *Science* **189:**63.

Kay, R. 1978. Molar structure and diet in extant cercopithecidae, in: *Development, Function, and Evolution of Teeth* (P. Butler and K. Joysey, eds.), pp. 309–339, Academic Press, New York.

Lavelle, C., Shellis, R., and Poole, D. 1977. *Evolutionary Changes to the Primate Skull and Dentition.* C. C. Thomas, Springfield, Illinois.

Le Gros Clark, W. E. 1963. *The Antecedents of Man,* 2nd ed., Harper and Row, New York.

Leutenegger, W. 1971. Metric variability of the postcanine dentition in colobus monkeys. *Am. J. Phys. Anthropol.* **35:**91–100.

Leutenegger, W., and Kelly, J. 1977. The relationship of sexual dimorphism in canine size and body size to social, behavioral and ecological correlates in anthropoid primates. *Primates* **18:**117–136.

McNamara, J. A. 1974. An electromyographic study of mastication in the rhesus monkey (*Macaca mulatta*). *Arch. Oral Biol.* **19:**821–823.

McNamara, J. A. 1980. Functional determinants of craniofacial size and shape. *Eur. J. Orthodontics* **2:**131–159.

Moss, M. 1973. A functional cranial analysis of primate craniofacial growth, in: *Craniofacial Biology of the Primates* (M. Zingeser, ed.), Vol. 3, pp. 191–208, S. Karger, Basel.

Napier, J. R., and Napier, P. H. 1967. *A Handbook of Living Primates,* Academic Press, New York.

Pilbeam, D., and Gould, S. 1974. Size and scaling in human evolution. *Science* **186:**892–901.

Pilbeam, D., Meyer, G., Badgley, C., Rose, M., Pickford, M., Behrensmeyer, A., and Ibrahim Shah, S. 1977. New hominoid primates from the Siwaliks of Pakistan and their bearing on hominoid evolution. *Nature* **270:**689–695.

Portman, O. 1970. Nutritional requirements of nonhuman primates, in: *Feeding and Nutrition of Nonhuman Primates* (R. Harris, ed.), pp. 87–115, Academic Press, New York.

Post, D., Goldstein, S., and Melnick, D. 1978. An analysis of cercopithecoid odontometrics. II.

Relations between dental dimorphism, body size dimorphism and diet. *Am. J. Phys. Anthropol.* **49:**533–544.

Schuman, E., and Brace, C. L. 1955. Metric and morphologic variations in the dentition of the Liberian chimpanzee; comparison with anthropoid and human dentitions, in: *The Nonhuman Primates and Human Evolution* (J. Gavan, ed.), pp. 61–90, Wayne State University Press, Detroit.

Shea, B. T. 1983. Size and diet in the evolution of African ape craniodental form. *Folia Primatol.* **40:**32–68.

Sirianni, J. E., and Swindler, D. R. 1979. A review of postnatal craniofacial growth in Old World monkeys and apes, *Yearb. Phys. Antrhopol.* **22:**80–104.

Sirianni, J. E., and Van Ness, A. L. 1978. Postnatal growth of the cranial base in *Macaca nemestrina, Am. J. Phys. Antrhopol.* **49:**329–340.

Smith, R. J. 1981. Interspecific scaling of maxillary canine size and shape in female primates: Relationships to social structure and diet. *J. Hum. Evol.* **10:**165–173.

Snedecor, G., and Cochran, W. 1967. *Statistical Methods,* 6th ed., Iowa State University Press, Ames, Iowa.

Stockmann, V. W. 1979. Differences in the shape of the mandibles of African bovidae (Mamalia) in relation to food composition. *Zool. Jahrb. Syst.* **106:**344–373.

Swindler, D. R. 1976. *Dentition of Living Primates,* Academic Press, New York.

Swindler, D., Sirianni, J., and Tarrant, L. 1973. A longitudinal study of cephalofacial growth in *Papio cynocephalus* and *Macaca nemistrina* from three months to three years, in: *Craniofacial Biology of the Primates* (M. Zingeser, ed.), Vol. 3, pp. 227–240, S. Karger, Basel.

Tattersall, I. 1973. Cranial anatomy of the Archaeolemurinae (Lemuroidea, Primates). *Anthropol. Pap. Am. Mus. Nat. Hist.* **52**(1)**:**1–110.

Tobias, P. V. 1967. *Olduvai Gorge,* Cambridge University Press, Cambridge.

Wolpoff, M. 1978. Some aspects of canine size in the Australopithecines. *J. Hum. Evol.* **7:**115–126.

Wood, B. A. 1979. Models for assessing relative canine size in fossil hominids. *J. Hum. Evol.* **8:**493–502.

Allometric Scaling in the Dentition of Primates and Insectivores

12

PHILIP D. GINGERICH AND
B. HOLLY SMITH

Introduction

Size is probably the single most important determinant of body architecture, physiology, ecology, life history, and social organization in mammals. Morphological characteristics associated with each of these broadly defined aspects of structure and function can profitably be studied in relation to size, and none can be fully understood without considering size. Here we outline the relationship of tooth size to body size in frugivorous and folivorous primates. For comparison we shall also consider the relationship of tooth size to body size in insectivorous mammals.

Why study tooth size in relation to body size? There are at least three distinct ways that the relationship of tooth size to body size is important:

1. *Functional inference.* Physiological requirements of animals change in predictable ways as body size changes, and one way to study the functional significance of a characteristic like tooth size is to examine how it changes in relation to body size and coordinated physiological changes.

2. *Baseline comparison.* A clear understanding of the common or general relationship of tooth size to body size permits one to identify outliers that require different and special functional explanation.

PHILIP D. GINGERICH • Museum of Paleontology, University of Michigan, Ann Arbor, Michigan 48109. B. HOLLY SMITH • Center for Human Growth and Development, University of Michigan, Ann Arbor, Michigan 48109.

3. *Prediction of body mass.* Body size is a powerful predictor of diet and other life history parameters in living primates and other mammals, and tooth size can be used to estimate body size in fossils, providing access to a more complete reconstruction of the biology of extinct species than would otherwise be possible.

Geometry and Metabolism

Organisms are commonly described in terms of lengths, areas, and volumes. Taken singly or together, length, area, and volume (or weight) are the measures of size. These simple elements of different dimension are interrelated in complex geometric ways. Length can be measured in any one or a combination of two or three independent orthogonal directions. Two lengths are necessary to define an area, and three are required to define a volume. Change in one length affects both area and volume, and leads to a change of shape. In fact, any change in any single measure of size (length, area, or volume) leads to a change in shape. Shape remains constant only when all lengths are changed by equal proportions (not equal amounts), and even here areas and volumes change disproportionately.

As an example, imagine a morphological transformation preserving shape. This change will be *isometric,* requiring all linear dimensions to change by a constant proportion. If one length, say height, doubles, what are the consequences for breadth and depth? These too must double, as shown in Fig. 1. What about surface area? Surface area will not only double but, being of greater dimension, surface area will increase as doubling raised to the power of two. Volume will increase as doubling raised to the power of three. These powers and proportions hold no matter how complex the shape involved.

We can write a series of equations describing the relationships of length, area, and volume under the constraint of isometry, when shape is preserved:

$$\text{length } Y_1 = b_1(\text{length } X)^{1.0} \tag{1}$$

$$\text{area } Y_2 = b_2(\text{length } X)^{2.0} \tag{2}$$

$$\text{volume } Y_3 = b_3(\text{length } X)^{3.0} \tag{3}$$

where b_i is a constant of proportionality and the exponents k_i equal to 1.0, 2.0, and 3.0 are associated with length, area, and volume, respectively.

One additional comparison is important here, comparison of a volume to an area. Considering that length $X = [(\text{area } Y_2)/b_2]^{1/2}$ [Eq. (2)] and that length $X = [(\text{volume } Y_3)/b_3]^{1/3}$ [Eq. (3)], then

$$\frac{\text{volume } Y_3}{b_3} = \left(\frac{\text{area } Y_2}{b_2}\right)^{3/2}$$

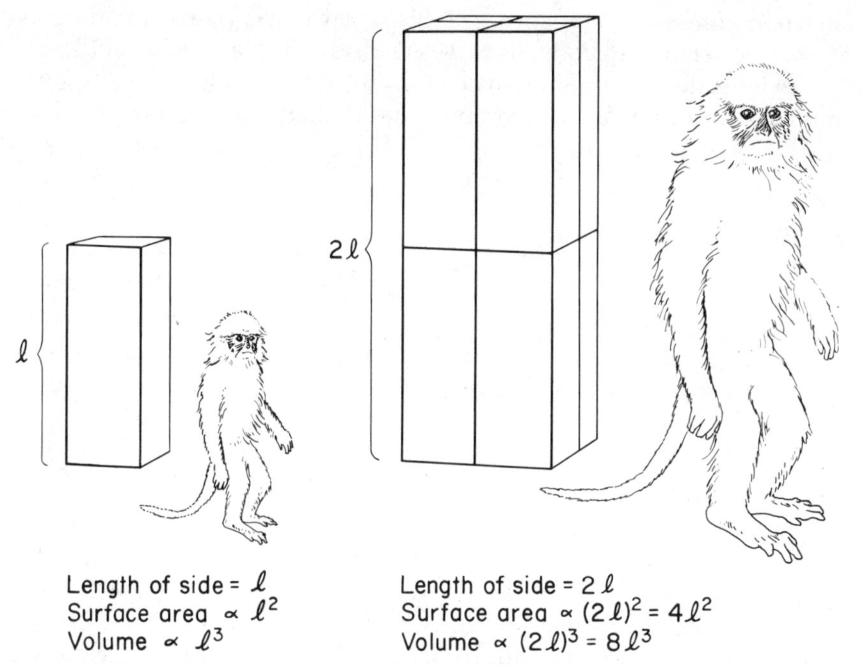

Fig. 1. Isometric transformation of simple or complex figures. Doubling all lengths l in figures at left leads to disproportionate increases in surface area and volume (or weight) in figures at right. To maintain shape at different sizes, all lengths must be changed by the same proportion, and area and volume will necessarily change as the second and third powers of this proportion.

and

$$\text{volume } Y_3 = \frac{b_3}{(b_2)^{3/2}} \left(\text{area } Y_2\right)^{3/2}$$

Consequently

$$\text{volume } Y_3 \propto (\text{area } Y_2)^{1.5} \tag{4}$$

Using a similar series of steps, but solving for area, one has

$$\text{area } Y_2 \propto (\text{volume } Y_3)^{0.67} \tag{5}$$

The exponents (or their inverses) 1.0, 2.0 (0.5), 3.0 (0.33), and 1.5 (0.67) are the powers associated with isometric or "geometric" scaling, preserving shape while the relationships of length, area, and volume are altered.

Nonisometric changes, involving changes in shape, are termed *allometric*. While there is a finite number of isometric relationships, allometry includes an infinite number of possibilities. For our purposes, only one allometric change is important. This is the allometry of heat production (basal metabolism) in eutherian mammals relative to body mass, which has been determined empirically to involve an allometric exponent k of 0.75 (Kleiber, 1932; Schmidt-Nielsen, 1975). Note that this metabolic exponent of 0.75 differs from all of the geometric exponents and their inverses discussed above. If, in examining tooth crown area in relation to body mass, we can distinguish a scaling coefficient of 0.75 from 0.67, then we might suspect that crown area is somehow responding to requirements of metabolism and not simply changing isometrically.

Methods

Quantitative study of size and shape requires careful consideration of the nature of the problem to be solved, and appropriate choices of variables to be compared, models of underlying relationship, and curve fitting techniques.

Choice of Variables

Body mass is usually used as a baseline in the study of primate body architecture, locomotion, substrate preference, ecology, faunal structure, home range size, life history, social organization, sexual dimorphism, and other parameters, but this is not invariably the case. Choice of a body size standard is often dictated by the standard employed in other studies to which comparison will be made. Occasionally it may be necessary to repeat allometric analyses with different body size standards to make them comparable to a diverse range of related studies. It is sometimes best to study the allometry of body architecture in relation to body length or limb length rather than body mass. Cranial allometry in a series of fossil species, for example, might best be studied relative to one or more cranial dimensions rather than body mass because body mass cannot be measured directly in fossil species. However, comparison with modern species for which body mass is known might make it advantageous to use predicted body mass in the fossils as a baseline. The choice of a body size standard must be appropriate for the overall objectives of any given study.

Choice of a measure or measures of tooth size also affects the results in studying tooth size allometry. Hylander (1975) used maxillary incisor breadth as a measure of tooth size in studying functional differences in the anterior dentition of folivorous and frugivorous cercopithecoids. Kay (1975*b*) measured total crown length and other detailed morphological characteristics of lower second molars to relate functional features of molar structure to diet in primates. Gould (1975) measured the sum of maxillary postcanine crown areas in studying the relationship of tooth size to metabolism and/or environmental grain. Each of these measures of tooth size is appropriate for a specific study, yet none can be regarded as representative for all possible problems of allometric scaling one might encounter.

Tooth crown area (crown length multiplied by width) is often used as a measure of tooth size, but crown area does not necessarily scale like crown length or crown width alone. Folivorous primates tend to have relatively longer and narrower crowns than do frugivorous primates. An index of crown length relative to width could be used to separate folivores from frugivores, but crown area combines length and width in a way unlikely to be sensitive to such dietary differences. This is advantageous in some situations and disadvantageous in others. Here again, the measure of choice must be appropriate for the problem under study.

Power Function Models

Power functions are customarily used to study the structural size relationships of various parts of organisms. The appropriateness of power functions in allometric studies was outlined years ago by Huxley (1932). We are interested not in "growth" but in "relative growth," difference not in *size* but in *proportion*. The general power function model used in studies of isometry and allometry is customarily written

$$Y = bX^k \tag{6}$$

where, for our purposes, X is body size, Y is tooth size, k is the exponent or power of body size (also referred to as the coefficient of allometry), and b is a constant. Equation (6) can be written in logarithmic form

$$\ln Y = k \ln X + \ln b \tag{7}$$

which is a linear equation of slope k and y intercept $\ln b$.

In a thoughtful analysis of allometry, Smith (1980) has argued that logarithmic transformation of raw data is inappropriate in many cases, and that linear functions are preferable to power functions for purposes of analysis. We disagree, for several reasons. As noted above, allometry is the study of difference not in absolute size but in proportion. In algebraic terms, a difference in one variable u with respect to another variable v can be written as du/dv, and a difference in proportion is this quantity du/dv considered with respect to u, i.e., $(1/u)\, du/dv$. It is a property of natural logarithms that

$$\frac{1}{u}\frac{du}{dv} = \frac{d \ln u}{dv} \tag{8}$$

Hence, differences in proportion, $(1/u)\, du/dv$, can be studied simply and directly by studying differences in logarithmically transformed original variables.

All of the equations relating length, area, and volume discussed above [Eqs. (1)–(5)] are power functions, as they must be when comparing characteristics of different dimension. Linear models are a special case of power function models that are only appropriate when comparing quantities of the same dimension. Power functions, as general models, include comparisons of quantities of the same dimension as a special case. Finally, as we shall show in the following analysis, power functions often fit actual data significantly better than do simple linear models.

Curve Fitting Techniques

In the Introduction we discussed three distinct objectives of a study of tooth size in relation to body size. The first two of these objectives, functional inference and baseline comparison, require quantification of the structural relationship of tooth size to body size. Error is inherent in both variables and, given some degree of correlation between variables, the principal or major axis best describes the slope or scaling relationship of the variables. The third objective, prediction of body size from tooth size, requires a different approach. Here tooth size is assumed to be known and error is inherent in only one variable, predicted body size. Least squares regression is explicitly designed for prediction problems, minimizing error in the dependent variable. Regression of body size on tooth size is appropriate in determining equations for predicting body size from tooth size.

Slopes calculated by regression are systematically lower than those calculated as principal or major axes, with the difference in calculated slopes increasing as the correlation between variables decreases. If regression is used to analyze structural relationship, the value of slopes (and allometric coefficients) will be systematically underestimated. The choice of methods for estimating slopes makes a difference, and this choice should be made with the differing objectives of analyzing structure versus making predictions clearly in mind.

Tooth Size and Body Size in Primates

Our objectives in studying primate tooth size scaling in relation to body size are several. We are interested to know whether the sizes of individual teeth, or upper and lower cheek teeth considered as a unit, scale isometrically or allometrically. This question is considered in relationship to the more specific problem of geometric versus metabolic scaling of tooth size. Second, we want to know the uniformity of tooth size–body size scaling in primates and to identify outliers that do not fit baseline scaling as defined by the majority of primate species. Finally, we want to know how to use tooth size to predict body size.

Metabolic scaling is defined in terms of body mass, and for this and other reasons we have used body weight or mass as our criterion of body size. Tooth size is measured as crown area—crown length multiplied by width—to avoid dietary grouping associated with tooth shape and to incorporate more information in our measure of tooth size than any single measurement of tooth size would yield. All weights and dental measurements were transformed logarithmically to facilitate analysis using power functions. Slopes of principal axes were used to calculate allometric coefficients in structural analyses of tooth size and body size, and regressions were used to predict body size from tooth size. Sources of data and our principal conclusions regarding allometric scaling of tooth size and the use of tooth size to predict body size have been published previously (Gingerich *et al.,* 1982). Here we review the general pattern of allometric scaling in the primate dentition.

Scaling of Individual Teeth

A simple scatter plot of tooth crown area and body weight is shown in Fig. 2A. This graph includes 43 species of folivorous and frugivorous primates, with males and females plotted separately. The distribution is curvilinear, approximating a parabola in shape, with slope decreasing as body size increases (solid line). Assuming a linear rather than curvilinear model for this distribution, the correlation of untransformed tooth size and body size measurements is 0.937. Transforming all measurements in Fig. 2A to logarithms yields the distribution shown in Fig. 2B. Here the distribution approximates a straight line. The correlation of transformed crown area and body mass measurements is 0.967, a correlation significantly higher than that for untransformed measurements ($p < 0.02$). The slope of the principal axis of the linear distribution in Fig. 2B is 0.638; this is the allometric coefficient describing how the crown area of the lower first molar of primates scales with respect to body mass. Correlations and slopes (allometric coefficients) for the remaining upper and lower teeth of folivorous and frugivorous primates are listed in Table 1.

All of the slopes listed in Table 1 are allometric in the sense that none of them is exactly equal to the coefficient of 0.67 expected for isometry of tooth crown area and body volume or mass. However, most of the slopes do not differ significantly from 0.67 in a statistical sense ($p < 0.05$). Of those that do differ significantly, six are greater than 0.67 and two are less than 0.67. Interestingly, there is a definite pattern of scaling coefficients corresponding to incisor, canine, and cheek tooth functional fields in the primate dentition (Fig. 3). Central cheek teeth have the lowest scaling coefficients, and teeth anterior and posterior to these have higher coefficients. Lower teeth as a rule have higher allometric coefficients than their counterparts in the upper dentition. The only notable exception is in the lower incisors, which have relatively low allometric coefficients for teeth in the lower dentition and relatively low coefficients by comparison with upper incisors.

The pattern of allometric scaling of individual teeth in the primate dentition is such that anterior and posterior cheek teeth become relatively larger

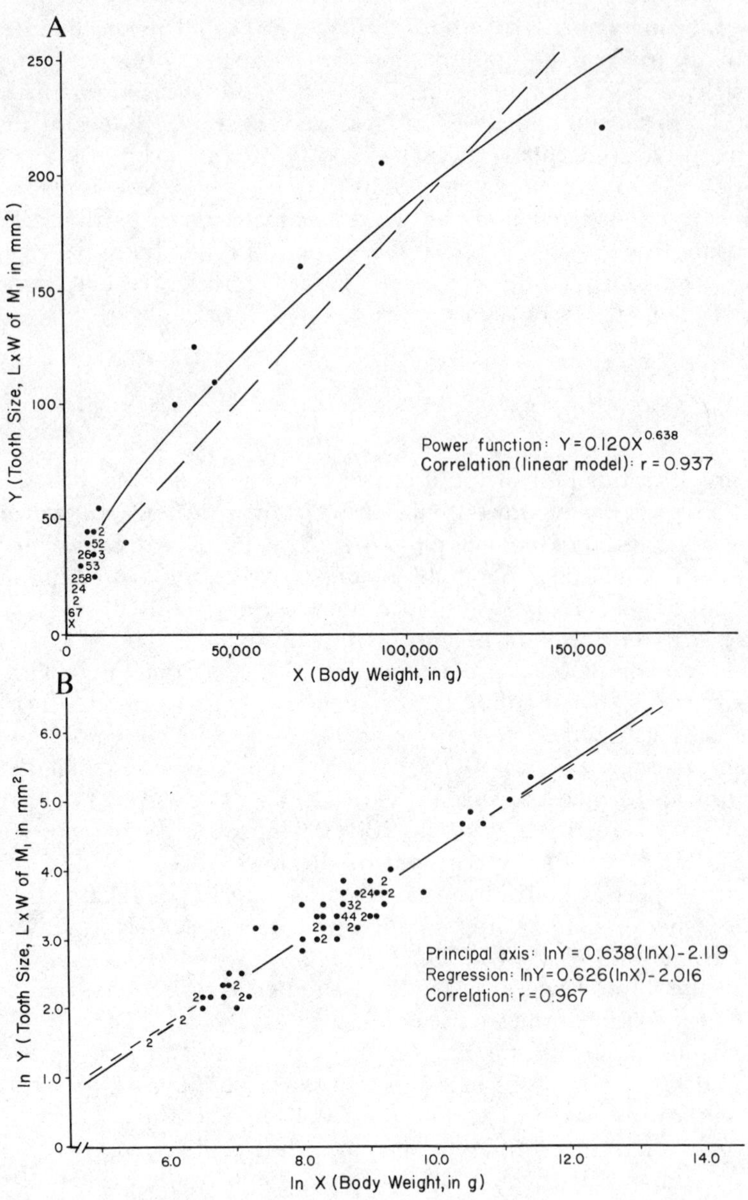

Fig. 2. Allometry of tooth size and body weight in 43 species of frugivorous and folivorous primates. Males and females are plotted separately. (A) Bivariate scatter with body weight on the abscissa and M_1 crown area on the ordinate. Note curvilinear relationship. A linear model applied to this distribution has a high correlation ($r = 0.937$) because most points are clustered at one end of the distribution. (B) Logarithmic transformation of the same distribution, showing a linear relationship of the transformed variables and a higher correlation ($r = 0.967$). The allometric coefficient k, the slope of the principal axis (solid line), is 0.638, and the ln$y$ intercept b is -2.119. Note how the use of regression (dashed line) to estimate slopes systematically underestimates k (regression $k = 0.626$ versus principal axis $k = 0.638$; this difference increases as the correlation decreases). Data from Gingerich *et al.* (1982).

Table 1. Correlation and Allometric Scaling of Tooth Crown Area and Body Weight in Frugivorous and Folivorous Primates[a]

Tooth position	Sample size $2N$	Correlation r	Slope k	95% Confidence interval for slope	Intercept $\ln b$
Individual upper teeth					
I^1	73	0.835	0.80	0.68–0.93	−3.70
I^2	73	0.858	0.75	0.65–0.86	−3.75
C^1	82	0.947	0.72	0.67–0.78	−2.71
P^2	27	0.902	0.70	0.57–0.84	−2.76
P^3	83	0.943	0.65	0.60–0.70	−2.48
P^4	83	0.934	0.59	0.54–0.64	−1.85
M^1	82	0.946	0.57	0.53–0.61	−1.27
M^2	83	0.945	0.68	0.63–0.73	−2.10
M^3	81	0.947	0.78	0.72–0.84	−3.19
Individual lower teeth					
I_1	70	0.854	0.72	0.62–0.83	−3.54
I_2	70	0.921	0.65	0.58–0.72	−2.85
C_1	75	0.882	0.79	0.69–0.89	−3.53
P_2	27	0.913	0.76	0.63–0.91	−3.15
P_3	83	0.954	0.78	0.72–0.83	−3.43
P_4	83	0.955	0.65	0.61–0.70	−2.56
M_1	83	0.967	0.64	0.60–0.68	−2.12
M_2	83	0.968	0.73	0.69–0.77	−2.72
M_3	81	0.947	0.80	0.74–0.86	−3.32
Sum of upper and lower cheek teeth					
P^2–M^3	78	0.949	0.62	0.57–0.67	−0.28
P_2–M_3	77	0.964	0.69	0.65–0.73	−0.91

[a] Number of species (N) includes males and females entered in analysis separately. Sources of data and confidence intervals are described in Gingerich *et al.* (1982).

with increasing body size. Lower incisors remain the same and upper incisors become relatively larger, while upper cheek teeth remain about the same and lower cheek teeth become relatively larger with increasing body size. Smaller primates have relatively smaller (narrower) lower cheek teeth and more V-shaped dental arcades, while upper and lower cheek teeth are more nearly equal in size (width) and the dental arcades are more U-shaped in larger primates.

Given the small number of tooth positions scaling with positive allometry (greater than isometry), and a strong tendency for these to be located at the ends or junctions of functional fields, it is unlikely that the few cases of positive allometry differing significantly from isometry reflect any influence of metabolic scaling. Geometric scaling (isometry) of the dentition as a whole would appear to require differential positive allometry at the ends of functional fields to compensate for negative allometry within them.

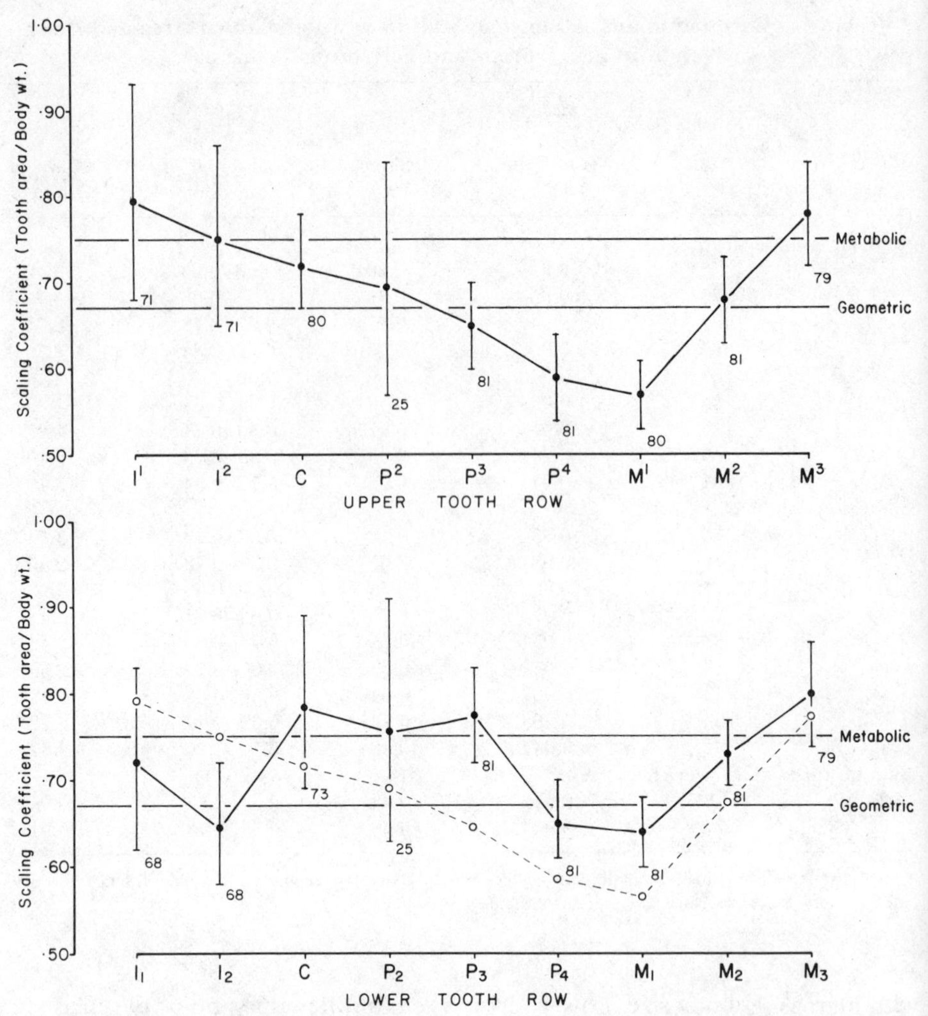

Fig. 3. Pattern of allometric scaling of crown area and body weight at individual tooth positions in the upper and lower dentition of primates. Solid circles are mean values of the allometric coefficient *k*. Vertical bars are 95% confidence intervals for means. Numerals are number of samples (2*N*), with males and females considered separately. Open circles and dashed lines show the pattern of scaling in the upper dentition (upper graph) superimposed on that in the lower dentition (lower graph). Note the low scaling coefficients for lower incisors (I_1, I_2) and upper and lower central cheek teeth (P^4_4, M^1_1). Note also the lower scaling values of upper cheek teeth compared to those of lower cheek teeth. Data from Table 1.

Scaling of Cheek Teeth as a Unit

One would anticipate that the size of cheek teeth considered as a unit might be most likely to reflect metabolic scaling because cheek teeth are involved directly in breaking down food for digestion, the digestive process providing the principal source of metabolic energy. The slopes (allometric coefficients) of upper and lower cheek teeth are listed in Table 1, and their positions relative to the geometric (isometric) and metabolic scaling models are shown graphically in Fig. 4. Neither differs significantly from geometric scaling ($p < 0.05$). Upper cheek teeth as a unit have a scaling coefficient of 0.62, which is considerably below geometric scaling. Lower cheek teeth are slightly positive allometrically, with a coefficient of 0.69, but this is much too low to suggest metabolic scaling.

Tooth Size and Body Size in Insectivores

Species of *Tarsius* stood out well above baseline in our original analysis of tooth size scaling in primates. *Tarsius* has relatively large cheek teeth for its body size. It is unusual among primates in being almost exclusively insectivorous (MacKinnon and MacKinnon, 1980). In an attempt to better understand scaling differences in *Tarsius,* we compiled tooth size and body size measurements for a broadly representative sample of mammalian insectivores, comprising 40 species of Insectivora in the five families Soricidae, Talpidae, Erinaceidae, Macroscelidae, and Tupaiidae. Zalambdodont species

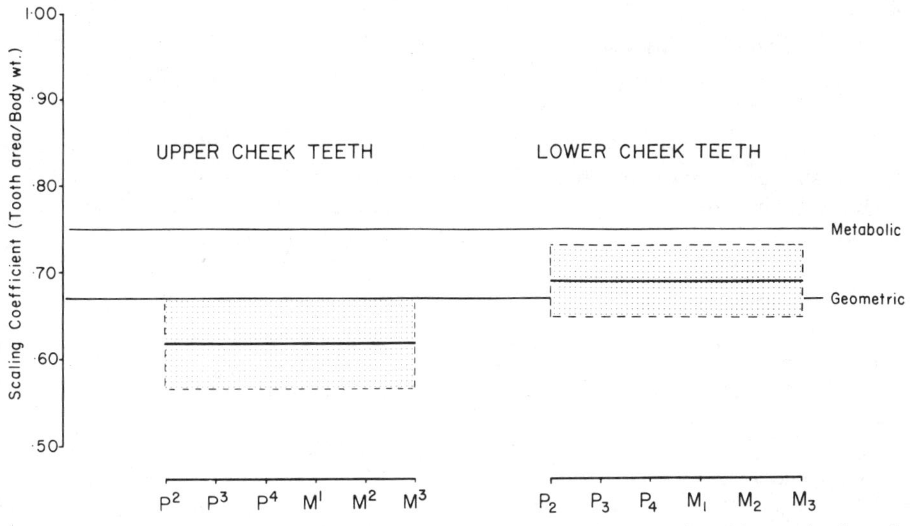

Fig. 4. Comparison of scaling coefficients k for upper cheek teeth as a unit and lower cheek teeth as a unit. Note that neither group differs significantly from geometric scaling (isometry). The scaling coefficient of each group is significantly below that predicted for metabolic scaling. Data from Table 1.

in the families Chrysochloridae, Solenodontidae, Tenrecidae, and Potamogalidae were also studied, but these species were excluded from analysis because of their unusually specialized dental morphology. We measured crown area (crown length multiplied by width) to characterize tooth size, and used body weight as a measure of body size. All measurements of upper and lower first molars were taken from original specimens. Body weights for individual specimens or species were taken from specimen labels or from the literature. Sexual dimorphism in insectivores is negligible, and we used five or more specimens of each sex to represent a species whenever possible. A full analysis of these data will be published elsewhere.

Upper first molars in Insectivora scale with an allometric coefficient k of 0.55 (95% confidence interval 0.50–0.60; Table 2). This is slightly less than the coefficient k of 0.57 in frugivorous and folivorous primates, and it is well below the geometric scaling coefficient of 0.67. We do not have measurements for other upper cheek teeth in insectivores, but, assuming the pattern in insectivores follows that in primates, coefficients for other cheek teeth are probably higher than 0.55. Upper cheek teeth as a unit may approach geometric scaling. However, the scaling value for M^1 alone is so far below the metabolic coefficient of 0.75 as to make metabolic scaling in the upper dentition very unlikely.

Lower first molars in Insectivora scale with an allometric coefficient k of 0.59 (95% confidence interval 0.54–0.64). This is again a little lower than that of frugivorous and folivorous primates, but probably not significantly so. It is well below the coefficient expected for geometric scaling, but lower cheek teeth as a unit may approach geometric scaling. The scaling value for M_1 alone is so far below metabolic scaling as to make metabolic scaling in the lower dentition very unlikely.

The principal difference between Insectivora and frugivorous and folivorous Primates is in tooth size intercepts. Insectivores have an intercept ln b

Table 2. Correlation and Allometric Scaling of Tooth Crown Area and Body Weight in Insectivores of the Families Soricidae, Talpidae, Erinaceidae, Macroscelidae, and Tupaiidae[a]

Tooth position	Sample size N	Correlation r	Slope k	95% Confidence interval for slope	Intercept ln b
Individual upper tooth					
M^1	40	0.966	0.55	0.50–0.60	−0.40
Individual lower tooth					
M_1	40	0.969	0.59	0.54–0.64	−0.95

[a]Zalambdodont insectivores are not included. Number of species N includes both males and females analyzed together. Full documentation to be published elsewhere.

well above that of primates (−0.40 versus −1.27 for M^1 and −0.95 versus −2.12 for M_1). Similar allometric coefficients coupled with greater tooth size intercepts mean that tooth size is greater in insectivores than it is in frugivorous and folivorous primates. This difference is illustrated graphically in Fig. 5, where solid figures represent insectivores, and half-shaded figures represent six species of Tupaiidae and six species of insectivorous primates (*Tarsius bancanus, T. borneanus, T. spectrum, T. syrichta, Galago demidovii,* and *Loris tardigradus*). *Microcebus murinus* is regarded as a frugivore (Kay, 1975*b*), but it falls very close to *Galago demidovii* and, like *G. demidovii* and *Loris tardigradus,* it appears to be intermediate between insectivores and primate frugivores/folivores. Species of *Tarsius* are unusual among primates in falling well above the general primate scaling axis and astride the insectivore scaling axis, a position consistent with their insectivorous habit.

Discussion

Many authors have studied the relationship of tooth size to body size in primates in recent years. Considering that most primates have between 30 and 40 teeth (which can be measured and combined in innumerable ways), and that body size is measured as, e.g., total body mass (weight), skull length, femur length, or in a variety of other ways at the convenience of the investigator, it is not surprising that published results cover the entire spectrum of reasonable possibility.

Pilbeam and Gould's survey of tooth size in primates and other mammals led them to conclude that larger species have relatively larger cheek teeth than one would expect if geometric similarity is maintained (Pilbeam and Gould, 1974; see also Gould, 1975). This positive allometry in tooth size approached the positive allometry of metabolism, leading Pilbeam and Gould to suggest a possible functional link between tooth size and metabolic rate in primates and other mammals. Kay (1975*a*) challenged this with data showing that tooth size usually scales isometrically (geometrically) rather than metabolically in homogeneous dietary groups. A major difference in the two studies involves the choice of a standard for body size. Pilbeam and Gould used skull length in six examples, body mass in two examples, and femur length in the final example. The only scaling coefficient that was significantly greater in a statistical sense than that predicted by isometry was scaling of total cheek tooth area against skull length in hystricognath rodents. Kay used body mass as a standard for body size, and found that none of the seven groups he examined had scaling coefficients significantly greater than predicted by isometry. Four of Kay's seven examples had scaling coefficients significantly lower than predicted by metabolic scaling. Total cheek tooth area was used in

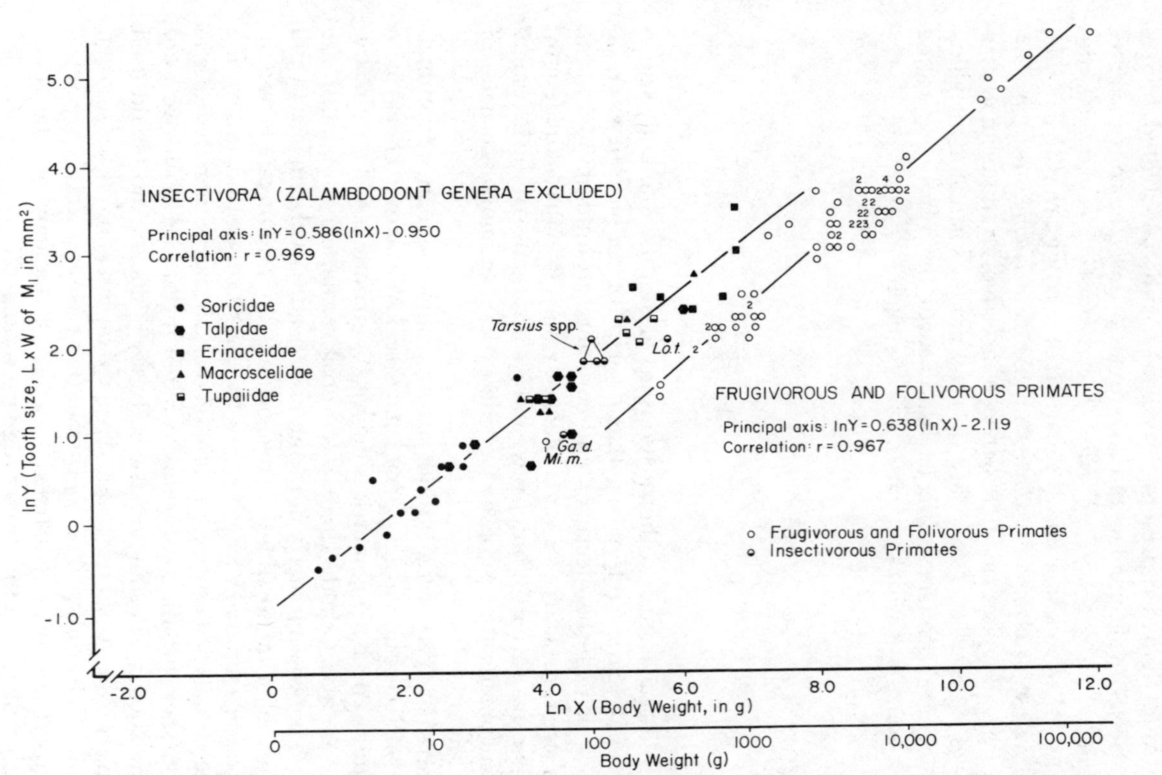

Fig. 5. Comparison of allometric scaling in 40 species of Insectivora (see Table 2) with that in 43 species of frugivorous and folivorous primates (Table 1). Note the similarity in allometric scaling coefficients (slopes) in both groups, and higher tooth size intercept in Insectivora. Insectivores have larger central cheek teeth than do frugivorous and folivorous primates at any given body size. *Tarsius* spp., *Microcebus murinus* (*Mi. m*), *Galago demidovii* (*Ga. d.*), and *Loris tardigradus* (*Lo. t.*) were not included in either analysis. They are plotted here to show the positions of insectivorous and partially insectivorous primates relative to Insectivora and frugivorous/folivorous primates.

two of these examples as a measure of tooth size, and the crushing surface area of lower second molars was used in the other two examples.

The apparent contradiction between Pilbeam and Gould's results and those of Kay can be resolved by noting the different standards of body size used in each study. Skull length, femur length, and body mass are highly correlated in mammals, but one does not scale isometrically with the other. Smith (1981) has shown this clearly for a broad range of anthropoid primates. Given that they are not the same, which standard is appropriate? In this particular case the answer is clear. All attempts to quantify the relationship of metabolism to body size involve scaling metabolic rate against body mass. The relationship of metabolic rate to skull length or to femur length has never been studied. Body mass is clearly the only appropriate standard if results are to be compared with metabolic scaling.

Different measures of tooth size will also yield different results. Crown area is based on two simple measurements, crown length and width, and gives a reasonably good measure of overall tooth size. Crown area alone cannot represent crown shape, crown height, or the size of particular functional features (cusps, crests, basins) that may be of interest. All of these are candidates for more detailed future study. Our work presented here shows that there are definite patterns of overall tooth size scaling that relate to position in the dentition and to diet.

Lower incisors (I_1, I_2) and central cheek teeth (P^4_4, M^1_1) have the lowest scaling coefficients. In the upper dentition these are well below geometric scaling and in the lower dentition they are approximately equal to geometric scaling. Upper incisors and teeth at the anterior and posterior limits of the cheek tooth series exceed geometric scaling and have scaling coefficients approaching or even exceeding metabolic scaling. Cheek teeth as a unit appear to scale geometrically, and it seems unlikely that "metabolic" coefficients in the anterior and posterior cheek teeth actually reflect any relationship to metabolism. Rather, teeth at the ends of cheek tooth units have high scaling coefficients because they are in areas where there is space for teeth to be enlarged differentially with increasing body size. High coefficients in marginal cheek teeth compensate for low coefficients in central cheek teeth in maintaining overall geometric scaling.

Upper cheek teeth scale with lower coefficients than their counterparts in the lower dentition. This may, again, be a reflection of the lower jaw being less constrained by surrounding structures, permitting its teeth to be differentially enlarged at progressively larger body sizes. Insectivorous primates and other mammals have larger cheek teeth than do frugivorous and folivorous primates, a difference that must be related in some way to mastication of animal prey.

Scaling of cheek tooth size against body weight in Primates and Insectivora involves coefficients clustered below or spanning the range expected for geometric scaling (isometry). A few coefficients reach the range to be expected in metabolic scaling, but these are in peripheral teeth of integrated functional units that themselves scale geometrically. Our results are clearly at variance with those of Pilbeam and Gould (1974, 1975) and Gould (1975), and they support Kay (1975*a*) in showing that tooth size in primates scales geometrically, not metabolically, with respect to body mass.

ACKNOWLEDGMENTS

We thank Drs. Philip Myers and Larry Heaney (University of Michigan Museum of Zoology), Sydney Anderson (American Museum of Natural History), and Michael Carleton (National Museum of Natural History) for access to skulls and body weights of insectivores. All figures were drawn by Karen Klitz. This work was supported by a grant from the National Science Foundation (BNS 80-16742).

References

Gingerich, P. D., Smith, B. H., and Rosenberg, K. 1982. Allometric scaling in the dentition of primates and prediction of body weight from tooth size in fossils. *Am. J. Phys. Anthropol.* **58:**81–100.

Gould, S. J. 1975. On the scaling of tooth size in mammals. *Am. Zool.* **15:**351–362.

Huxley, J. S. 1932. *Problems of Relative Growth,* Methuen, London.

Hylander, W. S. 1975. Incisor size and diet in anthropoids with special reference to Cercopithecidae. *Science* **189:**1095–1098.

Kay, R. F. 1975*a*. Allometry and early hominids. *Science* **89:**61–64.

Kay, R. F. 1975*b*. The functional adaptations of primate molar teeth. *Am. J. Phys. Anthropol.* **43:**195–216.

Kleiber, M. 1932. Body size and metabolism. *Hilgardia* **6:**315–353.

MacKinnon, J., and MacKinnon, K. 1980. The behavior of wild spectral tarsiers. *Int. J. Primatol.* **1:**361–379.

Pilbeam, D. R., and Gould, S. J. 1974. Size and scaling in human evolution. *Science* **186:**892–901.

Pilbeam, D. R., and Gould, S. J. 1975. Allometry and early hominids. *Science* **189:**61–64.

Schmidt-Nielsen, K. 1975. Scaling in biology: The consequences of size. *J. Exp. Zool.* **194:**287–307.

Smith, R. J. 1980. Rethinking allometry. *J. Theor. Biol.* **87:**97–111.

Smith, R. J. 1981. On the definition of variables in studies of primate dental allometry. *Am. J. Phys. Anthropol.* **55:**323–329.

Tooth Size–Body Size Scaling in a Human Population

13

Theory and Practice of an Allometric Analysis

MILFORD H. WOLPOFF

Introduction

The intent of this chapter is to examine the question of whether and how tooth size scales to body size in a living human population, and to determine the relationship that such intrapopulational scaling might have to the explanation and general understanding of the tooth size–body size relationship. In the process of this examination, focus is brought to the concept of allometry, its basis in fact and theory, the question of whether it is a descriptive or an explanatory device, and its potential for modeling the tooth size–body size relationship.

The relation of tooth size to body size in primates is of interest for a variety of reasons. The examination of this relationship has been used to predict body size from tooth size in fossil primates (Gingerich, 1977; Gingerich *et al.*, 1982; Kay and Simons, 1980; R. J. Smith, this volume; Zhang, 1982), to ascertain whether this relation can be explained by geometric scaling

MILFORD H. WOLPOFF • Department of Anthropology, University of Michigan, Ann Arbor, Michigan 48109.

or metabolic scaling (Creighton, 1980; Gingerich *et al.*, 1982; Goldstein *et al.*, 1978; Gould, 1975*a;* Kay, 1975*a,b;* Pilbeam and Gould, 1974, 1975; Pirie, 1978; Smith, 1981*b*), and to determine whether various fossil species show dental differences that can be explained as the necessary consequences of size differences (Garn and Lewis, 1958; Corruccini and Henderson, 1978; Pilbeam and Gould, 1974, 1975; Wolpoff, 1974, 1978, 1982; Wood and Stack, 1980). In primates, some sort of scaling relating tooth size to body size would usually seem to be an obvious phenomenon (Gould, 1975*a*), since larger individuals must eat more. Yet, several decades of literature have failed to result in either a reasonable explanatory model or even a set of consistent relationships; indeed, it remains unclear whether a significant relationship between tooth size and body size in humans can be claimed at all (Garn and Lewis, 1958; Garn *et al.*, 1968; Filipson and Goldson, 1963; Henderson and Corruccini, 1976; Perzigian, 1981), since various studies have drawn opposite conclusions about this issue.

The hominids are the best studied of all primates. One consequence of the resulting embarrassment of knowledge about hominids is the conclusion that the tooth size–body size relation has changed significantly over the course of hominid evolution. This indicates that there is not necessarily a single relationship between tooth size and body size in primates, even when the species compared are closely related. This fact has combined with other factors to result in the confusing and contradictory claims published about the tooth size–body size relationship. These other factors include the variety of measures of body size that have been used in different studies, the problem of whether height or weight is the most relevant variable (and the related question of whether estimators can be substituted for direct measures of the variables), the confusion of interspecific and intraspecific (or populational) variation, and the procedures used to determine the effects of scaling. While the decisions between the choices implicit in each of these factors can be justified, the resultant lack of consistency and presence of contradictions in the literature should not be surprising.

Which Question?

Fundamental to all of these problems is the question of the question. The question of whether or not there is a relationship between tooth size and body size within a population is a different problem than whether or not a constant relationship exists between closely related species (such as in an ancestral–descendant sequence or within an adaptive radiation). This in turn differs from whether or not a consistent relationship exists within a higher taxonomic category such as an order. These are different problems because in each case the tooth size–body size relationship in question has a different cause.

Thus, scaling relationships have become the object or basis for several different types of questions. One class of questions often asked is about the regularities of scaling in higher order taxa, since regularities can give insight into general biological phenomena. A rather different class of questions is about scaling as an explanation of differences; to what extent are differences in shape or disproportions in size a consequence of differences in size? A third class of questions related to both of these is about deviations, or residuals; to what extent do differences in shape or disproportions in size deviate from the consequences of size? Finally, there are questions that are essentially predictions about differences that are expected to result from variation in size. Each of these sets of questions requires different approaches and can also involve different techniques.

The Problem of Allometry

> Allometry then is the study of size and its consequences.
>
> (Gould, 1966, p. 587)

Allometry and the Consequences of Size

While allometry may have come to refer to the study of the consequences of size, it has traditionally also come to refer very specifically to the use of a power curve (or a linear fit of the log-transformed data) to describe and compare these consequences (Sprent, 1972; Smith, 1980). Use of the power curve was a main focus of Huxley's (1932) *Problems of Relative Growth.* Most contemporary discussions of this landmark volume focus on Huxley's discovery that the power function seems to describe a very wide range of size-based relationships. However, Huxley's identification of the power function was more than an empirical generalization. Huxley (1932, pp. 4–8) developed a theoretical basis for the use of this curve. In particular, he showed that a power curve relation can be expected to describe the phenomenon of scaling as a consequence of the hypothesis that the rates of growth of the dependent and independent variables are each proportional to the number of cells (i.e., weight) already present [for a binary model of this process see Katz (1980)]:

$$dx/dt \propto X \qquad \text{and} \qquad dy/dt \propto Y$$

or

$$dx/dt = \alpha X \qquad dy/dt = \beta Y$$

where α and β are the proportionality constants. The simultaneous solution of these two equations gives the log form of the power function:

$$\log Y = k \log X + \log b$$

where $k = \beta/\alpha$ and b is the constant of integration.

The fact is that the allometric (i.e., power) curve is almost never applied where the above assumptions actually describe the growth process. Attempts to posit alternative theoretical explanations for the allometric curve (Robb, 1929; Lambert and Teissier, 1927; Teissier, 1960; Serra, 1958; Rosen, 1967; Günther, 1975; Savageau, 1979) have not been convincing, and Huxley himself (Reeve and Huxley, 1945) later came to accept the lack of a theoretical model for allometry. Thus, as presently conceived, allometry seems to be a descriptive model and not a conceptual one (Gould, 1966, pp. 597–598; Kac, 1969). Whether or not data are very accurately fit by an allometric curve is usually not examined (Smith, 1980). When this accuracy is examined, it is as often as not left wanting (Wolpoff, 1967; R. J. Smith, this volume). I contend that this should not be surprising. This problem is especially relevant to the tooth size–body size relationship; body growth rates are anything but constant throughout the growth period (Harrison *et al.*, 1977, pp. 305–307), while the growth of tooth germs does not fit the model underlying the derivation of the allometric equation. There clearly are mathematical alternatives to the synonymy of allometric curve with power curve (Gould, 1966; Sprent, 1972), or for that matter with any two-dimensional curve (Hursh, 1976; Bookstein, 1977), that has become traditional in allometric studies. Actually, Huxley himself realized that the power curve was only an approximation, even under the assumptions he was willing to make (Lande, this volume).

Interesting, Huxley was not the first to use the power curve in describing scaling phenomena, and other authors of the time [reviewed by Gould (1966)] provided differing justifications for its use. The historic importance of Huxley's considerations about the relative growth process stems from his belief that relative growth could be explained by the individual growth processes, and that this phenomenon was the cause of observed regularities in scaling. Thus, the power function was given meaning far beyond its ability to describe (i.e., fit) data. Much of the subsequent development of allometric theory can profitably be seen in the context of the failure of these ideas.

One implication of the use of the power function (i.e., the power curve) to assess allometry follows from how poorly its assumptions reflect actual growth processes. Contrary to Huxley's (1924, 1932) original contention, the relative growth curve of an individual (or the curve describing individuals of various ages in a population) is not necessarily reflected in the relative size relations of adults in the population (Cock, 1966, pp. 148–151; Cheverud, 1982). This intrapopulational growth curve was first called heterogony (Huxley, 1932), and later heterauxesis (Huxley *et al.*, 1941). Gould (1975*b*, p. 258) also suggests that the intrapopulational adult allometric curve "may be a consequence (in part) of the fundamental ontogenetic relationship," but this suggestion is unsupported. Especially in the case of the tooth size–body size relation (but probably in other instances as well) the contention that intraspecific allometry is the direct result of growth is very unlikely. Teeth do not grow at all after they erupt (there is no growth over the period that the teeth can be measured for an allometric study), while body growth spans a long period and is characterized by variable rates (Harrison *et al.*, 1977, pp. 305–307). Thus, it is unreasonable to expect the scaling of tooth size to be explicable in terms of relative growth, in which the scaling exponent is said to be the ratio of the two growth rates (Huxley, 1932). The identity of these two different scaling relationships (represented by the growth curve and the curve reflecting adult sizes) was a presumption that Huxley made because he reasoned that the first was the cause of the second. By arguing that the growth process is described by the allometric model, it was possible to conclude that different sized adults in a population represented different stopping points along the same growth curve. It is now understood that the situation is somewhat more complex. Static intraspecific allometry seen in an adult series may be a consequence of growth, but probably not in a manner that allows the intraspecific curve to be predicted from or otherwise meaningfully related to the ontogenetic growth curve (Cheverud, 1982; Shea, 1983).

A second implication that follows from the above involves the relation of intraspecific and interspecific allometry. Once again Huxley (1932, pp. 212–224) seemed to regard these as the same phenomena. The distinction he made (Huxley *et al.*, 1941) was between the ontogenetic series within a population (heterauxesis) and comparisons between populations, races, species, or genera of either ontogenetic series or adult series (allomorphosis). It has since become clear that relative size relations in adult series are not the same interspecifically and intraspecifically (Gould, 1966, 1975*b*). The logic underlying the contention of similar causality in interspecific and intraspecific allometry [e.g., von Bertalanffy and Pirozynski (1952); Gould (1975*b*) and especially Gould (1977); Alberch *et al.* (1979); McNamara (1982); Shea (1983)] is that larger species can evolve from smaller ones by continuing or systematically altering the growth process, or that smaller species evolve from larger ones by stopping the growth process earlier in ontogeny. Presumably, as a larger species evolves from a smaller one (using the first case as an example), its scaling becomes an extension of the intraspecific allometric curve of the smaller species, since it is the larger individuals (i.e., the top part of the allometric curve) of the smaller species that are promoted by selection, either as adults or during ontogeny. A similar explanation proposed by Shea relates the interspecific allometric curve to the ontogenetic curve through an argument of "genetic convenience" (Shea, 1983, p. 52):

> If simple size increase or decrease is selectively favored, it may be most quickly and easily produced in the descendant population via exaggeration of the growth and developmental mechanisms which produce size differences within the ancestral population.

Some evidence supporting these contentions has been gleaned from the evolutionary interpretation of von Baer's law (Gould, 1977), which states that species are more likely to resemble each other in their early developmental stages than they are in later stages of development. Von Baer's law could be a result of species evolution due to extensions of the ontogenetic growth curve, or evolution as a consequence of subtle alterations of this curve early in development leading to major differences between the adult forms.

However, there are alternative explanations, and more than one reason to suppose that evolutionary change is generally not an extension of ontogenetic change. Von Baer's law, for instance, is alternatively perceived as a consequence of the fact that genes acting later in the developmental process "are likely to produce small, quantitative variations that are the basis for most morphological evolution in higher animals" (Lande, this volume). This is particularly likely to be the case in mammals (and especially in primates), where parental investment and delayed maturation shield a species' young from selection.

Another problem in the contention that phylogenetic change is constrained to following ontogenetic curves is in the fact that the scaling coefficient (i.e., the magnitude of the power curve exponent or of the linear slope of the log-transformed data) is usually, although not invariably (Giles, 1966; Gould, 1974), less in the intraspecific case than it is in interspecific scaling (Clutton-Brock and Harvey, 1979; Gould, 1966, 1975*b;* Kaplan and Salthe, 1979; Smith, 1981*c;* Wood, 1979). In the calculation of a species' intraspecific allometric curve, the extremes are of great importance in determining the slope. However, species change over time usually involves small shifts in the mean, and if these shifts respond to changing balancing selection or to genetic drift (Lande, 1976), the extremes may contribute nothing to the process: "The characteristics of extreme individuals do not determine the future of the lineage" (Kaplan and Salthe, 1979, p. 683). Thus, size increase over time does not necessarily follow an ontogenetic or an intraspecific curve, sensitive as these curves are to the extreme values, and (*contra* Gould, 1977; Shea, 1983) phylogenetic size increase may not even be under the control of any ontogenetic process (Cheverud, 1982). The relation between intraspecific and interspecific allometry has been far from clear.

Finally, even further confusing the issue, there are two different reasons why allometry might describe the relationship between variables (Gould, 1974) when this relationship is a consequence of size. These might apply in any of the cases discussed above, and have proven to be very difficult to distinguish, when attempts to distinguish them are made (the question is often not addressed). Allometry may characterize a bivariate relationship when selection acting to increase or decrease size (or, in the intraspecific case, when the growth of one variable) causes correlated change in the other variable. Alternatively, however, allometry can also characterize two variables when each responds to an independent source of selection, or when each responds independently to the same source of selection. Each of these alternative patterns also results in correlated change, either as a consequence of correlation between the sources of selection, or because of the identity of these sources. An example of this might be changes in tooth size that are required to meet the changing dietary resources necessary for larger animals. The new food resources required by larger body size might invoke selection for larger tooth size, or alternatively they might lead to selection changing tooth form.

In fact, then, three different scaling phenomena are subsumed in the concept of allometry: the relative growth of individuals (ontogenetic scaling), the scaling of adults in a population or species (intraspecific scaling, or static adult allometry), and the scaling reflected by comparing the means of different species (interspecific scaling, static allometry of adult means, or evolutionary allometry). The differences between these phenomena were discussed at some length by Gould (1966, 1975*b*). However, given the rejection of Huxley's assumption that they represent the same thing, and the potential confusion between causal correlation and noncausal correlation in the bivariate relationship, the three scaling phenomena have seemed unrelated to each other. Until recently, no model of either growth or of the evolutionary process has united the three in a predictable way.

Power Functions and Other Scaling Phenomena

If power functions are not theoretically acceptable as valid descriptors of relative size relations because they do not reflect relative growth, what other justifications underlie their continued use? Three answers to this question have been offered, each involving the use of power functions to test whether other scaling phenomena underlie relative size relationships.

One of these is geometric scaling, the change in relative size and proportions that comes only as a consequence of geometry (Gould, 1971; Schmidt-Nielsen, 1975). Geometric scaling posits a constant and consistent size increase for all parts of an organism. Linear measures, for instance, all change to the same proportion. However, surface areas and volumes cannot be expected to change at the same proportion as linear measures, because of the geometric relations of surfaces and volumes. Surface areas should change in proportion to the square of linear measures, and volumes in proportion to the cube. Geometric scaling thus provides a model for determining expected power curve exponents; for instance, linear tooth dimensions as a function of weight (i.e., a volume measure) should scale with an exponent of 0.33, and tooth areas as a function of weight should scale with an exponent of 0.67. Deviations from these expectations would then reflect deviations from the expectations of geometric scaling.

The basis and derivation of the geometric model seem quite straightforward (Gould, 1966, p. 591). Theoretical arguments supporting it derive from Teissier's (Lambert and Teissier, 1927) "theory of similitude"—a theory which now seems more empirical than theoretical (Brody, 1945), in spite of various attempts to elaborate on it (Günther, 1975; Economos, 1979; Lindstedt and Calder, 1981). These attempts seem to involve multiplication of empirical results by an arbitrary constant to allow the data to fit the theoretical expectations.

In dealing with the tooth size–body size relation the geometric model clearly utilizes the power curve as a description of scaling and not as a cause of scaling. Thus, for the tooth size–body size relation to conform to the model of geometric allometry, three hidden assumptions must be met. Conversely, if the relation deviates from this model, one or more of these assumptions are violated:

1. Dietary requirements increase in proportion to weight.
2. The lifelong resistance of a tooth crown to occlusal wear is proportional to a measure of its volume (occlusal length multiplied by breadth—i.e., area—multiplied by height, or occlusal area multiplied by enamel thickness). The proportionality of a body volume measure with a volume measure of tooth size is required by the contention of the geometric model that the occlusal surface area scales with an exponent of 0.67 when compared with body weight.
3. Food type is the same for all adult sizes.

These assumptions, then, provide the explanation for data fitting the expectations of the geometric scaling model; that is, the expected exponent of 0.67 for the scaling of occlusal area to body weight and the exponent of 0.33 scaling dental dimensions to body weight.

There is an alternative model related to the geometric model. This is the area–volume isometry model. As a replacement for the second assumption above, occlusal area is predicted to scale in direct proportion to body weight (Kay, 1975*b*). If tooth longevity (the lifetime resistance of a tooth to wear) is proportional to occlusal surface area instead of to the volume of a tooth, or if occlusal area is found to be functionally related to the daily amount of food required to maintain a given body size, the scaling exponent of occlusal area as a function of weight would be 1.00. This would no longer strictly be a geometric scaling model, even though the contention of area–volume proportionality could be interpreted to reflect the consequences of geometric scaling. This model might be expected to apply in species that spend an unusually high portion of the day eating, and/or species with continuously erupting teeth. The area–volume isometry model is a variant of the geometric model, changing the second hidden assumption.

A third scaling model, metabolic scaling, is based on the rejection of the first assumption of the geometric model. Following from the pioneering work of Keibler (1932), a considerable amount of research [reviewed in Schmidt-Nielsen (1975)] indicates that the interspecific relationship of various measures of metabolism to body weight scales with a power curve exponent of 0.75. This relation may actually be viewed as a consequence of geometric scaling since heat loss is a function of the body surface area–body volume relation, and as discussed above in geometric scaling, areas increase according to the 0.67 power of volume. Thus, larger mammals need not produce the same relative amount of heat to keep a constant core temperature.

The 0.75 exponent for metabolic scaling to body weight began as an empirical observation. McMahon (1973) attempted to derive a theoretical basis for it through an argument based on the contention that the elastic response of bone to bending and buckling underlies body proportions. His reasoning was that since bone is less well adapted to resist noncompressive forces than it is to resist compressive forces, the former should provide the limiting factors for the limb length–limb diameter relationship (McMahon, 1975). Unfortunately, however, more recent studies of the scaling of limb lengths and limb diameters to body weight (Alexander *et al.*, 1979) do not support the contention that elastic constraints are operative. Because the elastic constraint model does not seem to accurately explain the observed relations of limb length as a function of body weight, the extension of this model for explaining metabolic scaling is unsupported. This seemed to leave the observed 0.75 metabolic scaling exponent without explanation (Wilkie, 1977).

However, recently there have been several other attempts to derive the empirical metabolic exponent from other theoretical considerations. Thus, Heusner (1982) contends that the interspecific exponent is a statistical artifact—a consequence of connecting a series of lower intraspecific slopes. Heusner's analysis is marred by a number of problems, including the claim that these intraspecific slopes parallel each other (actually, the slope exponents he reports for 11 mammalian samples range from 0.48 to 0.91; four dog samples alone range from 0.55 to 0.69). However, the fundamental difficulty is in the assumption that the intraspecific relations (even if they all approximated 0.67 as claimed) can explain the interspecific slope.

An intriguing argument has been developed by Economos (1979) and elaborated by Linstedt and Calder (1981). Economos' development of the argument proceeds from the assumption that the basal metabolic rate (BMR) is proportional to the total blood flow to the tissues. Blood flow, in turn, is a function of heart rate (HR) and the stroke volume (SV) of the heart:

$$\mathrm{BMR} \propto \mathrm{HR} \times \mathrm{SV}$$

Since it is a volume measure, stroke volume is proportional to body mass:

$$\mathrm{SV} \propto M$$

However, experimental evidence (Guyton *et al.*, 1973; Linstedt and Calder, 1981; Calder, 1981) shows that heart rate is inversely proportional to the fourth root of mass for a surprisingly wide range of species:

$$\mathrm{HR} \propto M^{-0.25}$$

Thus, substituting, we find

$$\mathrm{BMR} \propto M \times M^{-0.25}$$

$$\mathrm{BMR} \propto M^{0.75}$$

which is the Keibler exponent! It appears that the observed interspecific consistency in metabolic scaling may be derivable from purely physiological considerations, although these considerations themselves are empirically rather than theoretically based.

The geometric, area–volume isometry, and metabolic models each provide a prediction for the expected power curve exponent. However, none addresses the more fundamental question of why the power curve is the appropriate vehicle for examining the consequences of size in the first place, apart from the obvious convenience of having a predictable exponent. Moreover, other problems involving geometric and metabolic scaling have become evident in the attempts to ascertain the tooth size–body size relationship. Some of the interspecific studies within the primates have resulted in scaling exponents for the power curve that were interpreted to be unable to distinguish the metabolic from the geometric model [Gould (1975*a*) and Goldstein *et al.* (1978); for a similar problem in distinguishing the metabolic from the geometric model in brain–body weight allometry see Armstrong (1982)], while others seem to support only the interpretation of geometric scaling (Kay, 1975*a,b;* Gingerich *et al.*, 1982). Finally, the three models are based on the interspecific case; it is not clear how these models might be expected to apply to intraspecific variation.

The Metabolic Model Misapplied

Tests of the metabolic model have been attempted in a number of tooth size–body weight studies, examining the relation of occlusal area to body weight in interspecific series. In every case (Gould, 1975*a*, p. 352; Kay, 1975*a*, p. 63; Kay, 1975*b*, p. 202; Gingerich *et al.*, 1982, p. 82) the expected scaling exponent for the power curve relating occlusal area to body weight was said to be 0.75, higher than the expectation of the 0.67 exponent given for the geometric model. This apparently accepted comparison of the metabolic and geometric models is incorrect.

The geometric scaling model rests on the three assumptions discussed above. In order to predict an area–volume relation between occlusal surface area and body weight (i.e., an exponent of 0.67), one must assume that *the tooth volume is proportional to weight.* This is the second assumption. The metabolic model is based on an alteration of the first assumption, since it assumes that food requirements increase with body weight raised to the 0.75 power (the metabolic scaling exponent), rather than linearly with body weight. The question is, how does tooth size relate to the dietary requirements? If it is occlusal surface area that relates proportionally with dietary requirements (i.e., metabolism), then one would expect occlusal area to scale to body weight with an exponent of 0.75. This is the exponent that has been traditionally given as the expected scaling of occlusal area to body weight under the metabolic model. *However, a proportionality between occlusal surface area and dietary requirements is not the assumption of the geometric model* (the model with which the metabolic model is compared). Instead, the assumption of the geometric model is that *the tooth volume is proportional to weight.* For a valid comparison of the metabolic and geometric models, both exponents should be derived from the same assumption about which measure of tooth size is proportional to dietary requirements. For most primates the assumption that the selection acting on tooth function is based on its longevity (and thus is proportional to its volume) is more reasonable than the assumption that only occlusal area responds to increased dietary intake (as might be expected in *Equus*). Indeed, given the proposal that metabolic rates are related to longevity (Linstedt and Calder, 1981), the assumptions of the geometric model make it the most reasonable framework for comparisons with metabolic scaling.

Thus, let us assume that tooth volume is proportional to dietary intake (as in the geometric model), and that dietary intake increases in proportion with metabolism.

> If tooth volume is proportional to metabolic requirements ($V \propto M$), and occlusal area is proportional to (tooth volume)$^{0.67}$ ($A \propto V^{0.67}$), and metabolic requirements are proportional to (weight)$^{0.75}$ ($M \propto W^{0.75}$), then occlusal area will be proportional to (body weight)$^{0.67 \times 0.75 = 0.5}$ ($A \propto W^{0.5}$).

The expectation of the metabolic model as applied to interspecific variation is that occlusal area will scale to body weight raised to the 0.5 power. If the metabolic model is to be compared with the geometric model, the relevant predictions are exponents of 0.5 for the former compared with 0.67 for the latter. The comparison of a 0.75 metabolic prediction with a 0.67 geometric prediction mixes two different sets of underlying assumptions and is therefore incorrect.

Interspecific versus Intraspecific Scaling

There is probably no more serious problem in the interpretation of allometry and the understanding of scaling than that of the relation of interspecific and intraspecific scaling phenomena. Although everybody seems to recognize that these are different, they are constantly confused with each other. For instance, in one place Gould (1975*b*, pp. 277–278) asserts:

> Of all types of scaling, the interspecific curve is by far the worst to consult for information about evolutionary mechanisms. . . . [T]he allometry of adults within a single population represents the correlated variability upon which evolution works. We can determine the rate of increase of organs and parts with body weight and predict the proportions that might accompany larger size if size alone is the object of selection. If evolutionary sequences scale along an intraspecific curve, we may wish to conclude that ancestors and descendants are animals of the same basic design expressed at different sizes.

Yet, in a paper published the same year (Pilbeam and Gould, 1975, p. 64), there is a contradictory claim. The statement is in response to the assertion that the larger incisor size in *Homo habilis* (descendant species) when compared with *Australopithecus africanus* (ancestral species) could be a consequence of larger body size in *Homo habilis,* since anterior teeth correlate best with body size in *Homo sapiens:*

> The relationship between anterior tooth size and femur length or body weight in *Homo sapiens* [intraspecific correlations] is irrelevant to a study of variation between species.

Similarly, after discussing the fact that interspecific brain–body weight allometry has a 0.67 exponent for a wide range of species, while the intraspecific exponent is invariably between 0.2 and 0.4, Pilbeam and Gould (1974, p. 895) contend:

> A sequence of closely related animals differing in size but not in function or 'grade' of evolution should yield a brain–body plot with a slope between 0.2 and 0.4. The highest slope that can be justified in making a claim of functional equivalence is 0.67.

These brain–body weight slopes are quite different in both magnitude and in derivation (the first is intraspecific and the second interspecific). Can they both mean "functional equivalence"?

The relationship between interspecies and intraspecies allometry is confusing because it has until recently been unexplained. The fact that these are actually different scaling phenomena raises a contradiction that has never been discussed, let alone resolved. Consider the two assertions that have been made. First, intraspecific allometry represents the evolutionary mechanism, and ancestral–descendant species that follow a single intraspecific curve demonstrate the consequences of size. From intraspecific allometry it is possible to "determine the rate of increase of organs and parts with body weight and predict the proportions that might accompany larger size if size alone is the object of selection" (Gould, 1975*b,* p. 278). Second, interspecific allometry "provides scaling criteria for the functional morphologist . . . [and] can be used to predict blueprints for adaptation . . . [but] does not represent the path travelled by any organism towards that adaptation" (Gould, 1975*b,* p. 277). The contradiction can be seen in the case when intraspecific scaling is less than interspecific scaling (as usually is true) and the ancestors and descendants are animals of the same basic design expressed at different sizes. Either this simply can never be the case, or the intraspecific and interspecific scaling must be the same, for how can a structure in a larger descendant species be functionally equivalent and lie along a higher sloped interspecific curve, but reach this state by size expansion along a lower sloped intraspecific curve? If only one of the curves represents the pattern of functional equivalence, which one would it be? The argument for the intraspecific curve is that it reflects the basis of and the results from the actual evolutionary mechanism of size increase. However, the interspecific curve presumably reflects general morphological patterns (when the curve accurately fits the data). If these are not patterns of scaling [as workers such as Jerison (1973) contend about the seemingly pervasive brain–body weight exponent of 0.67], what are they? In an ancestral–descendant sequence of increasing size, the description of the path traveled cannot differ from the description of the steps along the way. Yet, if scaling along an ancestral–descendant sequence is fundamentally different from the general pattern of interspecific scaling, one can reasonably question where this general pattern comes from, since in many cases the interspecific data include ancestral–descendant sequences (e.g., Radinsky, 1970).

One answer to these questions has been provided by Smith (1981*c*, p. 296), arguing that "neither intraspecific, interspecific, or ontogenetic allometry is related to evolutionary mechanisms." He contends that intraspecific allometry is unrelated to evolutionary mechanism under the punctuated equilibria hypothesis because (p. 296) "it is axiomatic that the direction of speciation is unrelated to intraspecific variation or selection." At the same time, under the assumption of gradualism he cites Lande's (1979) "demonstration" that the path of evolution also is not necessarily related to intraspecific variation. As for interspecific allometry, Smith restates Gould's contention that this relationship does not represent the pathway by which adaptations are attained [although see Kaplan and Salthe (1979), pp. 682–684]. Smith concludes that allometry is usually an empirical description.

Although this conclusion seems to parallel my own, following from the contention that the power curve relation apparently does not have the theoretical basis Huxley initially attributed to it (see above), I cannot fully agree with Smith's reasoning. My reading of Lande (1979) is somewhat different. In his discussion of brain–body size allometry, Lande not only confirms the contention that allometry exists as a real phenomenon to be explained, he also provides a theoretical explanation for it. Lande (1979, p. 414) argues:

> Under some simple conditions on the pattern of phenotypic and genetic covariation within populations, selection only on body size, certain types of multivariate selection, and random genetic drift in a stochastic phylogeny are each expected to produce allometric evolution, i.e., straight lines or linear regressions on logarithmic coordinates. . . . [T]he short-term differentiation of brain and body sizes in very closely related mammalian forms resulted either from directional selection mostly on body size with changes in brain sizes largely a genetically correlated response, or from random genetic drift.

Lande's proposed mechanism for allometry is twofold. First, when selection acts on only one variable, the observed intraspecific allometric relations are a consequence of genetic covariation (the result of a genetic coupling, or pleiotropy between the variables). When interspecific allometry follows this intraspecific pattern it reflects predictable (allometric) responses to selection on the independent variable (the gradualism model for change) or predictable changes resulting from random drift (Smith's interpretation of the punctuational model, citing Stanley). Second, interspecific deviation from the intraspecific allometric pattern reflects differences in selection acting separately on both the independent and the dependent variables. The differences in selection could be a consequence of scaling (i.e., the same form expressed at different sizes) that was prevented by the phenotypic correlation responding to the genetic coupling, or a consequence of fundamental adaptive changes in the relationship between the two variables. In this arguement, Lande provides insight into an important relationship between the interspecific and intraspecific phenomena.

Lande's Explanation for Allometry

Using quantitative genetics theory, Lande (1979, and this volume) derived a theoretical basis for the allometric relation, and provided a series of expectations governing the conditions under which intraspecific and interspecific allometric slopes will or will not approximate each other. His reasoning proceeds from the simplest case of populational change over time, in which an independent and dependent variable are genetically correlated, and selection on the independent variable results in changes in the dependent variable. If $\bar{Z}_X$ is the mean value for the independent variable (the variable under selection), S_X is the selection differential, the heritability of Z_X is $h_X{}^2$, the phenotypic variance is $\sigma_X{}^2$, and G_{XX} is the additive genetic variance, the formula for the one-generation change in the mean of Z_X is

$$\Delta\bar{Z}_Z = (G_{XX}/\sigma_X{}^2)S_X$$
$$= h_X{}^2 S_X$$

If the mean value for the dependent variable is $\bar{Z}_Y$, the heritability of Z_Y is $h_Y{}^2$, γ_{xy} is the correlation of the additive genetic values for Z_X and Z_Y, and the covariance of the additive genetic values is G_{YX}, the formula for the correlated response over one generation for the dependent variable is

$$\Delta\bar{Z}_Y = (G_{xy}/\sigma_X{}^2)S_x$$
$$= h_x h_y \gamma_{xy}(\sigma_y/\sigma_x)S_x$$

From these formulae, Lande argues that under the assumption that the phenotypic variances are roughly constant and independent of the mean values $\bar{Z}_X$ and $\bar{Z}_Y$ (this assumption can be met through the log transform of the variables, since the standard deviation of the natural logarithm of a variable is approximately equal to the coefficient of variation of the raw measurements), the slope of the line connecting the changing values of Z_Y as a function of Z_X can be derived from the above equations:

$$m = G_{YX}/G_{XX}$$
$$= \gamma_{XY}\ (\sigma_Y/\sigma_X)\ (h_Y/h_X)$$

The selection differential S_X has canceled, and therefore has no effect on this slope. As long as the relative values of the additive genetic covariances do not change much, this relation will remain linear for the log-transformed variables, and therefore will be allometric. Thus, Lande shows that when there is additive genetic correlation between two variables, and selection acts on one of them, an allometric relation between these variables can be expected to characterize populational change over time.

When there is random genetic drift, Lande argued that a combination of pleiotropic mutations and a stochastic phylogeny (a large number of independent lineages derived from a common ancestor showing random branchings and extinctions) will result in an interspecific regression slope for the surviving lineages that is exactly the same as the slope for correlated change over time with selection acting on one variable derived above:

$$m = G_{YX}/G_{XX}$$

Thus, the same predictable allometric relationship is shown to result from both selection acting on one of two genetically correlated (i.e., pleiotropic) characters, and from random genetic drift in a stochastic phylogeny when the pleiotropic characters are adaptively neutral.

The slope relating two genetically correlated variables that characterizes populational change over time, derived from the two cases discussed above, can be compared with the slope relating the phenotypic expression of two variables within a population. By definition, the least mean square regression through an intraspecific series of individuals relating the variables z_Y and z_X is given by the formula

$$m = \rho_{YX}\sigma_Y/\sigma_X$$

where ρ_{YX} is the correlation between the phenotypic expressions of the variables.

These two slope formulae can be compared to define the conditions relating the intrapopulational (static adult) and the interpopulational (population change over time) allometric slopes. By extension, they relate the intraspecific and interspecific (populational change over sufficient time to represent a new species) slopes. Under either the assumption of genetic correlation and selection on only one character or of random genetic drift, these slopes will be similar when and as long as the heritabilities are roughly the same, and both the genetic and phenotypic covariation patterns remain similar. Phenotypic covariance is the sum of the genetic covariance and the environmental covariance. Therefore, in comparing intraspecific and interspecific relations, "consistency of the phenotypic covariation is unlikely without consistency of genetic covariation if the latter is a substantial component of the former" (Lande, 1979, p. 405). It follows that under relatively stable conditions, maintaining similarity of environmental covariation, as long as there is continued similarity of genetic covariation selection on one character, or drift, will produce interspecific variation following the intraspecific allometric curve.

However, under these conditions the relation between the variables is a consequence of the genetic covariation [a point also made by Kaplan and Salthe (1979), pp. 682–683]. This relation may not be, and in fact probably is not, functional in an adaptive sense. Thus, a fallacy in Gould's reasoning is that while intraspecific allometry (and the cases in which interspecific allometry is roughly the same) reflects equivalence, this equivalence is a consequence of genetic covariation, and not a consequence of functional interrelationship. It is not "the same basic design expressed at different sizes" (Gould, 1975*b*, p. 278), except for those cases when the genetic correlation happens to result in phenotypically correlated change that maintains the same functional relationship between the two variables. The fact that intraspecific allometric slopes are commonly lower than interspecific slopes (see below, pp. 292–293) indicates that this correspondence is usually not the case. Another implication of Lande's reasoning is that this similarity in allometric relationships requires that the species be closely related, but not necessarily in the form of a phylogenetic sequence.

In sum, when interspecific allometry approximates intraspecific allometry in closely related forms, it is likely that two conditions are met [Lande (1979, p. 412), in discussing brain–body weight allometry]:

> First, closely related forms have usually diverged by selection mainly on body size or by random genetic drift with an allometric slope equal to the genetic regression of brain weight on body weight within populations; and second, there is a similarity in the slopes of the genetic and phenotypic regressions of brain weight on body weight within populations.

Since this situation has been observed and reported, it follows that there *is* such a thing as allometry. Denial of the concept because the path of evolution is not *necessarily* related to intraspecific variation (see below, p. 285) is the fallacy in Smith's reasoning.

In fact, the cases in which interspecific allometry substantially differs from intraspecific allometry are of equal if not greater interest. Any significant alteration of the environmental component of the covariance could lead to the divergence of the interspecific relation. Moreover, there are two conditions under which a divergent interspecific allometric relation could become the object of evolutionary change: when functional equivalence in the adaptive sense is maintained, and when the adaptive relation between the variables changes in a regular way.

The maintenance of functional equivalence is an alternative way of looking at Gould's "same basic design expressed at different sizes," relying, however, on adaptation rather than on form in determining what constitutes "sameness." Both concepts are somewhat more slippery than may appear obvious. For instance, ascertaining whether a morphological complex expressed at a larger size reflects the same basic design presupposed knowledge of *how* size effects form. This is different than knowing *whether* size effects form, since there are a variety of reasons why such an effect might occur. Only some of these are due to genetic covariation; others may be adaptive responses, and others yet the result of changing environmental covariance. The problem is which of these (or which combination) can be taken to represent changes in form that are not changes in design?

There is a parallel problem in the concept of functional equivalence. For instance, phyletic size increase may not require a change in the dietary resource base. Descendant populations may persist by utilizing more of the same resources that supported the smaller ancestral forms. In this case, if tooth size expansion associated with the size increase was no more or less than sufficient to stand up to the increased dental functions, the change might be considered functional equivalence. However, it is alternatively possible that phyletic size increase may lead to a different pattern of resource base utilization, or a change in the resource base utilized. Changes such as these can also be viewed as consequences of size increase, and the concomitant changes in the dentition might also be viewed as representing functional equivalence since they may reflect the only way that an organism with the same basic design can persist at a larger size. In my view, functional equivalence can be said to no longer characterize size-related changes only when these changes go beyond the requirements of what is size related. Thus, functional equivalence could involve a change in design, and these two different descriptions of changing form should be held separate.

I believe it is important to also distinguish this use of "functional equivalence" from a closely related concept, "the consequences of changes in size alone." While these are often held to be synonymous, the above discussion indicates that they need not be. Changes that are the "consequence of size alone" can refer to the phenomenon described sbove (i.e., the adaptive, structural, or functional consequence), but they can also refer to the genetic consequence (as Lande seems to use the phrase). One of the main points of this discussion is that the adaptive and genetic consequences of size almost always come to diverge from each other, leading to the separation of interspecific allometry from intraspecific allometry.

It is precisely because the effects of genetic covariance may not result in functional equivalence that intraspecific allometry may break down, as the two variables come under separate selections in order to maintain functional equivalence. As Lande puts it, again regarding brain–body weight allometry (Lande, 1979, p. 412):

> On a time scale typical for subspeciation or speciation, the coevolution of brain and body sizes occurs mainly through natural selection on body size and/or genetic drift, while on a longer time scale typical for diversification for higher taxonomic levels, natural selection is able to adjust brain size to accumulated changes in body size.

Thus, to maintain functional equivalence over a range of differing body sizes, in most cases selection must ultimately come to act on the two variables separately. According to Lande, species that actually represent the expression of the same functional design at different sizes may be the result of differential survival or extinction, since species that come to substantially deviate from the functional relationship "usually do not persist or speciate as well as those near the regression line" (Lande, 1979, p. 412) representing functional equivalence. Alternatively, a similar gradualist explanation is possible, viewing balancing selection as maintaining functional equivalence while size changes continue because of selection acting against individuals who deviate significantly from the functional pattern that is adaptive.

The other possibility for evolutionary deviation from the intraspecific pattern occurs when selection comes to promote a new functional relationship between the variables. The situation is similar in that selection is required to act separately on the two pleiotropic characters. The results, however, may be quite different and may not necessarily be characterized by allometric scaling at all. An example of such a deviation can be found in the case of posterior tooth size reduction in *Homo:* for most of the Pleistocene the teeth reduce without any significant body size change and therefore this reduction cannot be characterized as scaling. The other extreme can be found in Kurtén's (1955) analysis of relative paracone height in two species of *Ursus*. In this instance, a marked size difference between the brown bear and the cave bear results in no difference in relative paracone height even though both intraspecific curves show an allometric exponent greater than 1.0 for the cusp height–crown length relation.

Both cases of evolutionary deviation from intraspecific allometry—maintenance of the same functional pattern and development of a new pattern—have several things in common. Selection comes to act individually on the characters in question, and it follows that the resulting mathematical relation between the characters is a direct consequence of their adaptive relation. However, this pattern of interspecific variation, unlike the intraspecific case, need not be log–log linear (i.e., be accurately fit by a power function); there are examples in which allometry (as represented by log–log linearity) does not characterize the pattern of interspecific relationships. Once again *contra* Gould, this does not mean that interspecific allometry is inexplicable or without evolutionary interest. To the contrary, *those cases in which interspecific allometry can be shown* are of great evolutionary interest precisely because they may reflect a persistent adaptive pattern. Moreover, such cases provide a clear indication of evolutionary mechanisms. They reflect the pathway through which the consistent (or changing) adaptation was reached through either the differential extinction of species deviating from the pattern promoted by selection or through the differential reproduction and survivorship of individuals not deviating far from the mean relationship promoted by selection. Thus, interspecific allometry need not uniquely characterize all phylogenetic sequences in order to validly reflect both the evolutionary adaptation and the pathway by which it was established in those cases when phylogenetic sequences *are* allometric.

Finally, both patterns of deviation from intraspecific allometry have a genetic mechanism in common. In both cases, divergence from the intraspecific pattern is constrained by the additive genetic covariance for the variables. This limitation probably is irrelevant in many cases. However, when the directions of selection acting on the two variables are of sufficient difference to force the existing additive genetic correlation to affect the responses of the variables, the additive genetic variance must be lowered, or uncoupled, in order to allow the independent responses that are the object of selection. Without this uncoupling, interspecific allometry would be held to a relation that becomes increasingly inadaptive because of changes in one variable caused by the genetic correlation with the other. This phenomenon has been observed when particularly intense selection is applied to one feature, as for instance in dog breeding (Wright, 1977). Lande suggests that the brain–body size relation in hominids has been genetically decoupled in this way (Lande, 1979, p. 413):

> In an organism with population genetic parameters similar to those estimated here for mice, any increase in brain size without a relatively large change in body size . . . would require antagonistic selection for larger brains . . . *and for smaller bodies* . . . to counteract an excessive correlated increase in body size from selection on the brain.

An interesting implication is that when a lowering of the additive genetic covariance is required, it can be predicted that the genetic uncoupling lowers the phenotypic correlation by lowering the genetic correlation, and therefore the intraspecific least mean squares regression slopes will come to be lower than the interspecific slopes as a necessary consequence. Thus, because the changes in hominid tooth size were as great as the changes in brain size (although in the opposite direction), it is quite possible that during the evolution of the hominids, a functional coadaptation of the tooth–body size relation could only be maintained by reducing the tooth size–body size genetic correlation, thereby allowing each variable to respond to different selection. After all, for most of the span of tooth size change, average body size did not change at all. Paradoxically, in this case (as in hominid brain size evolution) the maintenance of a functional link between these variables requires a partial releasing of the genetic link between them.

One consequence of this model provides an explanation for an oddity that has been noted during phylogenetic dwarfing. When the process of evolving larger body size requires a genetic decoupling and reduced additive genetic correlation, and a consequent lowering of the intraspecific allometric slope for some size-related variables, subsequent selection to reduce body size may result in a different evolutionary progression for these variables. For instance, as in the case of tooth size, an intraspecific slope of lesser (than interspecific) magnitude might have evolved to maintain functional equivalence. However, subsequent body size reduction leading to correlated changes in tooth size along the (then lower) intraspecific slope would result in teeth that were more than large enough for functional equivalence. Tooth size (following the lower intraspecific slope) would reduce much more slowly than body size, without selection to disturb the correlated change of the intraspecific relationship with its lower genetic covariance. Differential selection would not be expected to disturb the intraspecific pattern in order to maintain functional equivalence because the more slowly reducing dentition would always be large enough. As long as, apart from size, the primary adaptation remained the same, the dwarfed descendant species would be expected to have relatively larger tooth size than similarly sized ancestral species, or similarly sized related species lacking the phylogenetic history of size expansion. Instances of dwarfed forms following this pattern, with relatively large teeth compared with either ancestors or collateral relatives of similar body size, are known in hippopotamus [Gould (1975*a*), but see Prothero and Sereno (1982)] and elephant (Maglio, 1972) species. *Contra* Gould (1975*a*, pp. 360–361), these cases are explicable; functional equivalence is maintained insofar as the teeth are as large as (actually larger than) necessary, a consequence of the fact that the low rate of dental reduction, following the intraspecific allometric curve but applied to populational change, did not invoke differential selection separately on tooth and body size.

In sum, Lande's explanation for allometry suggests that when interspecific and intraspecific allometric slopes correspond, this is probably the consequence of either selection on only one of the variables or of random genetic drift. The resulting allometric pattern reflects the magnitude of the additive genetic covariation, with the environmental covariation superimposed. Divergence of the interspecific slope from the intraspecific slope can occur for a variety of reasons, some of which only reflect changing environmental covariation. When such divergence has a genetic basis, it is a direct consequence of selection acting on each of the variables separately. If a divergent allometric pattern of interspecific variation persists, it could reflect either the maintenance of the same adaptive pattern expressed at different sizes, or the development of a new adaptive relationship between the variables. Under some conditions, continued evolution of the adaptive scaling relationship requires a reduction in the genetic covariation, with the implication of a smaller phenotypic correlation between the variables and a lowered intraspecific allometric exponent. Interpretation of the adaptive basis for particular cases of interspecific allometry is a problem for the investigator. However, it is the interspecific pattern that may correspond to a similar adaptation expressed at different sizes. In such cases the interspecific pattern of variation directly reflects the mechanism through which the similar adaptation was reached across the size range.

Implications of Different Interspecific and Intraspecific Tooth–Body Size Slopes in Primates

Lande's discussion is based on the brain–body weight phenomenon, in which the interspecific and intraspecific allometric slopes differ dramatically and consistently (Jerison, 1973). This relation of interspecific and intraspecific slopes may also characterize the tooth size–body size allometry in at least some primate species. Wood (1979) examined the relationship between interspecific and intraspecific tooth–body size allometry in four nonhuman primate taxa and in a human sample. Generally, Wood's results seem to show that the interspecific correlations and regression slopes are higher than the intraspecific slopes. This indicates that selection generally promotes a different tooth–body size relation than the genetic covariance provides, and might reflect some degree of genetic decoupling in these living species.

According to Lande's model, genetic decoupling is a response to selection acting separately on tooth size and body size, either maintaining the same functional relationship between these or systematically altering it as an adaptive response. The adaptive pattern underlying the reduced additive genetic covariation responsible for the lower intraspecific slopes, if the reduction in fact exists, remains to be demonstrated for Wood's primate data. Moreover, in general Wood's results are difficult to interpret because he used estimators rather than direct measures of body size (Smith, 1980, 1981*a*), and because he used the reduced major axis technique to calculate his regression slopes (see next subsection). However, one interesting implication of Wood's data is that the intraspecific pattern is unlikely to be useful in predicting one variable from the other, even in the sample from which the equations were derived, *let alone* in a sample of fossil primates such as Wood attempts (for further discussion of this problem, see R. J. Smith, this volume).

A Digression about Regression

Assuming that a power curve is a reasonably justified means of expressing a scaling relationship, the problems surrounding the mathematical expression of the scaling relationship do not cease. This is because there are different techniques for fitting a straight line to log-transformed data. A number of studies have shown that the regression slopes for biological data can vary dramatically when different curve fitting approaches are compared (Kuhry and Marcus, 1977; Sokal and Rohlf, 1981; Wolpoff, 1982). The two basic approaches to line fitting are probably best summarized by Sokal and Rohlf (1981; see also Kuhry and Marcus, 1977). Specific techniques based on each have been used in various studies of the tooth size–body size relation, and unfortunately several of these studies are not specific about which technique was used. It makes a difference.

The model I approach, using the least mean square technique, provides a predictive generalization about the linear dependence of the dependent variable on the independent one. The technique is often used to examine the specifics of a suspected functional relationship. It provides a means of examining causality, as long as one is willing to make the assumption that variation in the independent variable causes variation in the dependent variable. Conversely, if one is unwilling to make this assumption, it can provide the rationale for seeking an independent cause for the covariation of the regression variables.

In Lande's development of his allometry model, a strong case for using the least mean squares regression approximation is made (Lande, 1979, p. 404). The basis for this case is simply that the genetic model for intraspecific allometry can be used to directly derive the least mean squares regression slope for the log-transformed data. Lande's comparison of the expected formulas for the intraspecific slope and the interspecific slope approximated by a least mean squares regression results in the only major insight into the relation between intraspecific and interspecific allometry that has been achieved since Huxley's formulations.

However, it has been argued that the least mean squares technique (model I approach) is not valid for examining the question of underlying phenomena because it does not take all sources of covariation into account. Thus, when the question is not one of prediction but rather of underlying biological relationship, a model II regression may be required (Sokal and Rohlf, 1981, pp. 491–496). Model II regressions are said to be appropriate when variation in each of the variables is relevant, perhaps because of a common underlying source of variation in both or the presence of error in the measurements of both. Conversely, however, a model II regression is not appropriate for prediction and may only poorly describe a functional relationship when the independent variable is a functional consequence of the dependent variable.

The two model II regressions that have most often been used in allometric studies are the reduced major axis (or correlation surface) and the major axis (or principal axis) techniques. Of these, the major axis regression has been found to be the more desirable. Two other model II regression techniques, Bartlett's method and the covariance ratio method (Kuhry and Marcus, 1977), have also been suggested as appropriate but have never been widely used in allometric studies relating tooth size and body size.

Reduced major axis was the first of these to be applied to allometry (Kermack and Haldane, 1950). The slope of the regression is estimated by the ratio of the standard deviations of the log-transformed variables, and thus is the geometric mean of the least mean square regression of X on Y, and of Y on X. It is related to the least mean square regression slope, but will always be larger because it is equal to this regression slope divided by the correlation coefficient. The undesirable properties of the reduced major axis approach have been reiterated by Jolicoeur (1975), Gould (1975*b*), Kuhry and Marcus (1977), Wolpoff (1982), and others. Simply put, the main objections are two. First, because the slope is the ratio of the standard deviations, it responds to the separate variabilities but not to the covariation of the dependent and independent variables. For instance, if the dependent variable is by its nature the more variable of the two, the regression slope will be greater than 1.0 regardless of the actual relation of the variables. Second, at very low correlations the regression slope has no meaning; the ratio of the standard deviations can be quite high in this case while the least mean square slope, which is this ratio multiplied by the correlation, may not be significantly different from 0.0. In such a case, the reduced major axis slope is obviously a poor reflection of the biological relationship.

The major axis regression is not without its own problems. It requires that both variables are in the same unit of measurement, although Kuhry and Marcus (1977, p. 204) argue that this condition does not apply to the allometric equation, where "the scale becomes entirely arbitrary, since the variances are standardized by the logarithmic transformation." Under conditions of lower correlation this technique, like the reduced major axis, can give rather high (and misleading) slopes. After all, the major axis is the eigenvalue that accounts for the maximum amount of variation, and its slope is unlikely to have meaning if much of this variation is nonbiological, or random. Kuhry and Marcus (1977) report a case in which excessive nonrelated variation in the dependent variable causes the method to break down and give misleading results. However, because the major axis technique is covariance dependent, it is more desirable than the reduced major axis.

In sum, regression analyses of allometric relationships are only appropriate when a significant correlation describes the covariance of the variables in question. The model I approach is valid for questions about predictability (questions of the sort, Can larger bodied primates be expected to have larger teeth because of their larger size? Specifically, how do the teeth scale as a function of body size?). It has been claimed that a model II regression is the appropriate vehicle to examine biological relationship (questions of the sort, Is there a general relation between tooth size and body size in primates? Does this relation better fit the geometric or the metabolic model?). However, all of the model II regression approaches that have been used carry theoretical and usually also practical problems, and under certain reasonable conditions lead to bizarre results. Instead, a number of authors dealing with the tooth size–body size relation (as well as others) have argued that the least mean squares regression slopes are more appropriate for this purpose (Goldstein *et al.*, 1978; Smith, 1981*b;* Wolpoff, 1978, 1982). Lande's derivation of expected allometry coefficients for this regression technique from population genetics theory provides the most powerful argument for accepting the least mean squares regression as the most appropriate mathematical approximation for studying allometry.

Tooth Size Allometry in Humans

> Investigations of the strength and nature of the relationships between tooth size and body size in primates are a relatively neglected area of primate research.
>
> (Wood, 1979, p. 187)

Tooth Size and Body Height

Virtually all intraspecific studies of tooth size relative to body size in humans have related tooth size to stature, or to a linear estimator of stature such as femur length, cranial length, or the length of the cranial base. Of the studies relating linear dental dimensions to actual stature, the Filipson and Goldson analysis (1963) of 110 male army conscripts resulted in a lack of significant correlation between incisor and canine width, and stature. Similarly, the Anderson *et al.* (1977) study of 118 males and 102 females sampled at age 16 from the Burlington Growth Centre resulted in a complete lack of significant correlations between anterior tooth widths and body height for the male and female groups considered separately. Two other studies (summarized below in Table 5) considering the dimensions of all of the teeth have examined the tooth size—stature relation, both focusing on the question of predictability of one from the other, both considering males and females separately, and both using the strength of the correlation coefficient as the criterion of analysis (Garn *et al.*, 1968; Henderson and Corruccini, 1976). Although correlations differed from tooth to tooth, these analyses both concluded that the magnitude of the relationship is so low that predictability is impossible. As Henderson and Corruccini (1976, p. 94) put it: "inferring body size of fossil hominids from tooth size alone is unwarranted on the basis of these results." Similarly, Garn *et al.* (1968, p. 131) conclude that there is "no generalization that would assure the larger-bodied fossil species larger teeth as a natural consequence."

Whether or not one accepts Smith's (1980, 1981*a*) contention that variables estimating body size cannot be substituted for direct measures of body size in an allometric analysis, the conclusions of the more numerous studies relating tooth size to estimates of stature such as femur length or cranial dimensions are exactly the same as the results reported above; the correlations are either not significantly different from zero, or are of very low magnitude.

In sum, studies examining the scaling of tooth size to body height within human populations have provided uniformly negative results. Some correlations exist, but at a low magnitude that explains too little of the covariance to allow anything like the accurate predictability of one from the other. The conclusion drawn in these studies is that body height differences are unlikely to explain tooth size differences in fossil or living hominids. The only examination of this relationship using populational means (Garn and Lewis, 1958) reaches the same conclusion.

Tooth Size and Body Weight

Until now, there has been only one study relating intraspecific scaling of tooth size to actual body weight in humans, and none in any other of the higher primates. The Anderson *et al.* (1977) study using anterior tooth widths in the male and female samples taken from the Burlington Growth Centre records shows positive and mostly significant correlations between these dental dimensions and body weight at 16 years of age for the male sample, but smaller and generally not significant correlations for the female sample. The pattern of higher and more often significant correlations in the male sample parallels the pattern for tooth size–height correlations published by Garn *et al.* (1968) and Henderson and Corruccini (1976). In all cases, correlations with weight were much higher than correlations with height.

The most closely related analysis for nonhuman primates is Lauer's (1975), comparing tooth size correlations with stature estimates (limb lengths) and with weight estimates (limb cross sections and volumes) in a troop of provisioned rhesus monkeys. Lauer concluded that the correlations with stature estimates were higher than the correlations with weight estimates (in fact, when considered for each sex separately they differ little from the stature correlations reported for humans in the Garn *et al.* and Henderson and Corruccini studies), citing in explanation the fact that body weight can differ considerably over the adult lifespan of an individual, and differ between individuals for nongenetic reasons. The extent to which substitution of variables estimating stature and weight for direct measures of stature and weight also affects this conclusion is unknown.

While less than the correlations with height, the weight correlations for the combined sex sample were quite high in these primates and in all cases but one they were significant at the 0.005 level. Consideration of the sexes separately showed generally higher correlations in the male sample than in the female sample for both height and weight estimates.

What Remains Unknown

Human studies have been intraspecific [excepting Pilbeam and Gould's (1974) attempt to discuss tooth size allometry over the course of human evolution] and with only one exception have focused on the relation of tooth size to stature (or to a measure of stature). Studies relating tooth size directly to stature are limited to answering one question, relying on the correlation coefficient to ascertain the strength of the relationship in order to determine whether one of the variables can be accurately predicted from the other. The only attempts to determine the actual pattern of intraspecific scaling (i.e., the allometric exponent) are those of Wood (1979) and Perzigian (1981), but these both rely on indirect stature estimates. The conclusion that can be drawn from the correlation studies with actual (i.e., unestimated) statures is that a low level of relationship exists between tooth size and stature (too low for accurate predictability), and that the correlations are higher in the male sample than they are in the female sample (none of these studies examined the combined sex sample).

The only study utilizing weight (Anderson *et al.*, 1977) is limited because of two factors: (1) weights were all taken at 16 years, probably not the end of growth for all the individuals involved (Harrison *et al.*, 1977); (2) only the breaths of the anterior teeth were examined. This study concluded that weight was a better correlate with anterior tooth breadths than height, and (as above) the correlations were stronger in males. No attempt to ascertain scaling was made.

Thus, there is much left to know about the tooth size–body size relation in humans. The preliminary suggestion that weight is a better correlate with tooth size than height must be verified for the entire dentition. This would make sense, given the role of the teeth in masticating food, and if true it could help explain the low correlations of tooth size with height and measures of height. However, weight measures that accurately reflect the adult condition must be used. Intraspecific scaling of the various size measures for all of the teeth directly with either stature or weight is yet to be examined and compared with interspecific scaling. The scaling relations in males, females, and a combined sex sample have not been compared, and the difference between male and female correlations has not been explained [Smith (1981*c*) argues that the sample range can affect the magnitude of the correlation coefficient, but the effect is very small, the direction of the influence is opposite for the dependent and independent variable ranges, and the difference between male and female ranges is invariably too little for this influence to be significant]. Finally, an acceptable explanation for tooth size scaling is yet to be established either intraspecifically or interspecifically, and the related question of what (if any) conditions allow predictability is yet to be convincingly answered.

To help meet these goals, a new study using a living population of Australian Aborigines is reported here.

The Yuendumu Sample Data

Yuendumu is an Australian Aboriginal settlement, mostly of the Wailbri. This Commonwealth Government settlement is in the Northern Territory, about 285 km northwest of Alice Springs. The population of Yuendumu has cooperated in an intensive longitudinal study involving many aspects of growth, morphology, and adaptation that has spanned the last several decades (Barrett *et al.*, 1965; Brown and Barrett, 1973; Barrett, 1976; and references therein).

Growth data, including multiple observations of height and weight, and a full set of dental metrices from accurate dental casts were obtained for a sample from Yuendumu consisting entirely of Wailbri, from Dr. T. Brown of the University of Adelaide. The dental measurements were taken on casts that generally did not show dentin exposure, and consequently represented individuals at a young enough age to avoid the effects of interproximal attrition on mesiodistal length (Richards and Brown, 1981; Molnar *et al.*, 1983). The growth data for each individual were examined to find the age at which height stopped increasing, and body weight was recorded at this age. The purpose of this procedure was to minimize the environmental component in the variance of weight. Pregnant females and females who recently gave birth were omitted. The resulting data set is comprised of dental measurements, maximum adult height, and adult weight at the age that adult height was attained for 45 females and 21 males.

Methods

These data were used to examine the relation of tooth size to body size in the Yuendumu Wailbri sample. All determinations were separately established for the male, female, and combined sex samples. The measures of tooth size examined include the length and breadth dimensions of each tooth, occlusal areas for the canines and postcanine teeth, and several dental sums (not including M3 because the sample size is too small). Relation to height and weight was determined using the product-moment correlation with the small-sample modification suggested by Kendall and Stuart (1961). Scaling of tooth size to the two measures of body size utilized least mean squares regressions on the log-transformed data, for reasons discussed above. Regression slopes were only reported for cases in which the correlation coefficient was significant at least at the 0.10 level.

The Relation of Weight to Height

The weights recorded as described above were found to scale quite near the cube of height (Table 1), using a least mean squares regression to determine a power curve fit. The correlations were moderately high. Thus, weight and height have an isometric relationship, insofar as the scaling exponent for weight as a function of height is close to the prediction of the geometric model. However, the correlations are not high enough to allow the assumption of a one-to-one relationship between these size variables along a regression line relating them. Therefore, one cannot be validly substituted for the other in an allometric analysis, in spite of their geometric relationship (Smith, 1980). It follows that if weight is the primary causal variable for tooth size scaling, the relation of tooth size to height can be expected to be of lesser magnitude since height is not a completely accurate predictor of weight. The scaling of tooth size to estimators of height, such as limb lengths, should be of even lesser magnitude. Conversely, these expected relations can be used to ascertain whether weight is indeed the primary causal variable by comparing the relations of tooth size to weight and to height.

Interestingly, the scaling exponent for females is somewhat less than for males, and less than for the virtually identical combined sex sample. The similarity of the combined sex (i.e., total sample) curves to the male curves is also of some interest; it cannot be accounted for by a numerical predominance of males in the combined sample since there are only half as many males as there are females. When constrained to fit a cubic curve (Table 1), the prediction errors [comparison of the predicted to actual values; see Wolpoff (1982)

Table 1. Yuendumu Weight–Height Relation[a]

	Both sexes	Males	Females
N	99	37	62
r	0.790	0.611	0.600
Power curve	$W = 0.0019H^{3.343}$	$W = 0.0030H^{3.258}$	$W = 0.0468H^{2.712}$
Cubic curve	$W = 0.0125H^3$	$W = 0.0127H^3$	$W = 0.0094H^3$

[a]Weight W in grams, height H in centimeters.

and Smith (1980)] were exactly the same as for the power functions. Constants for the cubic equations are closer in magnitude, but still show that females scale at a slightly lower rate. This is not an inevitable relation in humans. In an adult sample of 81 female and 93 male Efé Pygmies (data provided by the Harvard Ituri Project), the power curve exponents for the weight–height relation were lower than for the Yuendumu Wailbri, and males and females were about equally different in magnitude (male exponent 2.24, female exponent 2.59). In this case, however, male scaling was at a lower magnitude than female scaling. There clearly are populational differences in both the magnitude and the male–female relationship of the scaling of weight to height in adults.

Correlation Analysis of the Yuendumu Data

Tables 2–4 show the correlations of the dental variables to weight and to height for the Yuendumu Wailbri female, male, and combined sex samples. For the male sample the correlations with weight are almost invariably greater than the correlations with height (one exception each for length and area, two for breadth). The combined sex sample shows a similar pattern, with the weight correlations exceeding height for 13 out of 16 lengths and breadths each, and for 10 out of 12 areas. However, for the female sample the situation is less consistent. Weight correlations exceed height for only five out of 16 lengths, but for 11 out of 16 breadths. In the occlusal areas, the weight correlations are greater only five out of 12 times. A binomial test shows that as a whole the female pattern of how many times weight correlations exceed height correlations is not significantly different from random; yet, the interesting fact remains that the height correlations predominate for lengths and areas, while weight correlations predominate for breadths.

In the summed dimensions, weight correlations are invariably higher than height correlations for the combined sample and for the males (in which only one height correlation, with the sum of the maxillary anterior lengths, is significant even at the 0.10 level). However, in the females height correlations are higher in all of the combinations of occlusal area sums (all of these are significant at the 0.05 level). Weight correlates more highly in the anterior dimension sums. Since the summed dimensions more likely reflect dental function, it is possible that females actually do maintain a somewhat different basis for their tooth size–body size relationship. Because the Anderson *et al.* (1977) study only examined the anterior breadths, while its conclusions are confirmed by this work, they are somewhat misleading.

Comparison of breadth correlations with length correlations provides only some consistent results. For instance, in the anterior teeth of the male sample both weight and height correlations with length are invariably greater than with breadth. However, these comparisons are mixed for the female and combined sex samples. In general, cases in which breadth correlations are greater than length correlations are about equal to cases of the reverse. For the male and the female samples, breadth correlations exceed length correla-

Table 2. Correlation of Dental Dimensions with Weight and Height for Yuendumu Females[a]

	Tooth length		Tooth breadth		Tooth area		
	Weight	Height	Weight	Height	Weight	Height	Sample size
I^1	0.441	0.219[b]	0.368	0.284			45
I^2	−0.002	−0.058	0.074	0.020			41
C	0.114	0.152	0.229[b]	0.173	0.203[b]	0.189	45
P^3	0.200[b]	0.214[b]	0.218[b]	0.115	0.234[b]	0.189	45
P^4	0.296	0.269	0.248	0.156	0.303	0.238[b]	44
M^1	0.200[b]	0.362	0.327	0.300	0.282	0.359	45
M^2	0.243[b]	0.358	0.194	0.330	0.239[b]	0.372	45
M^3	−0.005	−0.080	0.172	0.295	−0.013	0.088	19
I_1	0.278	0.156	0.180	0.152			45
I_2	0.244	0.098	0.239[b]	0.171			45
C	0.122	0.161	0.220[b]	0.212[b]	0.197[b]	0.213[b]	45
P_3	0.105	0.236[b]	0.322	0.198[b]	0.236[b]	0.233[b]	45
P_4	0.223[b]	0.229[b]	0.260	0.229[b]	0.270	0.258	44
M_1	0.192	0.232[b]	0.275	0.303	0.266	0.306	45
M_2	0.320	0.334	0.297	0.398	0.333	0.395	45
M_3	0.253[b]	0.207	0.122	0.239[b]	0.214	0.242[b]	35
Maxillary sums							
I^1–C	0.131	0.023	0.155	0.083			41
C–M^2	0.232[b]	0.309			0.261	0.295	44
I^1–M^2	0.168	0.173					42
P^3–M^2					0.266	0.311	41
P^4–M^2					0.269	0.340	44
Mandibular sums							
I_1–C	0.243[b]	0.153	0.228[b]	0.193			45
C–M_2	0.255	0.308			0.297	0.322	44
I_1–M_2	0.285	0.272					44
P_3–M_2					0.307	0.336	44
P_4–M_2					0.314	0.351	44

[a] Underlined correlations significant at $p(r = 0) \leq 0.05$.
[b] $p(r = 0) \leq 0.10$.

tions for 17 out of the 32 possible linear comparisons. However, in the combined sex sample, the breadth correlations are greater 26 out of 32 times.

Unlike the other two reported studies of tooth size relative to stature (Table 5), males generally do not have higher correlations with stature than females in the Yuendumu Wailbri sample. In fact, the male correlations only exceed the females eight out of 16 times for both length and breadth. While this might appear to be random, in most cases when the male correlations are higher, it is in the anterior portion of the toothrow. Conversely, in most cases when the female correlations are higher it is in the posterior of the toothrow. This anterior–posterior contrast can be clearly seen by comparing instances when correlations are significant; these also tend to be anterior for the males and posterior for the females. There is no equivalent pattern for the other

Table 3. Correlation of Dental Dimensions with Weight and Height for Yuendumu Males[a]

	Tooth length		Tooth breadth		Tooth area		
	Weight	Height	Weight	Height	Weight	Height	Sample size
I^1	0.572	0.348	0.286	0.309[b]			21
I^2	0.338[b]	0.264	0.295	0.194			20
C	0.453	0.404	0.245	0.076	0.425	0.294	21
P^3	0.420	0.209	0.471	0.358	0.453	0.271	21
P^4	0.178	0.008	0.523	0.258	0.400	0.144	21
M^1	0.343[b]	0.070	0.581	0.253	0.467	0.161	21
M^2	0.248	−0.003	0.335[b]	0.102	0.332[b]	0.063	21
M^3	0.754	0.437	0.874	0.644	0.927	0.759	7
I_1	0.454	0.248	0.416	0.176			21
I_2	0.678	0.287	0.468	0.242			20
C	0.484	0.378	0.241	0.114	0.433	0.309[b]	21
P_3	0.185	0.005	0.562	0.180	0.417	0.096	21
P_4	0.515	0.135	0.543	0.131	0.599	0.149	21
M_1	0.231	−0.155	0.735	0.303[b]	0.510	0.057	21
M_2	0.441	0.239	0.456	0.123	0.495	0.212	21
M_3	−0.163	−0.250	0.120	−0.153	−0.008	−0.234	12
Maxillary sums							
I^1–C	0.538	0.399	0.376	0.277			20
C–M^2	0.394	0.162			0.449	0.189	21
I^1–M^2	0.454	0.233					21
P^3–M^2					0.435	0.162	
P^4–M^2					0.417	0.126	
Mandibular sums							
I_1–C	0.640	0.364[b]	0.428	0.184			20
C–M_2	0.455	0.143			0.545	0.173	21
I_1–M_2	0.530	0.196					21
P_3–M_2					0.542	0.141	21
P_4–M_2					0.560	0.149	21

[a] Underlined correlations significant at $p(r = 0) \leq 0.05$.
[b] $p(r = 0) \leq 0.10$.

groups that have been studied (Table 5). In occlusal areas, male correlations with height exceed females only four out of 12 times.

The comparisons of sexes for weight correlations are quite different. Males exceed females 14 out of 16 times for lengths and breadths, and 11 out of 12 times for areas.

In sum, the following can be concluded from the analysis of correlations.

1. *Length and breadth.* Considering all of the teeth, on the average, length and breadth correlate about equally well with either weight or height separately in both the male and the female samples, although in the combined sex sample, breadth correlations tend to be higher. However, length correlations are invariably higher in the anterior teeth of the males. For the posterior teeth, breadth may provide a slightly higher correlation, but the difference is not great. This observation, of course, would be considerably different if the teeth measured showed significant interproximal wear.

Table 4. Correlation of Dental Dimensions with Weight and Height for Yuendumu Total Sample[a]

	Tooth length		Tooth breadth		Tooth area		
	Weight	Height	Weight	Height	Weight	Height	Sample size
I^1	0.582	0.422	0.537	0.545			66
I^2	0.263	0.225[b]	0.283	0.234[b]			61
C	0.398	0.394	0.377	0.338	0.444	0.417	66
P^3	0.267	0.184	0.352	0.265	0.334	0.244	66
P^4	0.293	0.241	0.450	0.324	0.413	0.314	65
M^1	0.369	0.369	0.559	0.470	0.491	0.444	66
M^2	0.311	0.314	0.444	0.459	0.408	0.416	66
M^3	0.071	−0.116	0.515	0.419	0.392	0.235	24
I_1	0.416	0.303	0.434	0.371			66
I_2	0.521	0.322	0.444	0.372			65
C	0.483	0.465	0.384	0.385	0.489	0.480	66
P_3	0.203[b]	0.213[b]	0.495	0.353	0.391	0.313	66
P_4	0.400	0.289	0.424	0.315	0.459	0.335	65
M_1	0.379	0.309	0.518	0.424	0.504	0.412	66
M_2	0.416	0.364	0.535	0.526	0.515	0.479	66
M_3	0.213[b]	0.196[b]	0.318	0.378	0.288	0.296	44
Maxillary sums							
I^1–C	0.455	0.368	0.427	0.396			63
C–M^2	0.391	0.359			0.470	0.419	65
I^1–M^2	0.415	0.345					63
P^3–M^2					0.455	0.402	65
P^4–M^2					0.466	0.426	65
Mandibular sums							
I_1–C	0.533	0.410	0.443	0.394			66
C–M_2	0.475	0.412			0.542	0.464	65
P_3–M_2	0.521	0.420					65
P_4–M_2					0.529	0.442	65
					0.542	0.457	65

[a]Underlined correlations significant at $p(r = 0) \leq 0.05$.
[b]$p(r = 0) \leq 0.10$.

2. *Male and female.* The pattern of correlations by sex differ for height and weight. In the relation to height, unlike previously published studies, there is no tendency for male correlations to be higher than female correlations. In fact, correlations of female teeth with height are more often significant than male correlations (although perhaps because of the larger female sample size). When the female correlations are greater in magnitude, it is most often in the posterior half of the toothrow, and most higher male correlations are in the anterior half. For weight, however, male correlations are almost invariably higher and are more often significant.

3. *Height and weight.* As already mentioned, these two measures of body size provide different estimates of the tooth size–body size relation. In both linear dimensions and in area, weight correlations exceed height correlations in the great majority of comparisons for males and for the combined sample.

Table 5. Correlations[a] of Tooth Dimensions with Height

	Males						Females					
	Length			Breadth			Length			Breadth		
	Ohio Whites[b]	Australian[c]	American Blacks[d]	Ohio Whites	Australian	American Blacks	Ohio Whites	Australian	American Blacks	Ohio Whites	Australian	American Blacks
I^1	0.02	*0.35*	*0.32*	0.13	0.31	*0.32*	−0.03	0.22	−0.09	0.02	*0.28*	−0.07
I^2	*0.20*	0.26	0.22	*0.31*	0.19	*0.27*	*−0.24*	−0.06	0.22	−0.07	0.02	0.05
C	*0.30*	*0.40*	*0.33*	0.04	0.08	0.17	0.11	0.15	0.24	0.00	0.17	0.17
P^3	0.18	0.21	0.19	0.16	*0.36*	*0.24*	0.17	0.21	0.24	*0.37*	0.12	0.10
P^4	*0.23*	0.01	*0.22*	*0.29*	0.26	0.13	*0.33*	*0.27*	0.21	*0.42*	0.16	0.13
M^1	*0.27*	0.07	*0.28*	0.18	0.25	0.11	0.15	*0.36*	0.09	0.01	*0.30*	0.06
M^2	*0.27*	0.00	*0.22*	*0.50*	0.10	0.18	0.24	*0.36*	*0.34*	0.25	*0.33*	−0.07
I_1	0.24	0.25	*0.25*	0.16	0.18	*0.25*	0.20	0.16	0.06	−0.04	0.15	0.07
I_2	0.11	0.29	*0.26*	0.09	0.24	*0.32*	0.17	0.10	0.16	−0.06	0.17	0.09
C	0.11	*0.38*	*0.22*	0.17	0.11	*0.27*	0.06	0.16	0.04	−0.01	0.21	0.13
P_3	*0.37*	0.05	*0.28*	*0.23*	0.18	*0.19*	0.15	0.24	0.11	*0.28*	0.20	0.14
P_4	0.05	0.14	*0.24*	*0.22*	0.13	0.15	0.10	0.23	0.15	0.22	0.23	0.19
M_1	0.01	−0.16	*0.37*	0.05	0.30	*0.24*	0.21	0.23	0.17	0.19	*0.30*	0.09
M_2	0.19	0.24	0.18	*0.25*	0.12	*0.22*	−0.04	0.33	0.14	0.14	*0.40*	0.03

[a]Correlations reported in italics when $p(r = 0) \leq .05$.
[b]From Garn *et al.* (1968).
[c]This study.
[d]From Henderson and Corruccini (1976).

Weight correlations also exceed height correlations in most breadth comparisons in the female sample. However, in the other female comparisons the height correlations tend to be greater.

Discussion of the Correlations

In his discussion of allometry, Smith (1980) contends that correlations must be exceedingly high to meaningfully reduce the magnitude of residuals around a regression line. None of the correlations reported here are close to the magnitude required to make a significant difference; in all cases the residuals are high. These correlations, not unlike those reported in other studies, show only a general relationship between tooth size and body size. With the exceptions noted above, this relationship is stronger for weight than it is for height, and for the weight correlations the relation is stronger in males than it is in females. It is very unlikely that the range effect on the magnitude of the correlation coefficient reported by Smith (1981*c*) has any influence on these comparisons. In the comparison of sexes, the Yuendumu male and female range difference is small, while in the comparison of height and weight as independent variables, weight has a greater range. Smith's data analysis shows a negative effect of independent variable range increase on the magnitude of the correlation (Smith, 1981*c*, p. 293) when the range of both variables is taken into account. This interpretation of Smith's data would predict a lower correlation for Yuendumu weight, with its greater range. Instead, the correlations with weight reported here are generally higher than the correlations with height. For this reason, I contend that the differences between the weight and height ranges do not influence the comparison of the correlations.

In the males, the correlations for dental sums with weight range between about 0.4 and 0.6, certainly not insubstantial, and much higher than the correlations of tooth size with height reported in the earlier studies with generally negative conclusions. In some cases the tooth size–weight correlations are high enough to approach the magnitude of the height–weight correlation. For the females, however, the corresponding correlations with weight are only slightly better than half the male value.

Thus, the idea that body weight is the primary or causal variable underlying tooth size variation is not unambiguously confirmed, unless part of the female data is ignored. Body weight may indeed be causal, but it cannot be the unique cause of tooth size variation. There is enough difference between the height and weight correlations to support the proposition that correlations with height may reflect the relation of tooth size to weight, but the data for the females again suggest that the situation is probably more complex.

One way of viewing the female data results from the observation that the female correlations with stature are similar to those for the males, while the female correlations with weight are significantly less. This could be a consequence of the non-weight-related dietary requirements associated with pregnancy. Interestingly, in the only related studies, involving a different human group (Anderson *et al.*, 1977) and in a second case involving rhesus monkeys (Lauer, 1975), female correlations with weight (or with weight estimates) were also found to be lower than male correlations. Thus, regarding the female tooth–body size relation as reflecting less genetic linkage, as Henderson and Corruccini (1976) suggest, allows us to consider the possibility that this relation might be the consequence of a difference in the selection acting on those alleles affecting tooth size that are located on the sex chromosomes. Indeed, the comparison of male and female coefficients of variation for the tooth areas shows that with only one exception, P_4, the variation in female occlusal areas is lower than the male variation for the postcanine teeth of both jaws. It is possible that balancing selection results in a narrower range in female postcanine areas, perhaps a consequence of promoting sufficient occlusal area to support the increased and non-weight-related dietary demands during pregnancy. Other data supporting this possibility come from a study of cusp height reduction as a function of age, using this same sample (Molnar *et al.*, 1983). The absolute magnitudes of the correlations of cusp heights and age were found to be lower in females than in males, although all correlations were significant. This might be expected if an additional factor, such as the increased dietary demands of pregnancy, was disturbing an otherwise linear relation between cusp height loss and age. Finally, should these considerations be correct, one might expect females to have relatively larger teeth than do males, as an evolutionary consequence of the elevated masticatory demands placed on them. This expectation is fulfilled when postcanine areas are determined as a fraction of body weight. For instance, maxillary grinding area (sum of P^3 through M^2 areas) calculated as a ratio to body weight is 17% higher in the females than it is in the males (the difference is significant at the 0.001 level). In sum, there is some reason to suppose that this sex-related difference in life histories might have resulted in a greater independence of tooth size from body size in the Yuendumu Wailbri females, resulting in a narrower range of relatively larger postcanine teeth.

Yuendumu Tooth Size Scaling

Scaling of Yuendumu Wailbri tooth dimensions and occlusal areas as a function of height (Table 6) differs from tooth to tooth, but in a fairly consistent pattern. In the combined sex sample, the highest scaling exponents are in the anterior teeth (excepting the maxillary I^2 breadth exponent, which is low, and the mandibular M_3, which is exceptionally high, but perhaps as an artifact of its smaller sample size). Generally, the exponents are smallest in the premolars and then increase distally. Breadth exponents are almost always of greater magnitude than length exponents in both jaws.

Because only exponents from cases with correlations with a significance of at least 0.10 are reported, a full set of comparisons between the separate sex samples cannot be made. For each sex, the exponent magnitudes follow

Table 6. Power Curve Exponents for Dental Dimensions as a Function of Height[a]

	Combined sample			Males			Females		
	L	*B*	*A*	*L*	*B*	*A*	*L*	*B*	*A*
I^1	0.713	0.913	—	0.864	0.599	—	0.461	0.563	—
I^2	0.521	0.463	—	—	—	—	—	—	—
C	0.629	0.652	1.281	0.888	—	1.049	—	—	—
P^3	—	0.365	0.659	—	0.601	—	0.456	—	—
P^4	0.398	0.433	0.831	—	—	—	0.587	—	—
M^1	0.490	0.590	1.080	—	—	—	0.624	0.481	1.105
M^2	0.474	0.678	1.152	—	—	—	0.746	0.581	1.327
M^3	—	0.532	—	—	1.556	3.259	—	0.424	—
I_1	0.616	0.630	—	—	—	—	—	—	—
I_2	0.548	0.635	—	—	—	—	—	—	—
C	0.745	0.638	1.383	0.881	—	1.054	—	0.503	0.806
P_3	0.328	0.568	0.896	—	—	—	0.483	0.446	0.929
P_4	0.431	0.435	0.866	—	—	—	0.466	0.466	0.932
M_1	0.385	0.528	0.913	—	0.399	—	0.346	0.534	0.880
M_2	0.533	0.671	1.024	—	—	—	0.655	0.633	1.228
M_3	0.416	0.719	1.190	—	—	—	—	0.562	1.213
Maxillary sums									
I^1–C	0.566	0.629	—	0.803	—	—	—	—	—
C–M^2	0.456	—	1.009	—	—	—	0.528	—	0.923
I^1–M^2	0.429	—	—	—	—	—	—	—	—
P^3–M^2	—	—	0.967	—	—	—	—	—	0.971
P^4–M^2	—	—	1.042	—	—	—	—	—	1.073
Mandibular sums									
I_1–C	0.645	0.635	—	0.752	—	—	—	—	—
C–M_2	0.477	—	1.032	—	—	—	0.446	—	0.931
I_1–M_2	0.496	—	—	—	—	—	0.399	—	—
P_3–M_2	—	—	0.985	—	—	—	—	—	0.984
P_4–M_2	—	—	1.008	—	—	—	—	—	1.014

[a]Exponents are only reported when $p(r = 0) \leq 0.10$. Here *L* denotes length, *B* breadth, and *A* area.

the same pattern as in the combined sex sample. Both male and female scaling exponents are generally (although not invariably) close to the same size or even larger than the exponents in the combined sample. The pattern of relationships can perhaps best be seen in the dental sums. Exponents for the area sums are quite close to linearity with stature (i.e., 1.0). Sums of linear dimensions including most of the teeth scale at close to 0.5 (as might be expected, since linear functions approximate the square root of area functions), while linear sums of only anterior dimensions scale somewhat above this. For those comparisons that can be made, male scaling exponents are generally higher than female exponents for the stature relation.

From the discussion above, it might be expected that tooth size scaling relative to weight is the more relevant comparison (Table 7). In fact, however, the pattern of scaling is not dissimilar to that for height. In this case there are

Table 7. Power Curve Exponents for Dental Dimensions as a Function of Weight[a]

	Combined sample			Males			Females		
	L	*B*	*A*	*L*	*B*	*A*	*L*	*B*	*A*
I^1	0.225	0.207	—	0.272	—	—	0.201	0.160	—
I^2	0.145	0.123	—	0.189	—	—	—	—	—
C	0.140	0.167	0.306	0.189	—	0.290	—	—	0.186
P^3	0.094	0.108	0.202	0.193	0.143	0.336	0.089	—	0.178
P^4	0.105	0.130	0.235	—	0.178	0.251	0.127	—	0.216
M^1	0.109	0.155	0.264	0.120	0.165	0.285	0.074	0.113	0.187
M^2	0.103	0.142	0.246	—	0.109	0.190	0.104	—	0.178
M^3		0.117	0.131	0.138	0.178	0.347			
I_1	0.193	0.167	—	0.221	0.176	—	0.179	—	—
I_2	0.203	0.166	—	0.332	0.155	—	0.125	—	—
C	0.176	0.144	0.320	0.235	—	0.309	—	—	0.162
P_3	0.067	0.172	0.239	—	0.184	0.257	—	0.144	0.187
P_4	0.131	0.129	0.260	0.195	0.148	0.343	0.093	0.112	0.205
M_1	0.105	0.140	0.245	—	0.178	0.257	—	0.107	0.167
M_2	0.137	0.152	0.290	0.157	0.118	0.275	0.138	0.105	0.243
M_3	0.102	0.125	0.252	—	—	—	0.142	—	—
Maxillary sums									
I^1–C	0.159	0.149	—	0.218	0.115	—	—	—	—
C–M^2	0.109	—	0.248	0.126	—	0.261	0.081	—	0.170
I^1–M^2	0.114	—	—	0.155	—	—	—	—	—
P^3–M^2	—	—	0.239	—	—	0.256	—	—	0.172
P^4–M^2	—	—	0.248	—	—	0.239	—	—	0.175
Mandibular sums									
I_1–C	0.190	0.156	—	0.262	0.130	—	0.115	0.107	—
C–M_2	0.122	—	0.266	0.141	—	0.281	0.078	—	0.183
I_1–M_2	0.138	—	—	0.171	—	—	0.090	—	—
P_3–M_2	—	—	0.259	—	—	0.277	—	—	0.192
P_4–M_2	—	—	0.264	—	—	0.281	—	—	0.196

[a]Exponents are only reported when $p(r = 0) \leq 0.10$. Here *L* denotes length, *B* breadth, and *A* area.

more exponents to compare because weight correlations are generally higher than height correlations, and power curve exponents were only reported when the correlation was significant at the 0.10 level or better. Higher scaling exponents for tooth size as a function of weight are almost invariably found in the anterior teeth, while posterior to these the exponents are lower in the premolars and in some instances also the first molar, but increase distally. As in the height exponents, breadth scaling in the combined sex sample is almost always of greater magnitude than length scaling. Exponents for the separate sexes do not differ dramatically from the combined sex sample, and, also like the height exponents, males tend to be higher and females lower. However, there is an interesting reversal for the length and breadth magnitudes; in the separate sex samples, the breadths tend to scale at a lesser magnitude than the lengths. Considering the dental sums, area sums scale at well below 0.33

(males higher, females lower), and anterior linear sums scale higher than posterior linear sums.

Clearly, an attempt to generalize from these scaling data is desirable. However, an accurate explanation of the magnitudes and patterns for the scaling exponents is not immediately obvious. Let us examine what is *not* the case. These exponents clearly do not fit the predictions of the geometric model (expected exponent for tooth area as a function of weight = 0.67) or of the metabolic model (expected exponent for tooth area as a function of weight = 0.5). On the other hand, what is interesting about the exponents is how similar they are. Compared with the correlations, which reflect a different pattern for height and weight in the male and female samples, and which tend to be higher for the combined sample, the scaling exponents show a surprisingly similar pattern in the male, female, and combined sex samples. Thus, unlike the pattern of phenotypic correlations, the pattern of scaling comes close to being independent of sex.

Discussion of the Intraspecific Data

In contrast to the initial expectation that tooth dimension scaling may be a primary response to weight variation, these data suggest that instead it may well be that the tooth areas scale in linear proportion to stature. Two results of this study might support this contention: the similar correlations of males and females with height as contrasted with the markedly higher correlations with weight in the males, and the magnitude of the power curve exponents for occlusal areas, varying near 1.0 for the height curves. These two results could be a coincidence, since there is no reason to have expected the scaling of tooth areas in proportion to body height, and no *a priori* explanation of why it might be true. It is credible that some other scaling exponent is being approximated by these data, near to but not the same as the 1.0 for scaling to height. Moreover, the exponents are considerably below a 0.33 for scaling to weight, the exponential value one might expect from a proportionality between occlusal area and height. Yet, the scaling exponent for weight as a function of height is above 3.0 (Table 1), so that the scaling of height as a function of weight should be below 0.33, possibly accounting for this deviation. Without an explanatory model, explanations are limited only by the ingenuity of the explainer. However, the possibility remains that tooth areas scale in linear proportion to stature in this intraspecific sample.

Let us look at the data another way. Intraspecific scaling, according to Lande's model, is proportional to the square root of the phenotypic variance ratios, the square root of the heritabilities, and the correlation of the additive genetic values. Comparing the male and female samples, the similarities of pattern and of magnitude in the scaling of tooth size to both weight and height (when the geometric relation of height to weight is taken into account, as the height–weight relation indicates it must be) suggests that the additive genetic correlation and the two variance ratios are similar for males and females. Yet, the *phenotypic* correlations with weight are quite different in the male and female samples. These considerations suggest that the difference might lie at least in part in the environmental component of the phenotypic covariance of tooth size and weight, possibly higher than the environmental component of the phenotypic covariance of tooth size and height. The heritabilities reported for tooth dimensions are well below 1.0, leaving ample opportunity for a significant component of environmental covariation to account for the higher phenotypic covariation in the male tooth size–weight relationship. It may simply be that prior to adulthood male food intake has a higher variance than female food intake. Supporting this possibility is the fact that the coefficient of variation for height is higher in the female sample than in the males (2.8 compared with 2.6), but in contrast the weight coefficient of variation is higher in the males than in the females (13.7 compared with 12.1). The contrasting greater variability in weight could be a consequence of more variance in male food intake.

Another factor to be taken into account is the apparently greater extent of the genetic decoupling of the tooth size–body size relation in females, possibly required by the dietary changes associated with pregnancy and the long period of infant dependency that is characteristic of humans. This would have more effect on the weight correlations than on the height correlations if the environmental component of the phenotypic covariance for tooth size and body weight were lower in the females.

Should the above considerations be correct, a relatively elevated environmental component of the male phenotypic correlations could account for the higher correlations in the male tooth size–weight relationship [see Atchley and Rutledge's (1980) discussion of the conditions under which the phenotypic allometric correlation is a poor estimator of the additive genetic correlation]. If so, this would bring focus on the lower phenotypic correlations with height as the more accurate measure of the magnitude of the additive genetic covariance for tooth size and body weight. Moreover, additional importance is thus brought to the apparent linearity of the tooth area–body height scaling relation. One must conclude that for either measure of body size the additive genetic covariance must therefore be fairly low.

In sum, examination of intraspecific scaling in the tooth size of the Yuendumu Wailbri suggests the possibility that occlusal area scaling is approximately linear to height. Similarities in the patterns of weight and height scaling indicate that the elevated correlation of tooth size to body weight in the male sample may be a consequence of environmental covariation and may not accurately reflect the additive genetic correlation. This, instead, is better approximated by the phenotypic correlations with height. The allometric model suggested by these data clearly indicates a scaling of tooth size to body size, perhaps linearly with height, and an underlying additive genetic correlation of low magnitude. Since modern humans are the living end product of a three million-year evolutionary history of changing relations between tooth size and body size, the Yuendumu Wailbri data are consistent with Lande's model for a genetically decoupled allometric relationship, perhaps with a greater degree of decoupling for the females (with their long periods of gestation and infant dependency) as compared with the males. The remaining magnitude of additive genetic covariation is low, and the magnitudes of the tooth size–body size allometric exponents are lower than any of the adaptive models discussed above (i.e., geometric, metabolic, etc.) could account for. Instead, it is likely that variation in body size alone, perhaps primarily height variation, largely accounts for the observable pattern of Yuendumu Wailbri tooth size allometry.

Comparison with Interspecific Allometry

Within the last decade, some very useful information has been published regarding the interspecific scaling of tooth size with direct measures of body weight in primates. Results of the relevant studies are summarized in Tables 8 and 9. Scaling exponents are invariably much higher than the intraspecific allometry of the Yuendumu Wailbri, as Lande's model would predict. With one exception, scaling for the areas of the individual teeth, and for the area sums of the postcanine teeth, is markedly above the prediction of 0.5 for the metabolic model in the combined sex comparisons. Various values reported for this exponent are distributed above and below the 0.67 prediction of the geometric model. For instance, in the Gingerich *et al.* (1982) data (when transformed to least mean squares exponents; see Table 8) for individual teeth spanning a very wide taxonomic range, as many exponents are above the 0.67 value as are below. Kay (1975*a,b*) attempted to parcel out the effects

Table 8. Comparison of Intraspecific and Interspecific Tooth Area Allometry: Power Curve Exponents as a Function of Body Weight

	Intraspecific: Yuendumu	Interspecific: Primates[a]	Female primates[b]
C	0.30	0.70	0.56
P^3	0.20	0.63	—
P^4	0.24	0.57	—
M^1	0.26	0.55	—
M^2	0.25	0.65	—
M^3	0.13	0.75	—
C	0.32	0.71	—
P_3	0.24	0.75	—
P_4	0.26	0.63	—
M_1	0.25	0.63	—
M_2	0.29	0.72	—
M_3	0.25	0.77	—
P^3–M^3	0.17	0.60	0.63
P_3–M_3	0.24	0.67[c]	—

[a]From Gingerich *et al.* (1982), calculated from the least mean squares slopes of weight as a function of tooth area, and not the same as the major axis slopes given in their Table 2.
[b]From Smith (1981*a,b*).
[c]Kay (1975*a*) reports a slope of 0.59 for noncercopithecoid primates.

Table 9. Variation in Primate Interspecific Allometry Exponents

	Area of M_2	Area of P^3–M^3
Primates[a]	0.72	0.60
Cercopithecoids[b]	0.62	—
Males[c]	—	0.78
Females[c]	—	0.79
Noncercopithecoids[b]	0.65	—
Primate frugivores[b]	0.68	—
Primate folivores[b]	0.71	—
Primate insectivores[b]	0.47	—
African apes[d]	—	0.85

[a]From Gingerich *et al.* (1982); see footnote for Table 7.
[b]From Kay (1975*a*).
[c]From Goldstein *et al.* (1978), with the same results given by Gould (1975*a*).
[d]From Gould (1975*a*).

of dietary differences in his data, providing separate allometric exponents for primate frugivores, folivores, and insectivores (Table 9). Only in the latter case is the exponent considerably below the 0.67 value, perhaps better approximating the metabolic model. However, his data are for only one tooth and the implications of this deviation are not clear. In all, most of the data for combined sex samples seem to support a geometric model for the interspecific scaling of primate tooth size. This conclusion contradicts several previous assertions because, as discussed in the first part of this chapter, an invalid prediction of the metabolic model has been in use.

The other interesting exception to the geometric pattern for intraspecific allometry is closer to home, involving Gould's (1975*a*) data for the African apes (Table 9). The exponent determined for the maxillary posterior area sum as a function of weight is higher than reported for any other combination of primate taxa with both sexes represented. At 0.85, this exponent is considerably greater than could be expected from geometric scaling, let alone from metabolic scaling. Gould shows that adding the Asian orangutan to the data set does not significantly change the slope. Clearly, something distinctive is resulting in the tooth area scaling for the great apes. Unfortunately, what may be distinctive is the data base. In particular, the sources of the body weight data [see footnote 29, Pilbeam and Gould (1974)] are largely out of date, and in many cases report zoo weights rather than weights taken in the wild. Since the publication of Gould's paper, the first accurate weights for the two chimpanzee species in their natural habitat have become available, showing among other things that the weight of the so-called pygmy chimpanzee is not significantly smaller than for *Pan troglodytes* (Horn, 1979). Better (i.e., larger sample sizes indicative of natural populations) gorilla weights are also now available, and it is clear that the elevated allometric slope for the great apes reported by Gould should be reexamined.

In sum, with a few exceptions of unknown significance, interspecific allometry in most primates approximates the predictions of the geometric model for tooth size scaling. The magnitude of primate interspecific scaling is much higher than the intraspecific case examined here (the Yuendumu Wailbri), as could be expected from the uncoupling of genetic covariation in tooth and body size that allowed their separate evolutionary directions during the course of hominid evolution. As mentioned above, in the Pleistocene hominids this evolutionary pattern does not involve scaling, since the posterior teeth consistently reduce in size over a period that is not characterized by any average change in body size (Pleistocene changes in the anterior teeth are regionally specific, but also are not the result of scaling, for the same reason).

Predictions

The rendering of a prediction, usually for a fossil form in which one of the variables is unknown, or in order to determine whether tooth size is "bigger (or smaller) than it is supposed to be," has been the primary goal of many of the studies reported here. In some cases, I am in agreement with those authors who contend that such predictions are fundamentally impossible. However, the impossibility of making predictions is not an inevitability. For instance, if morphological analysis indicates that a primate species is not extraordinarily specialized, and if a large sample size is available for its dental variation, use of the geometric model for estimating average species weight may be justified [something like what Gingerich *et al.* (1982) attempted, although I suspect that all of the predictions in this paper are not equally valid]. In this event, the model itself (i.e., a tooth area–weight exponent of 0.67), and not the data's approximation of it, should be used.

In other cases, the situation is more complex. For instance, closely related species may scale along an intraspecific curve because selection has not separately affected the two variables or acted to disrupt the genetic covariance. Thus, their relation may not reflect the general scaling that could characterize a more diverse taxonomic set. For instance, even though there are a number of postcranial remains for *Australopithecus africanus,* and even if one made the assumption that *Australopithecus boisei* had larger postcanine teeth only as a consequence of larger body size, tooth size could not be used to accurately predict body size in this megadont australopithecine species because it is unclear whether the intraspecific or the interspecific allometric relation actually (i.e., historically) applied.

While it is interspecific allometry that provides the criterion for whether two species can be said to represent "the same functional design expressed at a different size," all interspecific patterns are not necessarily a consequence of functional similarity (this is the case for the most recent three species of the genus *Homo*). Conversely, as in the case above, a descendant species may actually be a larger version of an ancestral species (i.e., correlated changes were only in response to increasing body size), without the relationship between the means following the more general interspecific curve.

In situations when one has reason to suspect that tooth sizes in related species are functionally equivalent, the situation may be no less ambiguous. For example, even if the body sizes of the Swartkrans and Sterkfontein australopithecines were known (as they might be estimated from the postcranial remains), the fact is that none of the specific body size estimates pertain to specimens with dentitions. How could one verify the hypothesis that the differences in the dentitions of these two samples reflected functional equivalence? Intraspecific allometry for tooth size is undeterminable for both. Applying the intraspecific relation from a different species is unjustified without reason to expect that the same genetic covariance ratio can be assumed. It might be possible to use the geometric scaling model, but the results would be misleading if an intraspecific allometric relation actually characterized these samples.

Two different issues are involved in the question of predictability. First, there is the question of whether larger primates can be expected to have larger teeth *along with* their larger size. For this question, whether asked at the individual or the species level, the answer clearly seems to be yes. The second question is *how much* larger the teeth can be expected to become *as a consequence of* larger body size. This is quite a separate issue. There is probably no general statement that can be made about predictability and the determination of functional equivalence across differing sizes. I do not believe that these are impossible in principle, but they must be approached with great care and caution, and a healthy dose of skepticism.

Conclusions

There *is* a theoretical basis for the power curve representation of the allometric relation, derived by Lande from population genetics theory. His model provides a clear understanding of intraspecific allometry, and defines the conditions under which interspecific allometry will be similar or will differ.

Intraspecific allometry *does* exist for tooth size scaling in a living human population, the Yuendumu Wailbri. To understand the details of the tooth size–body size relationship, correlation and scaling must be considered separately. The magnitudes of the tooth size–body size correlations are misleading in this regard, since the evolutionary history of the hominids suggests that the additive genetic covariance should be low in surviving species. This is because tooth size has changed independently of body size throughout the Pleistocene. The low genetic covariance underlies the low phenotypic correlations with height (in fact, higher correlations with weight in Yuendumu Wailbri males seem to reflect a higher male environmental covariance for tooth size and weight). The postcanine teeth of the females have low correlations with the body size measures, are individually less variable for tooth size measurements, and are larger than male teeth relative to body size. It is suggested that these characteristics are a consequence of the periodic increases in dietary demands during pregnancy. Female postcanine teeth have responded to selection promoting a relatively larger size and narrower range of variation and are therefore more independent of body size. Yuendumu Wailbri tooth areas seem to scale linearly to body height, and the allometric scaling of tooth size (to both height and weight) is quite regular and independent of sex. Its magnitude is well below that of interspecific scaling for most primate taxa.

The interspecific pattern approximates geometric scaling. The observed pattern can be shown to generally be of greater magnitude than the expectations of the metabolic scaling model when the expected exponents of these two models are both calculated under the assumption of tooth volume–body weight proportionality (the expected exponent for the metabolic model has been improperly determined in previous studies comparing the two models for interspecific tooth size scaling). Neither the hominids nor possibly the African apes have conformed to the geometric relation during the course of their post-Miocene evolution. In sum, the primate tooth size–body size relation is a real phenomenon and reflects both the level of existing genetic covariation and the effects of evolutionary history on this level. Within human (and probably most other primate) populations, tooth size can be expected to scale to body size. Over the course of evolution for a lineage, or when comparing related primate species, a geometric pattern of allometric scaling may apply (this has proven to be quite common in the primates), or alternatively this pattern may be disrupted by differing selection acting separately on tooth size and/or on body size. Thus, the data reported here clearly show that in general, larger primate individuals or larger primate species can, on the average, be expected to have larger sized teeth. On the other hand, however regular the pattern of allometry that is observed, particular application of allometry to understanding or predicting specific aspects of the scaling relation in fossil forms is problematic, and must be approached from a position of maximum information and caution.

ACKNOWLEDGMENTS

I am greatly indebted to T. Brown and his co-workers for permission to use the Yuendumu Wailbri dental and anthropometric growth data for the allometric determinations provided in this chapter, and to R. C. Bailey, N. R. Peacock, and members of the Harvard Ituri project for permission to use Efé height and weight data. I thank R. G. Northcutt, K. Rosenburg, M. D. Russell, E. Vrba, and R. Wrangham for their helpful comments, and I am particularly grateful to R. J. Smith, R. Lande, and W. L. Jungers for the many suggestions they made to greatly improve this manuscript. This research was supported in part by NSF grant BNS 76-82729.

References

Alberch, P., Gould, S. J., Oster, G. F., and Wake, D. B. 1979. Size and shape in ontogeny and phylogeny. *Paleobiology* **5**:296–317.

Alexander, R. McN., Jayes, A. S., Maloiy, G. M. O., and Wathuta, E. M. 1979. Allometry of the limb bones of mammals from shrews (*Sorex*) to elephant (*Loxodonta*). *J. Zool. Lond.* **189**:305–314.

Anderson, D. L., Thompson, G. W., and Popovich, F. 1977. Tooth, chin, bone, and body size correlations. *Am. J. Phys. Anthropol.* **46:**7–12.

Armstrong, E. 1982. A look at relative brain size in mammals. *Neurosci. Lett.* **34:**101–104.

Atchley, R. W., and Rutledge, J. J. 1980. Genetic components of size and shape. I. Dynamics of components of phenotypic variability and covariability during ontogeny in the laboratory rat. *Evolution* **34:**1161–1173.

Barrett, M. J. 1976. Dental Observations on Australian Aborigines: Collected Papers and Reports 1953–1973, Faculty of Dentistry, University of Adelade, Adelade.

Barrett, M. J., Brown, T., and Fanning, E. A. 1965. A long-term study of the dental and craniofacial characteristics of a tribe of Central Australian Aborigines. *Austr. Dent. J.* **10:**63–68.

Bookstein, F. L. 1977. Orthogenesis of the hominids: An exploration using biorthogonal grids. *Science* **197:**901–904.

Brody, S. 1945. *Bioenergetics and Growth,* Reinhold, New York.

Brown, T., and Barrett, M. J. 1973. Dental and craniofacial growth studies of Australian Aborigines, in: *The Human Biology of Aborigines in Cape York* (R. L. Kirk, ed.), pp. 69–80, Australian Institute of Aboriginal Studies, Canberra.

Calder, W. A., III. 1981. Scaling of physiological processes in homeothermic animals. *Annu. Rev. Physiol.* **43:**301–322.

Cheverud, J. M. 1982. Relationships among ontogenetic, static, and evolutionary allometry. *Am. J. Phys. Anthropol.* **59:**139–149.

Clutton-Brock, T. H., and Harvey, P. H. 1979. Comparison and adaptation. *Proc. R. Soc. Lond. B* **205:**547–565.

Cock, A. G. 1966. Genetical aspects of metrical growth and form in animals. *Q. Rev. Biol.* **41:**131–190.

Corruccini, R. S., and Henderson, A. M. 1978. Multivariate dental allometry in primates. *Am. J. Phys. Anthropol.* **48:**203–208.

Creighton, G. K. 1980. Static allometry of mammalian teeth and the correlation of tooth size and body size in contemporary mammals. *J. Zool. Lond.* **191:**435–443.

Economos, A. C. 1979. On structural theories of basal metabolic rate. *J. Theor. Biol.* **80:**445–450.

Filipson, R., and Goldson, L. 1963. Correlation between tooth width, width of the head, length of the head, and stature. *Acta Odont. Scand.* **21:**359–365.

Garn, S. M., and Lewis, A. B. 1958. Tooth size, body size, and "giant" fossil man. *Am. Anthropol.* **60:**874–880.

Garn, S. M., Lewis, A. B., and Kerewsky, R. S. 1968. The magnitude and implications of the relationship between tooth size and body size. *Arch. Oral Biol.* **13:**129–131.

Giles, E. 1956. Cranial allometry in the great apes. *Hum. Biol.* **28:**43–58.

Gingerich, P. D. 1977. Correlation of tooth size and body size in living hominoid primates, with a note on relative brain size in *Aegyptopithecus* and *Proconsul. Am. J. Phys. Anthropol.* **47:**395–398.

Gingerich, P. D., Smith, B. H., and Rosenberg, K. 1982. Allometric scaling in the dentition of primates and prediction of body weight from tooth size in fossils. *Am. J. Phys. Anthropol.* **58:**81–100.

Goldstein, S., Post, D., and Melnick, D. 1978. An analysis of cercopithecoid odontometrics. I. The scaling of the maxillary dentition. *Am. J. Phys. Anthropol.* **49:**517–532.

Gould, S. J. 1966. Allometry and size in ontogeny and phylogeny. *Biol. Rev.* **41:**587–640.

Gould, S. J. 1971. Geometric similarity in allometric growth: A contribution to the problem of scaling in the evolution of size. *Am. Nat.* **105:**113–136.

Gould, S. J. 1974. The origin and function of "bizarre" structures: Antler size and skull size in the "Irish Elk," *Megaloceros giganteus. Evolution* **28:**191–220.

Gould, S. J. 1975*a*. On the scaling of tooth size in mammals. *Am. Zool.* **15:**351–362.

Gould, S. J. 1975*b*. Allometry in primates, with emphasis on scaling and the evolution of the brain, in: *Approaches to Primate Paleobiology* (Contrib. *Primatol.,* Vol. 5, F. Szalay, ed.), pp. 244–292, S. Karger, Basel.

Gould, S. J. 1977. *Ontogeny and Phylogeny,* Belknap Press, Cambridge.

Günther, B. 1975. On theories of biological similarity. *Fortschr. Exp. Theor. Biophys.* **19:**7–111.

Guyton, A. C., Jones, C. E., and Coleman, T. G. 1973. *Circulatory Physiology: Cardiac Output and Its Regulation,* W. B. Saunders, Philadelphia.

Harrison, G. A., Weiner, J. S., Tanner, J. M., and Barnicot, N. A. 1977. *Human Biology,* 2nd ed., Oxford University Press, Oxford.

Henderson, A. M., and Corruccini, R. S. 1976. Relationship between tooth size and body size in American Blacks. *J. Dent. Res.* **55:**94–96.

Heusner, A. A. 1982. Energy metabolism and body size. I. Is the 0.75 mass exponent of Keibler's equation a statistical artifact? *Respir. Physiol.* **48:**1–12.

Horn, A. D. 1979. The taxonomic status of the bonobo chimpanzee. *Am. J. Phys. Anthropol.* **51:**273–282.

Hursh, T. M. 1976. Multivariate analysis of allometry in crania. *Yearb. Phys. Anthropol.* **18:**111–120.

Huxley, J. S. 1924. Constant differential growth-ratios and their significance. *Nature* **114:**895–896.

Huxley, J. S. 1932. *Problems of Relative Growth,* Methuen, London.

Huxley, J. S., Needham, J., and Lerner, I. M. 1941. Terminology of relative growth-rates. *Nature* **146:**225.

Jerison, H. J. 1973. *Evolution of the Brain and Intelligence,* Academic Press, New York.

Jolicoeur, P. 1975. Linear regression in fishery research: some comments. *J. Fish. Res. Board Can.* **32:**1491–1494.

Kac, M. 1969. Some mathematical models in science. *Science* **166:**695–699.

Kaplan, R. H., and Salthe, S. N. 1979. The allometry of reproduction: An empirical view in salamanders. *Am. Nat.* **113:**671–689.

Katz, M. J. 1980. Allometry formula: A cellular model. *Growth* **44:**89–96.

Kay. R. F. 1975*a*. Allometry and early hominids. *Science* **189:**63.

Kay, R. F. 1975*b*. The functional adaptations of primate molar teeth. *Am. J. Phys. Anthropol.* **43:**195–216.

Kay, R. F., and Simons, E. L. 1980. The ecology of Oligocene African Anthropoidea. *Int. J. Primatol.* **1:**21–37.

Keibler, M. 1932. Body size and metabolism. *Hilgardia* **6:**315–353.

Kendall, M., and Stuart, A. 1961. *The Advanced Theory of Statistics,* Vol. 2, Hafner, New York.

Kermack, K. A., and Haldane, J. B. S. 1950. Organic correlation and allometry. *Biometrika* **37:**30–41.

Kuhry, B., and Marcus, F. 1977. Bivariate linear models in biometry. *Syst. Zool.* **26:**201–209.

Kurtén, B. 1955. Contribution to the history of a mutation during 1,000,000 years. *Evolution* **9:**107–118.

Lambert, R., and Teissier, G. 1927. Theorie de la similitude biologique. *Ann. Physiol.* **3:**212–246.

Lande, R. 1976. Natural selection and random genetic drift in phenotypic evolution. *Evolution* **30:**314–334.

Lande, R. 1979. Quantitative genetic analysis of multivariate evolution, applied to brain:body size allometry. *Evolution* **33:**402–416.

Lauer, C. 1975. The relationship of tooth size to body size in a population of rhesus monkeys (*Macaca mulatta*). *Am. J. Phys. Anthropol.* **43:**333–340.

Lindstedt, S. L., and Calder, W. A. III. 1981. Body size, physiological time, and longevity of homeothermic animals. *Q. Rev. Biol.* **56:**1–16.

McMahon, T. 1973. Size and shape in biology. *Science* **179:**1201–1204.

McMahon, T. 1975. Allometry and biomechanics: Limb bones in adult ungulates. *Am. Nat.* **109:**547–563.

McNamara, K. J. 1982. Heterochronic and phylogenetic trends. *Paleobiology* **8:**130–142.

Maglio, V. J. 1972. The evolution of mastication in the Elephantidae. *Evolution* **26:**638–658.

Molnar, S., McKee, J. K., and Molnar, I. 1983. Measurements of tooth wear among Australian Aborigines: 1. Serial loss of the enamel crown. *Am. J. Phys. Anthropol.* **61:**51–65.

Perzigian, A. J. 1981. Allometric analysis of dental variation in a human population. *Am. J. Phys. Anthropol.* **54:**341–345.

Pilbeam, D., and Gould, S. J. 1974. Size and scaling in human evolution. *Science* **186:**892–901.

Pilbeam, D., and Gould, S. J. 1975. Allometry and early hominids. *Science* **189:**64.

Pirie, P. L. 1978. Allometric scaling in the postcanine dentition with reference to primate diets. *Primates* **19:**583–591.

Prothero, D. R., and Sereno, P. C. 1982. Allometry and paleoecology of medial Miocene dwarf rhinoceroses from the Texas Gulf coastal plain. *Paleobiology* **8:**16–30.

Radinsky, L. 1970. The fossil evidence of prosimian brain evolution, in: *Advances in Primatology,* Vol. 1 (C. R. Norback and W. Montagna, eds.), pp. 209–224, Appleton-Century-Crofts, New York.

Reeve, E. C. R., and Huxley, J. S. 1945. Some problems in the study of allometric growth, in: *Essays on "Growth and Form" Presented to D'Arcy Wentworth Thompson* (W. E. le Gros Clark and P. B. Medawar, eds.), pp. 121–156, Oxford University Press, Oxford.

Richards, L. C., and Brown, T. 1981. Dental attrition and age relationships in Australian Aboriginals. *Arch. Phys. Anthropol. Oceania* **16:**94–98.

Robb, R. C. 1929. On the nature of hereditary size limitation. II. The growth of parts in relation to the whole. *Br. J. Exp. Biol.* **6:**311–324.

Rosen, R. 1967. *Optimality Principles in Biology,* Butterworth, London.

Savageau, M. A. 1979. Allometric morphogenesis of complex systems: Derivation of the basic equations from complex principles. *Proc. Natl. Acad. Sci. USA* **76:**6023–6025.

Shea, B. T. 1983. Phyletic size change and brain/body allometry: A consideration based on the African pongids and other primates. *Int. J. Primatol.* **4:**33–62.

Schmidt-Nielsen, K. 1975. Scaling in biology: The consequences of size. *J. Exp. Zool.* **194:**287–308.

Serra, J. A. 1958. A simple derivation of the equation of relative growth. *Rev. Port. Zool. Biol. Ger.* **1:**301–310.

Smith, R. J. 1980. Rethinking allometry. *J. Theor. Biol.* **87:**97–111.

Smith, R. J. 1981*a*. On the definition of variables in studies of primate dental allometry. *Am. J. Phys. Anthropol.* **55:**323–329.

Smith, R. J. 1981*b*. Interspecific scaling of maxillary canine size and shape in female primates: Relationships to social structure and diet. *J. Hum. Evol.* **10:**165–173.

Smith, R. J. 1981*c*. Interpretation of correlations in intraspecific and interspecific allometry. *Growth* **45:**291–297.

Sokal, R., and Rohlf, F. 1981. *Biometry,* 2nd ed., Freeman, San Francisco.

Sprent, P. 1972. The mathematics of size and shape. *Biometrics* **18:**23–37.

Teissier, G. 1960. Relative growth, in: *The Physiology of Crustacea* (T. H. Waterman, ed.), Vol. 1, pp. 537–560, Academic Press, New York.

von Bertalanffy, L., and Pirozynski, W. J. 1952. Ontogenetic and evolutionary allometry. *Evolution* **6:**387–392.

Wilkie, D. R. 1977. Metabolism and body size, in: *Scale Effects in Animal Locomotion* (T. J. Pedley, ed.), pp. 23–36, Academic Press, New York.

Wolpoff, M. H. 1967. A critique of the applicability of Huxley's "relative growth" curve to primate population data. *Am. J. Phys. Anthropol.* **27:**242.

Wolpoff, M. H. 1973. Posterior tooth size, body size, and diet in South African gracile australopithecines. *Am. J. Phys. Anthropol.* **39:**375–394.

Wolpoff, M. H. 1974. The evidence for two australopithecine lineages in South Africa. *Yearb. Phys. Anthropol.* **17:**113–139.

Wolpoff, M. H. 1978. Some aspects of canine size in the australopithecines. *J. Hum. Evol.* **7:**115–126.

Wolpoff, M. H. 1982. Relative canine size. *J. Hum. Evol.* **10:**151–158.

Wood, B. A. 1979. An analysis of tooth and body size relationships in five primate taxa. *Folia Primatol.* **31:**187–211.

Wood, B. A., and Stack, C. G. 1980. Does allometry explain the differences between "gracile" and "robust" australopithecines? *Am. J. Phys. Anthropol.* **52:**55–62.

Wright, S. 1977. *Evolution and the Genetics of Populations,* Vol. 3, *Experimental Results and Evolutionary Deductions,* University of Chicago, Chicago.

Yinyun, Z. 1982. Variability and evolutionary trends in tooth size of *Gigantopithecus blacki*. *Am. J. Phys. Anthropol.* **59:**21–32.

Comparative Energetics and Mechanics of Locomotion 14

How Do Primates Fit In?

NORMAN C. HEGLUND

Introduction

There is no doubt that the primates are a highly diverse and specialized order of animals: they range in adult size from the mouse-size *Microcebus* (approximately 60 g) to the pony-size *Gorilla* (approximately 200 kg); some walk, trot, and gallop quadrupedally, others walk and run bipedally, some brachiate, some hop, some even use their tails during locomotion. Yet, taken as a whole, animals of all classes represent a much larger size range, and have the same and more specializations for locomotion. In this chapter I will review some aspects of the energetics and mechanics of terrestrial (and arboreal) locomotion, and then compare in more detail both the primates as a group and a few of the more specialized primate species. Then it will be possible to compare the many different sizes, shapes, and modes of locomotion of the primates to those of the terrestrial vertebrates in general.

Energetics

The energy consumed by an animal during steady state locomotion is usually determined by measuring the rate of oxygen consumption $\dot{V}_{O_2}$ of the

NORMAN C. HEGLUND • Concord Field Station, Museum of Comparative Zoology, Harvard University, Cambridge, Massachusetts 02138.

animal. It is relatively easy to train a small animal to run on a treadmill enclosed in a plexiglass box or to train a large animal to run while wearing a mask, although sometimes it can be dangerous and always it requires a lot of patience. Room air is pulled from the mask or box through a flow meter, and all of the expired air is collected. An aliquot of this air is then analyzed for oxygen content relative to room air. The decrease in oxygen content times the flow rate through the mask or box is equal to the oxygen consumed by the animal [a small correction is usually used to compensate for such factors as the CO_2 produced by the animal; see Fedak *et al.* (1981) for details].

In order for oxygen consumption to be a valid measure of the total energy input to the animal, the measurements must be done when the animal has reached a "steady state" and is not deriving any net energy from anaerobic metabolism. To ensure that a steady state is achieved, each oxygen consumption measurement is made over a long enough interval, usually 15–30 min at each speed; this allows one to disregard any transient startup effects. To ensure that all of the energy is derived from aerobic sources during the steady-state measurements, in the studies reported here the R value (the rate of CO_2 production/the rate of O_2 consumption) was often measured. Also, the lactate level in the blood was measured before and after some high-speed runs. At even the highest speeds of locomotion reported here, the R values were always less than 1.0, and the net energy derived from anaerobic sources during the experiment, calculated from the energetic equivalent of the lactate that accumulated in the blood during the run, was always less than 2.0% of the total energy consumed during the run (Seeherman *et al.*, 1981). These procedures have been repeated at many different running speeds, primarily in this laboratory and at the University of Nairobi, on 62 avian and mammalian species over a size range from 7-g pigmy mice (*Baiomys taylori*) to 260-kg Zebu cattle (*Bos indicus*) (Taylor *et al.*, 1982; Taylor and Heglund, 1982).

Rate of Oxygen Consumption as a Function of Running Speed

With few exceptions, the rate of oxygen consumption increases linearly with speed of locomotion over a wide range of tread speeds (Fig. 1). For each animal this empirical relationship may therefore be expressed as a normalized linear equation of the form

$$\dot{V}_{O_2}/M_b = \text{Slope}\cdot\text{velocity} + \text{intercept} \tag{1}$$

where the mass-specific oxygen consumption rate $\dot{V}_{O_2}/M_b$ and the intercept (extrapolated zero-velocity cost) have the units of ml O_2/sec per kg of body mass, velocity is in m/sec, and the slope is in ml O_2/m per kg.

Oxygen Consumption as a Function of Body Size

As can be seen in Fig. 1, there is a decrease with increasing body size in both the slope and the y intercept of the relation between rate of oxygen

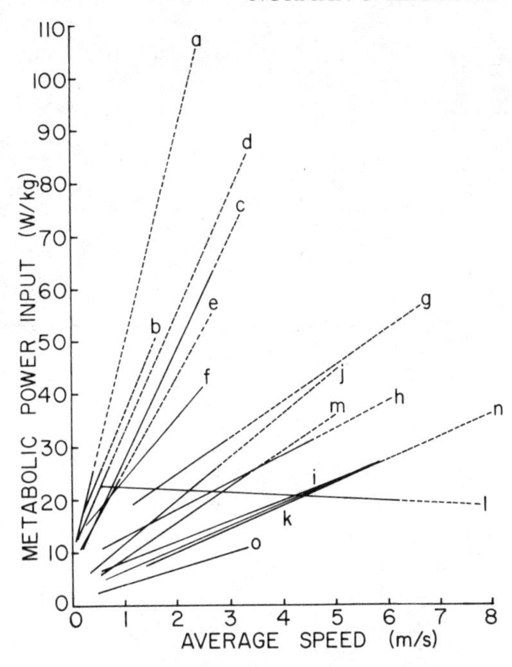

Fig. 1. Metabolic power input $\dot{V}_{O_2}/M_b$ (in W/kg of body mass) to animals during steady-speed terrestrial locomotion plotted against speed of locomotion. Metabolism was measured as oxygen consumption rate (see text for details) and converted to watts by the factor 20.1 J/ml O_2. The solid lines are the steady-state oxygen consumption measurements; the dashed lines extend the measurements to cover the entire speed range over which the mechanics of locomotion was measured. The animal key is: a, 0.035-kg kangaroo rat; b, 0.042-kg painted quail; c, 0.098-kg chipmunk; d, 0.112-kg kangaroo rat; e, 0.175-kg bobwhite; f, 0.186-kg ground squirrel; g, 2.5-kg spring hare; h, 3.6-kg monkey; i, 5.0-kg dog; j, 7.0-kg turkey; k, 17-kg dog; l, 20.5-kg kangaroo; m, 22.5 kg-rhea; n, 70-kg human; o, 73-kg ram.

consumption and speed. Figure 2 is a plot of the *y*-intercept values (top) and the slope values (bottom) versus body mass M_b on logarithmic coordinates for all of the animals measured, with the exception of the lion, the red kangaroo, and the waddling birds. Allometric equations can be used to describe these data. The equation (calculated by the method of least squares on the log-transformed data) for the *y*-intercept values is

$$y \text{ intercept} = 0.300 M_b^{-0.303} \tag{2}$$

where the *y* intercept has the units of ml O_2/sec per kg and M_b is in kg. The equation for the slope is

$$\text{Slope} = 0.533 M_b^{-0.316} \tag{3}$$

where the slope has the units of ml O_2/m per kg and M_b is in kg.

The data for the lion and kangaroo were not included in the line above because their oxygen consumption did not increase linearly over a wide range of speeds (Chassin *et al.*, 1976; Dawson and Taylor, 1973); the waddling birds were excluded because there was a large component to the cost of locomotion that was unique to these birds (Pinshow *et al.*, 1977).

Equations (2) and (3) can be combined into one general equation of the same type as Eq. (1); this permits us to estimate the rate of oxygen consumption of an animal solely on the basis of body mass and speed of locomotion *v*:

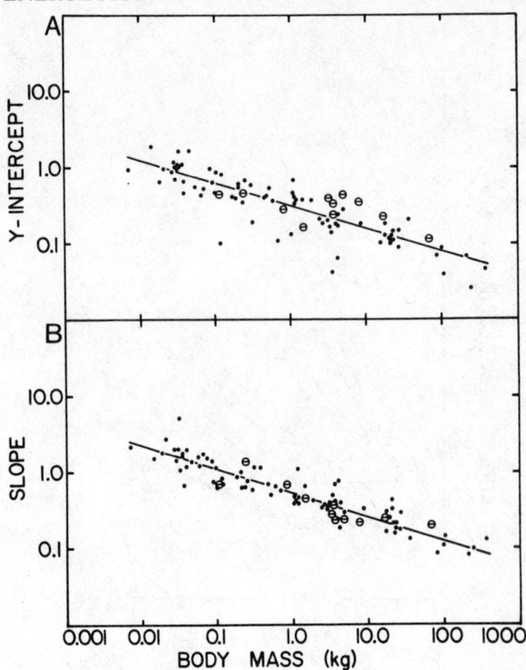

Fig. 2. The y-intercept values (top) and slope values (bottom) of mass-specific oxygen consumption rate versus speed for 73 avian and mammalian species plotted against body mass on logarithmic coordinates. The lines were calculated by the method of least squares on the log-transformed data and represent Eqs. (2) and (3) of the text. The open symbols represent the primate species and data listed in Table 2.

$$\dot{V}_{O_2}/M_b = 0.533 M_b^{-0.316}\, v + 0.300 M_b^{-0.303} \qquad (4)$$

where $\dot{V}_{O_2}/M_b$ has the units ml O_2/sec per kg, M_b is in kg, and v is in m/sec.

This equation appears to be quite general; in fact, if the animals are subdivided into different groups, the resulting subsets of the original data are not significantly different from the original line. Table 1, adapted from Taylor *et al.* (1982), gives the coefficients and exponents for Eq. (4) broken into the following groups: all animals except waddlers, red kangaroos, and lions (the original lines); all mammals except lions and kangaroos; all birds except waddlers; marsupials except red kangaroos; insectivores; artiodactyls; carnivores; rodents; and primates. None of the coefficients or exponents differs from those of Eq. (4) at the 95% confidence limits.

How Do the Primates Compare?

The coefficients and exponents for the primates are no different from any of the other groups listed in Table 1; but what about the individual primate species? Table II lists the available slope and y-intercept values [from Eq. (1)] for the primates; the data are for animals ranging in size from 0.12-kg tree shrews to 69-kg humans. In most cases, at a midrange speed of locomotion, the individual primate species agree very well with the generic value calculated on the basis of Eq. (4). The tree shrew and one individual of the

Table 1. The Coefficients, Exponents, and 95% Confidence Limits (in parentheses) for Eq. (4)[a]

Group	Coefficient a	Exponent b	Coefficient c	Exponent d
All animals except waddlers, red kangaroos, and lions	0.533 (0.502, 0.566)	−0.316 (−0.293, −0.339)	0.300 (0.268, 0.335)	−0.303 (−0.261, −0.346)
All mammals except lions and kangaroos	0.530 (0.496, 0.565)	−0.319 (−0.295, −0.344)	0.303 (0.268, 0.343)	−0.311 (−0.264, −0.357)
All birds except waddlers	0.566 (0.466, 0.688)	−0.285 (−0.208, −0.361)	0.279 (0.217, 0.358)	−0.246 (−0.149, −0.344)
Marsupials except red kangaroos	0.494 (0.423, 0.576)	−0.328 (−0.264, −0.391)	0.477 (0.323, 0.703)	−0.285 (−0.125, −0.445)
Insectivores	0.458 (0.244, 0.859)	−0.370 (−0.080, −0.660)	0.092 (0.017, 0.500)	−0.550 (−0.228, −1.330)
Artiodactyls	0.787 (0.552, 0.891)	−0.411 (−0.317, −0.506)	0.210 (0.093, 0.475)	−0.265 (−0.047, −0.483)
Carnivores	0.509 (0.417, 0.622)	−0.311 (−0.216, −0.407)	0.322 (0.202, 0.513)	−0.289 (−0.076, −0.521)
Rodents	0.483 (0.313, 0.746)	−0.364 (−0.211, −0.517)	0.463 (0.352, 0.609)	−0.157 (−0.061, −0.254)
Primates	0.523 (0.405, 0.674)	−0.298 (−0.171, −0.424)	0.345 (0.256, 0.465)	−0.157 (−0.009, −0.305)

[a]Here $\dot{V}_{O_2}/M_b = aM_b^{b}v + cM_b^{d}$; the mass-specific oxygen consumption rate $\dot{V}_{O_2}/M_b$ is in ml O_2/sec per kg; body mass M_b is in kg; and the speed v is in m/sec. This table is adapted from Taylor *et al.* (1982).

Table 2. Body Mass M_b, Slope, and y Intercept for Eq. (1)[a]

Animal	M_b, kg	Slope, ml O_2/m per kg	y Intercept, ml O_2/sec per kg
Tree shrew (*Tupaia glis*)	0.12	0.69	0.45
Bushbaby (*Galago senegalensis*)	0.24	1.47	0.48
Slow loris (*Nycticebus coucang*)	0.88	0.28	0.68
	1.55	0.17	0.48
Capuchin (*Cebus albifrons*)	3.34	0.28	0.42
Spider monkey (*Ateles geoffroyi*)	3.80	0.37	0.23
Stumped-tail macaque (*Macaca arctoides*)	5.10	0.25	0.43
Hamadryas baboon (*Papio hamadryas*)	8.50	0.24	0.36
Patas monkey (*Erythrocebus patas*)	3.80	0.25	0.35
Chimpanzee (*Pan troglodytes*)	17.5	0.25	0.22
Human (*Homo sapiens*)	68.8	0.20	0.12

[a]Here $\dot{V}_{O_2}/M_b$ = slope · v + y intercept, where v is in m/sec. This table is adapted from Taylor *et al.* (1982).

slow loris studied are low, and the bush baby, chimpanzee, and human are high. However, the data for the slow loris may be indicative of the variation between individuals; in this case, one individual was low by about 25% and the other high by perhaps 2%. On the basis of the available data reviewed here, it is clear that large variations in this cost are necessary for any interspecific arguments about adaptive specializations based on the metabolic cost of terrestrial locomotion.

Energetics of Specialized Locomotion

Functional morphologists have carefully studied many of the locomotory adaptations exhibited by different animals, usually for increasing speed or energy conservation. Three of the more highly specialized forms are the hopping kangaroo, the brachiating spider monkey, and the branch-walking slow loris. I will look at the energetics of each of these three locomotory specializations in turn.

The red kangaroo moves at low speed by what is called "pentapedal" locomotion; in this gait the animal swings both back legs forward while it is supported by the front legs and tail, and then swings the front legs and tail forward while supported by the hind legs to complete the cycle. This is an energetically very expensive way to move: the rate of oxygen consumption increases with speed much faster than would be predicted for a quadruped of the same size by Eq. (4). At the highest speed at which the animals used this gait, 1.7–2.0 m/sec, the oxygen consumption rate was approximately 2.3 times greater than for another animal of the same size [calculated from Dawson and Taylor (1973) and Eq. (4)].

At speeds above 2 m/sec the kangaroo hops, and the oxygen consumption rate actually decreases slightly with increasing speed. At about 5.0 m/sec the $\dot{V}_{O_2}$ of a comparably sized quadruped or biped would have increased to the point where the two animals would be consuming oxygen at the same rate. Only at speeds greater than 4.2 m/sec does the kangaroo realize any net energetic savings from its highly specialized form of locomotion.

Brachiation has long been thought to be an energetically inexpensive means of locomotion, primarily because of the pendulumlike characteristics of the movements. Parsons and Taylor (1977) tested this notion by measuring the oxygen consumption of two spider monkeys as they brachiated along a "rope-mill" (a treadmill-type device on which the animal suspends itself from a motor-driven rope) and walked or ran along a treadmill. They found that at all speeds brachiation involved greater energy expenditure than walking or running gaits in the same animals. However, as Parsons and Taylor point out, the advantages of an arboreal environment could easily outweigh any disadvantages in the energetic cost of moving in a straight line at a constant speed.

In the same study, Parsons and Taylor also measured the energetic cost of quadrupedal rope-walking in a loris. They trained a loris to walk both below and above the rope on the rope-mill. In this case, at all speeds, the rate of oxygen consumption was not significantly different, either for walking above or below the rope, than the rate that would be predicted by Eq. (4) for a terrestrial walker of the same size.

It would appear, at least for these animals, that their highly specialized forms of locomotion yield little, if any, benefits in terms of the energy cost of locomotion.

Mechanics

The study of the energetics of terrestrial locomotion has led to one intriguing conclusion: with very few exceptions, the energetic cost of locomotion increases linearly with speed and varies in a regular, simple way with body size, independent of the gait or method of movement. This is certainly a surprising result, and a mechanistic explanation would be very useful. Measurements of the mechanical power output required for steady-speed locomotion could then be compared to the measurements of metabolic power input.

Power is the rate at which energy is transferred. For example, the rate of oxygen consumption by an animal can be called the power input to the animal because the conversion between oxygen and energy is nearly independent of the foodstuffs being oxidized (20.1 J/ml O_2). In a mechanical sense, power is the rate at which work is performed. For example, if an animal lifts its center of mass, the work the animal performs is equal to the vertical distance of the lift times the body mass times the gravitational constant. The work performed is independent of the speed of the lift. On the other hand, the average power expended during the lift is equal to the work divided by the time it took to accomplish the work; therefore, the average power would be 10 times greater, for the same lift, if the lift were performed in ⅒ the time. As in the case of the metabolic measurements, we will only consider steady-speed measurements in our determinations of the average power output during terrestrial locomotion.

The mechanical power output of locomotion is often broken down into three components: (1) the power needed to overcome friction due to wind resistance, friction against the ground, and friction within the body; (2) the power needed to lift and accelerate the center of mass of the body within each step; and (3) the power needed to move the limbs and other body segments relative to the center of mass.

The Power Required to Overcome Friction

The power used to overcome wind resistance is small at all but the very highest speeds of terrestrial locomotion (Pugh, 1971). Humans present a relatively large frontal area to the air during running, but expend only about 2% of their metabolic power input to overcome wind resistance at 3 m/sec and only 8% at 8 m/sec [calculated from Cavagna and Kaneko (1977) and Hill (1927)]. The power used to overcome friction against the ground is zero unless the animal is slipping or deforming the ground (e.g., running on sand). The power required to overcome friction within the body is very difficult to estimate; certainly the joints are very well suited for minimizing friction. However, there are frictional losses as the viscera move within the body during running and losses within the muscles themselves. One estimate is that the power dissipated against friction within the body is less than 2% of the total power requirements [calculated from Cavagna *et al.* (1971) and Cavagna and Kaneko (1977)].

The Power Required to Lift and Accelerate the Center of Mass Each Step

It is often assumed that when an animal is moving on level ground at a constant speed, the body (or more properly, the center of mass) moves steadily and at a fixed height off the ground, and little additional energy input is required to maintain the average forward speed. However, the center of mass of the animal does move up and down, and accelerates and decelerates within each step. Thus, energy is required for the maintenance of the average forward speed and height of the center of mass. In a galloping 20-kg dog, for example, at 4.2 m/sec, the center of mass goes up and down 5 cm, and changes speed in the fore–aft direction 0.05 m/sec, each stride (Heglund, 1980).

These subtle but energetically important oscillations in the total energy of the center of mass are most conveniently measured by a force platform. This device resolves and measures the vertical and fore–aft horizontal forces exerted by an animal as it moves along the ground for one or more complete strides. From the measured forces the kinetic energy changes of the center of mass in both the vertical and forward directions can be calculated as a function of time. From the vertical kinetic energy changes the height changes, and therefore the gravitational potential energy changes, can be calculated as a function of time (Cavagna, 1975; Heglund, 1981). The total mechanical energy of the center of mass can then be calculated at each instant as the sum of the kinetic and potential energies; the work required to maintain the oscillations in the total energy of the center of mass each stride is calculated as the sum of the increments in the total energy curve. The average power required to maintain these oscillations in energy of the center of mass is the work per stride divided by the stride period.

There are two consequences of this method for calculating the total power required to maintain the observed oscillations in energy. First, summing the energy curves at each instant in time permits all possible transfers of energy to take place. For example, if an increase in the kinetic energy occurs simultaneously with a decrease in potential energy, the two tend to cancel each other out, resulting in a smaller oscillation in the total energy, and therefore smaller work and power expenditures. Second, by summing the increments in the total energy curve, we assume that all of the decrements in the curve result in energy that is degraded into heat and must be created *de novo* by the muscle. This assumption is not strictly true; some of the increments in the total energy can be due to elastic energy stored in the tendons and cross-bridges during a previous a decrement in the total energy (i.e., the energy was not degraded to heat but was instead stored in elastic elements).

Two Basic Mechanisms to Minimize the Cost of Maintaining Oscillations in the Mechanical Energy of the Center of Mass

While walking, animals used an exchange between the forward kinetic energy and the gravitational potential energy of the center of mass in a manner similar to a pendulum or an egg rolling end over end, in order to minimize the oscillations in the total energy of the center of mass (Cavagna *et al.*, 1977). Both bipeds and quadrupeds use this mechanism, but the effectiveness of the transfer varies with walking speed (Heglund, *et al.*, 1982*a*). At the optimal speed of walking (e.g., about 1.5 m/sec for 70-kg humans, about 0.2

m/sec for 0.15-kg quail) this transfer can be 75% of the energy changes that would have occurred had there been no transfer by the pendulum mechanism.

The second mechanism for minimizing the cost of maintaining the oscillations in the energy of the center of mass is storage and recovery of elastic strain energy in the muscles and tendons. During a bipedal run, bipedal hop, or a quadrupedal trot, the potential and kinetic energy changes of the center of mass are in phase, precluding any transfer between the two as occurs in a walk. However, the oscillations in energy of the center of mass in these gaits are very similar to the oscillations in energy of a bouncing ball. When the animal lands on the ground, part of the decrease in gravitational potential and kinetic energy of the center of mass may be stored in elastic strain energy of the muscles and tendons. Upon the subsequent takeoff, this elastic energy may be used to lift and reaccelerate the center of mass. This elastic energy recovery accounts for nearly 50% of the total energy oscillations in man (Cavagna *et al.*, 1964) and kangaroo (Alexander and Vernon, 1975).

The quadrupeds have a third gait, the gallop, which they use at high speeds of locomotion. The mechanism of galloping is different in large and small animals. Large animals at a slow gallop use a combination of the pendulum mechanism (up to 40% transfer) and the elastic bounce mechanism. As the speed of the gallop increases, the percentage transfer decreases to zero and the elastic nature of the gait increases. At the highest speeds of locomotion, the quadrupeds bounce off the back legs, have a short aerial phase, bounce off the front legs, and have a second, large aerial phase. In contrast, the small animals never achieve an appreciable transfer at the low-speed gallops. In addition, the stride cycle of a small animal differs in that the animal lands on the front legs and takes off from the back legs; if any elastic energy were stored in the front legs during landing it could not be used by the back legs during takeoff. (There is a possibility that some useful elastic storage and recovery can occur in the back of the small animals.) The mechanics of the gait not only precludes the use of elastic energy in the legs of the small animals, but the data also suggest that small animals may have relatively thicker tendons. The possible elastic storage in the tendons of a kangaroo rat is more than likely 14% or less as compared to the 50% or more in the red kangaroo (Biewener *et al.*, 1981).

The Power Required to Move the Limbs and Other Body Segments Relative to the Center of Mass

The third major component in the work of terrestrial locomotion is the power required for maintaining the oscillations in kinetic energy of the limbs and other body parts relative to the center of mass of the whole animal (Fedak *et al.*, 1982). For example, within each step cycle the foot is alternately moved forward and backward relative to the center of mass; this reciprocal motion implies that the kinetic energy of the foot must be alternately created and destroyed twice in each cycle. In actual fact the process is considerably more complicated because some of the translational kinetic energy can be converted to rotational kinetic energy and then back again, rather than having to be created anew twice each cycle.

The movements of the limbs are relatively gross and lend themselves nicely to a cinematographic analysis. High-speed films are made of the animal running on a treadmill. The limb and body segment positions on the film are digitized and entered into a computer. The animals are then sacrificed and frozen in a midstride position. The segments are dissected free and the location of the center of mass, the moment of inertia about the center of mass, and the mass of each segment are measured. From this information and the digitized films, it is possible to calculate the location of the composite center of mass (the center of mass of the whole animal) and the movements of each segment relative to the center of mass. The kinetic energy of each segment can then be calculated at each instant in time (i.e., each picture frame). The resulting kinetic energy versus time information is summed in a manner analogous to the force-plate analysis. However, in this case, transfers in energy are allowed only within a limb, not between limbs. Clear mechanisms exist for transfers within a limb (e.g., from a forearm to a hand), but it is not clear how kinetic energy of one limb could be transferred to another limb.

As in the previous analysis, the increases in the total rotational and translational kinetic energy of the body segments relative to the center of mass are summed for an integral number of strides to calculate the total work done during those strides to maintain the oscillations in kinetic energy. The work is then divided by the duration of the analyzed strides to obtain the average power output.

The animals utilize several mechanisms for minimizing the power required for the movements of the body segments. First is the well-documented decrease in the mass and lengthening of the distal segments in the cursorial animals (e.g., Hildebrand, 1974). While reduced limb segments undoubtedly decrease the oscillations in kinetic energy that would have occurred had the leg gone through the same motions without the decrease in mass, it is not clear that the animal actually reaps any net benefit. In an interesting study of the energetics of locomotion, Taylor *et al.* (1974) measured the oxygen consumption during treadmill running in three animals of the same total body mass and limb length but with very different limb masses and configurations. The total limb mass of a gazelle was two-thirds that of a goat and only one-half that of the cheetah, and the average distance from the pivot point to the center of mass of the extended limb in the gazelle was one-third that in a goat and one-ninth that in a cheetah (these measurements are only qualitative because the limb was assumed to be extended and rigid, when in fact the moments of inertia change considerably between the power and recovery phases of a stride cycle). Despite the large differences in limb configuration, they found no difference in oxygen consumption in these animals.

This apparent paradox between mechanical energy requirements and metabolic energy supply may be explained in part by a second energy-saving mechanism: the animals do not move their limbs and body segments in a reciprocal motion, but instead they often utilize a somewhat circular motion. Thus most segments do not stop moving relative to the center of mass at any point in their cycle. Some of the kinetic energy can then be conserved rather than recreated at the expense of metabolic energy.

Another possible energy-conserving mechanism is the storage and recovery of elastic energy in the tendons and muscles. To the extent that the movement of any segment is reciprocal, the mechanics of deceleration followed immediately by acceleration in the opposite direction is ideal for this metabolic energy-conserving mechanism. Studies of the electrical activity of the muscles (EMG) during locomotion have shown that the muscles are often active during the deceleration as well as acceleration phase of the cycle (Goslow *et al.*, 1981). This would allow elastic energy to be saved and recovered, although the extent to which this actually occurs has not been quantified.

All of these energy-conserving mechanisms do not mean that the cost of maintaining the oscillations in the kinetic energy of the body segments relative to the center of mass is insignificant to the total metabolic energy of the animal. Fedak *et al.* (1982) found that at high speeds this cost can often be more than half of the total work of locomotion. Thus the paradox is unresolved.

Total Mechanical Power Output as a Function of Speed and Size

The mechanical power required to maintain the oscillations in kinetic and potential energy of the center of mass of an animal increases linearly with running speed in all animals, independent of body size (Heglund *et al.*, 1982*a*), and can be expressed as a linear equation:

$$\dot{E}_{CM}/M_b = 0.685v + 0.072 \qquad (5)$$

where the mass-specific power $\dot{E}_{CM}/M_b$ is in W/kg of body mass and speed v is in m/sec.

The mechanical power required to maintain the oscillations in kinetic energy of the body segments relative to the center of mass increases as a power function of running speed in all animals, independent of body size (Fedak *et al.*, 1982), and can be expressed as

$$\dot{E}_{KE}/M_b = 0.478v^{1.53} \qquad (6)$$

where the mass-specific power $\dot{E}_{KE}/M_b$ is in W/kg of body mass and speed v is in m/sec.

We can simply add Eqs. (5) and (6) in order to calculate the total mechanical power required for terrestrial locomotion:

$$\dot{E}_{tot}/M_b = 0.478v^{1.53} + 0.685v + 0.072 \tag{7}$$

where the total mechanical power $\dot{E}_{tot}/M_b$ is in W/kg and speed v is in m/sec (Fig. 3).

Combining the equations in this way does not allow any transfers of energy between the center of mass and the kinetic energy relative to the center of mass or *vice versa* (Heglund *et al.*, 1982*b*). During an aerial phase no such transfers are possible; however, during the contact phase such transfers are possible but not probable. In order to evaluate the magnitude of the possible transfer of energy, selected strides from a few animals were analyzed by summing the E_{CM}/M_b and E_{KE}/M_b curves and then adding all the increments in the resulting total energy curve and dividing by the duration of the stride (Heglund *et al.*, 1982*b*). It was found that in a galloping quadruped (dog or chipmunk), Eq. (7) could result in a calculation of $\dot{E}_{tot}/M_b$ up to 34% too high; in other gaits and animals the difference between the two methods of calculating $\dot{E}_{tot}$ is much smaller.

Equation (7) is a very general relation; the mass-specific total mechanical power required during terrestrial locomotion is independent of the body size and increases curvilinearly with speed. This means that by knowing an animal's speed, one can calculate the mass-specific power output of the whole animal reasonably accurately.

Total Mechanical Power Output during Human Locomotion

There is a lack of quantitative data expressing total mechanical power output during terrestrial locomotion for primates; only humans have been

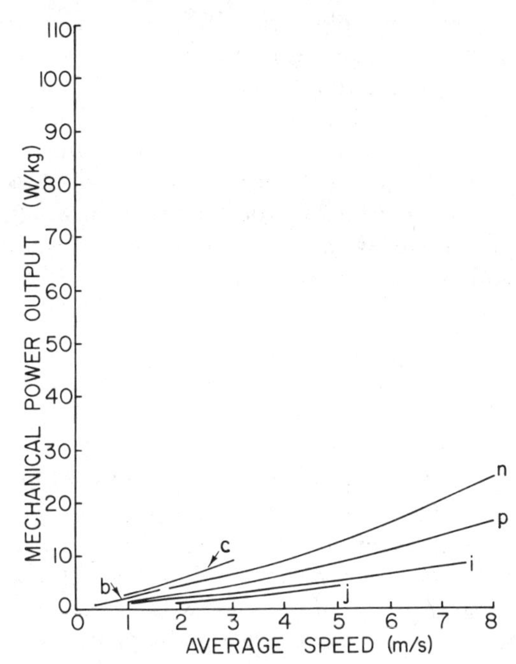

Fig. 3. Total mechanical power output $\dot{E}_{tot}/M_b$ versus speed for animals during steady-speed level locomotion. See text for details of how the measurements were made. The animal key is: b, 0.042-kg painted quail; c, 0.098-kg chipmunk; i, 5.0-kg dog; j, 7.0-kg turkey; n, 70-kg human; p, the average line calculated on the basis of Eq. (7) of the text.

studied extensively. Figure 3 shows that man is in the middle of the scatter of the data for the other animals that have been analyzed. It does not appear that the Hominidae have paid an inordinate price for their bipedalism; in fact, as we shall see, they are exceptionally efficient locomotors.

Efficiency of Locomotion

"Efficiency" is a term used in many different contexts and for that reason it is important to understand exactly how the term is meant to be interpreted. In a thermodynamic sense (the "true" definition), efficiency is the total work or power output divided by the total energy or power input. In this chapter we will use a modified version of this strict definition

$$\text{Efficiency} = \frac{\text{Mechanical power output}}{\text{Metabolic power input}} \tag{8}$$

This expression differs from the thermodynamic definition in that the total power input is the sum of the metabolic power input plus any elastically recovered energy (which we are unable to determine satisfactorily).

A striking result of the metabolism studies is that a 30-g quail, for example, will consume 13 times more energy to transport 1 g of its body mass a given distance than will a 100-kg ostrich. Yet from the mechanics studies we would expect both of these animals to perform about the same mechanical work per gram of body mass in order to travel that distance. Clearly, the efficiency of these two animals is very different.

Figure 4 plots the efficiency of locomotion as a function of speed for a quail, chipmunk, dog, turkey, and human. (These are the only animals for which the complete set of measurements necessary for calculating the efficiency have been made. In every case, the efficiency increases with speed, but the peak efficiency for the smallest animal (quail) is about 7% and the peak efficiency for the largest (human) is over 70%. A kangaroo attains an efficiency similar to that of the human even when the power output is just the $\dot{E}_{CM}$ power; surely kangaroos would attain even higher efficiencies if the $\dot{E}_{KE}$ were added to the power output.

Conclusions

From the data, it appears that efficiency is definitely size dependent, but probably independent of the number of limbs or mode of locomotion used by the animal. How is this possible? The size dependence is probably related to several different factors. First, as was previously mentioned, small animals have relatively thicker tendons and therefore are probably less able to utilize elastic storage and recovery of energy during steady-speed locomotion. In addition, at least for the quadrupeds, the small animals utilize a different gait

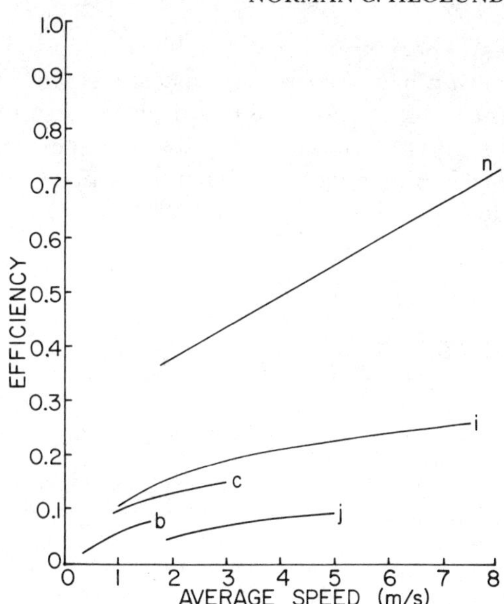

Fig. 4. The efficiency of locomotion plotted as a function of running speed. Efficiency is calculated as the ratio of mechanical power output divided by metabolic power input. The animal key is: b, 0.042-kg painted quail; c, 0.107-kg chipmunk; i, 5.0-kg dog; j, 6.4-kg turkey; and n, 70-kg human.

pattern during a gallop and as a result are unable to use what little elastic energy may be stored in the legs.

A second factor that may be responsible for the much lower peak efficiency in the small animals is the intrinsic velocity of their muscles. Intrinsic velocity is the velocity at which an unloaded muscle contracts and is related to the rate of myosin cross-bridge cycling. It is generally thought that one cross-bridge cycle costs one ATP. Close (1972) has shown that the intrinsic velocity of small-animal muscle is higher than that of homologous large-animal muscle. Thus we would expect that the metabolic cost of having the muscles "on" would be higher in the small animals than in the large animals.

Another factor that may be in part responsible for the lower peak efficiency in small animals is that they are turning their muscles on and off at a much higher rate. If this activation cost is the same per gram of muscle (this would seem to be reasonable if the concentration and gradient of Ca^{2+} are the same in large- and small-animal muscle), then the cost per gram per step would be independent of size. However, at a gallop, a 10-g mouse is taking nearly nine steps per second and a 600-kg horse is taking fewer than two steps per second (Heglund *et al.*, 1974); therefore, the mouse is consuming energy at about 4.5 times the rate of the horse just in activating the muscles. Is this an important factor? There is evidence that the cost of pumping Ca^{2+} may be as high as 30% of the total cost of an isometric twitch (Homscher *et al.*, 1972). The high efficiencies of the large animals must be due to very effective use of elastic energy, low intrinsic velocities, and relatively low stride frequency.

The final question to be answered is, why do all the animals of a given size seem to have the same oxygen consumption and efficiency, independent of locomotory specializations? Why is it, for example, that the highly cursorial and specialized kangaroo has the same peak efficiency as a human, who is most noncursorial? The answer probably lies in the fact that muscle and bone, the building blocks of the locomotory system, are basically the same in animals of roughly the same size. Rearrangement of these basic building blocks does yield some benefit to the animals involved, but not in terms of the energetics and efficiency of locomotion.

References

Alexander, R. McN., and Vernon, A. 1975. The mechanics of hopping by kangaroos (*Macropodidae*). *J. Zool.* **177:**265–303.

Biewener, A., Alexander, R. McN., and Heglund, N. C. 1981. Elastic energy storage in the hopping of kangaroo rats (*Dipodomys spectabilis*). *J. Zool.* **195:**369–383.

Cavagna, G. A. 1975. Force platforms as ergometers. *J. Appl. Physiol.* **39**(1)**:**174–179.

Cavagna, G. A., and Kaneko, M. 1977. Mechanical work and efficiency in level walking and running. *J. Physiol.* **268:**467–481.

Cavagna, G. A., Saibene, F. P., and Margaria, R. 1964. Mechanical work in running. *J. Appl. Physiol.* **19:**249–256.

Cavagna, G. A., Komarck, L., and Mazzoleni, S. 1971. The mechanics of sprint running. *J. Physiol.* **217:**709–721.

Cavagna, G. A., Heglund, N. C., and Taylor, C. R. 1977. Mechanical work in terrestrial locomotion: Two basic mechanisms for minimizing energy expenditure. *Am. J. Physiol.* **233**(5)**:**R243–R261.

Chassin, P. S., Taylor, C. R., Heglund, N. C., and Seeherman, H. J. 1976. Locomotion in lions: Energetic cost and maximum aerobic capacity. *Physiol. Zool.* **49**(1)**:**1–10.

Close, R. I. 1972. Dynamic properties of mammalian skeletal muscles. *Physiol. Rev.* **52:**129–197.

Dawson, T. J., and Taylor, C. R. 1973. Energetic cost of locomotion in kangaroos. *Nature* **246:**313–314.

Fedak, M. A., Rome, L., and Seeherman, H. J. 1981. One-step N_2-dilution technique for calibrating open-circuit VO_2 measuring systems. *J. Appl. Physiol.* **51**(3)**:**772–776.

Fedak, M. A., Heglund, N. C., and Taylor, C. R. 1982. Energetics and mechanics of terrestrial locomotion II: Kinetic energy changes of the limbs and body as a function of speed and body size in birds and mammals. *J. Exp. Biol.* **97:**23–40.

Goslow, G. E., Jr., Seeherman, H. J., Taylor, C. R., McCutchin, M. N., and Heglund, N. C. 1981. Electrical activity and relative length changes of dog limb muscles as a function of speed and gait. *J. Exp. Biol.* **94:**15–42.

Heglund, N. C. 1980. Mechanics of locomotion in primitive and advanced mammals, in: *Comparative Physiology: Primitive Mammals* (K. Schmidt-Nielsen, L. Bolis, and C. R. Taylor, eds.), pp. 213–219, Cambridge University Press, London.

Heglund, N. C. 1981. A simple design for a force-plate to measure ground reaction forces. *J. Exp. Biol.* **93:**333–338.

Heglund, N. C., Taylor, C. R., and McMahon, T. A. 1974. Scaling stride frequency and gait to animal size: Mice to horses. *Science* **186:**1112–1113.

Heglund, N. C., Cavagna, G. A., and Taylor, C. R. 1982*a*. Energetics and mechanics of terrestrial locomotion. III. Energy changes of the center of mass as a function of speed and body size in birds and mammals. *J. Exp. Biol.* **97:**41–56.

Heglund, N. C., Fedak, M. A., Taylor, C. R., and Cavagna, G. A. 1982*b*. Energetics and mechanics of terrestrial locomotion. IV. Total mechanical energy changes as a function of speed and body size in birds and mammals. *J. Exp. Biol.* **97:**57–66.

Hildebrand, M. 1974. *Analysis of Vertebrate Structure*, Wiley, New York.

Hill, A. V. 1927. The air resistance to a runner. *Proc. R. Soc. B.* **102:**380–385.

Homsher, E., Mommaerts, W. F. H. M., Ricchiuti, N. V., and Wallner, A. 1972. Activation heat, activation metabolism and tension related heat in frog semitendinosus muscles. *J. Physiol.* **220:**601–625.

Parsons, P. E., and Taylor, C. R. 1977. Energetics of brachiation versus walking: A comparison of a suspended and an inverted pendulum mechanism. *Physiol. Zool.* **50**(3)**:**182–188.

Pinshow, B., Fedak, M. A., and Schmidt-Nielsen, K. 1977. Terrestrial locomotion in penguins: It costs more to waddle. *Science* **195:**592–595.

Pugh, L. G. C. E. 1971. The influence of wind resistance in running and walking and the efficiency of work against horizontal or vertical forces. *J. Physiol.* **213:**255–276.

Seeherman, H. J., Taylor, C. R., Maloiy, G. M. O., and Armstrong, R. B. 1981. Design of the mammalian respiratory system: Measuring maximum aerobic capacity. *Resp. Physiol.* **44:**11–23.

Taylor, C. R., and Heglund, N. C. 1982. Energetics and mechanics of terrestrial locomotion. *Ann. Rev. Physiol.* **44:**97–107.

Taylor, C. R., Shkolnik, A., Dmi'el, R., Baharav, D., and Borut, A. 1974. Running in cheetahs, gazelles, and goats: Energy cost and limb configuration. *Am. J. Physiol.* **227**(4)**:**848–850.

Taylor, C. R., Heglund, N. C., and Maloiy, G. M. O. 1982. Energetics and mechanics of terrestrial locomotion. I. Metabolic energy consumption as a function of speed and body size in birds and mammals. *J. Exp. Biol.* **97:**1–21.

Body Size and Limb Design in Primates and Other Mammals

15

R. McNEILL ALEXANDER

Introduction

The aim of this chapter is to discover major differences of limb design between primates and other mammals. It is not concerned with qualitative features such as possession of nails or claws, but with quantitative ones, the dimensions of bones and muscles. It is not concerned with differences between primates, but looks for design features shared by primates as a whole, irrespective of body size and taxonomic group.

Most of the data come from investigations by Alexander *et al.* (1979, 1981). These investigations did not focus particularly on primates, but examined only six primate species (*Galago, Homo,* and four cercopithecoids). So small a group can hardly be representative of the order, but the design features that will be revealed are shared by all the members of this group and are probably characteristic of primates in general. The primates will be compared with about 30 other species of mammal, mainly rodents, insectivores, fissipeds, and artiodactyls. The body masses of the primates ranged from 0.6 to 50 kg, and those of the other mammals from 3 g (a shrew *Sorex*) to nearly 3 tons (an elephant, *Loxodonta*).

R. McNEILL ALEXANDER • Department of Pure and Applied Zoology, University of Leeds, Leeds LS2 9JT, England.

Bones

Figure 1 shows femur dimensions plotted (on logarithmic coordinates) against body mass. The continuous lines represent allometric equations calculated from the data for all taxa. The lines for femur length and femur diameter both have gradients of 0.36, indicating that femur length and diameter both tend to be proportional to (body mass)$^{0.36}$. Similar exponents were found for all the major limb bones: the mean values were 0.35 for lengths and 0.36 for diameters. These exponents are just a little greater than the exponent of 0.33 that would be found if mammals of different sizes were geometrically similar to each other. The exponents for lengths are much greater than predicted by McMahon's (1973) theory of elastic similarity, which requires lengths proportional to (body mass)$^{0.25}$ and diameters to (body mass)$^{0.375}$.

In Fig. 1, different symbols are used for different taxa. The graph for diameters shows no consistent peculiarity of any taxon, though one rodent (*Pedetes,* 2 kg) and two primates (*Papio,* 15 kg, and *Homo,* 64 kg) seem to have unusually thick femora for their body masses. In the graph for femur length, however, the primates and the bovids stand out from the other taxa. The femora of primates are consistently about 50% longer than those of other

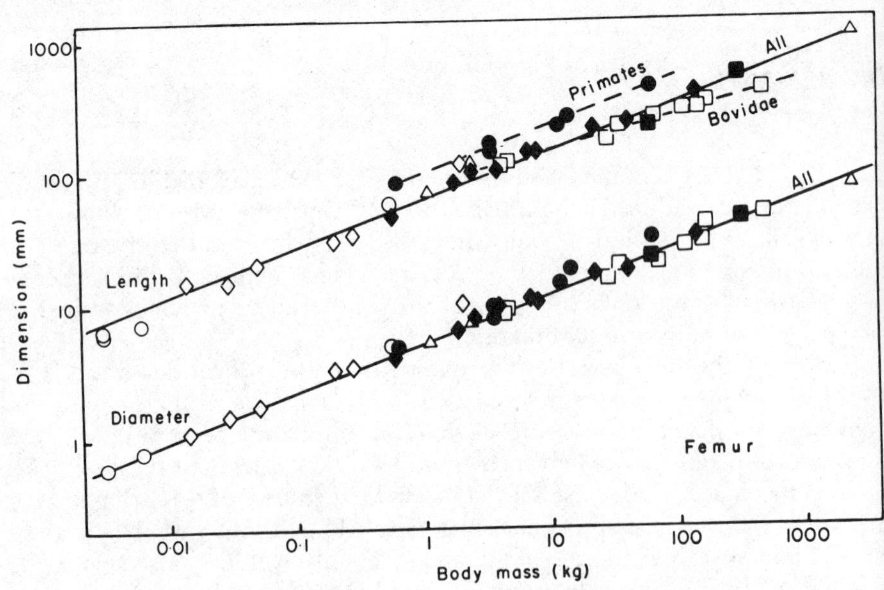

Fig. 1. A graph on logarithmic coordinates showing the lengths and diameters of the femurs of mammals plotted against body mass. (○) Insectivora; (●) Primates; (◇) Rodentia; (◆) Fissipedia; (□) Bovidae; (■) other Artiodactyla; (△) other orders (Lagomorpha and Proboscidea). The lines represent allometric equations fitted by least squares regression after transformation to logarithms. (The correlation coefficients are so high that model II regression methods would give only slightly different equations.) Continuous lines represent regression equations fitted to all the data and broken lines represent equations fitted to Primates only and to Bovidae only. [From Alexander *et al.* (1979), by permission of the Zoological Society of London.]

mammals of equal mass. The femora of small bovids have about the lengths predicted by the general allometric equation, but large bovids have unusually short femora for their body masses. The allometric exponents for bovid leg bone lengths are all lower than the exponents for mammals in general, averaging 0.24.

Figure 2 shows the same pattern for the humerus as for the femur. The humerus is generally about 40% longer in primates than in more typical mammals. The allometric exponent for bovid humeri is lower than the one for mammals in general.

Similar graphs for the tibia and ulna show that these bones also are longer in primates than predicted by the allometric equations for mammals in general. The sole exception is the human ulna, which has about the length predicted by the general equation (Alexander *et al.*, 1979). Man has shorter radii and presumably ulnae than great apes of equal mass (Aiello, 1981).

A different picture is presented by the metatarsals and metacarpals (Fig. 3). In both cases the lengths and diameters for primates are very close indeed to the predictions of the allometric equations for mammals in general. Bovids have their metatarsals and metacarpals fused to form cannon bones, which are much thicker and generally also much longer than the metatarsals and metacarpals of other mammals of equal mass. (The nonbovid artiodactyls included in Fig. 3 are *Camelus*, which has cannon bones, and *Phacochoerus*, which does not.)

Most of the medium to large mammals in our data set are primates, fissipeds, and bovids. The primates and bovids are the two groups that have

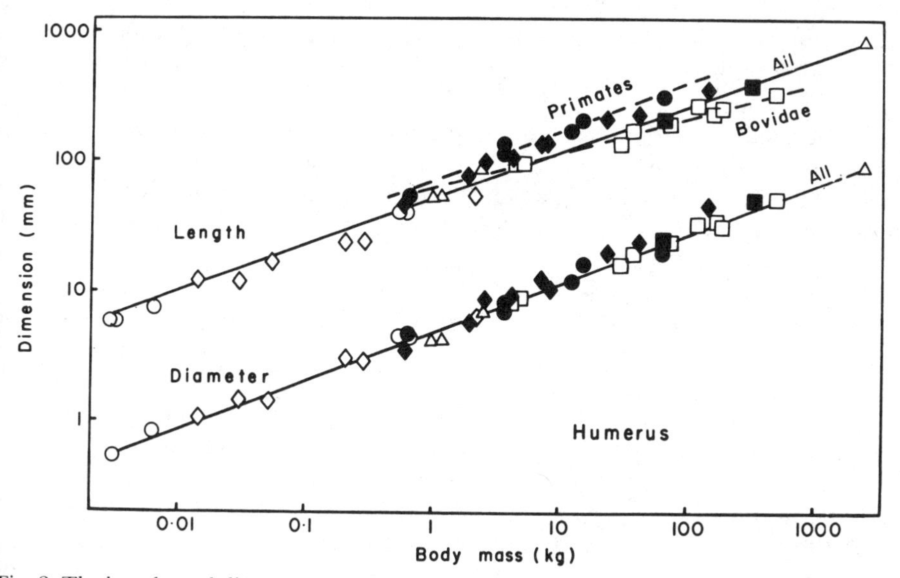

Fig. 2. The lengths and diameters of the humerus of mammals plotted against body mass. Other details as Fig. 1. [From Alexander *et al.* (1979), by permission of the Zoological Society of London.]

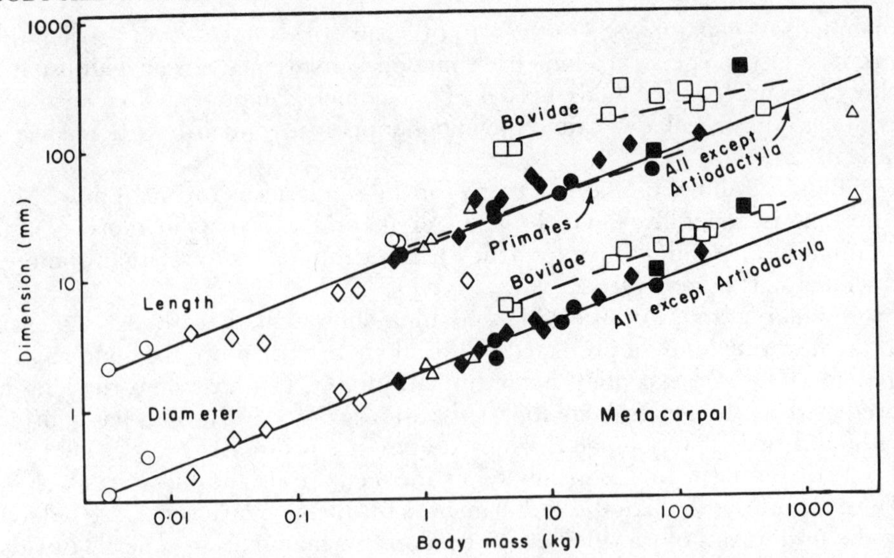

Fig. 3. The lengths and diameters of the metacarpals of mammals plotted against body mass. Other details as Fig. 1, except that the continuous lines represent allometric equations for all the data *excluding Artiodactyla.* [From Alexander *et al.* (1979), by permission of the Zoological Society of London.]

stood out from the others in this analysis. Most of the limb bone dimensions of fissipeds, however, are close to the predictions of the allometric equations for mammals in general. Table 1 shows mean bone lengths predicted for 10-kg members of the three taxa. This particular mass has been chosen because it is close to the geometric mean masses for the primate and fissiped samples (7 and 8 kg, respectively).

The totals in Table 1 exclude the phalanges, but give a fair indication of limb length. They show that a 10-kg primate and a 10-kg bovid could be expected to have much longer limbs than a 10-kg fissiped. The extra length is in the femur, tibia, radius, and ulna of primates, but in the metatarsals and metacarpals of bovids.

The general allometric equations for mammals (Alexander *et al.*, 1979) show that the lengths of the femur, humerus, ulna, and longest metacarpal all tend to be proportional to the 0.36 or 0.37 power of body mass. The lengths of the tibia and longest metatarsal, however, tend to be proportional to the 0.32 and 0.30 powers of body mass. Thus the crural index 100(tibia length)/(femur length) tends to decrease as body mass increases and the intermembral index 100(humerus + radius length)/(femur + tibia length) tends to increase. Reanalysis of the data shows that the crural index is proportional to (body mass)$^{-0.05 \pm 0.02}$. Also, a modified intermembral index, using ulna length in place of radius length (which was not measured) is proportional to (body mass)$^{0.02 \pm 0.01}$. The exponents were calculated by least squares regression after transformation to logarithms, as in Alexander *et al.* (1979), and are given with

Table 1. Limb Bone Lengths Calculated for 10-kg Mammals from the Allometric Equations of Alexander *et al.* (1979)

	Bone length, cm		
	Primates	Fissipedia	Bovidae
Femur	22	16	14
Tibia	19	15	18
Longest metatarsal	5	6	14
Total	46	36	47
Humerus	17	14	11
Ulna	17	15	15
Longest metacarpal	4	5	13
Total	38	36	40

95% confidence limits. The correlation coefficients are −0.72 and 0.61. *Pedetes* was omitted from the analysis because it is a biped with reduced forelimbs.

Similarly, Rollinson and Martin (1981) found that in primates, crural index tends to decrease and intermembral index to increase as body mass increases. Also, Aiello (1981) found that in anthropoids, tibia length tends to be proportional to (femur length)$^{0.88}$. The relationships between limb proportions and body size seen in the primates are similar to those seen in mammals in general.

Muscles

Figure 4 shows the masses and fiber lengths of the muscles that flex the wrist and fingers in a diverse selection of mammals. It shows differences between taxa at least as sharp as those that have already been demonstrated from bone dimensions. These particular muscles have two to three times the mass and two to three times the muscle fiber length in primates as in fissipeds of equal body mass. In bovids they have only half the mass and half or less the fiber length as in fissipeds of equal body mass. Primates presumably need long muscle fibers, capable of large length changes, to operate their very mobile hands. In contrast, bovids probably resemble camels, in which changes of muscle fiber length seem to be much less important than elastic extension of tendons in governing the movements of distal forelimb joints (Alexander *et al.*, 1982).

Data for other muscles are shown in Table 2. The extensors of the ankle (e.g., gastrocnemius) and the flexors of the hind digits tend to be larger and longer fibered in primates than in fissipeds of equal mass. The long fibers are presumably important for the mobility of the foot, just as the long fibers of

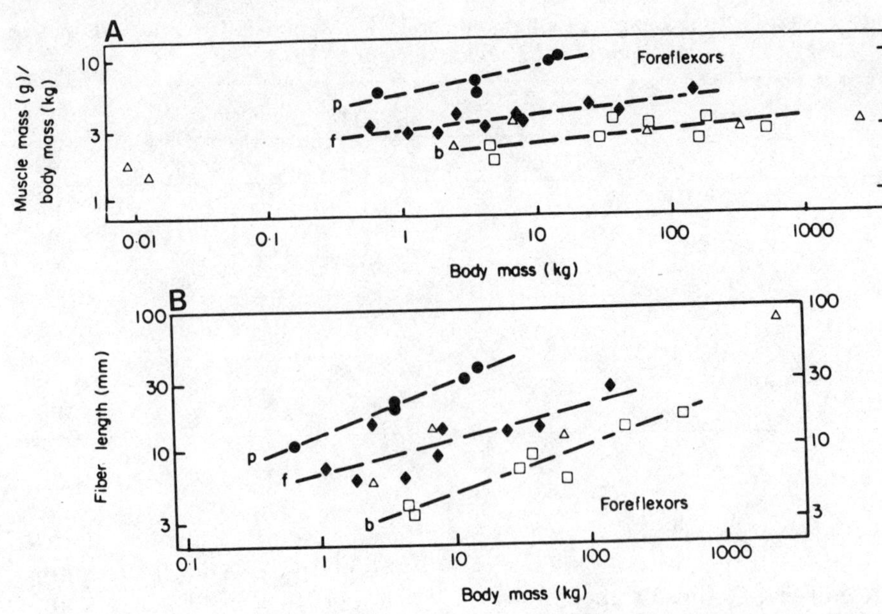

Fig. 4. Graphs on logarithmic coordinates showing properties of the flexor muscles of the wrist and of the digits of the fore foot, plotted against body mass. The upper graph shows the total mass of these muscles and the lower one the harmonic mean length of their muscle fibers. (●) Primates; (◆) Fissipedia; (□) Bovidae; (△) other mammals. The broken lines represent allometric equations fitted as for Fig. 1 for (p) Primates, (f) Fissipedia, and (b) Bovidae. [From Alexander *et al.* (1981), by permission of the Zoological Society of London.]

the corresponding forelimb muscles seem to be important for the mobility of the hand.

Table 2 also seems to show that primates tend to have rather large quadriceps muscles for their body masses.

The maximum force a muscle can exert can be expected to be propor-

Table 2. Mass *m* and Fiber Length *l* of Muscles Calculated for 10-kg Mammals from the Allometric Equations of Alexander *et al.* (1981)[a]

	Primates		Fissipedia		Bovidae	
	m, g	*l*, mm	*m*, g	*l*, mm	*m*, g	*l*, mm
Hamstrings and adductor femoris	260	81	260	85	390	99
Quadriceps	190	36	110	32	150	36
Gastrocnemius, soleus, and plantaris	71	25	51	15	49	8
Deep digital flexors of hind foot	24	25	14	14	19	15
Triceps	110	36	100	40	63	43
Flexors of wrist and fore digits	85	31	37	12	23	5

[a]The fiber lengths *l* are weighted harmonic means for the groups of muscles.

tional to its physiological cross-sectional area, which is proportional to (muscle mass)/(muscle fiber length). Estimates of the relative magnitudes of the forces exerted by the muscles described in Table 2 can therefore be obtained by dividing the mass estimates by the fiber length estimates. This method indicates that homologous muscle groups in 10-kg primates and fissipeds would be about equally strong, except that the quadriceps would be stronger in the primates. The ankle extensors and the flexors of the wrist and fore digits of a 10-kg bovid would be stronger than their homologues in primates and fissipeds. These muscles may be required to exert very large forces because the long metatarsals and metacarpals of bovids tend to make them work with small mechanical advantages.

Conclusions

Primates and bovids generally have longer limbs than typical mammals of equal body mass. Primates have the extra length in the femur, tibia, humerus, and ulna, but bovids have it in the metatarsals and metacarpals. As body mass increases, the crural index tends to decrease and the intermembral index tends to increase, both in mammals in general and in primates in particular.

The extensor muscles of the ankle and the flexors of the wrist and digits tend to be larger in primates than in other mammals of equal mass, and to have longer muscle fibers. In bovids, most of these muscles have exceptionally short muscle fibers. Primates apparently need muscles with long fibers to operate their very mobile hands and feet, but the movements of the distal joints of bovid limbs depend largely on tendon elasticity rather than changes of muscle fiber length.

References

Aiello, L. C. 1981. The allometry of primate body proportions. *Symp. Zool. Soc. Lond.* **48:**331–358.

Alexander, R. McN., Jayes, A. S., Maloiy, G. M. O., and Wathuta, E. M. 1979. Allometry of the limb bones of mammals from shrews (*Sorex*) to elephant (*Loxodonta*). *J. Zool. Lond.* **189:**305–314.

Alexander, R. McN., Jayes, A. S., Maloiy, G. M. O., and Wathuta, E. M. 1981. Allometry of leg muscles of mammals. *J. Zool. Lond.* **194:**539–552.

Alexander, R. McN., Maloiy, G. M. O., Ker, R. F., Jayes, A. S., and Warui, C. N. 1982. The role of tendon elasticity in the locomotion of the camel (*Camelus dromedarius*). *J. Zool. Lond.* **198:**293–313.

McMahon, T. A. 1973. Size and shape in biology. *Science,* **179:**1201–1204.

Rollinson, J., and Martin, R. D. 1981. Comparative aspects of primate locomotion, with special reference to arboreal cercopithecines. *Symp. Zool. Soc. Lond.* **48:**377–427.

Body Size and Scaling of Limb Proportions in Primates

16

WILLIAM L. JUNGERS

Introduction

Extant terrestrial mammals span an enormous size range, from tiny pygmy shrews weighing in at less than 5 g to massive African elephants tipping the scales at over 2500 kg (Eisenberg, 1981). Despite such marked differences in adult body mass, Alexander *et al.* (1979) report that the lengths and diameters of limb bones in a shrew-to-elephant size series scale surprisingly close to geometric similarity; i.e., linear dimensions are almost proportional to $(\text{mass})^{1/3}$. Although adult members of the order Primates cannot match the overall range of body sizes in mammals, the size distribution of living primate species is still quite impressive, ranging from dwarf galagos and mouse lemurs (~60–70 g) at one end of the spectrum to male gorillas (>200 kg in some individuals) at the other. Given the unexpected results of the Alexander *et al.* (1979) study, it is reasonable to wonder if the linear dimensions of the long bones of adult primates also conform to the expectations of geometric similarity. A few minutes of casual observation at the zoo or natural history museum would probably suffice to allow one to reject this null hypothesis for primate limb proportions. Galagos and tarsiers simply do not resemble gibbons or gorillas very much with regard to their respective bodily proportions. Locomotor differences between these extremes seem equally obvious and

WILLIAM L. JUNGERS • Department of Anatomical Sciences, School of Medicine, State University of New York, Stony Brook, New York 11794.

appear to be correlated with differences in limb proportions. If we choose galagos and humans as the endpoints of our comparative size series in primates, however, we might not be so quick to discount the possibility of geometric similarity in limb proportions. In fact, limb bone allometry in just such a series (galago, colobine, cercopithecines, and human) is very close to geometric similarity (isometry), with the average exponent for the lengths of all long bones equal to 0.34 (Alexander *et al.*, 1979). Clearly, the composition of the species included in an allometric comparison will strongly influence one's conclusions about body size, limb proportions, and skeletal allometry in primates.

The gross correspondence between overall body proportions and locomotor behavior in primates has been recognized since Mollison's (1911) classic study on "Die Körperproportionen der Primaten." Schultz's (1930, 1933, 1937, 1954) heroic efforts added enormous amounts of quantitative information about variation in the body proportions of primates, but the implications of his findings were frequently more taxonomic than functional in nature. Somewhat later, Erikson (1963) attempted to relate taxonomic variation in proportions specifically to locomotor differences among groups of New World monkeys; he found a reasonably good fit between postcranial proportions and his "locomotor types" of springers, climbers, and brachiators (also see Hershkovitz, 1977). Efforts to extend his findings to other anthropoids created unexpected complications, so much so that Erickson ultimately despaired of making generalizations and concluded that numerous "disharmonious" combinations of locomotor characteristics could coexist in any given species. Nevertheless, differences in body proportions formed the essence of Napier's (Napier and Napier, 1967; Napier and Walker, 1967) locomotor classification, and subsequent authors continued to seek for ways to link locomotion and primate proportions in a causal manner (e.g., Stern and Oxnard, 1973; Ashton *et al.*, 1975; Jouffroy and Lessertisseur, 1978, 1979). Although some functional relationships could be easily discerned (e.g., relatively long arms in brachiators versus relatively long legs in leapers), as Schultz had discovered earlier, taxonomic similarities and differences proved to be very serious complicating factors. Size differences among taxa were also generally recognized as another set of confounding factors, but Biegert and Maurer's (1972) study of body proportions and allometry in catarrhine primates represented the first explicit attempt to directly assess the functional significance of covarying patterns of body size and postcranial dimensions in a large group of primates. In his consideration of locomotor adaptations in prosimian primates, Walker (1974) also recognized that interpretations of body proportions are seriously complicated by allometric differences. The recent study by Rollinson and Martin (1981), although not allometric in design, correctly notes that limb proportions in primates "are influenced by an intricate combination of body size and locomotor specialization" (pp. 408–409) and concludes that it is vital to include an explicit consideration of body size in any analysis of primate locomotor evolution.

Allometric analysis of the locomotor skeleton has the potential to partially unravel the complex relationships among body size, overall proportions, and positional behavior in adult primates (Biegert and Maurer, 1972; Jungers, 1978, 1979, 1980, 1984*a;* Aiello, 1981*a,b;* Dykyj, 1982; Steudel, 1982). Much remains to be discovered about the causal relationships between body size and locomotor adaptation in primates, and this study hopes to contribute to that understanding by employing the allometric approach in a systematic analysis of limb dimensions and limb proportions across the entire primate order. It must be stressed, however, that allometry *per se* is not an explanation, but rather a "descriptive tool of gross morphology" (Giles, 1956). To demonstrate specific scaling trends is only the first step in the analysis; it is the allometric patterns themselves that require explication.

Materials and Methods

Skeletal Sample and Body Weights

A total of 1083 fully adult primate skeletons were measured in osteological collections in the United States and Europe. Adult status was judged by fusion of the epiphyses of the long bones; "dental adults" were therefore excluded if they did not meet this criterion. More than 99% of this sample is composed of wild-collected specimens; nonpathological individuals that died in zoological parks are included in rare cases (e.g., three specimens of the patas monkey, *Erythrocebus patas*). Table 1 lists the 90 different primate taxa analyzed in this study and indicates the number of skeletal specimens measured for each group. The sample includes 34 strepsirhine taxa and 56 haplorhine taxa (one species of *Tarsius,* 18 platyrrhine species, 22 cercopithecoid species, and 15 hominoid taxa).

Body weight values for each of the taxa (and sex in many cases) are also indicated in Table 1. These values are compiled from numerous sources that are noted in the footnote to the table. Wild-collected body weights were always given priority. Those skeletal samples with an asterisk are composed totally of individuals with associated body weights; these body weights are therefore sample-specific. Skeletal samples and body weights are separated by sex in sexually dimorphic taxa.

Analytical Methods

Relative Limb Proportions

Four standard anthropometric indices were constructed to assess relative limb proportions in primates:

Table 1. Skeletal Sample and Body Weights of the Primate Species Included in This Study[a]

Species	Number of adult skeletons	Body weight, g	Species	Number of adult skeletons	Body weight, g
Microcebus rufus	7	60	*Ateles geoffroyi*	3*	7,730
Microcebus murinus	20	70	*Miopithecus talapoin* ♂	7	1,380
Mizra coquereli	1	330	♀	9	1,120
Phaner furcifer	2	440	*Cercopithecus neglectus* ♂	8	6,320
Cheirogaleus major	10	450	♀	8	3,960
Hapalemur griseus	20	880	*Cercopithecus aethiops* ♂	10	5,370
Lemur mongoz	14	2,025	♀	7	3,600
Lemur macaco	7	2,410	*Cercopithecus cephus* ♀	5	2,880
Lemur catta	17	2,670	*Cercopithecus nictitans* ♂	6	6,500
Lemur rubriventer	4	2,350	*Cercopithecus ascanius* ♂	7	4,070
Lemur fulvus rufus	4	2,685	*Erythrocebus patas* ♂	3	11,100
Lemur fulvus collaris	7	2,505	♀	5	5,900
Lemur fulvus albifrons	11	2,295	*Cercocebus albigena* ♂	11	8,980
Varecia variegata	17	3,800	♀	8	6,400
Lepilemur leucopus	22	545	*Macaca nemestrina* ♂	6	10,210
Avahi laniger	16	920	♀	6	6,350
Propithecus verreauxi	14	3,780	*Macaca fascicularis* ♂	6*	4,930
Propithecus diadema	13	6,500	♀	6*	3,130
Indri indri	16	10,000	*Mandrillus sphinx* ♂	6	26,900
Daubentonia madagascariensis	11	2,800	*Papio hamadryas* ♂	9	21,300
Loris tardigradus	16	275	*Papio anubis* ♂	11	25,100
Arctocebus calabarensis	22	265	♀	6	14,100
Nycticebus coucang	27	920	*Papio cynocephalus* ♂	4	22,800
Perodicticus potto	20	1,150	♀	6	12,350
Galagoides demidoff	18	70	*Colobus verus*	3	4,280
Galagoides thomasi	3	110	*Colobus guereza* ♂	10	10,100
Galagoides zanzibaricus	2	150	♀	10	8,040
Galagoides alleni	7	295	*Colobus polykomos* ♀	4	7,700
Galago senegalensis	7	215	*Presbytis rubicunda* ♂	6	6,190
Galago moholi	14	160	♀	6	5,680
Galago elegantulus	17	275	*Presbytis cristata* ♂	6*	6,930
Galago inustus	1	210	♀	6*	5,950

Otolemur garnettii	9	760	*Presbytis obscura* ♂	7	7,540
Otolemur crassicaudatus	11	1,150	♀	8	6,080
Tarsius bancanus	3	125	*Presbytis aygula* ♂	5	6,680
Cebuella pygmaea	13	135	*Nasalis larvatus* ♂	6	20,370
Callithrix jacchus jacchus	4*	310	♀	6	9,820
Leontopithecus rosalia	10	680	*Hylobates klossi*	11	5,830
Saguinus fuscicollis	4	360	*Hylobates lar agilis*	7	5,740
Saguinus oedipus	5	450	*Hylobates lar carpenteri*	30	5,520
Saguinus midas midas	1*	570	*Hylobates lar muelleri*	19	5,730
Aotus trivirgatus	3*	1,220	*Hylobates lar albibarbis*	7	6,010
Callicebus moloch	9	1,070	*Hylobates hoolock*	9	6,800
Saimiri sciureus ♂	11*	960	*Hylobates syndactylus*	17	11,050
♀	9*	790	*Pan paniscus* ♂	10	45,000
Pithecia pithecia	4*	1,800	♀	10	33,200
Chiropotes satanus	4*	2,980	*Pan troglodytes schweinfurthii* ♂	15	43,000
Cebus apella ♂	8*	3,330	♀	11	33,200
♀	8*	1,940	*Pan troglodytes troglodytes* ♂	20	60,000
Cebus albifrons ♂	9*	3,260	♀	20	47,400
♀	5*	2,220	*Gorilla gorilla gorilla* ♂	21	169,500
Alouatta caraya ♂	4*	8,280	♀	18	71,500
♀	3*	5,410	*Gorilla gorilla beringei* ♂	7	159,200
Alouatta seniculus ♂	3*	7,880	♀	8	97,700
♀	5*	5,500	*Pongo pygmaeus pygmaeus* ♂	10	81,700
Alouatta palliata pigra ♂	1*	11,590	♀	12	37,800
♀	3*	6,290	*Pongo pygmaeus abelli* ♂	3	77,500
Lagothrix lagothricha ♂	8	8,770	♀	6	37,700
♀	4	5,740	*Homo sapiens* ♂	9*	68,230
			♀	9*	55,000

[a] An asterisk denotes that body weight is specific to the skeletal sample (i.e., all specimens have associated body weights). Body weights of the remaining nonhuman primates are compiled from the following sources: Miller, 1903; Lyon, 1908*a,b,* 1911; Hrdlicka, 1925; Pocock, 1927; Schultz, 1944, 1973; Grzimek, 1956; Bramblett, 1967; Fooden, 1963, 1976; Schaller, 1963; Rahm, 1967; Cousins, 1972; Gautier-Hion, 1975; Rothenfluh, 1976; Charles-Dominique, 1977; Hershkovitz, 1977; Petter *et al.*, 1977; Fleagle, 1978; Doyle and Martin, 1979; Niemitz, 1979c; Olson, 1979; Charles-Dominique *et al.*, 1980; Wrangham and Smuts, 1980; Stephan *et al.*, 1981; Coolidge and Shea, 1982; Hoage, 1982; Tattersall, 1982; Dechow, 1983; Jungers, 1984*a*; Jungers and Susman, 1984; Groves, Fleagle, Mittermeier, Oates, Susman, Tattersall, personal communications; records of the Duke University Primate Facility; records of the following museums: American Museum of Natural History; Museum of Comparative Zoology, Harvard University; National Museum of Natural History; Museum of Zoology, University of Michigan; British Museum (Natural History); Powell-Cotton Museum; Vandebroek's field notes, Lovain-La-Neuve. Body weights of the human specimens (all of European descent) are from the Cleveland Museum of Natural History.

1. Intermembral index: $\frac{\text{Humerus length} + \text{radius length}}{\text{Femur length} + \text{tibia length}} \times 100$

2. Humerofemoral index: $\frac{\text{Humerus length}}{\text{Femur length}} \times 100$

3. Brachial index: $\frac{\text{Radius length}}{\text{Humerus length}} \times 100$

4. Crural index: $\frac{\text{Tibia length}}{\text{Femur length}} \times 100$

In addition to these four commonly used indices, two additional ratios were calculated to further gauge the proportional differences in forelimb and hindlimb lengths:

5. Forelimb length (mm)/[body weight (g)]$^{1/3}$

6. Hindlimb length (mm)/[body weight (g)]$^{1/3}$

These indices in the smallest and largest members of closely related taxa were compared in order to evaluate whether or not significant proportional changes occur within and between the limbs with increasing size (e.g., *Colobus verus* versus *Nasalis larvatus* within colobines). By restricting these pairwise comparisons to closely related species (within a family or subfamily), much of the "phylogenetic noise" inherent in comparisons of higher level taxonomic groups can be eliminated. Proportional differences that are observed are therefore probably more fundamentally linked to body size differences (e.g., Jungers, 1979). Taxonomic groups compared in this manner include cheirogaleids, lemurids, indriids, galagids, lorisids, callitrichids, cebids, colobines, cercopithecines, hylobatids, and African apes.

Body Weight as the Size Variable

One of the most important steps in any study of size and scaling is the selection of an appropriate size variable that can be employed consistently throughout the allometric analyses at all levels of interest. All conclusions concerning isometry (geometric similarity) or the departure from isometry (allometry in the strictest sense) are valid only in reference to this specified size variable (Mosimann and James, 1979; R. J. Smith, 1980, 1981*a;* Jungers and German, 1981; Steudel, 1981; Holloway and Post, 1982). Body weight is regarded as the single most appropriate and meaningful size variable in this study for a variety of reasons. Mass is easily measured with considerable accuracy, and because the density of almost all animals is approximately 1.0, body weight (or mass) is an excellent measure of total volume, "a fact which greatly simplifies many functional considerations" (Schmidt-Nielsen, 1977, p. 4). A test of the null hypothesis of geometric similarity itself requires that the

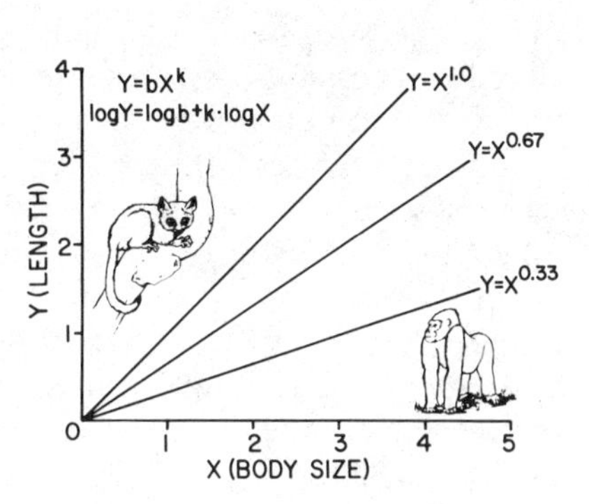

Fig. 1. Schematic bivariate plot (on logarithmic coordinates) of body size versus length. Given the volumetric nature of overall body size (body weight or body mass), linear dimensions in different-sized animals are geometrically similar when the exponent (slope) is equal to 0.33. Positive allometry and negative allometry obtain when exponential values are greater or less than 0.33, respectively. The exponential value of 0.67 represents the expectations for geometric similarity for measurements of area. Geometric similarity of other variables measured in units of volume, weight, or mass is indicated by an exponent of 1.0.

size variable bear this type of direct relationship to volume (Fig. 1). More importantly, most theoretical and empirical scale effects in animal locomotion are a primary function of body weight (Cartmill, 1974*b;* McMahon, 1975; Schmidt-Nielsen, 1975; Pedley, 1977; Alexander *et al.,* 1979, 1981). A meaningful test of theoretically inspired hypotheses or a truly commensurate comparison to other empirical studies requires that body weight be employed as the size variable.

Substitute measures for body weight (e.g., skeletal trunk length, femur length, or midshaft diameter of a long bone) often scale nongeometrically with body weight and can lead, therefore, to erroneous interpretations of skeletal allometry and proportionality. In addition, a substitute variable that scales close to isometry with body weight across a "mouse-to-elephant" size series will not necessarily exhibit an isometric relationship at lower taxonomic levels of analysis; comparisons of allometric results among different groups is further complicated as a consequence. In sum, Lindstedt and Calder (1981, p. 2) have argued cogently that "Body mass, more than any other single descriptive feature, is the primary determinant of ecological opportunities, as well as of the physiological and morphological requirements of an animal."

Bivariate Allometry and Regression Models

The widespread use and general popularity of the power formula for allometric relationships (cf. Huxley, 1932)

$$Y = bX^k$$

and its log-transformed counterpart

$$\log Y = k \log X + \log b$$

are easily appreciated. Although mathematically simple, the power function remains "the most useful descriptive tool for understanding the evolution of size" (Lindstedt and Calder, 1981, p. 2). Numerous analytical and computational advantages of this bivariate allometry model have been pointed out by various authors (Kuhry and Marcus, 1977; Wood, 1978; Harvey, 1982). An empirical fit of data to a straight line on logarithmic coordinates may be the ultimate argument for its use in many studies (R. J. Smith, 1980; Lande, 1979, and this volume); however, theoretical justification for its application come from many sources (Jungers, 1984*b*). For example, in numerous biomechanical analyses of size-related locomotor phenomena, predictive bivariate power functions have been derived from both theoretical considerations and empirical observations (e.g., Rashevsky, 1948; McMahon, 1973, 1975, 1977; Prange, 1977; Taylor, 1977; Alexander, 1980). The legitimate place of the bivariate power function in modern biology can also be traced to simplifying assumptions in the dimensional analysis of biological similarity (Lambert and Teissier, 1927; Günther, 1975; Günther and Morgado, 1982; Economos, 1982). Of equal importance, the power function is usually interpretable in biological terms, and the individual equations provide for an orderly reduction of quantitative observations to "formal statements from which patterns and parallels emerge. Ultimately we seek to identify the physical constraints and causes that explain the empirical exponents" (Calder, 1981, p. 301).

Opinions are rather polarized as to which line fitting technique is most appropriate for allometric analysis. Reasonable arguments can be made in favor of both model I (least squares) and model II (major axis, reduced major axis, Bartlett's three-group) regression methods. In model II both X and Y variables are presumed to have associated error terms and to be invertible; i.e., no assumptions are made about the dependence of either variable. In model I error variance is taken into account in the Y variable only, and X is regarded as an independent variable affecting the values of the Y variable. This last assumption is most compatible with the view that body size is a primary determinant of numerous biological characteristics, including body proportions. In other words, the X (size) and Y (e.g., limb length) terms are not logically invertible. Moreover, when the correlation coefficient between a pair of variables is reasonably high, regression models tend to produce almost identical results and have little, if any, impact on interpretations. In those cases where the choice of regression technique largely determines one's conclusions, the logical strength and biological importance of the size relationship itself should be questioned. For the sake of comparison and completion, both least squares and major axis models are employed throughout the present study.

Because another important goal of this study is to examine the pattern of departures (residuals) from various allometric baselines, the "predictive" model of least squares regression is clearly the most appropriate method (Sokal and Rohlf, 1981; Jungers, 1982, 1984*a*). The deviations of selected species from different regression lines (e.g., suspensory forms from quadrupedal baselines) allow one to assess both the direction and degree of departure from some common trend; the nature and magnitude of proportional divergence (or convergence) can be evaluated in different taxa (e.g., gibbons and spider monkeys). Following R. J. Smith (1980), positive and negative percentage deviations are computed as

$$\frac{\text{Observed value} - \text{predicted value}}{\text{Predicted value}} \times 100$$

The species that will be examined in this fashion include:

1. "Prosimians" compared to all quadrupedal taxa and to strepsirhines as a whole: *Loris tardigradus, Daubentonia madagascariensis, Lepilemur leucopus, Propithecus verreauxi, Galago senegalensis,* and *Tarsius bancanus.*
2. "Brachiators" compared to all quadrupedal taxa and to cercopithecoids as a whole: *Ateles geoffroyi, Hylobates lar carpenteri, Hylobates syndactylus.*
3. Humans (Caucasian males and females) and orangutans (*Pongo pygmaeus pygmaeus* and *Pongo pygmaeus abelli*) compared to all quadrupedal taxa and to African apes.

This approach is not intended to be equivalent to the "criterion of subtraction" concept advocated by Gould (1975; also see Sweet, 1980). Rather, all regression lines in this study are empirically derived, and points lying on or near a given line should not necessarily be regarded as "functionally equivalent." Some change in function and adaptation can accompany highly correlated changes in size, even among closely related taxa (R. J. Smith, 1980).

Sample Composition

Scaling parameters will be examined independently for various taxonomic assemblages in order to assess the generality of "interspecific" allometric trends in primates and to gauge the impact of different natural groupings on functional inferences. Such subdivisions are necessary because primates are a very heterogeneous group behaviorally, morphologically, and phylogenetically. Taxonomic groups at the level of family or subfamily approach quite closely to what Davis (1962, p. 505) has convincingly described as the "ideal situation" to study size-related or allometric phenomena, namely "comparisons between closely related organisms with similar habits and behavior, but differing substantially in size."

One must exercise extreme caution in interpreting the results of all "broad allometric" (R. J. Smith, 1980) studies as if the animals differed in size only. This observation is not intended to diminish the great significance of "mouse-to-elephant" allometric analyses (e.g., Alexander *et àl.*, 1979, 1981; Fedak and Seeherman, 1979). Such broad-spectrum studies serve to identify basic organizing principles in animal biology and allow one to test theoretically inspired "laws" of scaling. However, without simultaneous allometric analysis of more homogeneous taxonomic (and functional) subgroups, these exercises in broad allometry can actually obscure other important adaptive differences among species. In other words, allometric analysis of taxonomic and functional subunits over a narrower size range complements the broader perspective with a much finer grain resolution of size-related differences among closely related animals. This analytical strategy allows one to minimize the impact of very disparate adaptive and/or phylogenetic histories, and thereby to isolate other proportional and morphological changes that are related more fundamentally to body size differences.

Three different levels of broad interspecific allometry are explored in this study: nonhuman primates as a whole, strepsirhines, and haplorhines (without humans). Scaling trends of the limbs are also documented for a variety of more homogeneous taxonomic assemblages: cheirogaleids, lemurids, indriids, galagids, lorisids, callitrichids, cebids (with and without *Ateles*), cercopithecoids, cercopithecines, colobines, hylobatids, and African apes.

Results

Relative Limb Proportions (Table 2)

With only one exception—the cheirogaleids—interlimb proportions (as assessed by the intermembral index) of small and large members of closely related taxa exhibit significant differences. This index is identical in *Microcebus rufus* and *Cheirogaleus major.* The most consistent trend is for the intermembral index to increase significantly with increasing body size (lemurids, indriids, galagids, callitrichids, cebids, colobines, cercopithecines, hylobatids, and African apes). Only in the lorisids does this index decrease with increasing size (from angwantibos and slender lorises to slow lorises and pottos). If this index were contrasted between *Otolemur crassicaudatus* and virtually any member of the galagids (e.g., *G. senegalensis* at 51.5 or *G. moholi* at 53.9) other than *Galagoides demidoff* (or the closely related *G. thomasi*), the size-related differences in interlimb proportions in this family would be much more pronounced.

The size-related pattern of the humerofemoral index tends to parallel the trend seen in the intermembral index, but this alternative index of interlimb proportions is not as redundant with the intermembral index as Howell (1944) originally believed, at least not in primates. Small species-to-large species contrasts of the humerofemoral index are insignificant not only in cheirogaleids, but also in galagids, lorisids, and callitrichids. In the remaining cases, this index increases with increasing body size.

The brachial index also tends to exhibit size-related differences within primate families and subfamilies. These differences are significant in all but three cases, the indriids, lorisids, and hylobatids. This index decreases as size increases in most of the remaining groups: cheirogaleids, lemurids, galagids, cebids, and African apes. Relative radius length increases with size, however, in callitrichids and in both subfamilies of the cercopithecids.

Significant size-related differences in the crural index of closely related taxa are present in all but one group, the African apes (Jungers, 1984*a*). In the majority of the remaining cases, this index decreases with increasing body

Table 2. Comparisons of Limb Proportions between the Smallest and Largest Members of Closely Related Taxa[a]

Group	Intermembral index $\bar{x}$ (SD)		Humerofemoral index $\bar{x}$ (SD)		Brachial index $\bar{x}$ (SD)		Crural index $\bar{x}$ (SD)		Forelimb length (mm) / [body weight (g)]$^{1/3}$	Hindlimb length (mm) / [body weight (g)]$^{1/3}$
Cheirogaleids										
Microcebus rufus (N = 6)	71.5 (0.6)	NS	72.5 (0.9)	NS	118.7 (1.9)	$p < 0.001$	121.9 (3.6)	$p < 0.001$	10.19	14.28
Cheirogaleus major (N = 10)	71.5 (1.0)		72.9 (2.0)		97.0 (2.9)		100.9 (3.1)		10.62	14.85
Lemurids										
Hapalemur griseus (N = 20)	64.4 (2.5)	$p < 0.001$	60.7 (3.0)	$p < 0.001$	110.4 (3.4)	$p < 0.001$	98.7 (2.2)	$p < 0.001$	12.69	19.79
Varecia variegata (N = 17)	72.2 (1.7)		70.6 (1.9)		96.9 (1.8)		92.4 (1.8)		12.91	17.89
Indriids										
Avahi laniger (N = 16)	57.5 (1.2)	$p < 0.001$	48.8 (1.5)	$p < 0.001$	120.7 (2.5)	NS	87.3 (1.6)	$p < 0.02$	14.25	24.80
Indri indri (N = 16)	64.3 (1.5)		54.4 (1.7)		122.9 (4.2)		88.8 (1.7)		13.75	21.40
Galagids										
Galagoides demidoff (N = 18)	67.5 (1.7)	$p < 0.01$	66.1 (2.0)	NS	114.2 (3.6)	$p < 0.001$	109.8 (2.6)	$p < 0.001$	13.25	19.60
Otolemur crassicaudatus (N = 11)	69.5 (2.1)		65.8 (1.6)		105.0 (2.8)		94.1 (1.3)		12.59	18.09
Lorisids										
Arctocebus calabarensis (N = 21)	89.0 (1.3)	$p < 0.01$	84.2 (2.1)	NS	102.2 (3.3)	NS	91.2 (2.3)	$p < 0.001$	18.28	20.60
Perodicticus potto (N = 20)	87.5 (2.2)		85.6 (3.7)		100.4 (3.7)		95.9 (3.1)		13.72	15.68
Callitrichids										
Cebuella pygmaea (N = 13)	83.2 (1.6)	$p < 0.001$	88.6 (1.6)	NS	90.6 (1.8)	$p < 0.001$	103.2 (1.3)	$p < 0.05$	12.32	15.01
Leontopithecus rosalia (N = 10)	86.7 (2.4)		89.2 (1.4)		99.9 (3.8)		105.5 (3.3)		14.06	16.10
Cebids										
Saimiri sciureus (N = 9)[b]	79.1 (1.2)	$p < 0.001$	82.4 (1.7)	$p < 0.001$	94.2 (1.5)	$p < 0.001$	102.5 (1.4)	$p < 0.001$	14.20	17.96
Lagothrix lagothricha (N = 8)[c]	97.9 (1.1)		99.8 (1.7)		89.9 (2.3)		93.7 (2.9)		15.47	15.84
Colobines										
Colobus verus (N = 3)	80.1	*	81.5	*	92.2	*	95.6	*	14.11	17.61
Nasalis larvatus (N = 6)[c]	93.6 (1.9)		87.3 (1.9)		104.8 (2.3)		90.9 (0.8)		16.12	17.22
Cercopithecines										
Miopithecus talapoin (N = 8)[b]	84.6 (1.3)	$p < 0.001$	85.2 (1.2)	$p < 0.001$	99.5 (1.4)	$p < 0.001$	101.1 (0.9)	$p < 0.001$	14.42	16.98
Papio anubis (N = 11)[c]	97.4 (2.2)		90.1 (2.0)		104.9 (2.5)		89.7 (1.8)		15.69	16.13
Hylobatids										
Hylobates lar carpenteri (N = 30)	129.7 (2.7)	$p < 0.001$	115.0 (2.7)	$p < 0.001$	110.8 (2.6)	NS	86.9 (2.1)	$p < 0.005$	28.21	21.76
Hylobates syndactylus (N = 17)	147.0 (3.3)		128.5 (4.1)		110.0 (2.8)		84.8 (2.3)		25.91	17.63
African apes										
Pan paniscus (N = 10)[b]	102.7 (2.1)	$p < 0.001$	97.6 (2.0)	$p < 0.001$	91.6 (1.2)	$p < 0.001$	82.1 (1.2)	NS	17.00	16.49
Gorilla gorilla (N = 21)[c]	115.6 (2.1)		117.2 (2.9)		80.4 (2.4)		82.9 (2.3)		14.39	12.46

[a]Intermembral index = [(humerus length + radius length)/(femur length + tibia length)] × 100; humerofemoral index = (humerus length/femur length) × 100; brachial index = (radius length/humerus length) × 100; crural index = (tibia length/femur length) × 100. Significance of mean differences indicated; NS signifies that $p > 0.10$. An asterisk indicates that sample size is too small to test for significance of differences.

[b]Females only.

[c]Males only.

size: cheirogaleids, lemurids, galagids, cebids, colobines, cercopithecines, and hylobatids. Only in indriids, lorisids, and callitrichids does relative tibia length increase in larger members of each group.

Simultaneous comparisons of relative forelimb length and relative hindlimb length (the last two columns of Table 2) between closely related species yield a pattern of interlimb proportions essentially the same as that embodied in the contrasts of intermembral indices. However, this type of comparison also allows one to isolate the limb that is changing most with increasing size. For example, in lemurids it becomes apparent that the increasing intermembral index from *Hapalemur* to *Varecia* is due to relative reduction of hindlimb length in *Varecia*. Relative reduction occurs in both forelimbs and hindlimbs of *Indri* in comparison to *Avahi*, but the relative reduction is greater in the hindlimb; hence the higher intermembral index in *Indri*. Within the lorisids relative reduction also occurs in both limbs, but in this case it is the forelimb that changes most and leads to somewhat lower intermembral indices in the larger species. By contrast, both limbs of the callitrichids are relatively elongated in the largest species, *Leontopithecus;* the relatively greater elongation of the forelimb results in a higher intermembral index in comparison to *Cebuella*. In cebids, relative forelimb elongation occurs in conjunction with relative hindlimb reduction. In colobines and cercopithecines slight to moderate relative hindlimb reduction is coupled with substantial relative forelimb elongation. In hylobatids and African apes, relative reduction can be seen in both forelimbs and hindlimbs, but hindlimb reduction is more pronounced. The impact of these disparate trends on the intermembral index in cebids, colobines, cercopithecines, hylobatids, and African apes is similar, nevertheless; the forelimb increases in length relative to the hindlimb in all five of these groups as body size increases.

Scaling of Interlimb Proportions (Table 3)

As Fig. 2 illustrates, the overall trend among nonhuman primates is for the intermembral index to increase with increasing body size. It is also apparent from this plot that hylobatids and orangs possess unusually high intermembral indices for their respective body sizes, whereas indriids are characterized by especially low values for their size. The allometric basis of this general trend is summarized in Table 3. Within nonhuman primates as a whole, forelimb length scales in a positive allometric fashion ($k > 0.33$) with body size, whereas hindlimb length is characterized by negative allometry ($k < 0.33$). Allometric trends within haplorhines as a group parallel those of primates as a whole; i.e., positive forelimb allometry coupled with negative hindlimb allometry. Although the intermembral index does not increase with body size increases within strepsirhines as a whole (e.g., *Microcebus rufus* at 71.5 versus *Indri indri* at 64.3), forelimb length does scale slightly faster than does hindlimb length in this group. Strepsirhine forelimb length exhibits

Table 3. Scaling of Forelimb Length and Hindlimb Length in Nonhuman Primates[a]

Group	Forelimb length					Hindlimb length				
	b	k (SE of k)		MA	r	b	k (SE of k)		MA	r
Nonhuman primates	9.76	0.386	(0.008)	0.389	0.971	24.00	0.296	(0.006)	0.297	0.974
Strepsirhines	11.79	0.345	(0.015)	0.348	0.971	18.37	0.336	(0.016)	0.339	0.964
Haplorhines	11.81	0.368	(0.012)	0.372	0.954	27.30	0.281	(0.008)	0.282	0.966
Cheirogaleids	9.12	0.366	(0.022)	0.367	0.995	11.99	0.381	(0.030)	0.382	0.991
Lemurids	12.41	0.336	(0.019)	0.337	0.988	33.14	0.255	(0.024)	0.256	0.970
Indriids	15.31	0.317	(0.048)	0.319	0.978	37.90	0.267	(0.031)	0.268	0.987
Galagids	12.44	0.335	(0.021)	0.336	0.984	25.12	0.293	(0.029)	0.295	0.963
Lorisids	59.04	0.123	(0.018)	0.123	0.978	61.56	0.136	(0.018)	0.136	0.981
Callitrichids	9.33	0.394	(0.047)	0.397	0.972	11.36	0.402	(0.058)	0.406	0.960
Cebids (with *Ateles*)	9.43	0.392	(0.025)	0.396	0.966	30.04	0.266	(0.022)	0.268	0.944
Cebids (without *Ateles*)	10.90	0.372	(0.017)	0.373	0.984	33.38	0.251	(0.018)	0.252	0.960
Cercopithecoids	10.79	0.373	(0.014)	0.375	0.977	20.52	0.318	(0.017)	0.321	0.951
Cercopithecines	10.99	0.372	(0.014)	0.373	0.984	19.91	0.318	(0.016)	0.320	0.974
Colobines	8.78	0.395	(0.048)	0.404	0.926	30.72	0.278	(0.055)	0.287	0.836
Hylobatids	40.68	0.284	(0.093)	0.295	0.807	165.50	0.092	(0.066)	0.094	0.526
African apes	48.75	0.230	(0.021)	0.231	0.966	111.97	0.148	(0.022)	0.148	0.916

[a] $Y = bX^k$ (X in grams, Y in millimeters); SE of k denotes the standard error of the least squares estimate of k; MA is the major axis estimate of k; r is the correlation coefficient.

slightly positive allometry, whereas hindlimb length is isometric with body size. A comparison restricted to cheirogaleids and indriids only would suggest very different scaling trends (i.e., positive allometry in *both* limbs) because both the forelimbs and hindlimbs of cheirogaleids are relatively short (cf. last two columns of Table 2).

Within more homogeneous taxonomic assemblages of strepsirhines, a number of quite disparate interlimb allometric patterns are evident. Both

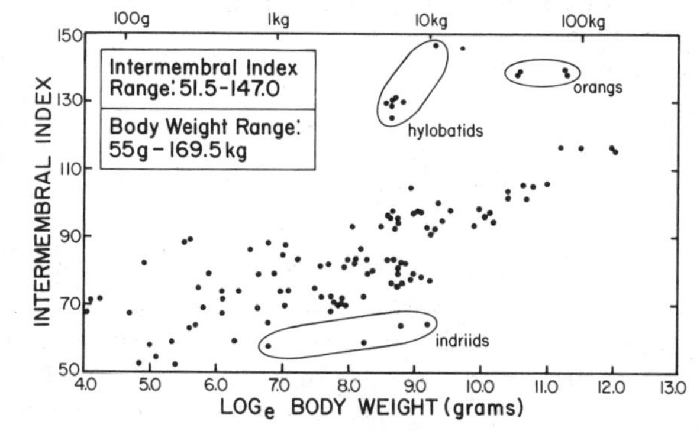

Fig. 2. Relationships between body weight and intermembral index in adult nonhuman primates. The overall trend of relative forelimb elongation with increasing body size is readily apparent. At least three groups of outliers are also obvious; hylobatids and orangutans possess unusually high intermembral indices, whereas indriids exhibit relatively low ratios of forelimb length to hindlimb length.

forelimb and hindlimb lengths exhibit positive allometry in cheirogaleids, but the hindlimb scales slightly faster than does the forelimb. In contrast to cheirogaleids, the forelimbs of both lemurids and galagids scale isometrically, whereas their hindlimbs are characterized by moderately to relatively strong negative allometry. Both indriids and lorisids are characterized by negative allometry of both limbs, but details of interlimb allometry in the lorisids are very different from those seen in the other two groups. Pronounced negative allometry occurs in both the forelimb and hindlimb of lorisids, slightly more in the forelimb than in the hindlimb. In indriids and galagids, the negative allometry observed in the forelimb is relatively slight, especially in comparison to the stronger negative scaling seen in their hindlimbs (but which does not approach the condition seen in the lorisids).

The clawed New World monkeys, the callitrichids, exhibit positive allometry of both forelimbs and hindlimbs. Cebids (both with and without *Ateles*) are characterized instead by positive allometry of the forelimbs in conjunction with negative allometry of the hindlimbs. Scaling trends of the limbs of both cercopithecoids as a whole and within cercopithecines alone are very similar; positive allometry is seen in the forelimb, whereas the hindlimb exhibits only slightly negative allometry. Within colobines, the positive allometry observed for the forelimb is more pronounced than that seen in the cercopithecines; the negative allometry of the colobine hindlimb is also more exaggerated. The lower correlation coefficients seen for both limbs of colobines in comparison to cercopithecines possibly suggest that the allometric trends in the extremities of leaf-eating Old World monkeys contain relatively more "noise"; in other words, substantial variation in limb length occurs at similar body weights in different colobine species. This interpretation is complicated by the fact that correlation coefficients are sensitive to the absolute ranges of X and Y variables (narrower ranges lead to lower correlations), and the colobine range of values is somewhat constricted in comparison to cercopithecines (R. J. Smith, 1981*b*). The forelimb of hylobatids scales negatively with body weight, as does the hindlimb. However, relative hindlimb reduction from gibbon to siamang is especially pronounced (despite the relatively weak association between size and hindlimb length across hylobatids). Both forelimbs and hindlimbs of African apes are also characterized by negative allometry, with the exponent of the hindlimb significantly less than that of the forelimb.

Scaling of Intralimb Proportions (Table 4)

Only the more natural groupings of closely related taxa are considered in this part of the analysis (i.e., families and subfamilies). Elements of the forelimb will be considered first. The positive allometry noted for the cheirogaleid forelimb is due to strong positive allometry of humerus length in conjunction with isometry of radius length. Positive allometry of the humerus combine

Table 4. Scaling of Limb Bone Lengths in Nonhuman Primates[a]

Group	Humerus				Radius				Femur				Tibia			
	b	k (SE of k)	MA	r	b	k (SE of k)	MA	r	b	k (SE of k)	MA	r	b	k (SE of k)	r	
Cheirogaleids	3.64	0.402 (0.020)	0.403	0.996	5.67	0.331 (0.027)	0.332	0.990	4.69	0.420 (0.029)	0.420	0.993	7.53	0.346 (0.033)	0.347	0.987
Lemurids	4.69	0.369 (0.029)	0.371	0.979	8.12	0.305 (0.017)	0.306	0.989	14.83	0.272 (0.027)	0.273	0.966	18.58	0.238 (0.022)	0.239	0.971
Indriids	6.87	0.322 (0.033)	0.322	0.990	8.48	0.313 (0.066)	0.315	0.958	20.76	0.264 (0.028)	0.264	0.989	17.22	0.271 (0.035)	0.271	0.984
Galagids	5.57	0.348 (0.020)	0.349	0.987	7.08	0.317 (0.022)	0.319	0.981	11.17	0.319 (0.041)	0.323	0.940	13.92	0.271 (0.018)	0.271	0.983
Lorisids	25.33	0.145 (0.018)	0.146	0.984	35.06	0.098 (0.029)	0.098	0.921	32.93	0.129 (0.024)	0.129	0.967	28.87	0.142 (0.021)	0.142	0.979
Callitrichids	5.41	0.377 (0.044)	0.378	0.973	3.94	0.415 (0.056)	0.419	0.965	5.65	0.400 (0.061)	0.405	0.957	5.73	0.403 (0.056)	0.407	0.962
Cebids (without *Ateles*)	5.64	0.373 (0.018)	0.375	0.981	5.25	0.371 (0.017)	0.372	0.983	14.46	0.273 (0.019)	0.275	0.962	19.42	0.228 (0.018)	0.229	0.955
Cercopithecines	6.23	0.357 (0.011)	0.358	0.989	4.79	0.388 (0.018)	0.390	0.977	8.47	0.339 (0.014)	0.341	0.981	11.61	0.298 (0.017)	0.300	0.964
Colobines	5.27	0.374 (0.036)	0.378	0.953	3.72	0.414 (0.069)	0.434	0.873	15.90	0.278 (0.053)	0.287	0.484	15.41	0.273 (0.057)	0.282	0.823
Hylobatids	15.82	0.305 (0.108)	0.322	0.783	24.97	0.267 (0.081)	0.275	0.829	64.49	0.128 (0.075)	0.132	0.604	100.67	0.062 (0.062)	0.063	0.406
African apes	16.46	0.272 (0.028)	0.274	0.960	39.23	0.181 (0.021)	0.181	0.951	55.95	0.156 (0.023)	0.156	0.921	56.98	0.137 (0.025)	0.138	0.890

[a] b, k, SE, MA, and r as in Table 3.

with slightly negative allometry of the radius to produce forelimb isometry in both lemurids and galagids. The negative allometry exhibited by the indriid forelimb is due more to the negative scaling of radius length than to allometry of humerus length; the latter is almost isometric with body size. The strong negative allometry seen in the lorisid forelimb is due to negative scaling of both humeral and radial elements, but the negative allometry of the radius is very pronounced.

The increase in the value of the brachial index and the overall positive allometry of the callitrichid forelimb are due to positive allometry in both the humerus and radius, especially so in the latter element. Within cebids (without *Ateles*), the positive allometry characteristic of the forelimb is due to almost equivalent scaling trends in both humerus and radius lengths. In both cercopithecines and colobines the radial element contributes more to the positive allometry of forelimb length than does the humeral element; both elements do scale positively, however, in both groups. Negative allometry of the radius influences overall forelimb negative allometry in hylobatids much more than does the allometry seen in the humerus (which is also negative). The substantial decline in the brachial index with size and the overall negative allometry of the forelimb in African apes is the result of especially negative scaling of the radius in conjunction with moderately negative allometry of humerus length.

Positive allometry of the cheirogaleid hindlimb is due primarily to the strong positive scaling of the femur; tibia length shows only slightly positive allometry. The much lower crural index of *Cheirogaleus major* in comparison to *Microcebus rufus* or *M. murinus* is a reflection of these differing allometric rates of the hindlimb elements. Both femur and tibia lengths scale negatively in lemurids, but the trend is stronger in the tibia (again resulting in lower crural indices in the larger species such as *Varecia*). The slight increase in the crural index of the *Avahi* to *Indri* size series results from negative allometry in both femoral and tibial elements, with the tibial exponent only slightly greater than the femoral. Moderately negative scaling of the femur coupled with much stronger negative allometry of tibia length results in overall negative allometry of the hindlimb and lower crural indices with increasing size in galagids. Both hindlimb elements scale negatively to a marked degree in lorisids, but the crural index increases with size because this negative trend is slightly less exaggerated in the tibia.

Positive scaling of the hindlimb in callitrichids is due to roughly equal (positive) allometric trends in femoral and tibial lengths. In contrast, the negative allometry of the cebid hindlimb results from moderately negative scaling of femur length in conjunction with more pronounced negative allometry of tibia length (and hence the decrease in crural indices with increasing size). The femur of cercopithecines is virtually isometric, whereas the tibia exhibits negative allometry. In colobines both femoral and tibial elements scale negatively. In both of the cercopithecoid groups the difference in femoral versus tibial allometry results in lower crural indices in the larger species of each group. The very strong negative allometry noted for the hylobatid hindlimb is related to negative allometry of both bony elements but especially to the marked negative scaling of tibia length. The pronounced negative allometry of the entire hindlimb observed in African apes is due to strong negative scaling of both femoral and tibial lengths.

Percentage Prediction Errors in Selected Taxa

The forelimbs and hindlimbs of six prosimian species are plotted in Fig. 3, using their percentage deviations from quadrupedal primate baselines (A and C) and from strepsirhine baselines (B and D) as bivariate coordinates. Three of the four "leapers" possess forelimbs shorter than predicted for their body size (i.e., prediction errors are negative with respect to both forelimb baselines): *Lepilemur leucopus, Propithecus verreauxi* (sifaka), and *Galago senegalensis* (galago). *Tarsius bancanus,* the fourth leaper par excellence, deviates from its strepsirhine counterparts by possessing forelimbs that are slightly longer than expected (prediction errors are positive). *Daubentonia madagascariensis* (the aye-aye) possesses forelimbs approximately as long as expected for a strepsirhine of its size (−2.2%), but which are relatively short compared to all quadrupedal primates (−12.9%). The forelimbs of *Loris tardigradus* (the slender loris) are relatively very long in comparison to either all quadrupeds (+35.5%) or to strepsirhines alone (+43.2%).

All four leaping prosimians are characterized by positive predition errors for their hindlimbs. *Tarsius* possesses much longer hindlimbs than expected (+40.0%, +37.5%), as does *Galago senegalensis* (+22.9%, +19.1%). The hindlimbs of the sifaka are relatively long compared to all quadrupedal species (+20.8%) but only moderately elongated when contrasted to strepsirhines alone (+9.4%). *Loris tardigradus,* a species not well known for its jumping prowess, has hindlimbs that are relatively as long as or longer than those of

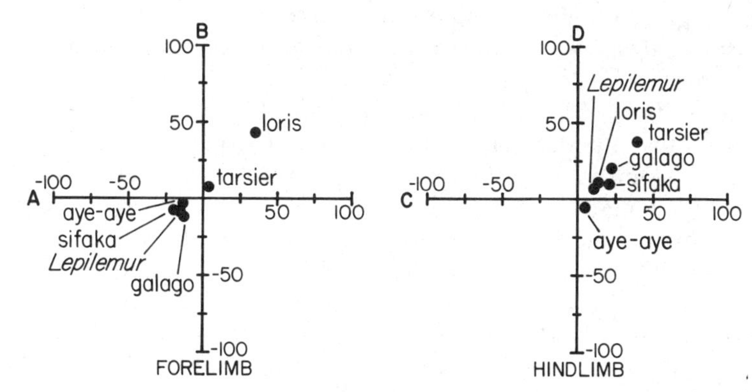

Fig. 3. Percentage prediction errors of the limbs for six prosimian taxa with respect to baselines of quadrupedal primates as a group (A, C) and to all strepsirhines (B, D). The species represented in this fashion include *Tarsius bancanus* (tarsier), *Loris tardigradus* (loris), *Daubentonia madagascariensis* (aye-aye), *Propithecus verreauxi* (sifaka), *Galago senegalensis* (galago), and *Lepilemur leucopus.*

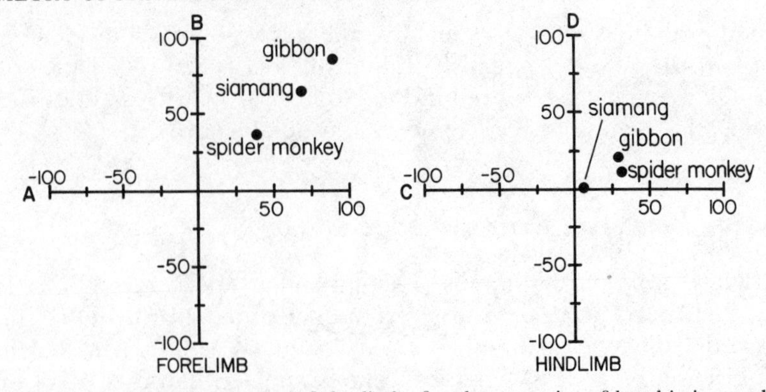

Fig. 4. Percentage production errors of the limbs for three species of brachiating anthropoids with respect to baselines of quadrupedal primates as a group (A, C) and to cercopithecoids (B, D). Gibbon refers to *Hylobates lar carpenteri*, siamang to *Hylobates syndactylus*, and spider monkey to *Ateles geoffroyi*.

Lepilemur (+12.6%, +8.6% versus +11.5%, +5.7%). The hindlimb of the aye-aye is either relatively long (+4.5% from quadrupedal line) or relatively short (−4.7% from strepsirhine line), depending on the baseline of comparison.

As one might expect for those primate species particularly adept at brachiation and other bimanual positional activities, the gibbon (*Hylobates lar carpenteri* in this case), the siamang, and the spider monkey (*Ateles geoffroyi*) all possess forelimbs that are much longer than expected for their respective sizes (Fig. 4). This elongated condition is most pronounced in the gibbon (+88.8%, +85.5%) and least marked in the spider monkey (+38.5%, +36.1%). The baselines in this set of comparisons are quadrupedal primates as a whole (A, C) and all cercopithecoids (B, D). The relatively long arms of both gibbons and spider monkeys tend to obscure the fact that their respective hindlimbs are also relatively long (e.g., gibbon values are +28.5%, +21.0%). The hindlimb of the siamang is slightly longer than expected when compared to all quadrupeds, but it is almost exactly as long as one would predict for a cercopithecoid of its body size.

In the contrast of orangutans and humans (Fig. 5), the baselines of comparison are quadrupedal primates as a whole (A, C) and the African apes alone (B, D). Compared to the African ape trend, orangutan forelimbs are relatively quite long (ranging from +14.7% for male *Pongo pygmaeus pygmaeus* to +20.2% for female *Pongo pygmaeus abelli*). Relative to the expectations of the quadrupedal primate baseline, however, it is only the female orangs that possess especially long forelimbs (+22.3% and +23.2% for *pygmaeus* and *abelli*, respectively). Humans possess relatively short forelimbs with respect to both baselines: females (−14.2%, −11.8%), males (−12.7%, −7.5%). Orangutan hindlimbs are somewhat shorter than expected (e.g., male *Pongo pygmaeus abelli* at −6.4%) relative to the African ape baseline but very short compared to all quadrupeds (e.g., male *pygmaeus* at −21.1%). Humans exhibit the reverse condition, with relatively very long hindlimbs regardless of the comparative baseline (Jungers, 1982).

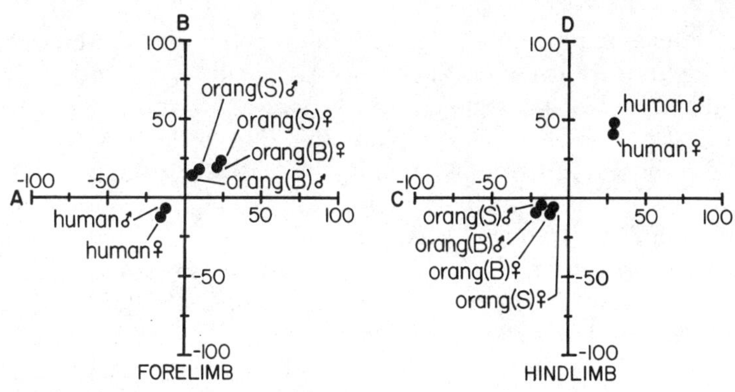

Fig. 5. Percentage prediction errors of the limbs for humans and orangutans with respect to baselines of quadrupedal primates as a group (A, C) and to African apes (B, D). The humans are all of European descent. Borneo (*Pongo pygmaeus pygmaeus*) and Sumatra (*Pongo pygmaeus abell*) orangutans are treated separately (B and S, respectively).

Discussion

Overall Allometric Trends in Primate Limb Proportions

As Fig. 2 and the results of Table 3 indicate, interlimb proportions change with increasing size in nonhuman primates such that relative forelimb length increases at a faster rate than relative hindlimb length. In other words, there is a pervasive trend for the intermembral index to increase with increasing body size (also see Biegert and Maurer, 1972; Rollinson and Martin, 1981; Dykyj, 1982). Aiello's (1981*b*) bivariate plots also demonstrate that forelimb length is related in a positive allometric fashion to hindlimb length in anthropoid primates, and Biegert and Mauer (1972) argued earlier that both forelimb and hindlimb lengths of catarrhine primates scale positively with size (skeletal trunk length) but forelimb length increases relatively faster than does hindlimb length. Trunk length itself is now known to scale negatively with body mass (Aiello, 1981*b;* Jungers, 1984*a*) and conclusions about the specifics of limb allometry in catarrhines have been modified accordingly [i.e., forelimb length is virtually isometric, whereas hindlimb length scales negatively with body mass (Jungers, 1984*a*)]. The reason forelimb length scales positively relative to hindlimb length in higher primates (haplorhines in this study) and for nonhuman primates as a whole is because forelimb length scales positively with body size, whereas hindlimb length scales negatively. The largest primates generally possess absolutely and relatively longer forelimbs than smaller primates, but their hindlimbs are relatively shorter (ignoring cheirogaleids for the moment, which possess relatively short forelimbs *and* hindlimbs).

Whereas Howell (1944) and Hildebrand (1952, 1961) essentially dismiss the intermembral index as a nearly "worthless" tool in the functional analysis of the mammalian locomotor skeleton [e.g., "no association is apparent between the intermembral index and habits or body size" in didelphid marsupials (Hildebrand, 1961, p. 247)], one can argue instead that the size-related increase in this index in primates follows logically and causally from basic biomechanical principles related to climbing. The order Primates is fundamentally an arboreal radiation (Cartmill, 1972, 1974*a*), and all primates with access to trees remain adept climbers. Due to their characteristic lack of claws (excluding callitrichids and selected strepsirhines), the negotiation of large vertical supports by primates entails special problems relating to the generation of adequate levels of manual and pedal friction (Cartmill, 1974*b*, 1979). This basic problem leads to seriously conflicting mechanical constraints: the necessity to generate adequate pedal friction to avoid slipping during clinging and climbing versus the desirability of minimizing the magnitude of rotatory torques away from the support at each contact point. The "solution" to this dilemma, including its allometric corollary, is for larger, clawless primates within a size series to possess forelimbs that increase in length relatively faster than their hindlimbs (Cartmill, 1974*b*; Jungers, 1977, 1979, 1984*a*). Negative allometry of the hindlimb in particular allows a larger individual to "lean away" from the support in order to optimize pedal friction and stability without inordinately increasing the moment arm of body weight (Jungers, 1978), which will tend "to pluck the animal's extremities away from the surface of the support" (Cartmill, 1979, p. 506). In sum, the mechanical prerequisites of climbing in primates should promote the observed increase in intermembral index with increasing body size in order to maintain *functional competence* in scansorial activities. Clearly, different phylogenetic starting points in different primate taxa and disparate adaptive strategies in different groups introduce appreciable noise into this size-related trend, but the overall fit of the observed results to the predictions of the model are compelling. Just how good this fit is in more homogeneous primate assemblages will also be considered group by group, but first it is necessary to consider competing explanations for the general trend.

Following Napier (1967), Rollinson and Martin (1981) believe that the trend for increasing intermembral indices with increasing body size is probably "related to the interaction between primate body size and branch size in an arboreal habitat" (p. 413). Small-bodied quadrupedal primates will tend to progress above the branch because even with side-to-side oscillations in the position of their center of gravity, it will usually remain above the support plane upon which they are moving. These lateral deviations in the center of gravity are relatively greater in larger bodied primates moving quadrupedally on similarly sized supports (the center of gravity is more likely to fall out of the support plane) and result in greater instability; hence, below-branch suspensory locomotion is preferable. Although these constraints on stability of arboreal travel probably account for the frequent trend to move the center of gravity relatively closer to the support in larger arboreal quadrupeds (Grand, 1968; Fleagle, 1977; Jungers, 1979; Jungers and Fleagle, 1980; Thorington and Heaney, 1981), especially by relative shortening of the distal segment of each limb (lower brachial and crural indices), such an explanation does not account for relative elongation of the forelimb in those groups in which larger members are no more (or possibly less) suspensory than their smaller counterparts (e.g., lemurids, indriids, galagids, cercopithecines, hylobatids, and African apes). Rollinson and Martin (1981) were well aware that this explanation lacked generality, and noted that a different functional basis must be invoked to explain the same trend in terrestrial cercopithecines (see pp. 372–373).

Aiello (1981*a*) has speculated that relative hindlimb reduction and an increasing intermembral index with size might impart some biomechanical advantage in terrestrial locomotion (e.g., in African apes) if these proportional changes are "connected with greater weight support on the forelimbs" (p. 85). Although it is not entirely clear what is meant by "more efficient weight carriage and movement in terrestrial locomotion" (p. 79), which is supposedly due to this alteration in interlimb proportions, the data on weight distribution between forelimbs and hindlimbs in primates contradict the suggestion that higher intermembral indices will necessarily lead to more weight being borne by the forelimbs during terrestrial locomotion. Forceplate data demonstrate that during quadrupedal walking, the forelimb support factors (the vertical impulse acting on a limb divided by the product of the subject's weight and the gait cycle duration) as well as the forelimb contribution to propulsion do *not* increase with relatively longer forelimbs in primates (Kimura *et al.*, 1979; Reynolds, 1981). For example, the fraction of body weight supported by the forelimbs during walking is roughly similar in *Lemur, Macaca, Erythrocebus, Ateles,* and *Pan* (all less than 50%) despite grossly different interlimb proportions. In addition, it should be noted that with increasing forward speed, both peak vertical forces on the forelimb and the forelimb support factors actually *decrease* in chimpanzees, presumably due to the activity of extrinsic muscles of the hindlimb (Reynolds, 1981).

Fleagle (this volume; Fleagle and Mittermeier, 1980) offers an alternative "ecological" explanation for increasing intermembral indices with size in Surinam monkeys wherein leaping activities (and relatively long hindlimbs) characterize the smaller species but suspensory behaviors (and relatively long forelimbs) are predominant among the larger species. A decrease in the frequency of leaping activities is presumed to be causally linked to a relative decrease in hindlimb length, whereas an increase in suspensory positional behaviors should be accompanied by relatively longer forelimbs. It should be pointed out, however, that relative hindlimb length [as a ratio of (body weight)$^{1/3}$] is longer in *Ateles geoffroyi* (20.02) than in *Saimiri* (females at 17.96) and almost as long as in *Pithecia pithecia* (21.58). Mittermeier's (1978; Mittermeier and Fleagle, 1976) study of the locomotor behavior of *Ateles geoffroyi* indicates that leaping remains an important component of locomotion during travel (11.4% of bouts). The relatively long arms of *Ateles* are clearly associated with an increase in all types of suspensory activities, especially brachiation, but also including vertical climbing of the nature inherent in Cartmill's model. When *Ateles* climbs relatively large supports, "the monkey extends its arms as far as possible around the trunk, presses its feet against the support, and moves up in a cross extension pattern" (Mittermeier, 1978, p. 167). The point here is that the interlimb proportions of *Ateles* (as well as those of *Lagothrix* and *Alouatta*) conform exceedingly well to the climbing hypothesis; i.e., vertical climbing is facilitated by their higher intermembral indices. Moreover, a decrease in the frequency of leaping behaviors is not correlated with relative hindlimb reduction in cheirogaleids (*Cheirogaleus major* leaps less frequently but has relatively longer hindlimbs than does *Microcebus*). Conversely, even when leaping remains an integral part of the locomotor repertoire of the largest member of a closely related group of species (e.g., *Indri indri* within the Indriidae or *Varecia variegata* within the Lemuridae), relative hindlimb reduction can and does occur in order to maintain a high level of competence or to improve performance in vertical climbing (see below in the group-specific discussions).

Arboreal mammals with claws form a natural control group against which the predictions of the vertical climbing model can be assessed. Because clawed mammals that live in the trees do not face the same mechanical constraints as primates with regard to friction and effective support diameter (Cartmill, 1974*b*), a size-related increase in intermembral index is not a prerequisite in order to maintain competence in climbing. It is perhaps not surprising, then, that Hildebrand (1961) discerned no size-related trend in interlimb proportions in didelphid marsupials; rather, "species that differ most in body proportions tend to be of similar body size" (p. 249). Jungers (1980) discovered that larger bodied sloths tended to possess lower intermembral indices than smaller species. The excellent study by Thorington and Heaney (1981) is especially relevant here. Despite significant differences in the body size of different species of tree squirrels, the intermembral index failed to exhibit any consistent size-related trend (e.g., compare *Microsciurus* at 0.77 to *Ratufa* at 0.74–0.76). It is also interesting to note that equivalent levels of negative allometry occurred in both forelimbs and hindlimbs, and that both the crural and brachial indices decreased with increasing body size in this same size series. These trends serve to move the large animals' center of gravity relatively closer to the support. As expected, the evidence on arboreal, clawed mammals does not reveal a size-related tend in interlimb proportions that is analogous to that seen in primates.

It should be stressed that Cartmill's model is specific for *arboreal mammals*. Other circumstances and biomechanical factors could conceivably also promote an increase in intermembral index with increasing size in a terrestrial context. As judged from the study by Alexander *et al.* (1979), the intermembral index appears to increase with size in both fissiped carnivores and bovids. Allometric details of this trend differ in the two groups, however; positive allometry dominates the proportional changes in fissipeds, whereas negative allometry prevails in all limb segments of the bovids. The negative allometric trends in the limbs of this specific group of bovids may partially account for the overall decrease in maximum running speed from the small Thompson's gazelle (81 km/h) to the large African buffalo (57 km/h) (Garland, 1983). The assumption here is that high-speed terrestrial travel is facilitated by relative elongation of the extremities, especially the distal segments, inasmuch as the resultant mechanical advantages (velocity ratios) would favor speed over force output (J. M. Smith and Savage, 1956; Alexander, 1968; Hildebrand, 1974). In view of the overall positive allometry of the limbs in Alexander's fissiped group, one might therefore expect an improved cursorial adaptation in the largest members of the group. Unfortunately, data on maximum velocity are available (Garland, 1983) for only the three largest species of Alexander's fissiped sample (*Canis familiaris, Crocuta crocuta,* and *Panthera leo*), but it seems highly probably that maximum speed in the very short-limbed species such as the ferret and mongoose is less than that of the larger members of this sample-specific group. Regardless, the speed-related argument does not address why the forelimb should necessarily scale faster than the hindlimb in terrestrial assemblages. Although the allometric basis of the interlimb trend cannot be extracted from Hildebrand's (1952) study of body proportions of the Canidae, the intermembral index again exhibits an overall tendency to increase with increasing body size (i.e., the slope of intermembral index on body mass is positive and the correlation is significant at 0.70). There is also no discernible relationship between intermembral index and differential body weight distribution on the forelimb versus hindlimb in terrestrial mammals [data summarized by Rollinson and Martin (1981)]. At the moment we lack a unifying biomechanical theory that accounts for this prevalent trend in nonprimate terrestrial mammals.

Group-Specific Aspects of Body Size and Limb Proportions

Cheirogaleids

As noted previously, cheirogaleids as a group possess the relatively shortest forelimbs and hindlimbs of the nonhuman primates. Only *Gorilla* is characterized by a hindlimb that is relatively shorter. Both limbs become relatively longer as body size increases from *Microcebus* to *Phaner* and *Cheirogaleus major,* with hindlimb length increasing slightly faster than forelimb length (*Phaner* possesses the lowest intermembral index at 67.4). The elongated tarsus of cheirogaleids serves to lengthen the hindlimb considerably (Jouffroy and Lessertisseur, 1978, 1979). When this element is added to femur and tibia length, however, the hindlimb of most cheirogaleids is still relatively short in comparison to most other strepsirhines and many haplorhines. Because the calcaneus and navicular of *Microcebus* are relatively longer than those of *C. major,* this composite measure of hindlimb length suggests some degree of tradeoff such that relative hindlimb length in *Microcebus* is only slightly less than that in *C. major.* One cannot appeal to enhanced leaping abilities to explain the relatively longer hindlimbs of *C. major* in comparison to *Microcebus* because *C. major* leaps less frequently (Petter *et al.,* 1977; Walker, 1979; Tattersall, 1982). Such an explanation probably does pertain to *Phaner.* They are frequent and adept leapers (Petter *et al.,* 1971; Tattersall, 1982) and possess the absolutely and relatively longest hindlimb and tarsus of the cheirogaleids.

Phaner is able to climb on very large vertical tree trunks without relative forelimb elongation because of their clawlike nails (Petter *et al.*, 1971; Martin, 1972). The previously reported negative allometry of limb length in cheirogaleids (Jungers, 1978) was with respect to trunk length; this latter measure is now known to be relatively long in this strepsirhine family. This once again points out the problems inherent in using this type of substitute measure for body weight. Relative trunk length varies greatly among different primate taxa and this variation itself is clearly of locomotor significance (Cartmill and Milton, 1977; Preuschoft, 1978; Jungers, 1984*a*).

Lemurids

The size-related trends in limb proportions documented for the Lemuridae are an excellent fit to the expectations of the climbing model. The intermembral index increases with body size from *Hapalemur* to *Varecia* via positive allometry of the forelimb and negative allometry of the hindlimb; forelimbs become absolutely and relatively longer, whereas the hindlimbs become relatively shorter. *Contra* Jungers (1979), *Varecia* remains a frequent and agile leaper despite relative reduction of hindlimb length (Petter *et al.*, 1977; Tattersall, 1982). The higher intermembral index of *Varecia* should enhance this species' ability to climb on vertical supports, and theory is matched nicely by practice in his case because the frequency and capability of climbing such supports is known to increase in *Varecia* (Petter *et al.*, 1977). Relative hindlimb reduction in *Varecia* may also increase the stability of quadrupedal travel on horizontal branches.

Indriids

Vertical clinging, climbing, and leaping are the staples of indriid positional behavior (Petter *et al.*, 1977; Walker, 1979; Tattersall, 1982). Allometric changes in interlimb proportions (increasing values of the intermembral index) from *Avahi* to *Indri* serve to preserve locomotor and postural competence on vertical supports as body size increases. In addition, by virtue of these scaling trends *Indri* is known to be able to utilize vertical supports of much larger diameter than is either *Avahi* or *Propithecus* (Petter *et al.*, 1977). These proportional modifications in indriids are effected by slightly negative allometry of forelimb length coupled with considerably stronger negative allometry of hindlimb length. Despite this relative reduction in the hindlimb length, *Indri* still possesses one of the longest hindlimbs (relative to body size) among all primates (cf. Table 2) and remains a frequent and spectacular leaper (Petter *et al.*, 1977; Tattersall, 1982). In both lemurids and indriids, size-related proportional alterations related to the maintenance of adequate performance in vertical climbing and clinging appear to have a higher adaptive priority than the preservation of complete functional equivalence in leaping ability. The proportional endpoints and the allometric pathways in each of these two Malagasy families are not identical because the structural starting points and overall locomotor repertoires are not the same. The allometric details of this analysis should replace those published in Jungers (1979).

Galagids

Interlimb scaling in galagos was seen to be similar in many respects to that observed in lemurids, i.e., virtual isometry of forelimb length coupled with negative allometry of hindlimb length. Without *Galagoides demidoff* and *G. thomasi,* however, the scaling patterns would recall those of the cebids; the forelimb exponent would be 0.361 (r = 0.984) and the hindlimb exponent would be 0.237 (r = 0.975). The increase in intermembral index from *G. demidoff* to *O. crassicaudatus* is significant but slight; the difference between *G. moholi* (53.9) or *G. senegalensis* (51.5) is much more pronounced. Relative length of the tarsus is almost as great in *G. senegalensis* as it is in *G. demidoff;* the relative length of the tarsus of *O. crassicaudatus* is reduced in comparison to all other galagos, but is relatively longer than in all other primates (including *Phaner*) except *Tarsius.* Interlimb allometry makes for a nice functional story in a *G. moholi* or *G. zanzibaricus* to *O. crassicaudatus* size series, but is much more complicated when the low end of the range includes *G. demidoff* and *G. thomasi.* Although leaping remains an important activity in both *G. demidoff* and *O. crassicaudatus* (Charles-Dominique, 1977; Doyle and Bearder, 1977), these two groups are reported to be the galago species most committed to quadrupedalism. Climbing does become a more important component of the greater bush baby's locomotor repertoire (Doyle and Bearder, 1977), and average substrate diameter does increase (Nash, 1983), but horizontal supports are still preferred by *O. crassicaudatus* (Doyle and Bearder, 1977; Nash, 1983). The relatively lower center of gravity afforded by the negative scaling of the galago hindlimb may serve to increase the stability of quadrupedal travel on horizontal supports by *Otolemur* (see above discussion of lemurids). *Galago senegalensis,* with its relatively very long hindlimb, is a superb leaper (Doyle and Bearder, 1977); the same relationship holds for *G. zanzibaricus* (Nash, 1983). *Galago elegantulus* can feed on the exudates of relatively large tree trunks by virtue of its clawlike nails (Charles-Dominique, 1977).

Lorisids

Compared to the size-related changes in limb proportions seen in all other taxonomic assemblages of primates, those observed in the lorisids are the most deviant. Relatively extreme reduction occurs in both limbs between the endpoints of the size dichotomy of *Arctocebus/Loris* versus *Nycticebus/Perodicticus.* For their size, the forelimbs of the slender loris and the angwantibo are exceptionally long; only hylobatids and female orangutans can match them. The hindlimbs of these small-bodied lorisids are also relatively long; in fact, they match the condition seen in a number of leaping primates (e.g., the sifaka, *Lepilemur, Hapalemur*). Although the hindlimbs are relatively shorter in the potto and slow loris, they are nevertheless relatively longer than in most cheirogaleids, some callitrichids, and the largest great apes. The forelimbs of these large-bodied lorisids, although very short in comparison to their small-bodied counterparts, are relatively as long as, or longer than, the forelimbs of cheirogaleids, lemurids, *Otolemur,* sifakas, and some callitrichids. Improved stabilization in quadrupedalism is probably gained because the center of gravity in the potto and the slow loris is moved toward the horizontal supports by relative reduction of both limbs rather than by relative shortening of the hindlimb alone. The angwantibo and slender loris both prefer the lower levels of the forest, a location dominated by lianes, foliage, and fine branches (Subramonian, 1957; Jewell and Oates, 1969; Sabater Pi, 1972; Charles-Dominique, 1977; Petter and Hladik, 1970). Such a habitat necessitates frequent bridging behavior between discontinuous but adjacent supports; a relatively small-bodied primate with relatively long extremities is well suited to such a niche. *Perodicticus* and *Nycticebus* are both found consistently higher up in the canopy and prefer to move on supports of larger average diameter (Jewell and Oates, 1969; Charles-Dominique, 1977; D'Souza, 1974). These larger bodied lorisids seek out continuous pathways whenever possible and will take detours to avoid bridging between discontinuous supports; bridging does occur but it is relatively rare.

Lepilemur and Tarsius

Both sportive lemurs and tarsiers are known to be dedicated leaping prosimians (Charles-Dominique and Hladik, 1971; Petter *et al.,* 1977; Niemitz, 1979*a,b*). Both possess low intermembral indices (59 and 52, respectively) and relatively long hindlimbs. The hindlimb of the tarsier is exceptionally long, however, even if the very long tarsus is excluded from consideration. The forelimb of the tarsier is also relatively long for a leaper (Fig. 3), a relationship that is obscured by their very long hindlimbs (Jouffroy and Lessertisseur, 1978, 1979). Relative hindlimb length in *Lepilemur* is very similar to that of *Hapalemur griseus,* but is shorter than in most other habitual leapers. In fact, small lorisids and small hylobatids possess relatively longer hindlimbs than do sportive lemurs.

Callitrichids

Because marmosets and tamarins possess claws rather than nails, they are not faced with the same mechanical constraints during climbing or clinging to tree trunks that confront most primates with nails. Interlimb scaling trends in callitrichids, as might be expected, then, do not conform to the climbing model detailed above. Both forelimbs and hindlimbs exhibit positive allometry, with the hindlimb exponent slightly greater than that of the forelimb. Despite this general trend, the intermembral index of the largest species

Leontopithecus rosalia is significantly greater than that of *Cebuella pygmaea* (see below). Within the six species considered in this study, the least squares regression of intermembral index on body weight has a slope of 0.003 (R^2 = 0.017); overall, therefore, intermembral index tends to be independent of body size in callitrichids. Although *Cebuella* has been characterized as a "neotropical vertical clinger and leaper" (Kinzey *et al.*, 1975), its hindlimb is relatively very short. Only cheirogaleids (except *Phaner*) and male gorillas possess relatively shorter hindlimbs than the pygmy marmoset. The relatively longer forelimb of *Leontopithecus* (and especially the forearm) together with its relatively long hands and fingers may be related in an adaptive sense less to locomotion and more to a foraging strategy involving extensive probing for and extracting of insects in a manner analogous to *Dactylopsila* (Hershkovitz, 1977; Thorington, 1982). Regardless, the clawed cheiridia of callitrichids allow all of them to move upon and cling to very large tree trunks (Kinzey *et al.*, 1975; Fleagle and Mittermeier, 1980; Garber, 1980). By virtue of powerful claws on all digits (except the hallux), the aye-aye (*Daubentonia*) exhibits a scansorial ability similar to the callitrichids; the claws permit them to use very large tree trunks in climbing and clinging, and to adopt feeding postures impossible for most other primates to maintain (Petter and Peyrieras, 1970; Hershkovitz, 1977).

Cebids

Although specific details of interspecific scaling of the limbs in Cebidae (*sensu* Hershkovitz, 1977) are altered somewhat depending on the inclusion or exclusion of *Ateles* in the analysis, the overall allometric trends observed in cebid limb proportions remain quite similar. The intermembral index increases from *Samiri* to *Lagothrix* or to *Ateles* due to positive scaling of the forelimb in conjunction with negative allometry of hindlimb length. As was noted previously, size-related proportional changes of this nature fit the predictions of the climbing model exceedingly well. Dykyj (1982) has recently reached the same conclusion independently from a different data base; as she notes, the observed allometry serves "to facilitate climbing and suspensory adaptation at larger body sizes" (p. V). Dykyj's comprehensive analysis also demonstrates that a consideration of scaling trends based on Rosenberger's (1979) taxonomic revision of most of the noncallitrichid platyrrhines into the "Atelidae" (*Aotus, Callicebus,* pithecines, and atelines) leads one to essentially the same functional conclusions about size-related changes in limb proportions. *Lagothrix,* according to Hershkovitz (1977, p. 47) is "more a climber and less a leaper or brachiator" than is *Ateles;* and differences in relative limb proportions correlates well with these behavioral observations. Both the forelimb and hindlimb of *Ateles* are relatively longer than those of *Lagothrix.* Among the cebids only *Pithecia* can boast relatively longer legs than *Ateles.* As was suggested for lemurids and galagids, the negative allometry of hindlimb length in cebids (without *Ateles*) not only facilitates climbing by the larger species such as *Alouatta* and *Lagothrix,* but also serves as the most effective way to move a large-bodied primate's center of gravity relatively closer to the base of support (even as relative forelimb length increases); stability during pronograde travel on horizontal supports should be improved as a consequence (Napier, 1967; Jungers and Fleagle, 1980).

Cercopithecids

The high intermembral index of *Nasalis larvatus,* the largest of extant colobines, results from positive allometry of forelimb length in conjunction with negative scaling of hindlimb length. The degree of relative hindlimb reduction in *Nasalis* is substantially greater than a comparison to *Colobus verus* alone would suggest; compared to other species of *Colobus* (e.g., *C. guereza* or *C. polykomos*) or *Presbytis* (e.g., *P. rubicunda* or *P. aygula*), the hindlimb of *Nasalis* is relatively quite short (but still longer than many cercopithecines; see below). The absolute and relative increase in the length of the radial element contributes more to the observed positive scaling of forelimb length in colobines than does the humeral element. Locomotor information on *Nasalis* is vague at best (Kern, 1964; Kawabe and Mano, 1972), but this species does seem to frequent both the trees and the ground; it has also been said to possess a greater "terrestrial tendency" (Kawabe and Mano, 1972) than most colobines. Even if *Nasalis* climbs in the manner described for *Colobus guereza* as "shinnying" up a tree trunk [wherein the hands are placed around the trunk and the hindlimbs are flexed (Mittermeier and Fleagle, 1976); also see the hand and foot postures adopted on more or less vertical supports by *C. guereza* in Rose (1979)], climbing is still facilitated by its relative limb proportions. For the same reasons as those noted for cebids, the relative hindlimb reduction in *Nasalis* may also contribute to better stability during pronograde travel on large horizontal branches. The functional basis of the relatively short hindlimbs of *C. verus* (in comparison to virtually any colobine except *Nasalis* or male *P. obscura*) is not immediately obvious.

The talapoin monkey-to-olive baboon size series also exhibits a significant increase in the intermembral index due to positive forelimb allometry coupled to nearly isometric scaling of hindlimb length. The large-bodied cercopithecines, the baboons, tend to be much more terrestrial than the smallest species (Jolly, 1970; Gautier-Hion, 1971, 1975), and the observed pattern of limb allometry can be interpreted as an adaptive response to this shift in habitat preference (Jungers, 1984*a*). Longer, subequal limbs are biomechanical adaptations for greater speed in a terrestrial context, and the cercopithecines appear to be converging on the proportional pattern seen in other cursorial mammals (as discussed previously). Although an appeal to the climbing model would appear to be uncalled for in order to account for the increase in relative forelimb length in cercopithecines (cf. Aiello, 1981*a;* Rollinson and Martin, 1981), it should be noted that such interlimb proportions do preserve an adequate facility for climbing by baboons. As Rose (1977) notes for the rare occasions during which olive baboons did climb trees: "Upward climbing in trees and on fenceposts almost always involves the diagonal limb sequence used in walking, with the hands grasping round the support and the feet braced against it" (p. 79). This description is very similar to that cited previously for vertical climbing by *Ateles* (Mittermeier and Fleagle, 1976; Mittermeier, 1978), and recalls the description of gorilla climbing by Bingham (1932, p. 6).

Hylobatids

The intermembral index increases with size within the lesser apes due primarily to dramatic relative hindlimb reduction in the siamang. Despite the moderately negative allometry of hylobatid forelimb length, the forelimb of the siamang is still much longer for its body size than that of the smallest great ape (*Pan paniscus*). As Tuttle (1972) has noted, the exceptional length of the hylobatid forelimb has obscured the fact that the hindlimb of all lesser apes (except the siamang) is also very long and muscular. Considering the frequency of hindlimb-dominated locomotor behaviors such as leaping and bipedalism (Carpenter, 1940; Fleagle, 1976; Gittins, 1983), well-developed hindlimbs are probably to be expected. No doubt the very long forelimbs of gibbons are the result of a specialized adaptation to bimanual locomotion (Tuttle, 1975; Cartmill and Milton, 1977; Preuschoft and Demes, 1983). However, the size-related changes in forelimb to hindlimb proportions also accord well with the postulated demands of the vertical climbing hypothesis. Perhaps the hylobatid allometric trends represent selection for both long forelimbs and hindlimbs, with the mechanical constraints related to climbing superimposed (Jungers, 1984*a*).

Great Apes and Humans

The scaling of interlimb proportions in African apes also conforms closely to the theoretical expectations derived from the biomechanics of climbing in large, clawless mammals (Jungers, 1978, 1983, 1984*b;* Shea, 1981). Due to negative allometry of forelimb length together with much more pronounced negative scaling of hindlimb length, the already quite high intermembral index of bonobos (~100) increases with body size to a significantly higher value in gorillas (~116). A high level of competence in climbing is maintained as a consequence of these size-related change in proportions. Allometric trends are virtually identical if the sexes are analyzed separately, but it is also clear that males and females of a given species tend to share very similar limb proportions even when sexual dimorphism in body size is considerable (Jungers, 1984*a*). Ecological advantages in foraging strategies are said to accrue to this allometric pattern of changing limb proportions in great apes (MacKinnon, 1971; Kortland, 1975), and even the largest male gorillas are frequent and adept climbers in pursuit of desirable food items (Bingham,

1932; Schaller, 1963; Deschryver in Stern, 1976; Goodall, 1977). The exaggeration of this trend in *Pongo* (a much higher intermembral index due to the combined effects of pronounced relative forelimb elongation and relative hindlimb reduction) underscores the selective importance of such bodily proportions in a highly aboreal, large-bodied ape.

As discussed previously, this pattern of relative forelimb elongation in African apes does not redistribute their weight to their forelimbs (*contra* Aiello, 1981*a*). Given the cranially directed glenoid fossae of African apes, one can argue that their forelimbs are poorly suited for a weight-bearing role in quadrupedal progression (Roberts, 1974; Reynolds, 1981; *contra* Vrba, 1979). The further redistribution of weight to the hindlimb with increasing speed in chimpanzees supports such an interpretation (Reynolds, 1981; Jungers and Susman, 1984). Whether increasing body size in African apes forced the largest members (gorillas) to spend more time on the ground or allowed them to exploit new terrestrial resources with less concern for predators (or both), size-related alterations in their interlimb proportions have served to perserve their climbing capabilities. Climbing remains a critical (*sensu* Prost, 1965), if not most frequent, locomotor habit of all nonhuman hominoids, including the African apes and orangutans (Fleagle, 1976). Although the forearm of gorillas is relatively short, the functional analogy to graviportal species (longer proximal segments, shorter distal ones) seems rather farfetched (Roberts, 1974; Buschang, 1982).

In comparison to all pongids and the siamang (and many other nonhylobatid catarrhines), humans are characterized by relatively very long hindlimbs (Jungers, 1982). Although humans appear to converge on the proportions of leapers in this respect, hindlimb elongation has been selected for in the human lineage to improve our striding bipedal form of gait. Compared to the relatively short hindlimb seen in the earliest australopithecines, relative elongation permits increasing velocity of gait at only a slight increase in energy cost; increased speed is achieved by increased stride length rather than by an increase in the number of gait cycles per unit time (cadence). Longer hindlimbs probably also increase the maximum attainable velocity of terrestrial bipeds (Jungers, 1982). Because the forelimbs of modern humans are freed from a primary locomotor role (except for balance), the departure of human interlimb proportions from the climbing model is not terribly surprising. Although the forelimbs of the earliest hominids were also relatively short compared to extant pongids of comparable body size (Jungers, 1982; Lovejoy, 1983), interlimb proportions in these ancestors (as judged by the high humerofemoral index) would still greatly improve climbing abilities relative to those capabilities characteristic of the modern human condition. Numerous other morphological features of the postcranial skeleton of these fossil hominids attest to the importance of climbing in their locomotor repertoire (Tuttle, 1981; Stern and Susman, 1983). It should be emphasized, however, that an adaptation to bipedalism is clearly driving numerous changes in their locomotor skeleton. For example, relatively short forelimbs in comparison to pongids of the same body size is one mechanism to more favorably reposition the center of gravity in a biped (Preuschoft, 1978); the reduction of the distance between their acetabulum and sacroiliac articulation is another (Lovejoy, 1975). Relative elongation of the hindlimbs and numerous other anatomical examples of biomechanical fine tuning for bipedalism in later hominids (Stern and Susman, 1983) indicate that the bipedal adaptation of these earliest australopithecines was still far from being perfected.

ACKNOWLEDGMENTS

I wish to thank the museum directors and curators of primate skeletal material for their kind cooperation and valuable assistance: Dr. C. Smeenk at the Rijksmuseum van Natuurlijke Historie; Dr. F. K. Jouffroy at the Laboratoire d'Anatomie Comparée; Dr. F. E. Thys van den Audenaerde and D. Meirte at the Tervuren Museum; Prof. J. J. Picard (Laboratoire d'Embryologie) at Louvain-la-Neuve; D. R. Howlett and L. Barton at the Powell-Cotton Museum; Prue Napier and Paula Jenkins at the British Museum of Natural History; Dr. R. Thorington at the National Museum of Natural History; Dr. F. A. Jenkins, Jr., and Maria Rutzmoser at the Museum of Comparative Zoology, Harvard University; Barbara Becker and Dr. P. Hershkovitz at the Field Museum of Natural History; Dr. P. Myers at the Museum of Zoology, University of Michigan; Dan Russell at the American Museum of Natural History; and Dr. W. Kimbel at the Cleveland Museum of Natural History. I also wish to express my sincere appreciation to Luci Betti for the graphics and Joan Kelly for her patience and expertise in typing this manuscript. This research was supported by NSF grants BNS 8041292, 8119664, and 8217635.

References

Aiello, L. C. 1981*a*. Locomotion in the Miocene Hominoidea, in: *Aspects of Human Evolution* (Symposia for the Study of Human Biology, Vol. 21, C. B. Stringer, ed.), pp. 63–97, Taylor and Francis, London.

Aiello, L. C. 1981*b*. The allometry of primate body proportions. *Symp. Zool. Soc. Lond.* **48:**331–358.

Alexander, R. McN. 1968. *Animal Mechanics.* University of Washington Press, Seattle.

Alexander, R. McN. 1980. Forces in animal joints. *Engin. Med.* **9:**93–97.

Alexander, R. McN., Jayes, A. S., Maloiy, G. M. O., and Wathuta, E. M. 1979. Allometry of limb bones of mammals from shrews (*Sorex*) to elephant (*Loxodonta*). *J. Zool. Lond.* **189:**305–314.

Alexander, R. McN., Jayes, A. S., Maloiy, G. M. O., and Wathuta, E. M. 1981. Allometry of the leg muscles of mammals. *J. Zool. Lond.* **194:**539–552.

Ashton, E. H., Flinn, R. M., and Oxnard, C. E. 1975. The taxonomic and functional significance of overall body proportions in Primates. *J. Zool. Lond.* **1975:**73–105.

Biegert, J., and Maurer, R. 1972. Rumpfskelettlänge, Allometrien und Körperproportionen bei catarrhinen Primaten. *Folia Primatol.* **17:**142–156.

Bingham, H. C. 1932. Gorillas in a natural habitat. *Carnegie Inst. Wash. Publ.* **426:**1–66.

Bramblett, C. 1967. The Skeleton of the Darajani Baboon, Ph.D. Diss., University of California, Berkeley.

Buschang, P. H. 1982. The relative growth of the limb bones for *Homo sapiens*—as compared to anthropoid apes. *Primates* **23:**465–468.

Calder, W. A., III. 1981. Scaling of physiological processes in homeothemic animals. *Annu. Rev. Physiol.* **43:**301–322.

Carpenter, C. R. 1940. A field study in Siam of the behavior and social relationships of the gibbon (*Hylobates lar*). *Comp. Psychol. Monogr.* **16:**1–212.

Cartmill, M. 1972. Arboreal adaptations and the origin of the order Primates, in: *The Functional and Evolutionary Biology of Primates* (R. H. Tuttle, ed.), pp. 97–122, Aldine-Atherton, Chicago.

Cartmill, M. 1974*a*. Rethinking primate origins. *Science* **184:**436–443.

Cartmill, M. 1974*b*. Pads and claws in arboreal locomotion, in: *Primate Locomotion* (F. A. Jenkins, Jr., ed.), pp. 45–83, Academic Press, New York.

Cartmill, M. 1979. The volar skin of primates: Its frictional characteristics and their functional significance. *Am. J. Phys. Anthropol.* **50:**497–510.

Cartmill, M., and Milton, K. 1977. The lorisiform wrist joint and the evolution of "brachiating" adaptations in the Hominoidea. *Am. J. Phys. Anthropol.* **47:**249–272.

Charles-Dominique, P. 1977. *Ecology and Behavior of Nocturnal Primates: Prosimians of Equatorial West Africa,* Columbia University Press, New York.

Charles-Dominique, P., and Hladik, M. 1971. Le *Lepilemur* du sud de Madagascar: Ecologie, alimentation et vie sociale. *Terre Vie* **25:**3–66.

Charles-Dominique, P., Cooper, H. M., Hladik, A., Hladik, C. M., Pages, E., Pariente, G. F., Petter-Rousseaux, A., Petter, J. J., and Schilling, A. 1980. *Nocturnal Malagasy Primates,* Academic Press, New York.

Coolidge, H. J., and Shea, B. T. 1982. External body dimensions of *Pan paniscus* and *Pan troglodytes* chimpanzees. *Primates* **23:**245–251.

Cousins, D. 1972. Body measurements and weights of wild and captive gorillas, *Gorilla gorilla. Zool. Gart. N. F. Leipzig* **41:**261–277.

Davis, D. D. 1962. Allometric relationships in lions vs domestic cats. *Evolution* **16:**505–514.

Dechow, P. C. 1983. Estimation of body weights from craniometric variables in baboons. *Am. J. Phys. Anthropol.* **60:**113–123.

Doyle, G. A., and Bearder, S. K. 1977. The galagines of South Africa, in: *Primate Conservation* (Prince Ranier and G. H. Bourne, eds.), pp. 1–35, Academic Press, New York.

Doyle, G. A., and Martin, R. D. 1979. *The Study of Prosimian Behavior,* Academic Press, New York.

D'Souza, F. A. 1974. A preliminary field report on the lesser tree shrew (*Tupaia minor*), in: *Prosimian Biology* (R. D. Martin, G. A. Doyle, and A. C. Walker, eds.), pp. 167–182, University of Pittsburgh Press, Pittsburgh.

Dykyj, D. 1982. Allometry of Trunk and Limbs in New World Monkeys, Ph.D. Diss., City University of New York.

Economos, A. C. 1982. On the origin of biological similarity. *J. Theor. Biol.* **94:**25–60.

Eisenberg, J. F. 1981. *The Mammalian Radiations. An Analysis of Trends in Evolution, Adaptation, and Behavior,* University of Chicago Press, Chicago.

Erikson, G. E. 1963. Brachiation in New World monkeys and in anthropoid apes. *Symp. Zool. Soc. Lond.* **10:**135–164.

Fedak, M. A., and Seeherman, A. J. 1979. Reappraisal of energetics of locomotion shows identical cost in bipeds and quadrupeds including ostrich and horse. *Nature* **282:**713–716.

Fleagle, J. G. 1976. Locomotion and posture of the Malayan siamang and implications for hominoid evolution. *Folia Primatol.* **26:**245–269.

Fleagle, J. G. 1977. Locomotor behavior and skeletal anatomy of sympatric Malaysian leaf-monkeys (*Presbytis obscura* and *Presbytis melalophos*). *Yearb. Phys. Anthropol.* **20:**440–453.

Fleagle, J. G., and Mittermeier, R. A. 1980. Locomotor behavior, body size, and comparative ecology of seven Surinam monkeys. *Am. J. Phys. Anthropol.* **52:**301–314.

Fooden, J. 1963. A revision of the woolly monkeys (genus *Lagothrix*). *J. Mammal.* **44:**213–247.

Fooden, J. 1976. Primates obtained in peninsular Thailand June–July, 1973, with notes on the distribution of continental Southeast Asian leaf-monkeys (*Presbytis*). *Primates* **17:**95–118.

Garber, P. A. 1980. Locomotor behavior and feeding ecology of the Panamanian tamarin (*Saguinus oedipus geoffroyi,* Callitrichidae, Primates). *Int. J. Primatol.* **1:**185–201.

Garland, T., Jr. 1983. The relation between maximal running speed and body mass in terrestrial mammals. *J. Zool. Lond.* **199:**157–170.

Gautier-Hion, A. 1971. L'ecologie du Talapoin du Gabon. *Terre Vie* **25:**427–490.

Gautier-Hion, A. 1975. Dimorphisme sexuel et organization social chez les Cercopithecines forestiers Africans. *Mammalia* **39:**356–374.

Giles, E. 1956. Cranial allometry in the great apes. *Hum. Biol.* **28:**43–58.

Gittins, S. P. 1983. Use of the forest canopy by the agile gibbon. *Folia Primatol.* **40:**134–144.

Goodall, A. G. 1977. Feeding and ranging behavior of a mountain gorilla group (*Gorilla gorilla beringei*) in the Tshibinda-Kahuzi Region (Zaire), in: *Primate Ecology* (T. H. Clutton-Brock, ed.), pp. 449–479, Academic Press, London.

Gould, S. J. 1975. Allometry in primates, with emphasis of scaling and the evolution of the brain, in: *Approaches to Primate Paleobiology* (Contrib. Primatol., Vol. 5, F. Szalay, ed.), pp. 244–292, S. Karger, Basel.

Grand, T. 1968. The functional anatomy of the lower limb of the howler monkey (*Alouatta caraya*). *Am. J. Phys. Anthropol.* **28:**163–182.

Grzimek, B. 1956. Mässe und Gewichte von Flachland-Gorillas. *Z. Säugetierkd.* **21:**192–194.

Günther, B. 1975. Dimensional analysis and theory of biological similarity. *Physiol. Rev.* **55:**659–699.

Günther, B., and Morgado, E. 1982. Theory of biological similarity revisited. *J. Theor. Biol.* **96:**543–559.

Harvey, P. H. 1982. On rethinking allometry. *J. Theor. Biol.* **95:**37–41.

Hershkovitz, P. 1977. *Living New World Monkeys* (*Platyrrhini*), Vol. 1, University of Chicago Press, Chicago.

Hildebrand, M. 1952. An analysis of body proportions in the Canidae. *Am. J. Anat.* **90:**217–256.

Hildebrand, M. 1961. Body proportions of didelphid (and some other) marsupials, with emphasis on variability. *Am. J. Anat.* **109:**239–249.

Hildebrand, M. 1974. *Analysis of Vertebrate Structure,* Wiley, New York.

Hoage, R. J. 1982. Social and physical maturation in captive lion tamarins, *Leontopithecus rosalia rosalia* (Primates, Callitrichidae). *Smithson. Contrib. Zool.* **354:**1–56.

Holloway, R. L., and Post, D. G. 1982. The relativity of relative brain measures and hominid mosaic evolution, in: *Primate Brain Evolution* (E. Armstrong and D. Falk, eds.), pp. 57–76, Plenum Press, New York.

Howell, A. B. 1944. *Speed in Animals,* Hafner, New York.

Hrdlicka, A. 1925. Weight of the brain and the internal organs in American monkeys. *Am. J. Phys. Anthropol.* **8:**201–211.

Huxley, J. S. 1932. *Problems of Relative Growth,* Methuen, London.

Jewell, P. A., and Oates, J. F. 1969. Ecological observations on the lorisoid primates of African lowland forest. *Zool. Afr.* **4:**231–248.

Jolly, C. J. 1970. The large African monkeys as an adaptive array, in: *Old World Monkeys; Evolution, Systematics and Behavior* (J. R. Napier and P. H. Napier, eds.), pp. 139–174, Academic Press, New York.

Jouffroy, F. K., and Lessertisseur, J. 1978. Etude ecomorphologique des proportions des membres des primates et specialement des prosimiens. *Ann. Sci. Nat. Zool. Paris* **20:**99–128.

Jouffroy, F. K., and Lessertisseur, J. 1979. Relationships between limb morphology and locomotor adaptations among prosimians: An osteometric study, in: *Environment, Behavior and Morphology: Dynamic Interactions in Primates* (M. E. Morbeck, H. Preuschoft, and N. Gomberg, eds.), pp. 143–181, Gustav Fischer, New York.

Jungers, W. L. 1977. Hindlimb and pelvic adaptations to vertical climbing and clinging in *Megaladapis,* a giant subfossil prosimian from Madagascar. *Yearb. Phys. Anthropol.* **20:**508–524.

Jungers, W. L. 1978. The functional significance of skeletal allometry in *Megaladapis* in comparison to living prosimians. *Am. J. Phys. Anthropol.* **19:**303–314.

Jungers, W. L. 1979. Locomotion, limb proportions and skeletal allometry in lemurs and lorises. *Folia Primatol.* **32:**8–28.

Jungers, W. L. 1980. Adaptive diversity in subfossil Malagasy prosimians. *Z. Morphol. Anthropol.* **71:**177–186.

Jungers, W. L. 1982. Lucy's limbs: Skeletal allometry and locomotion in *Australopithecus afarensis*. *Nature* **297:**676–678.

Jungers, W. L. 1984*a*. Scaling of the hominoid locomotor skeleton with special reference to the lesser apes, in: *The Lesser Apes: Evolutionary and Behavioural Biology* (H. Preuschoft, D. Chivers, W. Brockelman, and N. Creel, eds.), pp. 146–169, Edinburgh University Press, Edinburgh.

Jungers, W. L. 1984*b*. Aspects of size and scaling in primate biology with special reference to the locomotor skeleton. *Yearb. Phys. Anthropol.* **27** (in press).

Jungers, W. L., and Fleagle, J. G. 1980. Postnatal growth allometry of the extremities in *Cebus albifrons* and *Cebus apella:* A longitudinal and comparative study. *Am. J. Phys. Anthropol.* **53:**471–478.

Jungers, W. L., and German, R. Z. 1981. Ontogenetic and interspecific skeletal allometry in nonhuman primates: Bivariate versus multivariate analysis. *Am. J. Phys. Anthropol.* **55:**195–202.

Jungers, W. L., and Susman, R. L. 1984. Body size and skeletal allometry in African apes, in: *The Pygmy Chimpanzee: Evolutionary Biology and Behavior* (R. L. Susman, ed.), pp. 131–177, Plenum Press, New York.

Kawabe, M., and Mano, T. 1972. Ecology and behavior of the wild proboscis monkey, *Nasalis larvatus* (Wurmb), in Sabah, Malaysia. *Primates* **13:**213–228.

Kern, J. A. 1964. Observations on the habits of the proboscis monkey *Nasalis larvatus* (Warmb), made in the Brunei Bay Area, Borneo. *Zoologica* **49:**183–192.

Kimura, T., Okada, M., and Ishida, H. 1979. Kinesiological characteristics of primate walking: Its significance in human walking, in: *Environment, Behavior, and Morphology: Dynamic Interactions in Primates* (M. E. Morbeck, H. Preuschoft, and N. Gomberg, eds.), pp. 297–311, Gustav Fischer, New York.

Kinzey, W. G., Rosenberger, A. L., and Ramirez, M. 1975. Vertical clinging and leaping in a neotropical anthropoid. *Nature* **225:**327–328.

Kortland, A. 1975. Ecology and paleoecology of ape locomotion, in: *Proceedings from the Symposium of the 5th Congress of the International Primatological Society* (S. Kondo, M. Kawai, A. Ehara, and S. Kawamura, eds.), pp. 361–364, Japan Science Press, Tokyo.

Kuhry, B., and Marcus, L. F. 1977. Bivariate linear models in biometry. *Syst. Zool.* **26:**201–209.

Lambert, R., and Teissier, G. 1927. Theorie de la similitude biologique. *Ann. Physiol.* **3:**212–246.

Lande, R. 1979. Quantitative genetic analysis of multivariate evolution applied to brain:body size allometry. *Evolution* **33:**402–416.

Lindstedt, S. L., and Calder, W. A., III. 1981. Body size, physiological time, and longevity of homeothermic animals. *Q. Rev. Biol.* **56:**1–16.

Lovejoy, C. O. 1975. Biomechanical perspectives on the lower limb of early hominids, in: *Primate Functional Morphology and Evolution* (R. H. Tuttle, ed.), pp. 291–326, Mouton, The Hague.

Lovejoy, C. O. 1983. Locomotor Anatomy of *A. afarensis,* Paper presented at the Institute of Human Origins (Berkeley, CA), Conference on the Evolution of Human Locomotion, April 22, 1983.

Lyon, M. W. 1908*a*. Mammals collected in Western Borneo by Dr. W. L. Abbott. *Proc. U.S. Natl. Mus.* **33:**547–571.

Lyon, M. W. 1908*b*. Mammals collected in Eastern Sumatra by Dr. W. L. Abbott during 1903, 1906, and 1907, with descriptions of new species and subspecies. *Proc. U.S. Natl. Mus.* **34:**619–679.

Lyon, M. W. 1911. Mammals collected by Dr. W. L. Abbott on Borneo and some of the small adjacent islands. *Proc. U.S. Natl. Mus.* **40:**53–146.

MacKinnon, J. 1971. The orang-utan in Sabah today. *Oryx* **11:**141–191.

McMahon, T. A. 1973. Size and shape in biology. *Science* **197:**1201–1204.

McMahon, T. A. 1975. Using body size to understand the structural design of animals: Quadrupedal locomotion. *J. Appl. Physiol.* **39:**619–627.

McMahon, T. A. 1977. Scaling quadrupedal galloping: Frequencies, stresses, and joint angles, in: *Scale Effects in Animal Locomotion* (T. J. Pedley, ed.), pp. 143–151, Academic Press, New York.

Martin, R. D. 1972. Adaptive radiation and behavior of the Malagasy lemurs. *Philos. Trans. R. Soc. Lond. B* **264**:295–352.

Miller, G. S., Jr. 1903. Seventy new Malayan mammals. *Smithson. Misc. Coll.* **45**:1–73.

Mittermeier, R. A. 1978. Locomotion and posture in *Ateles geoffroy* and *Ateles paniscus. Folia Primatol.* **30**:161–193.

Mittermeier, R. A., and Fleagle, J. G. 1976. The locomotor and postural repertoires of *Ateles geoffroyi* and *Colobus guereza,* and a reevaluation of the locomotor category semibrachiation. *Am. J. Phys. Anthropol.* **45**:235–255.

Moismann, J. E., and James, F. C. 1979. New statistical methods for allometry with application to Florida red-winged blackbirds. *Evolution* **33**:444–459.

Mollison, T. 1911. Die Körperproportionen der Primaten. *Morphol. Jahrb.* **42**:79–304.

Napier, J. R. 1967. Evolutionary aspects of primate locomotion. *Am. J. Phys. Anthropol.* **27**:333–342.

Napier, J. R., and Napier, P. H. 1967. *A Handbook of Living Primates,* Academic Press, London.

Napier, J. R., and Walker, A. C. 1967. Vertical clinging and leaping—a newly recognized category of locomotor behavior of primates. *Folia Primatol.* **6**:204–219.

Nash, L. T. 1983. Differential habitat utilization in two species of sympatric *Galago* in Kenya. *Am. J. Phys. Anthropol.* **60**:231.

Niemitz, C. 1979*a*. Outline of the behavior of *Tarsius bancanus,* in: *The Study of Prosimian Behavior* (G. A. Doyle and R. D. Martin, eds.), pp. 631–660, Academic Press, New York.

Niemitz, C. 1979*b*. Relationships among anatomy, ecology and behavior: A model developed in the genus *Tarsius,* in: *Environment, Behavior and Morphology: Dynamic Interactions in Primates* (M. E. Morbeck, H. Preuschoft, and N. Gomberg, eds.), pp. 119–137, Gustav Fischer, New York.

Niemitz, C. 1979*c*. Results of a field study on the western tarsier (*Tarsius bancanus borneanus* Horsfield, 1821) in Sarawak. *Sarawak Mus. J.* **27**:171–228.

Olson, T. R. 1979. Studies on aspects of the morphology and systematics of the genus *Otolemur* Coquerel 1859 (Primates: Galagidae), Ph.D. Diss., University of London.

Pedley, T. J. 1977. *Scale Effects in Animal Locomotion,* Academic Press, London.

Petter, J. J., and Hladik, C. M. 1970. Observations sur le domaine vital et la densité de population de *Loris tardigradus* dans les forets de Ceylon. *Mammalia* **34**:394–409.

Petter, J. J., and Peyrieras, A. 1970. Nouvelle contribution a l'etude d'un lemurien malgache, le aye-aye (*Daubentonia madagascarensis*). *Mammalia* **34**:167–193.

Petter, J. J., Schilling, A., and Pariente, G. 1971. Observations eco-ethologiques sur deux lemuriens malgaches nocturnes: *Phaner furcifer* et *Microcebus coquereli. Terre Vie* **25**:287–327.

Petter, J. J., Albignac, R., and Rumpler, Y. 1977. Mammiferes lemuriens (Primates Prosimiens). *Faune Madagascar* **44**:1–513.

Pocock, R. I. 1927. The gibbons of the genus *Hylobates. Proc. Zool. Soc. Lond.* **1927**:719–741.

Prange, H. D. 1977. The scaling and mechanics of arthropod exoskeletons, in: *Scale Effects in Animal Locomotion* (T. J. Pedley, ed.), pp. 169–183, Academic Press, New York.

Preuschoft, H. 1978. Recent results concerning the biomechanics of man's acquisition of bipedality, in: *Evolution* (Recent Advances in Primatology, Vol. 3, D. J. Chivers and K. A. Joysey, eds.), pp. 435–458, Academic Press, New York.

Preuschoft, H., and Demes, B. 1983. The biomechanics of brachiation, in: *The Lesser Apes: Evolutionary and Behavioural Biology* (D. Chivers, H. Peuschoft, W. Brockelman, and N. Creel, eds.), pp. 96–118, Edinburgh University Press, Edinburgh.

Prost, J. H. 1965. A definitional system for the classification of primate locomotion. *Am. J. Anthropol.* **67**:1198–1214.

Rahm, U. 1967. Observations during chimpanzee captures in the Congo, in: *Neue Ergebnisse der Primatologie* (D. Starck, R. Schneider, and H. J. Kuhn, eds.), pp. 195–207, Gustav Fischer, Stuttgart.

Rashevsky, N. 1948. *Mathematical Biophysics,* University of Chicago Press, Chicago.

Reynolds, T. R. 1981. Mechanics of Interlimb Weight Redistribution in Primates, Ph.D. Diss., Rutgers University, New Jersey.

Roberts, D. 1974. Structure and function of the primate scapula, in: *Primate Locomotion* (F. A. Jenkins, Jr., ed.), pp. 171–200, Academic Press, New York.

Rollinson, J., and Martin, R. D. 1981. Comparative aspects of primate locomotion with special reference to arboreal cercopithecines. *Symp. Zool. Soc. Lond.* **48:**377–427.

Rose, M. D. 1977. Positional behavior of olive baboons (*Papio anubis*) and its relationship to maintenance and social activities. *Primates* **18:**59–116.

Rose, M. D. 1979. Positional behavior of natural populations: Some quantitative results of a field study of *Colobus guereza* and *Cercopithecus aethiops*, in: *Environment, Behavior and Morphology: Dynamic Interactions in Primates* (M. E. Morbeck, H. Preuschoft, and N. Gomberg, eds.), pp. 75–93, Gustav Fischer, New York.

Rosenberger, A. L. 1979. Phylogeny, Evolution and Classification of New World Monkeys (Platyrrhini, Primates), Ph.D. Diss., City University of New York.

Rothenfluh, E. 1976. "Überprüfung der Gewichtsangaben adulter Primaten. Sem. Arbeit. Univ. Zürich.

Sabater Pi, J. 1972. Notes on the ecology of five Lorisiformes of Rio-Mumi. *Folia Primatol.* **18:**140–151.

Schaller, G. B. 1963. *The Mountain Gorilla,* University of Chicago Press, Chicago.

Schmidt-Nielsen, K. 1975. Scaling in biology: The consequences of size. *J. Exp. Zool.* **194:**257–308.

Schmidt-Nielsen, K. 1977. Problems of scaling: Locomotion and physiological correlates, *Scale Effects in Animal Locomotion* (T. J. Pedley, ed.), pp. 1–21, Academic Press, New York.

Schultz, A. H. 1930. The skeleton of the trunk and limbs of higher primates. *Hum. Biol.* **2:**303–438.

Schultz, A. H. 1933. Die Körperproportionen der erwachsenen catarrhinen Primaten, mit spezieller Berucksichtigung der Menschenaffen. *Anthropol. Anz.* **10:**154–185.

Schultz, A. H. 1937. Proportions, variability and asymmetries of the long bones of the limbs and clavicles in man and ape. *Hum. Biol.* **9:**281–328.

Schultz, A. H. 1944. Age changes and variability in gibbons. A morphological study on a population sample of a man-like ape. *Am. J. Phys. Anthropol.* **12:**1–129.

Schultz, A. H. 1954. Studien uber die Wirbelzahlen und die Körperproportionen von Halbaffen. *Z. Naturforsch. Ges. Zurich* **99:**39–75.

Schultz, A. H. 1973. The skeleton of the Hylobatidae and other observations on their morphology. *Gibbon and Siamang* **2:**1–54.

Shea, B. T. 1981. Relative growth of the limbs and trunk in the African apes. *Am. J. Phys. Anthropol.* **56:**179–201.

Smith, J. M., and Savage, R. J. G. 1956. Some locomotory adaptations in mammals. *J. Linn. Soc. (Zool.)* **42:**603–622.

Smith, R. J. 1980. Rethinking allometry. *J. Theor. Biol.* **87:**97–111.

Smith, R. J. 1981*a*. On the definition of variables in studies of primate dental allometry. *Am. J. Phys. Anthropol.* **55:**323–329.

Smith, R. J. 1981*b*. Interpretation of correlations in intraspecific and interspecific allometry. *Growth* **45:**291–297.

Sokal, R. R., and Rohlf, F. J. 1981. *Biometry,* 2nd ed., Freeman, San Francisco.

Stephan, H., Frahm, H., and Baron, G. 1981. New and revised data on volumes of brain structure in insectivores and primates. *Folia Primatol.* **35:**1–29.

Stern, J. T., Jr. 1976. Before bipedality. *Yearb. Phys. Anthropol.* **19:**59–68.

Stern, J. T., Jr., and Oxnard, C. E. 1973. Primate locomotion: Some links with evolution and morphology. *Primatologia* **4**(11)**:**1–93.

Stern, J. T., Jr., and Susman, R. L. 1983. The locomotor anatomy of *Australopithecus afarensis*. *Am. J. Phys. Anthropol.* **60:**279–317.

Steudel, K. 1981. Body size estimators in primate skeletal material. *Int. J. Primatol.* **2:**81–90.

Steudel, K. 1982. Allometry and adaptation in the catarrhine postcranial skeleton. *Am. J. Phys. Anthropol.* **59:**431–441.

Subramonian, S. 1957. Some observations on the habits of the slender loris. *J. Bombay Nat. Hist. Soc.* **54:**388–398.

Sweet, S. S. 1980. Allometric inference in morphology. *Am. Zool.* **20:**643–652.

Tattersall, I. 1982. *The Primates of Madagascar,* Columbia University Press, New York.

Taylor, C. R. 1977. The energetics of terrestrial locomotion and body size in vertebrates, in: *Scale Effects in Animal Locomotion* (T. J. Pedley, ed.), pp. 127–141, Academic Press, New York.

Thorington, R. W., Jr. 1982. Forelimb anatomy of marmosets and tamarins. *Int. J. Primatol.* **3:**340.

Thorington, R. W., Jr., and Heaney, L. R. 1981. Body proportions and gliding adaptations of flying squirrels (Petauristinae). *J. Mammal.* **62:**101–114.

Tuttle, R. H. 1972. Functional and evolutionary biology of hylobatid hands and feet. *Gibbon and Siamang* **1:**136–206.

Tuttle, R. H. 1975. Parallelism, brachiation, and hominoid phylogeny, in: *Phylogeny of the Primates: A Multidisciplinary Approach* (W. P. Luckett and F. S. Szalay, eds.), pp. 447–480, Plenum Press, New York.

Tuttle, R. H. 1981. Evolution of hominid bipedalism and prehensile capabilities. *Philos. Trans. R. Soc. Lond. B* **292:**89–94.

Vrba, E. S. 1979. A new study of the scapula of *Australopithecus africanus* from Sterkfontein. *Am. J. Phys. Anthropol.* **51:**117–130.

Walker, A. 1974. Locomotor adaptations in past and present prosimian primates, in: *Primate Locomotion* (F. A. Jenkins, Jr., ed.), pp. 349–381, Academic Press, New York.

Walker, A. 1979. Prosimian locomotor behavior, in: *The Study of Prosimian Behavior* (G. A. Doyle and R. D. Martin, eds.), pp. 543–565, Academic Press, New York.

Wood, B. A. 1978. Allometry in hominid studies, in: *Geological Background to Fossil Man* (W. W. Bishop, ed.), pp. 125–138, Scottish Academic Press, Edinburgh.

Wrangham, R. W., and Smuts, B. B. 1980. Sex differences in the behavioural ecology of chimpanzees in the Gombe National Park, Tanzania. *J. Reprod. Fertil. (Suppl.)* **28:**13–31.

Influence of Size and Proportions on the Biomechanics of Brachiation

17

HOLGER PREUSCHOFT
AND BRIGITTE DEMES

Introduction

In an earlier study (Preuschoft and Demes, 1984), we investigated the biomechanical advantages of the considerable arm length in brachiators—a well-known phenomenon. We showed that longer arms increase traveling velocity in arm-swinging locomotion. In addition, it provides a quicker start and reduces the muscle force necessary to lift the body, i.e., reduces the energetic cost of locomotion. This seems to provide a fine explanation for why gibbons have long arms. Here we ask why these animals are not larger, and why their arms are not even more elongated.

The principle of brachiation is based on a sequence of pendulumlike swings in which the suspending arm is the cord and the rest of the body the mass of the pendulum (Fig. 1; Tuttle, 1968; Kummer, 1970; Fleagle, 1974, 1977; Preuschoft and Demes, 1984). The greatest problem to solve is to obtain a secure handhold; body weight plus centrifugal force is countered by a

HOLGER PREUSCHOFT AND BRIGITTE DEMES • Arbeitsgruppe Funktionelle Morphologie, Ruhr-Universität Bochum, 4630 Bochum 1, Federal Republic of Germany.

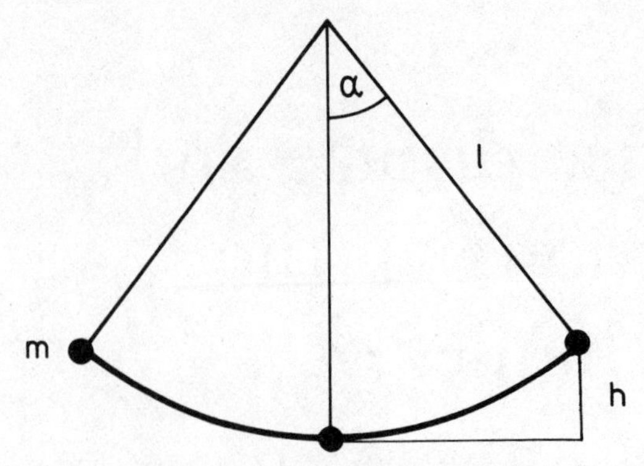

Fig. 1. Mathematical pendulum as a model of a brachiator. Here $h = 2l \sin^2(\alpha/2)$.

contraction of the finger flexors. All other joints can assume positions in which they are exposed to minimal torques (Kummer, 1970).

Body Mass

Increase of body mass m with dimensions enlarged isometrically also leads to an increase of the centrifugal force F_C that, with body weight (mg), tends to dorsiflex the fingers passively and requires balancing by a muscle force $F_A = (F_C + mg)$. The centrifugal force can be calculated from

$$F_C = mgv^2/l$$

with l the pendulum length, m the mass, g the gravitational acceleration, mg the weight (a force), and v the orbital velocity. The maximal orbital velocity is

$$v_{max} = (2gh)^{1/2}$$

where h is the vertical difference between the highest and the lowest points of the swinging body on its circular path (see Fig. 1). It can be expressed by

$$h = 2l \sin^2 (\alpha/2)$$

Consequently it follows that

$$v_{max} = [4gl \sin^2(\alpha/2)]^{1/2}$$

and

$$F_C = 4mg \sin^2(\alpha/2)$$

The increase of the reaction force follows linearly the increase of mass, provided the excursions of the swinging body remain constant (Fig. 2).

If we assume that a 6-kg hylobatid's forearm possesses a diameter of 6 cm, it has a cross-sectional area of 28.3 cm^2, of which about one-third is made up of the finger flexors. Given that the maximum muscle tension per unit area is 10 kp/cm^2 [this is a high value for trained persons; Schmidt-Nielsen (1977) assumes only 3–4 kp/cm^2], the available force of the finger flexors is roughly 95 kp. If the body mass is enlarged and the linear dimensions of the body increase isometrically, the diameter of the forearm and in consequence muscle force increases by

$$d = em^{1/3}$$

The proportionality factor e can be calculated from the values given above for a gibbon. The available muscle force of the finger flexors follows the parabolic curve F_M in Fig. 2. Although the true muscle force of the finger flexors may be underestimated by this rough approximation (morphological cross section smaller than physiological cross section because of pinnation, but assumed maximum tension per unit area high), the curve shows clearly that the reserve force, or safety margin, is relatively greater in a smaller than in a larger animal. With increasing mass, the available force runs more and more parallel to the forces necessary to balance the centrifugal force F_C plus body weight.

The lever arm of the reaction force applied to the fingers is assumed to be about three times the power arm of the flexor muscles at the proximal interphalangeal joints. Therefore, the relative torque of the finger flexors follows the curve $F_M/3$ in Fig. 2. This curve is well above the greatest reaction force in body masses between 5 and 20 kg, but then comes close to the reaction force and to an intersection at 30 kg. This indicates that small brachiators have no problems in keeping their fingers flexed against the torque of body weight plus centrifugal force. The larger they grow, however, the more difficult this task will be without (allometric) alteration of body shape, i.e., relative increase of power arm length or muscular strength. At weights greater than 30 kg, muscular strength no longer suffices to allow rapid, pendulumlike arm-swinging at wide excursions.

Figure 3 shows the relation between the necessary and the available muscle force at the interphalangeal joints: The force reserves decrease rapidly from gibbon size (5–7 kg) to siamang size (10–12 kg), showing a tendency to level out for the great apes. The curves make evident why continuous and rapid arm-swinging is restricted to the lesser apes.

Arm Length

Given this weight limitation, it seems possible to imagine an ape with arms elongated beyong the length in hylobatids without an increase in total weight.

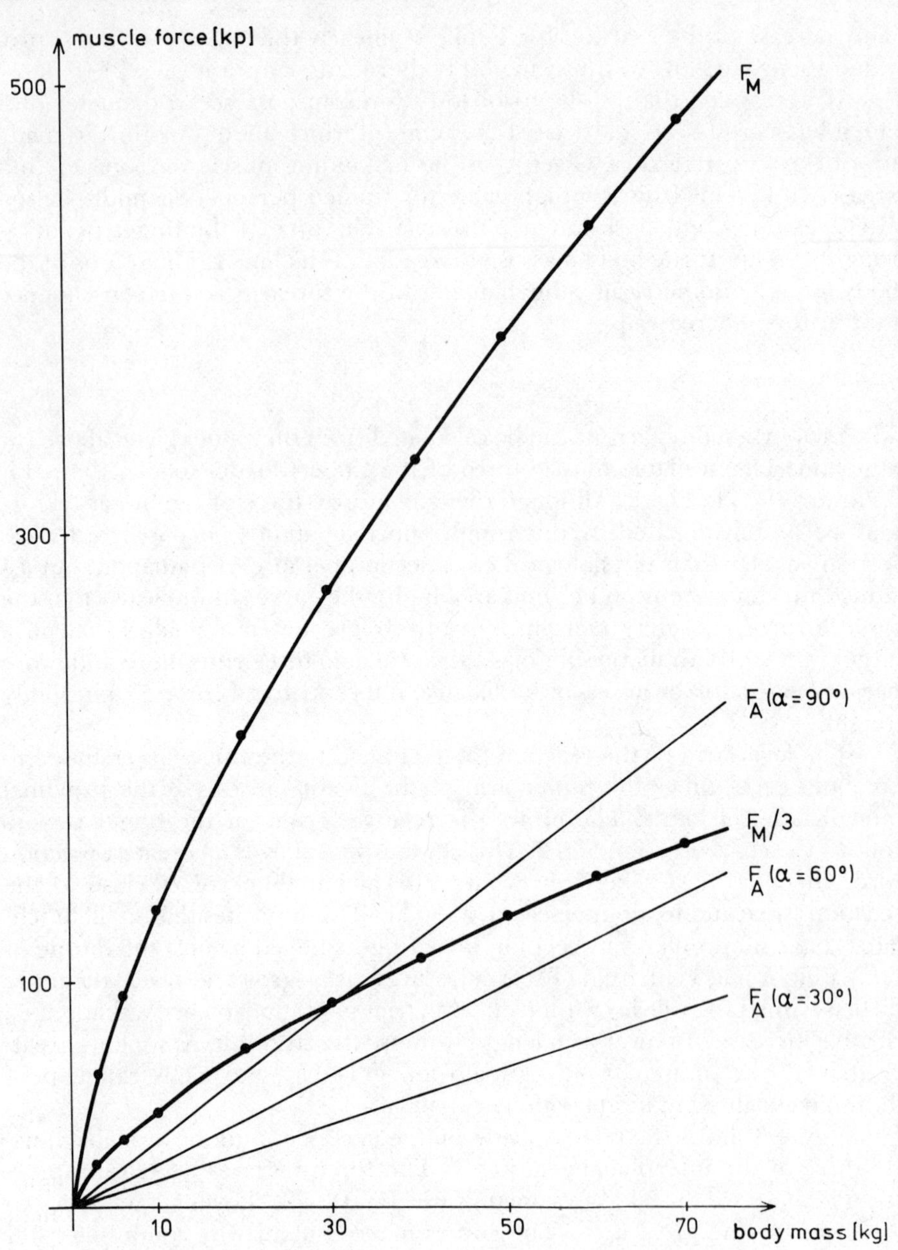

Fig. 2. Force of the finger flexors as a function of body mass. F_M is the available muscle force and F_A is the muscle force necessary for balancing the finger joints against centrifugal force plus body weight at three different excursion angles.

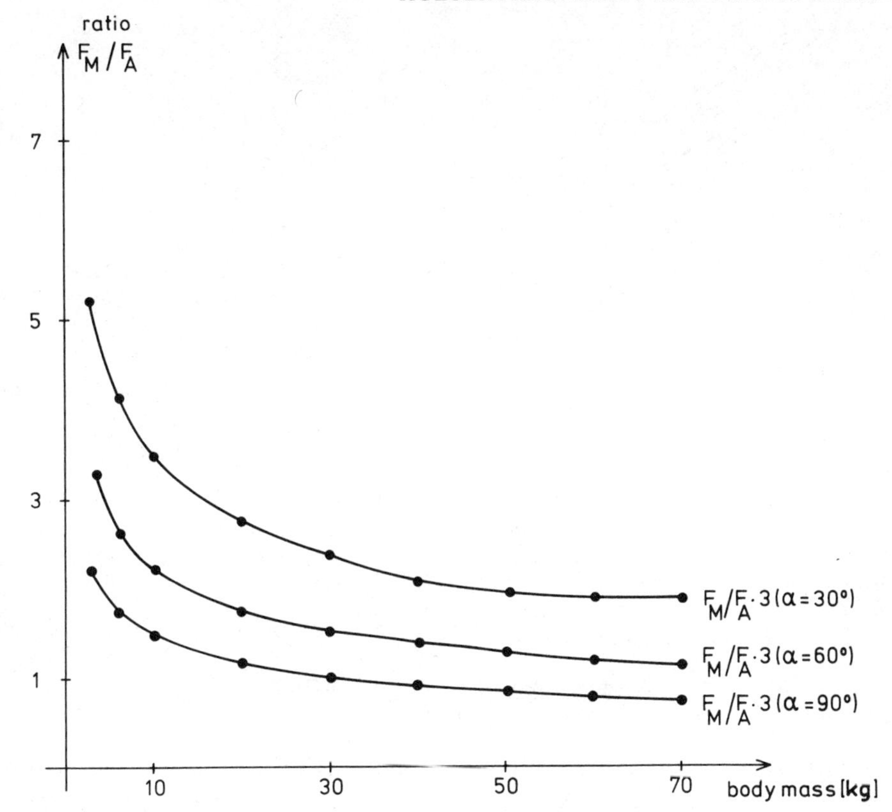

Fig. 3. Relation between available (F_M) and necessary (F_A) force of the finger flexors as a function of body mass. F_A has been multiplied by the factor three, because the body weight that has to be balanced by the muscles acts on a lever arm assumed to be three times longer than the power arm of the muscles. F_A also depends on the velocity of the swinging, which is determined, among other factors, by the angle of excursion α.

As has been detailed above, arm length does not influence the force that the finger flexors must produce to ensure the grip. A brachiator therefore may profit from the advantages of elongated upper extremities without the disadvantage of generating higher muscle forces, i.e., without an increase of the energetic cost of locomotion. One *limitation on arm length* seems to be posed by the space requirements of arm-swinging. To perform this mode of locomotion, an animal essentially needs a clearance between handhold and the next lower branch equal to arm length plus trunk length plus lower limb length. If this width is not offered by the three-dimensional meshwork of the branches, the swinging movement is disturbed or completely impossible.

A second limitation is that any increase of arm length inevitably entails an increase of mass of the arm. The upper extremity of an ape may be modeled by three cylinders, "hand," "forearm," and "upper arm" (Fig. 4). The muscle forces necessary to move the hand cylinder at the wrist joint and the hand plus forearm at the elbow joint can be calculated by

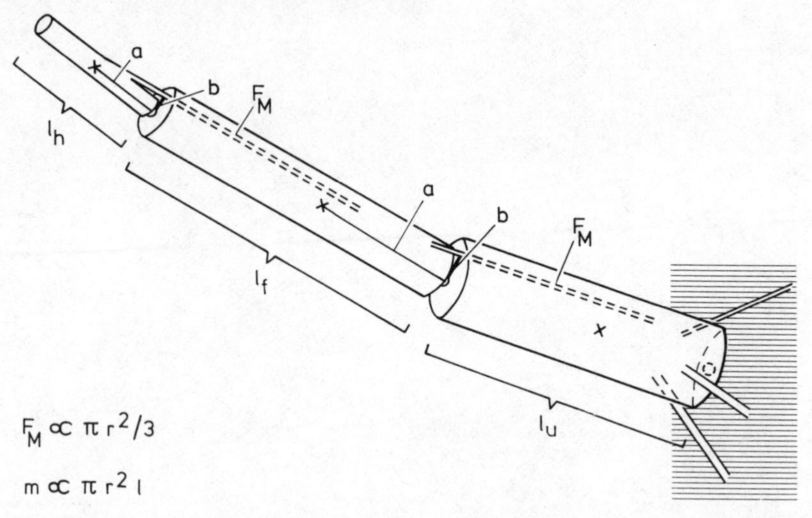

Fig. 4. Simplified model of a hylobatid arm. Hand, forearm, and upper arm are replaced by three cylinders. The cross sections of these cylinders are proportional to the muscle forces necessary to balance them against the acceleration of gravity (or weight force) or to move them against inertia. Cross-section and length determine the mass of the cylinders.

$$F_{\mathrm{M}} = \frac{(\text{weight } mg) \times (\text{load arm } a)}{\text{power arm } b}$$

We started our calculations with the dimensions of a siamang arm (hand length $l_{\mathrm{h}} = 15.5$ cm; forearm length $l_{\mathrm{f}} = 29.0$ cm; upper arm length $l_{\mathrm{u}} = 26.0$ cm). If the hand length is increased without an increase of mass by making the hand more slender, its center of gravity is shifted distally. The muscle forces necessary to balance the wrist joint increase by the same factor as hand length does, provided the lever arms of the muscles remain constant (Fig. 5).

This increase of muscle force influences the diameter of the forearm, as the cross-sectional area grows proportional to muscle force ($F_{\mathrm{M}} \propto \pi r^2/3$). If the forearm is elongated by the same factor as the hand, its mass thus grows proportional to the increase of length *and* diameter. The volumes as equivalents to mass increase by factors up to 2.25 (Fig. 5).

This heavy forearm plus unchanged hand must be kept in balance at the elbow joint. The forces that balance this joint act through lever arms increased by the same factor as forearm diameters, and grow by factors up to 2.75 (Fig. 5).

These forces again determine the enlargement of the cross-sectional area of the upper arm, which at the same time is subject to a length increase. The volume of the cylindrical upper arm therefore grows more rapidly than the mass of the forearm, namely by 4.13, nearly three times the length increase (Fig. 5).

The changes are more dramatic if a *kinetic* rather than a static situation is considered. Let us assume that a hylobatid is able to move its arm as fast as before while elongating it linearly. Mass inertia of the arm depends upon the

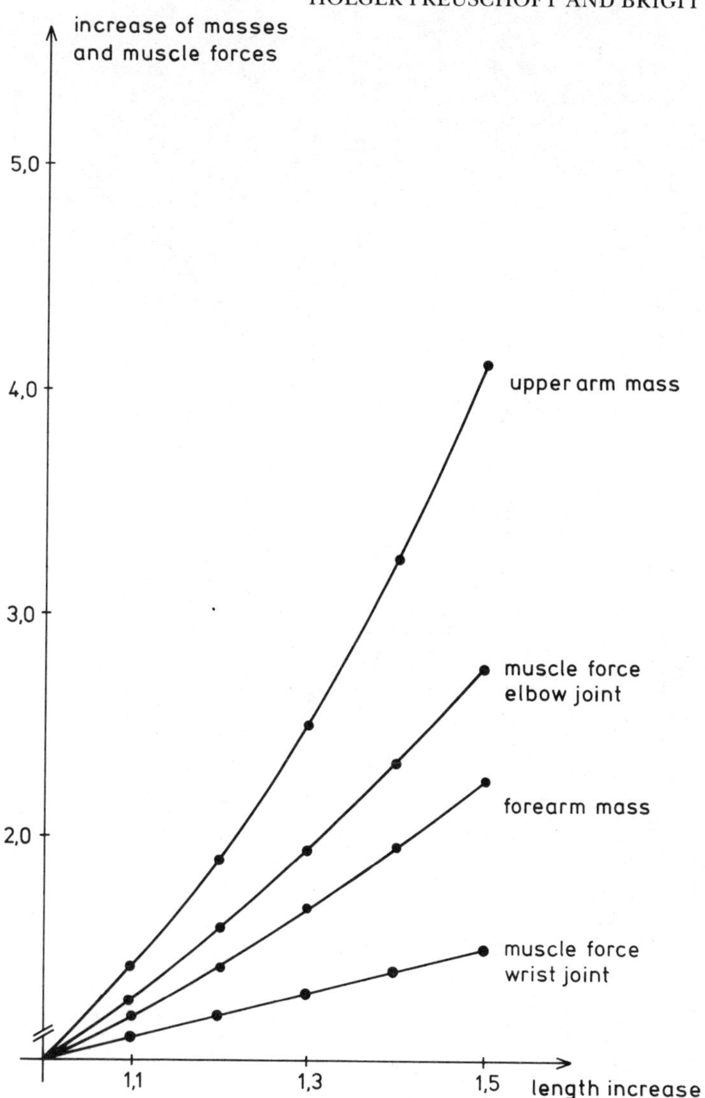

Fig. 5. Increase of mass and muscle force with length under static conditions. Length on the x axis refers to the pendulum length, that is, the distance between handhold and center of gravity. It is composed of arm length plus hand length minus the flexed middle and terminal phalanges plus distance between shoulder joint and center of gravity.

square of the distance between the axis of rotation and the segment center of gravity ($\theta \propto ma^2$). The forces necessary to move the elongated hand therefore increase parabolically (Fig. 6).

The growing muscular cross sections entail an increase of the forearm's volume and of the lever arms of muscles by the factor of length increase. This taken into consideration, the muscle forces necessary to move the forearm plus hand in the elbow joint increase more than in the forearm muscles. In

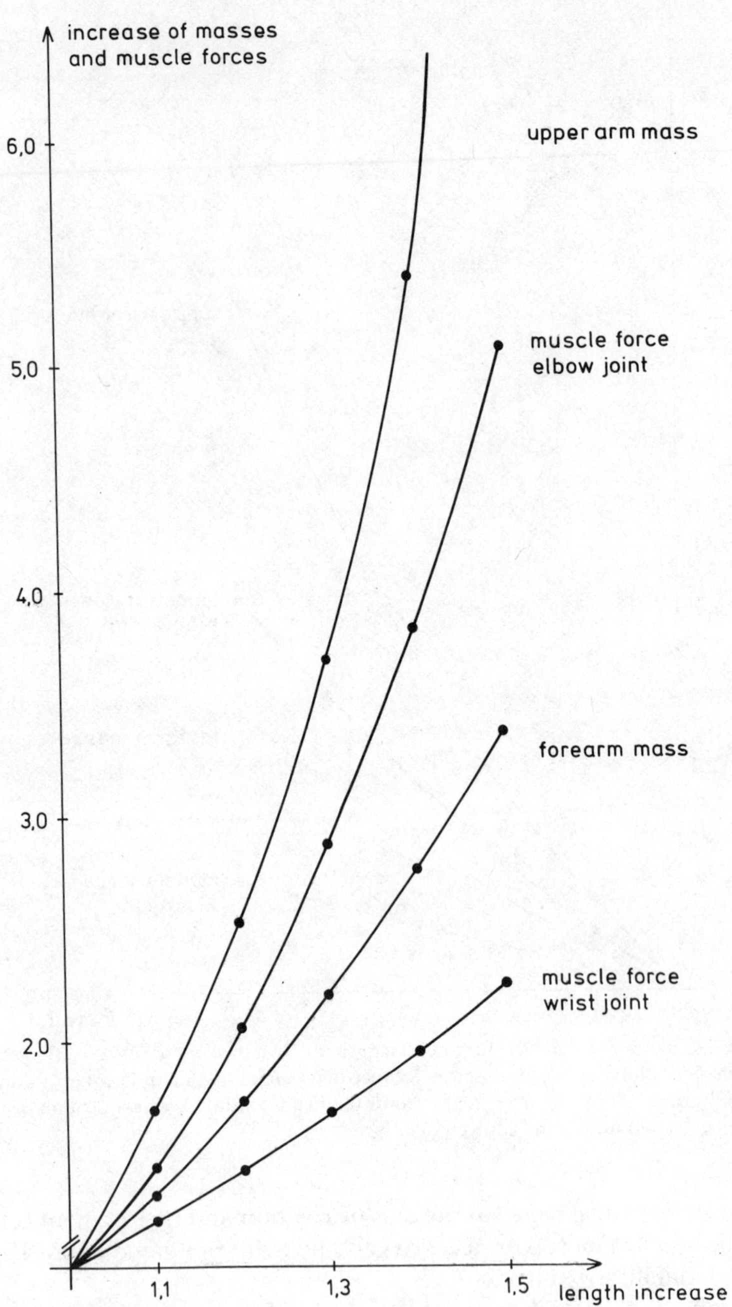

Fig. 6. Increase of mass and muscle force with length under kinetic conditions.

consequence, the mass of the upper arm increases rapidly with length (Fig. 6). If the elongated arm is to be moved as quickly as the shorter arm, a length increase by 1.1 leads to a volume (= mass) increase of the forearm by 1.3 and of the upper arm by 1.6. A length increase by 1.5 entails a growth of volume (= mass) by 3.4 in the forearm and by 7.6 in the upper arm.

This probably sets limitations to an elongation of the arms. In contrast to body mass, we were not able to define an absolute forelimb length that cannot be exceeded.

Bone Strength

An often discussed limitation for growth is bone strength. Body mass increases to the third power of linear dimensions and the resistance of bones, determined by their cross sections, only to the second power (Galilei, 1638; Schmidt-Nielsen, 1977; Kummer, 1972; Bühler, 1972). However, this holds true only for the compressive or tensile strength of bone, not for its bending strength. The moments of resistance against bending increase, like mass, by the third power of linear dimensions (Preuschoft and Weinmann, 1973; Alexander *et al.*, 1979*a*, 1981). In most animals, and definitely in most primates, it is not compressive, but bending stress that is most dangerous for the bones (Preuschoft, 1971, 1973*a,b*). Bending moments occur in all cases in which muscle forces and weight force do not fall in line with the long axis of a bone. This is true in all flexed positions of the joints.

It is a characteristic of gibbon arm bones that they are normally not exposed to high bending stresses, as the line of action of body weight runs along with the suspending arm (Kummer, 1970). They are, however, bent as soon as the arm is flexed. Arm flexion is a means to accelerate arm-swinging locomotion. By lifting the center of gravity, the body gains potential energy (Tuttle, 1968; Kummer, 1970; Fleagle, 1974, 1977; Preuschoft and Demes, 1984). It is also a means to brake after phases of free flight.

We calculated the bending moments in the radius during flexion of the arm for an animal of constant body mass (m = 10 kg). For the sake of simplicity, the elbow flexors are assumed to be reduced to one muscle. Its origin is close to the shoulder joint, and its insertion at one-seventh of radius length. Introduction of more elbow flexors and of hand and finger flexors would reduce the stresses, but not change their pattern in principle (Pauwels, 1948; Kummer, 1970; Preuschoft, 1973*a*). The maximal bending moment in the radius occurs at the insertion of the remaining flexor muscle. It is determined by the component of the muscle force perpendicular to the bone's long axis ($F'_M = F_M \sin \pi$; see Fig. 7):

$$M_{B\max} = \frac{F_M' \, c/7(c - c/7)}{c}$$

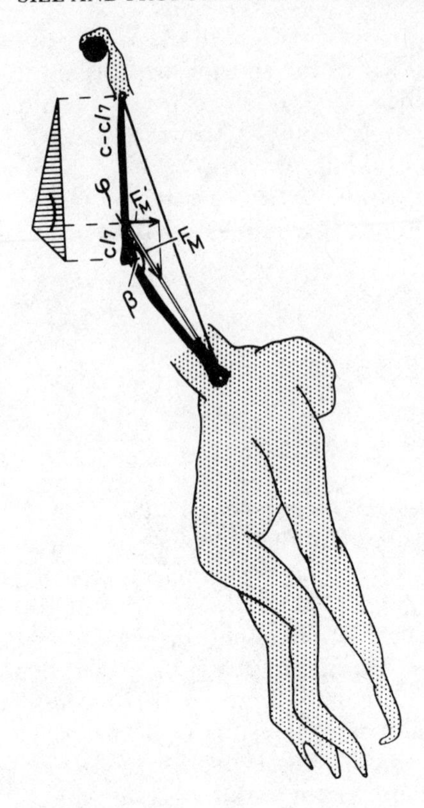

Fig. 7. Bending moments in the forearm of a hylobatid. Bending occurs if the arm is flexed in order to lift the body's center of gravity, for instance, between two swings. Only one elbow flexor (biceps) is shown. See text.

with F_M' the vertical component of muscle force and c the radius length (forearm length).

Because of the equilibrium of moments at the elbow joint (load times load arm equals power times power arm), the force of the flexor muscle can be calculated by

$$F_M = mg\ a/b$$

with mg the body weight, a the load arm (equal to the lever arm of the body mass), and b the power arm (equal to the lever arm of the elbow flexors).

If the elbow joint of a brachiator is flexed to constant angles, the bending moments in the forearm bones increase linearly with increased length (Fig. 8). The elongation of the arm bones evidently requires a remarkable increase of bending strength, especially to resist the high bending moments occurring at strong flexion (small angles between upper arm and forearm).

In reality, however, hylobatids avoid this dangerous bending of the arm bones. They tend in all phases of movement to keep their elbow joints extended or flexed to rather blunt angles. This is obvious in hoisting, but more so in arm-swinging locomotion. In order to reach a higher position in a tree,

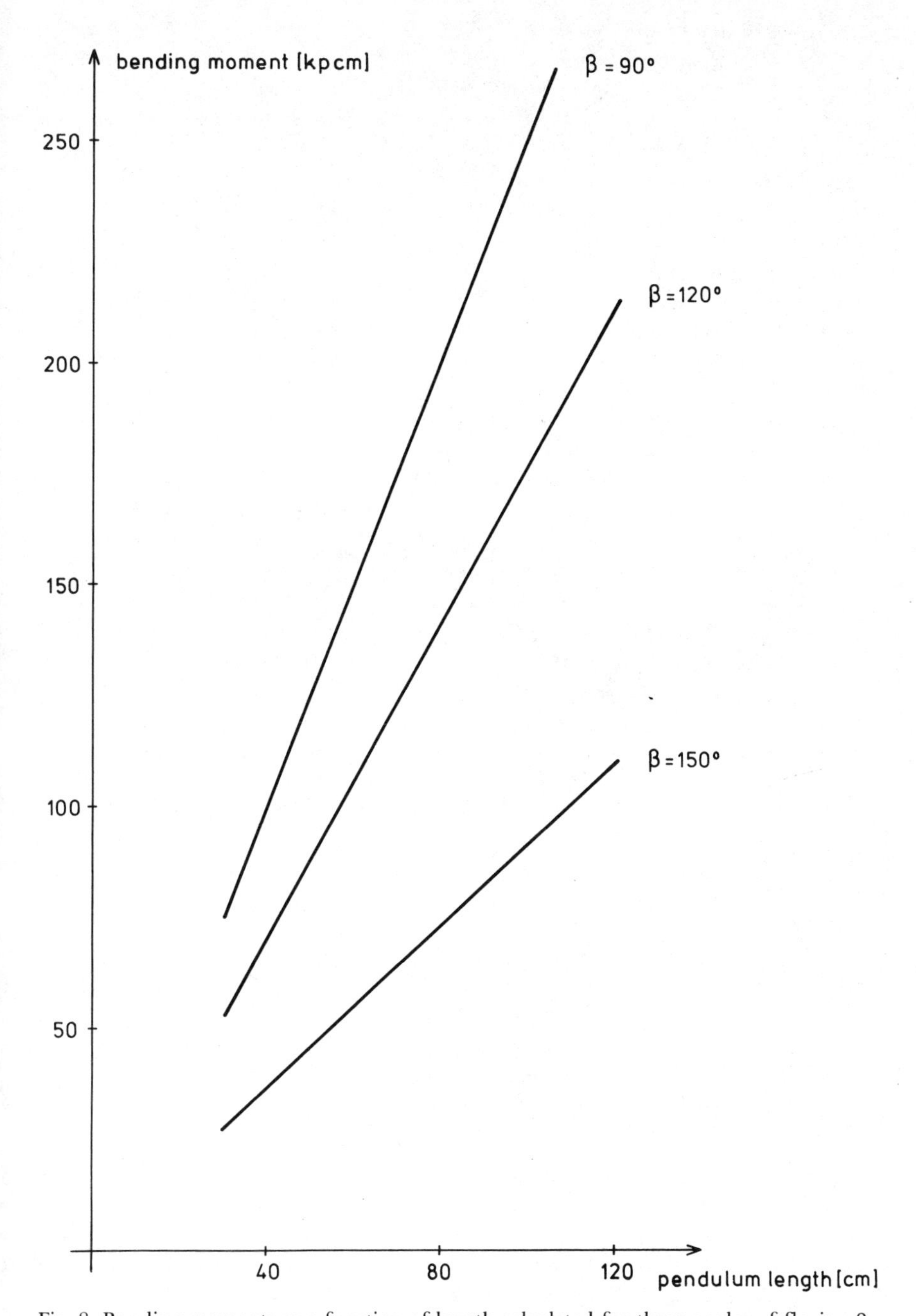

Fig. 8. Bending moments as a function of length calculated for three angles of flexion β.

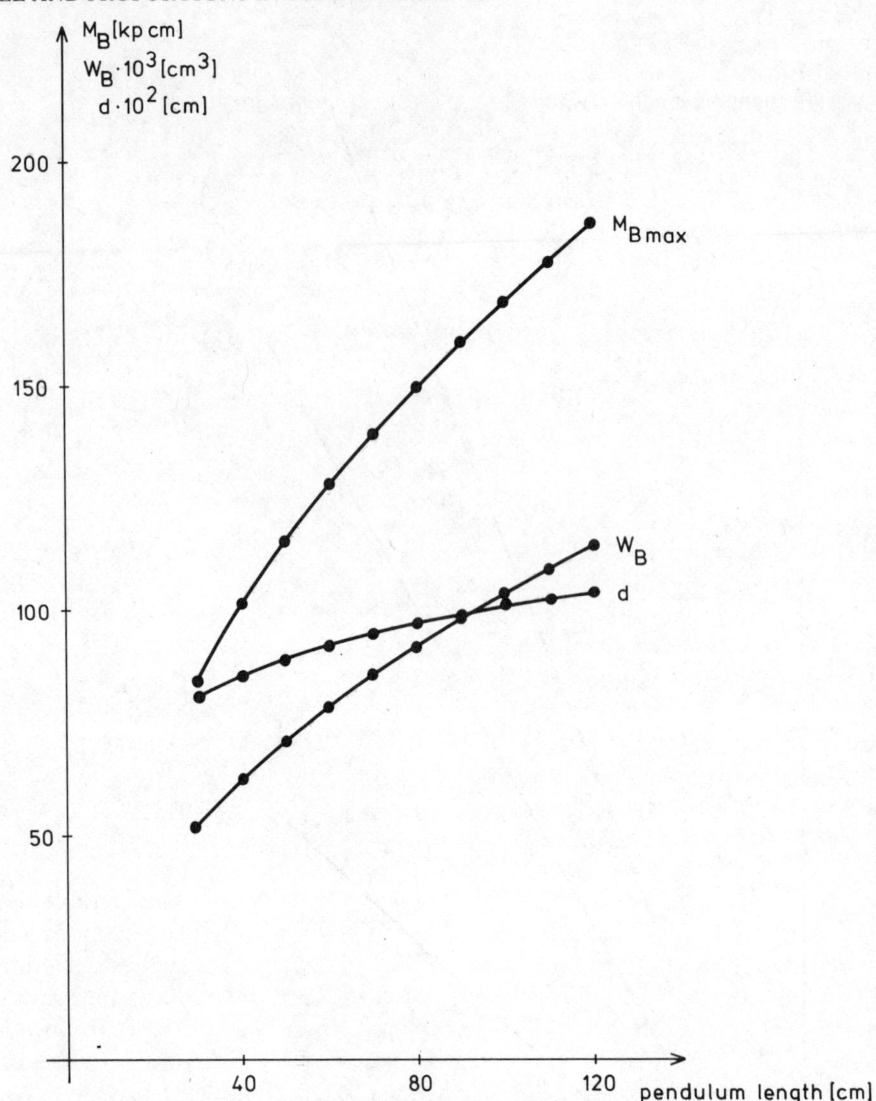

Fig. 9. Increase of maximum bending moment $M_{B\ max}$, moment of resistance against bending W_B, and bone diameter d with length. The moment of resistance and the bone diameter are calculated under the assumptions that bending stress is kept constant and that the bone is cylindrical.

they often make use of the kinetic energy stored in the swinging body. To gain the potential energy necessary for arm swinging, the body's center of gravity must be lifted to a given height. This requires only a slight flexion of the arm. As shown in an earlier study (Preuschoft and Demes, 1984), an animal with long forelimbs needs less muscle force than one with short arms of the same body mass to reach the same height of lift (equivalent to the gain of potential energy). It must not flex its arm to the same degree, and the ratio of load arm to power arm is more favorable. Although the muscle forces therefore decrease with arm length, the bending moments increase (Fig. 9). If the bending stresses are to remain constant, the resistance against bending must increase as well.

If the radius of a hylobatid is cylindrical in shape, the moments of resistance against bending of its cross sections are

$$W_B = \pi d^3/32 \approx d^3/10$$

with d the diameter.

From the moment of resistance, the increase of bone diameter necessary to keep the bending stress constant can be deduced:

$$d = (10W_B)^{1/3}$$

We started our calculations from a siamang, whose radius has a diameter of 0.95 cm at the insertion of the biceps, where the bending moments are at a maximum. The moment of resistance of a cross section at this point is 0.086 cm^3. The maximal bending moment for a siamang of 10 kg body mass who flexes its arm to reach a height of lift of 10 cm is ~140 kp cm. The bending stress, given by the ratio of bending moment to moment of resistance, is 1.628 kp/cm^2. If this value is to be maintained in animals with longer arms, the moment of resistance must increase; in animals with shorter arms, it can become smaller (Fig. 9). The curve labeled d in Fig. 9 shows how the diameter must change if the safety margin is to be kept constant.

Although the increase of bone diameter with length is considerable, it is less than length increase itself. The *relative* diameter therefore decreases with arm length (Fig. 10). That means that the longer the arm, the more slender the radius can be without an increase of bending stresses. The same should hold true for the other arm bones, because the bending stresses follow the same function in all cases. The relative decrease of diameter with length is not only a consequence of the slighter flexion necessary to reach a given height of lift in animals with long forelimbs. A similar result is obtained if a constant angle of flexion at the elbow joint is assumed. This is shown for an angle of $\beta = 120°$ in Fig. 10 (dotted line).

To conclude, bending strength does not seem to be a limiting factor for arm length. In contrast, our calculations have shown that elongation of the forearm bones permits relatively shorter diameters. To put it the other way round: Because hylobatids possess long upper extremities, they are able to avoid high bending stresses and consequently may have gracile arm bones.

Mass was assumed to be constant for these last calculations. If mass is enlarged by the same factor as arm length raised to the third power, the ratio of bone thickness to bone length increases (broken line in Fig. 10). This increase is slight, however, so that an isometric growth of total body size

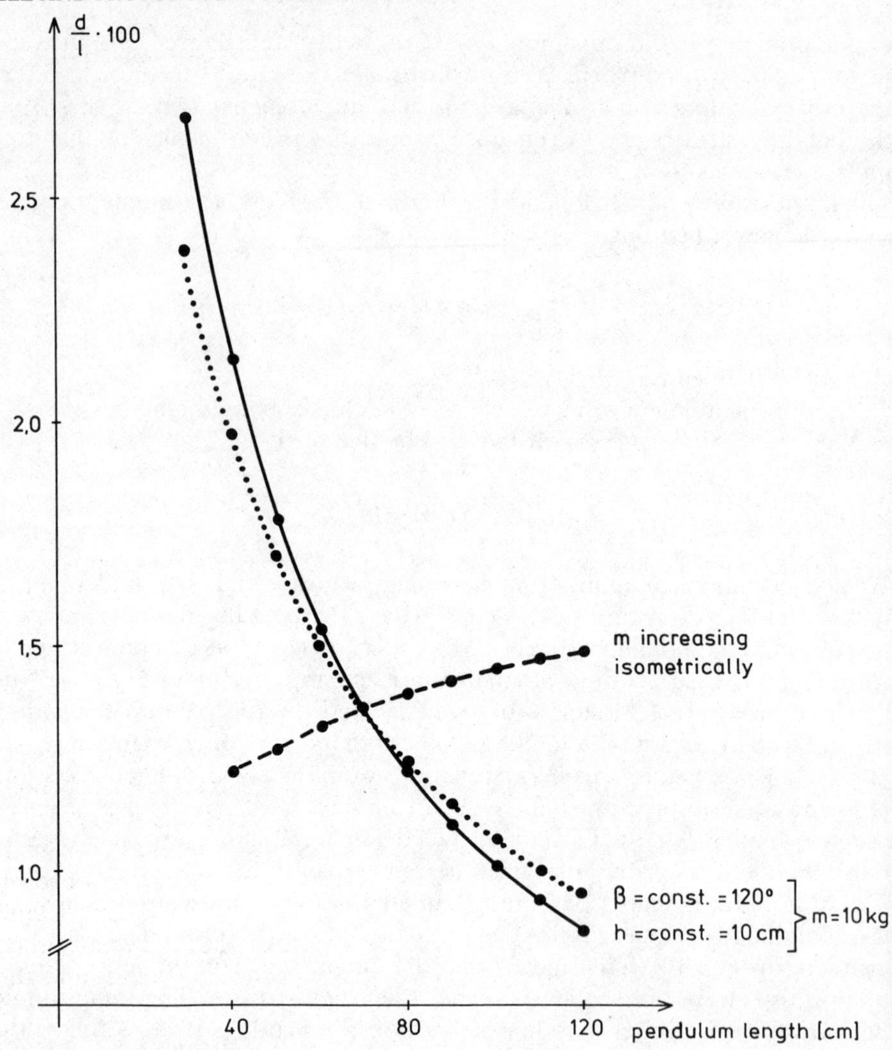

Fig. 10. Bone diameter in percent of length, as a function of length. The solid line indicates the relative bone diameter necessary to keep bending stress constant if the arm is elongated without an increase of body mass and the elbow is flexed to a degree to reach a constant height of lift (h = 10 cm). The dotted line shows the same for a constant angle of flexion at the elbow joint (β = 120°). The dashed line indicates the increase of bone diameter necessary to keep the bending stress constant if body mass is increased isometrically with arm length.

(which includes mass) does not require a much faster (allometric) growth of bone diameter in order to maintain identity of strength.

Our theoretical predictions fit rather well with the relations between body mass, bone diameter, and bending moments based on empirical data given by Alexander *et al.* (1979*b*, 1981). If we insert our calculated bending moments into an allometric equation of the form $M_B = am^x$, we get a proportionality to (body mass)$^{1.18}$. Alexander *et al.* (1981) measured muscle masses and lever arms in 37 mammalian species and calculated for these animals the maximum bending moments. They turned out to be proportional to (body mass)$^{1.2}$, a value in reasonable agreement with our result. Under the assumption of constant bending stresses we predict bone diameters growing proportional to (body mass)$^{0.39}$. Alexander *et al.* (1979) found diameters of corresponding bones in different mammals to be proportional to (body mass)$^{0.36}$.

According to Jungers (1984), arm length *within* hylobatids increases more or less isometrically with body weight from the smaller gibbons to the larger siamang. Therefore on the basis of our calculations we expect a slightly positive allometry of relative bone diameters with body weight. Jungers (personal communication) confirms that antero-posterior diameters of the humerus (i.e. diameters in the plane of bending) indeed scale slightly positive with body weight. However, according to Schultz (1953) and Knussmann (1967) the relation between girth and length of the arm bones is roughly the same for gibbons and siamangs. This means that relative bone diameters decrease with body weight. Jungers, on the other hand (personal communication), found a negative allometry for mediolateral diameters (vertical to the plane of bending). This partial difference between empirical data and theoretical predictions may be due to the narrow data base as well as to yet unobserved behavioral differences.

Preuschoft and Weinmann (1973) tried to explain the negative allometry of relative thickness of the arm bones with body weight within the hylobatids by a more rapid increase of strength over weight. This does not seem to hold true. In addition, the authors varied the external reaction force (*Gp*, which is an equivalent to body mass), while maintaining limb dimensions constant. This procedure led to the assumption of a linear growth of the bending moments with body weight. Instead, the authors should have varied limb dimensions as the third root of the external force acting against the extremity, as has been done in this chapter. By doing so, they would have found the nonlinear growth of bending moments with increased weight.

Summary

1. Body mass of brachiators is limited by the force of the finger flexors. Muscle force does not increase to the same degree with mass as does the centrifugal force plus body weight that tend to dorsiflex the fingers. Necessary and available forces intersect at a body mass of 30 kg. Assuming a security factor of three, siamangs seem to represent the upper limit of body weight for brachiators.

2. Arm length of brachiators is limited by an inevitable corresponding mass increase, as the muscle forces necessary to move the elongated arm grow by disproportionally high factors.

3. Since long arms reduce bending stresses of the arm bones in arm-swinging locomotion, bone strength is not the limiting factor for length of forelimb segments. In contrast, long-limbed animals may possess more slender arm bones than short limbed ones without a reduction of strength, provided that body mass is constant.

4. In an isometrically enlarged brachiator, the diameters of forearm bones should grow proportional to (body mass)$^{0.39}$ in order to guarantee the same strength. This gives a biomechanical explanation for the relation between bone diameter and body mass that has been found in a very similar form for different mammal species on the basis of empirical data (Alexander *et al.*, 1979*b*).

5. Siamangs show the same relation of arm length to bone diameter as the smaller gibbons do. The predicted values for relative bone thickness are slightly greater than those found empirically.

References

Alexander, R. McN., Maloiy, G. M. O., Hunter, B., Jayes, A. S., and Nturibi, J. 1979*a*. Mechanical stresses in fast locomotion of buffalo (*Syncerus caffer*) and elephant (*Loxodonta africana*). *J. Zool. Lond.* **189:**135–144.

Alexander, R. McN., Jayes, A. S., Maloiy, G. M. O., and Wathuta, E. M. 1979*b*. Allometry of the limb bones of mammals from shrews (*Sorex*) to elephant (*Loxodonta*). *J. Zool. Lond.* **189:**305–314.

Alexander, R. McN., Jayes, A. S., Maloiy, G. M. O., and Wathuta, E. M. 1981. Allometry of leg muscles of mammals. *J. Zool. Lond.* **194:**539–552.

Bühler, P. 1972. Sandwichstrukturen der Schädelkapseln verschiedener Vögel—Zum Leichtbauprinzip bei Organismen. *Mitt. Inst. Leichte Flächentragwerke Univ. Stuttgart IL* **4:**40–51.

Fleagle, J. G. 1974. Dynamics of a brachiating siamang *Hylobates* (*Symphalangus*) *syndactylus*. *Nature* **248:**259–260.

Fleagle, J. G. 1977. Brachiation and biomechanics: The siamang as example. *Malay. Nat. J.* **30**(1):45–51.

Galilei, G. 1914. *Dialogues Concerning Two New Sciences* (H. Crew and A. DeSalvio, transl.), Macmillan, New York (original publication, 1638).

Jungers, W. L. 1984. Scaling of the hominoid locomotor skeleton with special reference to the lesser apes, in: *The Lesser Apes, Evolutionary and Behavioural Biology* (H. Preuschoft, D. Chivers, W. Brockelman, and N. Creel, eds.), pp. 146–169, Edinburgh University Press, Edinburgh.

Knussmann, R. 1967. *Humerus, Ulna und Radius der Simiae*, S. Karger, Basel.

Kummer, B. 1970. Die Beanspruchung des Armskeletts beim Hangeln. Ein Beitrag zum Brachiatorenproblem. *Anthropol. Anz.* **32:**74–82.

Kummer, B. 1972. Biomechanics of bone: Mechanical properties, functional adaptation, in: *Biomechanics—Its Foundations and Objectives* (Y. C. Fung, N. Perrone, and M. Anlicker, eds.) pp. 237–271, Prentice-Hall, Englewood Cliffs, New Jersey.

Pauwels, F. 1948. Die Bedeutung der Bauprinzipien des Stütz- und Bewegungsapparates für die Beanspruchung der Röhrenknochen. *Z. Anat. Entwicklungsgesch.* **114:**129–166. Reprinted in: Pauwels, F. 1980. *Biomechanics of the locomotor apparatus*. Springer-Verlag, Berlin.

Preuschoft, H. 1971. Body posture and mode of locomotion in early Pleistocene hominids. *Folia Primatol.* **14:**209–240.

Preuschoft, H. 1973*a*. Functional anatomy of the upper extremity, in: *The Chimpanzee,* Vol. 6, pp. 34–120, S. Karger, Basel, and University Park Press, Baltimore.

Preuschoft, H. 1973*b*. Body posture and locomotion in some East African Miocene Dryopithecinae. *Hum. Evol.* **XI:**13–46.

Preuschoft, H., and Demes, B. 1984. Biomechanics of brachiation, in: *The Lesser Apes: Evolutionary and Behavioural Biology* (H. Preuschoft, D. Chivers, W. Brockelmann, and N. Creel, eds.), pp. 96–118, Edinburgh University Press, Edinburgh.

Preuschoft, H., and Weinmann, W. 1973. Biomechanical investigations of *Limnopithecus* with special reference to the influence exerted by body weight on bone thickness. *Am. J. Phys. Anthropol.* **38:**241–250.

Schmidt-Nielsen, K. 1977. Problems of scaling: Locomotion and physiological correlates, in: *Scale Effects in Animal Locomotion* (I. J. Pedley, ed.), pp. 1–21, Academic Press, London.

Schultz, A. H. 1953. The relative thickness of the long bones and the vertebrae in primates. *Am. J. Phys. Anthropol.* **11:**277–311.

Tuttle, R. H. 1968. Does the gibbon swing like a pendulum? *Am. J. Phys. Anthropol.* **29:**132.

Intraspecific, Interspecific, Metabolic, and Phylogenetic Scaling in Platyrrhine Primates

18

SUSAN M. FORD
AND ROBERT S. CORRUCCINI

Introduction

The relative scaling of body parts to body size is an important consideration in modern primatology. Allometric studies allow investigation of changes in shape that occur in conjunction with changes in body size separately from those independent of body size change, the latter often a result of significant new adaptations (Aiello, 1981; Gould, 1975).

Most current studies have concentrated on interspecific scaling in adult animals (e.g., Aiello, 1981; Alexander *et al.*, 1979; Biegert and Maurer, 1972; Bronson, 1981; Corruccini and Henderson, 1978; Gingerich, 1977; Goldstein, *et al.*, 1978; Hylander, 1975; Jungers, 1978, 1979; Kay, 1975; Pirie, 1978; Prothero and Sereno, 1982; Radinsky, 1967; Roth and Thorington, 1982; Smith, 1981; Steudel, 1981). However, several different kinds of allometry exist: ontogenetic or developmental scaling, static adult intraspecific scaling, and interspecific scaling both across a large range of species and within presumed lineages or distinct clades (evolutionary scaling). The rela-

SUSAN M. FORD AND ROBERT S. CORRUCCINI • Department of Anthropology, Southern Illinois University, Carbondale, Illinois 62901.

tionship among these different allometric patterns is unresolved. Some have suggested that interspecific scaling may represent a simple extension of ontogenetic scaling, perhaps of a common ancestral growth pattern (Bauchot and Stephan, 1969; Cock, 1966; Gould, 1966, 1975, 1977; Huxley, 1932; Jerison, 1973; Kurtén, 1954; Lumer, 1939; Marshall and Corruccini, 1978). As pointed out by Gould (1966, 1977) and Jungers (1979), this "hypermorphosis" (Gould, 1977) can be contrasted with interspecific scaling that does *not* follow the ontogenetic patterns of the included species or some ancestral species, but is the result of mechanical requirements of increased or decreased size to retain the functional efficiency of a particular biophysical relationship. These two types of interspecific scaling have been suggested for dwarfing lineages by Marshall and Corruccini (1978), corresponding to their "phenotypic" and "genotypic" dwarfing.

Lande (1979) strongly argues against any necessary correlation among ontogenetic and interspecific scaling, at least between brain weight and body weight. Shea (1981), however, stressed that the differences may in part be artificial, citing von Bertalanffy and Pirozynski (1952) and Dodson (1975) to show that developmental shape changes may themselves ultimately be the result of biomechanical function. Three independent factors influence the developmental growth and resulting adult proportions of different body parts: time of onset of growth, speed of growth, and duration of growth (Gould, 1977; Lumer and Schultz, 1947). As Aiello (1981) has indicated, alterations in any of these can lead to significant changes in ontogenetic and static adult allometry.

The relationship between static adult (intraspecific) allometry and other types of scaling is also unclear. Many have suggested that adult intraspecific allometry can be used to infer ontogenetic or developmental scaling (e.g., Corruccini and McHenry, 1979; Hersh, 1934; Lumer, 1939; McHenry and Corruccini, 1981; Reeve and Murray, 1942; Zihlman and Cramer, 1978). Thorington (1972, pp. 13 and 14) reasoned that, since all postinfant individuals are functioning in the same environment as adults are, "functionally important relationships between linear measurements [should be] the same or similar to the relationships that would be found in a sample of important relationships between linear adults of the same species." However, in testing this with gray squirrels, he found it to be true for some features but not for others. Shea (1981) also found significant differences between ontogenetic and adult intraspecific allometric scaling in hominoids. It has been argued that there is no reason to expect a correspondence between adult intraspecific variation and either developmental variation (Cock, 1966) or interspecific variation (Simpson, 1953), except where there is very little individual variation in ontogeny and small adults appear similar to early developmental stages of large adults of the same species (Gould, 1966, 1977; Lande, 1979).

In allometric studies, as well as in other studies of the functional and metabolic relationships between features of biological organisms, "body size" is a loosely used term. Many different variables have been equated with or used as estimators of "body size," including various skull dimensions (Corruccini, 1980; Corruccini and Ciochon, 1979; Corruccini and Henderson, 1978; Gould, 1975; Lavelle, 1974; Pilbeam and Gould, 1974; Pirie, 1978; Prothero and Sereno, 1982; Radinsky, 1967; Smith, 1981; Wood, 1979*a,b*), head and body length or stature in humans (Albrecht, 1980; Anderson *et al.*, 1977; Filippson and Goldson, 1963; Fooden, 1975; Garn and Lewis, 1958; Garn *et al.*, 1968; Henderson and Corruccini, 1976; Roth and Thorington, 1982), trunk or vertebral length (Aiello, 1981; Biegert and Maurer, 1972; Corruccini, 1978; Erikson, 1963; Henderson and Corruccini, 1976; Jungers, 1978, 1979; Lumer and Schultz, 1941, 1947; Schultz, 1937, 1938, 1953; Thorington, 1972), femur length (Gould, 1975; Lavelle, 1977; Wood, 1979*a,b*), and body weight (Aiello, 1981; Anderson *et al.*, 1977; Bronson, 1981; Gingerich, 1977; Goldstein *et al.*, 1978; Hylander, 1975; Jungers and Fleagle, 1980; Jungers and German, 1981; Lande, 1979; Prothero and Sereno, 1982; Roth and Thorington, 1982; Shea, 1981; Smith, 1981; and Wolpoff, 1973). This last, body weight, is frequently presumed "according to convention" (Roth and Thorington, 1982, p. 171) to be equivalent to "body size." Many have referred to body weight or mass as a "known [measure] of total size" (Jungers and German, 1981, p. 199) and selected other measures based on their correlation and isometric relationship with body weight (Corruccini, 1980; Corruccini and Henderson, 1978; Jungers, 1978, 1979; Smith, 1981), although Mosimann and James (1979) have pointed out that no one measure of size is sufficient to serve all purposes.

A readily available source of data to examine different types of allometry and including at least two possible measures of body size (body weight, trunk length) is information on specimen tags and in catalogues of museums. The National Museum of Natural History (USNM) now has this information stored on a computer for platyrrhines. These data are especially plentiful for platyrrhines because of the yellow fever study collections recently made in the wild. Platyrrhines are especially useful for this comparison because, in addition to providing developmental and static adult intraspecific samples, much recent work has strongly indicated that both increases and decreases (phyletic dwarfing) in body size have occurred in various platyrrhine lineages (Ford, 1980*a,b;* Garber, 1980; Leutenegger, 1973, 1980; Rosenberger, 1977, 1979). Therefore, suggested differences in interspecific allometric scaling of dwarfing lineages (Gould, 1975; Marshall and Corruccini, 1978; see also Ford, 1980*a;* Prothero and Sereno, 1982) can also be examined. These data are used here to investigate the relative value of different estimators of body size and the relationship between intraspecific adult and relative growth allometry and interspecific allometry of different platyrrhine clades.

Materials and Methods

The data used were those available in the National Museum of Natural History catalogue, repeated on specimen tags. The variables are the standard measurements taken by field collectors: body weight (BW); total length (TOTL); trunk length, head and body (TrL); tail length (TAILL); foot length, hind foot (FOOTL); and ear length (EARL). All measurements were taken externally on fresh specimens by the individual collector at the time of collection. Data on some variables were not available on every specimen; when only one of the three variables, TOTL, TrL, and TAILL was missing, it was computed from the other two using the formula TOTL = TrL + TAILL.

These data provide a sample of easily accessed measurements on an astoundingly large series of platyrrhine specimens, totalling 1522, and include information on features of importance in studies of various types of allometry as well as possible metabolic scaling, if ears and tails are being used as heat-dissipating organs (Rosenblum and Schwartz, 1982). There are some obvious limitations and potential problems inherent in use of these data. Primary is the fact that measurements were taken by different individuals using different instruments and, on occasion, perhaps employing nonstandard landmarks or techniques. This problem exists in all studies using body weight data from museum specimen tags. In addition, no detailed information on age is provided for many specimens. In these cases, specimens were

Table 1. Minimum Body Weight and Trunk Length for Adult Specimens

Species	Minimum adult body weight, kg	Minimum adult trunk length, m	Sources[a]
Ateles geoffroyi[b] ♂	5.00	0.40	1,3,4
♀	—	0.30	
Alouatta belzebel	4.00	0.41	(1,3,4)
Alouatta caraya	4.00	0.41	(1,3,4)
Alouatta fusca	4.00	0.41	(1,3,4)
Alouatta palliata[b]	4.00	0.41	5 (1,3,4)
Cebus apella[b]	1.80	0.36	1,4 (3,5)
Cebus capucinus[b]	1.80	0.36	1,3–5
Callicebus moloch	0.60	0.25	1,4 (5)
Callicebus personatus	0.60	0.25	(1,4,5)
Aotus trivirgatus[b]	0.45	0.24	1,3–5
Saimiri sciureus[b] ♂	0.55	0.22	1,3–5
♀	0.365	—	
Saguinus fuscicollis	0.25	0.175	2 (1,3,4)
Saguinus labiatus	0.30	0.22	2 (1,3,4)
Saguinus midas[b]	0.30	0.206	1,2,4 (3)
Saguinus mystax	0.40	0.22	2 (1,3,4)
Callithrix humeralifer	0.17	0.17	2 (1,3,4)
Callithrix jacchus[b]	0.17	0.17	1–4
Cebuella pygmaea	0.14	0.12	1,2

[a]All rely on values for specimens in the USNM collection identified as adult or juvenile. Sources listed are: (1) Gaulin and Konner (1977), (2) Hershkovitz (1977), (3) Kay (1975), (4) Napier and Napier (1967), (5) Smith (1981). Sources in parentheses give data for other species of that genus.

[b]Species used in both intra- and interspecific samples.

classified as juvenile or adult by their body weight and trunk length, based on published ranges for the species examined and values in the USNM data for identified adult and juvenile individuals. The ranges are given in Table 1. There is a clear bias toward adults in the sample, which, as noted in a similar study by Lumer and Schultz (1947), is unavoidable in data of this nature.

The sample for the intraspecific aspect of the study utilized the 456 (with body weight as the estimator of body size) and 943 (with trunk length as the estimator) specimens which comprised the largest intraspecies samples. They include eight species (sample size for specimens with body weight given; see also Tables 3–5): *Ateles geoffroyi,* 23 (11 adult); *Alouatta palliata,* 42 (35 adult); *Cebus apella,* 269 (185 adult); *Cebus capucinus,* 27 (20 adult); *Saimiri sciureus,* 25 (18 adult); *Aotus trivirgatus,* 26 (25 adult); *Saguinus midas,* 17 (all adult); and *Callithrix jacchus,* 27 (25 adult). Sexes were analyzed both separately and combined for the two sexually dimorphic species with the largest samples, *Cebus apella* and *Alouatta palliata.*

The interspecific sample included all species for which there were body weights for both males and females, and was limited to those adult specimens for which body weight was available. Eighteen species were used (sample sizes are given in Table 2): *Ateles geoffroyi, Alouatta belzebel, Alouatta caraya, Alouatta fusca, Alouatta palliata, Cebus apella, Cebus capucinus, Callicebus moloch, Callicebus personatus, Aotus trivirgatus, Saimiri sciureus, Saguinus fuscicollis, Saguinus labiatus, Saguinus midas, Saguinus mystax, Callithrix humeralifer, Callithrix jacchus,* and *Cebuella pygmaea.* Mean values for each species were determined in two different ways: first, by taking the mean of all specimens of that species, and

Table 2. Number of Specimens per Species for Interspecific Sample[a]

Species	Males	Females	Total
Ateles geoffroyi	6	5	11
Alouatta belzebel	4	6	10
Alouatta caraya	4	4	8
Alouatta fusca	5	4	9
Alouatta palliata	19	16	35
Cebus apella	95	90	185
Cebus capucinus	17	3	20
Callicebus moloch	2	3	5
Callicebus personatus	2	1	3
Aotus trivirgatus	10	15	25
Saimiri sciureus	9	10	19
Saguinus fuscicollis	1	2	4
Saguinus labiatus	1	1	2
Saguinus midas	10	7	17
Saguinus mystax	7	2	9
Callithrix humeralifer	2	2	4
Callithrix jacchus	11	14	25
Cebuella pygmaea	1	1	2

[a]Only specimens with data on body weight used; see text.

second, by computing a mean value for males and a mean value for females and then taking the average of those two values to represent the species. The second method controls for an overrepresentation of one or the other sex, which could otherwise skew the resulting means, especially in species that are sexually dimorphic. Results using the second method are labeled "averaged by sex" in Tables 3–5.

All values were log-transformed, with log of body weight divided by three to allow it to scale linearly with the other attributes. Pearson product-moment correlation coefficients r were computed. Allometric relationships between different pairs of variables were calculated using the log-transformed data.

There exists much controversy over the most appropriate method of estimating the allometry coefficient k of the allometry equation

$$Y = bX^k$$

(Jolicoeur, 1963, 1973, 1975; Kuhry and Marcus, 1977; Manaster and Manaster, 1975; Zar, 1968). A common method is the least squares regression (LS) of logarithmically transformed data, using the regression formula

$$\log Y = k \log X + \log b$$

However, this method is based on the assumption that Y is dependent on X (which is usually an estimator of body size) (Gould, 1975; Kuhry and Marcus, 1977). Use of a "best fit" method, such as the major axis (MA) and standard or reduced major axis (RMA) techniques, allows for residual variance in both X and Y and includes a consideration of the degree of correlation between the two variables, more closely reflecting most biological situations (Gould, 1975; Jolicoeur, 1963, 1968, 1973; Jungers, 1979; Kuhry and Marcus, 1977; Roth and Thorington, 1982; Thorington, 1972; Wood, 1979*a*). The LS method tends to give a lower estimate of k than any of the methods that consider error in both variables, departing more and more as the correlation coefficient r decreases (Corruccini, 1980; Gould, 1975; Jungers, 1979; Prothero and Sereno, 1982). Since here we are investigating the relationship between two variables of biological organisms, both of which exhibit variance and uncontrolled error, we prefer the reduced or standard major axis estimate of the allometry coefficient k (Kermack and Haldane, 1950). Both LS and RMA estimates are presented in Tables 3–5 but subsequent analysis and the following discussion centers on the standard or reduced major axis estimates (RMA).

Correlation and allometry coefficients were computed using standard SPSS programs (Nie *et al.*, 1975) for each species in the intraspecific developmental and adult samples and for several different subsets of the interspecific sample. In order to compare alternate methods of determining species means, as well as possible sex differences, coefficients were computed for the total sample of all 18 platyrrhine species using the overall species means,

means averaged by sex, means for male specimens only, and means for female specimens only.

Samples were also selected in order to test for possible differences between clades, in particular, a clade resulting from size increase versus one exhibiting size decrease. As discussed above, there is now substantial evidence to indicate that, contrary to Hershkovitz (1977), the callitrichids are phyletic dwarfs (Ford, 1980*a;* Garber, 1980; Leutenegger, 1973, 1980; Rosenberger, 1977, 1979). Due to the absence of any fossil record for ancestors of this group during the period of dwarfing, it is impossible to examine allometry coefficients from representatives of stages of the actual dwarfing event, as was done by Marshall and Corruccini (1978) and Prothero and Sereno (1982). However, extant callitrichids span a fairly large size range, from the relatively large Goeldi's monkey (*Callimico*) and slightly smaller *Saguinus mystax* (used in this study) to the diminutive pygmy marmoset (*Cebuella*). These species can be used as living representatives of different ancestral stages in the dwarfing event, interpreted as side branches of "arrested dwarfing" at progressively smaller sizes (Ford, 1980*a*). In this manner, a less perfect but alternative sample (in the absence of a fossil record) for examining allometric trends in dwarfing lineages can be studied.

This can be compared to coefficients of samples of progressively larger platyrrhines that are presumed to be the result of phyletic size increase, in this case, members of the Atelinae, including *Alouatta* (Ford, 1980*b*, 1981; Hershkovitz, 1977; Rosenberger, 1979). *Cebus* spp. and *Saimiri sciureus* are excluded from both of these samples; their phylogenetic affinities have long been a source of controversy [see reviews in Ford (1980*b*) and Rosenberger (1979, 1981)] and quite possibly represent independent, parallel cases of size increase and decrease, respectively (Ford, 1980*b*, 1981; Rosenberger, 1979). *Callicebus* and *Aotus* may come closest to representing the ancestral body size (and locomotor type) of all living platyrrhines (Ford, 1980*b*, 1981), and so are included in both the sample of species resulting from phyletic size increase (the atelines) and that from size decrease (the callitrichids). In addition, the seven callitrichid species used here were examined without *Aotus* and *Callicebus* spp. (the sample of ateline species was too small and unvarying to yield significant results alone).

RMA estimates of the allometry coefficient were compared between samples (interspecific and intraspecific) to determine if they were sampled from the same statistical universe (within 95% confidence limits) using the formula for the *z*-statistic given by Sokal and Rohlf (1981):

$$z = \frac{k_1 - k_2}{(S_1^2 + S_2^2)^{1/2}}$$

where k_i is the RMA estimate of the allometry coefficient for group i (species or interspecific sample) and S_i^2 is the standard error squared for group i, or:

$$S_i^2 = \frac{S_Y^2}{S_X^2}\left(\frac{1 - r^2}{N}\right)$$

with S_Y^2 and S_X^2 the standard deviation squared of Y and X variables, respectively, r is Pearson's r correlation coefficient, and N is the sample size of the group.

Results

Choice of Body Size Variable

Almost uniformly, low correlations are found between both body weight and trunk length and the other variables (see Tables 3–5). Only 11 correlation coefficients out of 60 in the adult intraspecific data are over 0.6 (three each with TOTL and TrL, one with TAILL, four with FOOTL), none greater than 0.83. Trunk length fares no better, with only 10 of 48 correlation coefficients over 0.6, all with TOTL. Somewhat higher correlations are found for the mixed age or developmental sample of each species. Of 50 correlations with body weight, 43 are greater than 0.6. Nine have values for r greater than 0.9: the correlations with TOTL and TrL in *Cebus apella* males, with TOTL in *Callithrix jacchus,* and with FOOTL in *Aotus,* and five correlations in *Alouatta palliata* combined and separate sex samples. Correlations are especially low between EARL and BW. They are generally markedly higher between TOTL and BW than between TrL and BW. Fifty percent of the 48 correlations between trunk length and the other variables (excluding BW) are greater than 0.6, six of these greater than 0.9 (five with TOTL, one with TAILL in *Alouatta palliata* males). Only 11 of 50 correlations with body weight are less than 0.9 in the interspecific sample, 10 of these between BW and EARL. Fourteen of 40 interspecific correlations with trunk length are less than 0.9, again 10 with EARL.

A number of explanations can be offered for these findings. The first is that the range of the data can affect correlation. Wood (1979*a*) and Steudel (1981) also found low correlations between body weight and other variables in an intraspecific study. Both they and recently Smith (1981) have suggested that one could expect correlations to improve as the study increased its scale (i.e., to interspecific comparisons), and Wood felt that one would get the lowest correlations in the narrowest samples, those of single-sex intraspecific comparisons. We do find overall that r increases as size range of the sample increases—from adult intraspecific to mixed age intraspecific to interspecific. In fact, the increase in average correlation coefficient from the intraspecific (mixed age) to the interspecific samples is far greater than Smith's (1981) postulated 16–19%, ranging from 43% to 79%. But even the larger correlation coefficients in the interspecific samples are not uniformly 0.95 or better. The correlations between TrL and BW, TAILL and TrL, FOOTL and TrL, and especially EARL and both BW and TrL are frequently lower than 0.95, in the case of EARL ranging from 0.76 to 0.85. Also, there are two striking exceptions to Wood's postulated trend. In the two species where single-sex samples were examined for the developmental or mixed-age samples, *Alouatta palliata* females (where X = BW) and *Cebus apella* males consistently showed stronger correlations than either combined-sex sample. This may indicate much greater consistency in one sex of the relationship between individual variables during growth and development (but not in static adult samples or across species with adult-only means).

A second possibility is that these values reflect the lack of a linear relationship between body size measures and the other variables. Ear length in particular has low correlations with BW or TrL. Individual species may vary as well. In a study of cross-sectional growth data for *Cebus capucinus* and *Ateles geoffroyi,* Lumer and Schultz (1947) found that their measure of body size exhibited far less of a linear relationship to other variables in *Cebus capucinus* than it did in *Ateles.* This is consistent with our findings in the mixed-age sample and may reflect a real difference in the growth patterns of these two species; however, it cannot be generalized to the other species of *Cebus* examined here, especially the *Cebus apella* male sample. Correlation coefficients for this species are generally equal to or greater than those for *Ateles geoffroyi.* The smaller bodied species in general exhibited the lowest correlation coefficients in the developmental (mixed age) data. Three marked exceptions are the unusually high correlations in *Callithrix jacchus* of TOTL to BW and EARL to both BW and TrL. This may reflect real deviations in the nature of the relationship between the variables examined.

But this does not explain the almost uniformly low correlations between presumed measures of "body size"—body weight and trunk length—and the frequent departure from isometry between these variables, exhibiting both negative and positive allometry. Smith (1981) points out that both criteria should be met for variables to be substituted for one another. Aiello (1981) has also found that trunk length and body weight are not isometric in anthropoids. Which brings us to the third possibility, that body weight may not be the best reflection of the enigmatic object of our quest—"body size." Others (Pilbeam and Gould, 1974; Gould, 1975; Corruccini and Henderson, 1978; Corruccini, 1980) have expressed some concern over the frequent necessity of relying on mean species weights reported in the literature, usually gathered from a number of sources. These data do not account for the marked individual variation that can occur (Wrazen and Wrazen, 1982). Only a few studies have had access to samples with weight data for the specimens examined (e.g., Jungers and Fleagle, 1980; Jungers and German, 1981; Shea, 1981; Roth and Thorington, 1982; and this study), and many are based on animals raised in captivity. A number of complicating factors can be implicated in the use of body weight data in some or all of these studies. Rarely, if ever, is information given with wild-killed or captive material on the time of day the animal died, whether it had a full, partially full, or empty stomach, how much blood or other tissue loss occurred before weighing (as from a bullet wound), and, if female, whether she was pregnant (and how much) or lactating. Yet all of these factors can greatly alter an animal's body weight over a brief period of time. In addition, captive animals are notorious for having large and somewhat uncontrolled amounts of body fat accumulations. And of course, with data from museum collections or compiled from the literature, there is the problem of interobserver error.

Aside from the issue of strength of correlation, measurement error is probably greater with dead-specimen body weights than with subsequently collected dimensions. The stability of a broad "mouse-to-elephant" regression is usually cited to assuage that concern. However, it now seems that inaccurate species weight means will have an effect, as such regression slopes may be unduly influenced by the endpoints, i.e., the "mouse" and the "elephant" (e.g., Susman and Creel, 1979).

Despite these complicating factors, an argument could be made that body weight data gathered by trained collectors on wild-caught specimens as part of a unified study would be a valid estimator of "body size." However, in a recent study on African squirrels, Roth and Thorington (1982) have shown that gross body weight is not always a good estimator of overall size, as some squirrels are disproportionately heavy animals. They suggest using lean eviscerated body weight, under the assumption that body fat or gut size is influencing gross body weight. Jungers and Fleagle (1980), in their study of two species of *Cebus,* also found one species to be markedly more heavy and stocky, but in this case it was suggested that the difference may be due to increased physical strength in one species related to foraging strategies, so lean eviscerated weight would still reflect body weight differences.

Clearly, as others have cautioned, choice of a measure of "body size" can greatly affect analyses of various types of allometry. Certainly, as most recently pointed out by Lindstadt and Calder (1981), body mass is a prime determinant of physiological and metabolic requirements of an animal and should be the size variable of choice in studies of purely metabolic scaling. As will be discussed, this may be particularly appropriate here in the study of relative ear and tail length. But we suggest that body weight may be less appropriate in studies of mechanical scaling than other measures of overall size, such as trunk length (it is unclear why total length shows an even stronger correlation with body weight than does trunk length). We support Aiello's (1981) suggestion that the combined analysis of body weight, trunk length, or some other measure of body size and other variables of interest may prove the most useful and informative, when feasible.

Intraspecific: Mixed Age or Developmental

Many of the allometry coefficients for species are not significantly different. A few are, and comparison with findings from other studies are particularly interesting. There have been only a few studies of ontogenetic or developmental allometry in platyrrhines reported, and only one using longitudinal data, that by Jungers and Fleagle (1980) on two species of *Cebus* raised in captivity. The others rely on cross-sectional data on wild-caught specimens, as in this study, and include one on *Saguinus oedipus* (Thorington, 1972) and a comparison of *Ateles geoffroyi* and *Cebus capucinus* by Lumer and Schultz (1947). Unfortunately, these studies concentrate primarily on different features, but Jungers and Fleagle do discuss relative foot length to body weight, and Lumer and Schultz both foot and tail length to trunk length. These results can be compared to the present study.

Separate Sex Samples

Males and females were examined separately as well as combined in two species, *Cebus apella* and *Alouatta palliata* (see Table 3). Jungers and Fleagle (1980) also used separate sex samples. They do not report on whether differences are significant, and their sample sizes are small. However, *Cebus apella* females (over 1100 days) scaled more slowly than males (lower RMA coefficients) for all features except foot, where females scaled more rapidly. *Cebus albifrons* females (over 1100 days) scaled more slowly than males for all features examined, including the foot. However, this was not true for *C. albifrons* older females (to 2000 days), which scaled the same as or more quickly than males for several features (humerus, radius, femur) but retained lower RMA coefficients for foot length to body weight. In our study, foot length in *C. apella* females scales more slowly than in males relative to trunk length, but has a greater value than males when measured relative to body weight. In neither case, however, are the differences significant to the 0.05 level, as may be the case for the Jungers and Fleagle data.

In a few cases, the differences between sexes or between one sex and the total sample are significant. *Cebus apella* females have significantly larger values than males for both TAILL and EARL to BW and TrL. For all of these except TAILL/TrL, females are also significantly different from the total *C. apella* sample. *Cebus apella* females have basically an isometric relationship between tail length and body size (measured either way), while males exhibit negative allometry, also seen in the combined-sex sample; that is, in *C. apella* males, tail length increases more slowly than body size with increasing age. It should be noted that tail length and body size (measured both ways) show a much poorer correlation in *C. apella* females. For this species, males may have a fairly strong and negatively allometric growth relationship for tails, while the relationship is much weaker during ontogeny in females but averaging as isometry. For ear length, *C. apella* males show a more isometric relationship during ontogeny (to both BW and TrL), while females exhibit marked positive allometry of EARL. While correlation coefficients are low for both sexes, they are lowest in females, again perhaps indicating a less linear relationship as well as a significantly different average relationship during ontogeny. The other differences in *C. apella* males and females are not significant, although in most cases the coefficients for females are larger than those for males.

Table 3. Comparison of Body Weight (BW) and Trunk Length (TrL) to Other Parameters: Intraspecific, Mixed Age[a]

Species	N, X = BW	N, X = TrL	Total length/BW			Trunk length/BW			Tail length/BW			Foot length/BW		
			r	LS	RMA	r	LS	RMA	r	LS	RMA	r	LS	RMA
Ateles geoffroyi	23	51	0.86	0.721	0.838	0.73	0.794	1.088	0.72	0.653	0.906	0.76	0.583	0.767
Alouatta palliata	42	98	0.92	0.801	0.871	0.84	1.003	1.194	0.88	0.647	0.735	0.89	0.920	1.034
Alouatta palliata ♂	22	38	0.86	0.623	0.724	0.78	0.763	0.978	0.84	0.506	0.602	0.90	0.578	0.642
Alouatta palliata ♀	21	61	0.95	0.954	1.004	0.85	1.119	1.316	0.94	0.835	0.888	0.93	1.187	1.276
Cebus apella	269	289	0.83	0.620	0.747	0.86	0.774	0.900	0.63	0.491	0.779	0.77	0.437	0.567
Cebus apella ♂	139	152	0.90	0.676	0.751	0.90	0.807	0.897	0.78	0.564	0.723	0.81	0.460	0.568
Cebus apella ♀	130	137	0.66	0.548	0.830	0.75	0.750	1.000	0.38	0.388	1.021	0.56	0.323	0.577
Cebus capucinus	27	84	0.82	0.635	0.774	0.61	0.583	0.956	0.65	0.716	1.101	0.51	0.467	0.916
Saimiri sciureus	25	194	0.75	0.702	0.936	0.83	0.836	1.007	0.66	0.607	0.920	0.82	0.643	0.784
Aotus trivirgatus	26	115	0.85	1.089	1.281	0.87	0.959	1.102	0.82	1.203	1.467	0.91	0.956	1.051
Saguinus midas	17	22	0.65	0.799	1.229	0.42NS	0.988	—	0.65	0.600	0.923	0.81	0.642	0.793
Callithrix jacchus	27	52	0.93	0.923	0.992	0.50	0.834	1.668	0.84	0.972	1.157	0.57	0.683	1.198

Species	Ear length/BW			Total length/TrL			Tail length/TrL			Foot length/TrL			Ear length/TrL		
	r	LS	RMA	r	LS	RMA	r	LS	RMA	r	LS	RMA	r	LS	RMA
Ateles geoffroyi	0.55	0.443	0.805	0.89	0.686	0.771	0.66	0.492	0.745	0.55	0.779	1.416	0.38	0.259	0.682
Alouatta palliata	0.46	0.432	0.939	0.96	0.815	0.849	0.84	0.658	0.783	0.70	0.655	0.936	0.42	0.244	0.581
Alouatta palliata ♂	0.41NS	0.376	—	0.98	0.876	0.894	0.91	0.771	0.847	0.89	0.695	0.781	0.40	0.302	0.755
Alouatta palliata ♀	0.46	0.492	1.070	0.95	0.803	0.845	0.81	0.653	0.806	0.62	0.658	1.061	0.39	0.202	0.518
Cebus apella	0.45	0.428	0.951	0.92	0.761	0.827	0.67	0.561	0.837	0.68	0.484	0.712	0.39	0.386	0.990
Cebus apella ♂	0.46	0.392	0.852	0.95	0.795	0.837	0.78	0.619	0.794	0.71	0.512	0.721	0.42	0.369	0.879
Cebus apella ♀	0.31	0.380	1.226	0.85	0.695	0.818	0.47	0.449	0.955	0.52	0.326	0.627	0.26	0.327	1.258
Cebus capucinus	0.29NS	0.429	—	0.66	0.476	0.721	0.31	0.275	0.887	0.40	0.309	0.772	0.26	0.367	1.411
Saimiri sciureus	0.17NS	0.275	—	0.85	0.750	0.882	0.57	0.571	1.002	0.76	0.499	0.657	0.22	0.267	1.214
Aotus trivirgatus	0.65	0.477	0.734	0.87	0.656	0.754	0.42	0.336	0.800	0.51	0.274	0.537	0.24	0.201	0.837
Saguinus midas	0.46NS	0.570	—	0.77	0.492	0.639	0.19NS	0.130	—	0.46	0.174	0.378	0.24NS	0.238	—
Callithrix jacchus	0.56	1.068	1.907	0.58	0.316	0.545	−0.04NS	−0.027	—	0.09NS	0.053	—	0.64	0.662	1.034

[a]LS, Least squares estimate of allometry coefficient. RMA, Reduced major axis estimate of allometry coefficient. NS, p not significant.

In *Alouatta palliata,* most allometry (RMA) coefficients for females are also larger than those for males, but in only three cases are the differences significant. Two include the scaling of foot length to both BW and TrL. In neither case is the female value significantly different from the combined-sex sample, and both approximate isometry. The *A. palliata* males have much lower coefficients, indicating that foot length scales negatively to body size during growth and increases much more slowly than it does in females. (Correlations are fairly high in all cases.) Just the opposite is true for ear length. The correlation between EARL and BW in males is not significant, but for females the relationship approximates isometry. However, the relationship between EARL and TrL in females is highly negative (the opposite of that exhibited by *C. apella* females), and significantly more negative than that seen for *A. palliata* males (which also exhibit negative allometry of ear length to trunk length). Increases in ear length keep pace with increases in body weight during growth but not with increases in body (trunk) length, especially in females.

Species of Cebus

The two species of *Cebus* examined here (*C. apella* and *C. capucinus*) can be compared for similarities or differences in allometric relationships during growth, and also to the study of data on foot length to body weight in the study of *C. apella* and *C. albifrons* by Jungers and Fleagle (1981) and on foot and tail length to trunk length in the study of *C. capucinus* by Lumer and Schultz (1947). The differences in allometry coefficients for TOTL to body size (BW or TrL) or TrL to BW in the two species are not significant. FOOTL to TrL differences are also not significant. Jungers and Fleagle (1981) note that foot length scales more slowly than the rest of the hindlimb in most primates examined, perhaps due to the importance of grasping in the neonate, and they also have coefficients indicating negative allometry to body weight in both *C. apella* (0.333 for 2000-day males) and *C. albifrons* (0.436 for 2000-day males). Negative allometry of FOOTL to body size in *C. apella* is also seen here, even more strongly negative to body weight ($k = 0.567$) than to TrL ($k = 0.712$). This, however, cannot be generalized to the third species for which there is information, *C. capucinus.* Although FOOTL scales negatively to TrL, as was also reported by Lumer and Schultz (their value for LS was 0.552), FOOTL scales almost isometrically to BW, and is significantly different (0.01 level) from the equivalent value for *C. apella* (both combined- and single-sex samples). The same is true of TAILL—differences between the two species for TAILL to TrL are not significant, but are significant (0.01 level) for TAILL to BW. Here again *C. apella* (combined and males alone) has a lower value than *C. capucinus,* in which TAILL and BW scale isometrically. However, tails of *C. apella* females scaled to BW as in *C. capucinus.*

Jungers and Fleagle (1981) found coefficients in *C. apella* were consistently lower than in *C. albifrons* and that *C. apella* is always heavier and stockier. *Cebus capucinus* is also more slender and longer limbed than *C. apella* (Moynihan, 1976), and shows similarly less negative growth relationships of both tail length and foot length to body weight compared to *C. apella* (interestingly, this is not the case for trunk length or total length; the value of *r* for EARL/BW is not significant for *C. capucinus,* so no comparison can be made). Moynihan suggests that *C. albifrons* and *C. apella* may have extremely similar foraging strategies. However, Jungers and Fleagle, based on an unpublished study by Janson (1975), suggest there may be distinct differences in distances traveled daily and foraging techniques of the two species. This would place different selective pressures on the two species, thus giving adaptive value to their alternative body builds. In particular, *C. apella* appears to have a more active and aggressive foraging strategy. A similar difference may exist with *C. capucinus. Cebus apella* often eats extremely hard palm nuts, which must be pounded against a hard surface and broken open; *C. capucinus* has not been seen to exhibit similar nut-cracking behavior or to eat these palm nuts (Hernandez-Camacho and Cooper, 1976; Klein and Klein, 1976; Moynihan, 1976; Struhsaker and Leland, 1977). The difference does indeed seem to be one of increased body bulk in *C. apella,* as tail and foot (and presumably other aspects of the extremities) scale approximately the same in *C. apella* and *C. capucinus* relative to trunk length; the difference is in scaling to body weight. In this case, a comparison using different measures of body size indicates that extremities do not increase more slowly to body size in *C. apella,* but rather, that body weight appears to increase more rapidly than other features. It may be expected that scaling of extremities to trunk length will not be significantly different in the other *Cebus* species from that seen in *C. apella* and *C. capucinus,* but that coefficients of scaling to body weight for *C. nigrivittatus* will be intermediate between those of *C. apella* and the other two species, as it is somewhat intermediate in body weight (Moynihan, 1976) and perhaps in the rigor of its foraging strategy.

The last trait, ear length, cannot be compared between species relative to body weight, as the correlation coefficient for *C. capucinus* is not significant. But EARL scales significantly higher relative to TrL in *C. capucinus,* exhibiting positive allometry, compared to the isometric relationship seen in *C. apella* combined-sex and male samples. The *C. apella* females fall intermediate between the two species combined samples.

Other

The only prior study on other platyrrhines that examines some of the same features is that of Lumer and Schultz (1947). They found that both foot and tail length scaled negatively to trunk length in *Ateles geoffroyi* and *Cebus capucinus,* and that LS coefficients for *Cebus* were always lower than those for *Ateles.* Our LS coefficients for these two species on these comparisons give the same results, although for TAILL/TrL, *C. apella* has a larger LS value than *Ateles* or *C. capucinus.* However, when coefficients are computed using the

RMA method, a very different picture emerges. Both *Cebus* species have higher values than *Ateles* for TAILL/TrL, although all are negative and the differences are not significant (0.05 level). Also, *Ateles* shows very strong positive allometry of FOOTL/TrL, significantly (0.001 level) different from the negative scaling of both *Cebus* species. As our values for LS reveal similar relationships (though quite different absolute values) as those of Lumer and Schultz, who also reported low correlations, one could assume that the RMA method applied to their data would have led to results similar to ours.

Looking across all species examined for each trait, some general trends can be seen. (While *Saguinus midas* is included in the tables for comparison, the sample is actually all adults, so it will be included only in later discussion.) Trunk length is fairly isometric to BW in most species, with the striking exceptions of positive allometry of TrL in *Alouatta palliata* females (slightly so in the total *A. palliata* sample) and especially in *Callithrix jacchus,* and negative allometry in *Cebus apella* total and male-only samples. TOTL, TAILL, and FOOTL scale negatively to TrL in most species. This is most marked for TOTL/TrL in *Callithrix jacchus,* followed by *A. geoffroyi, C. capucinus,* and *Aotus;* for TAILL/TrL in *A. geoffroyi, A. palliata* (total sample), *C. apella* males, and *Aotus;* and for FOOTL/TrL in *C. apella* all samples, *Aotus,* and *Saimiri.* There are only a few exceptions to this negative scaling: isometry of TAILL/TrL in *Saimiri* and *C. apella* females and of FOOTL/TrL in *A. palliata* total and female samples, and marked positive scaling of FOOTL/TrL in *A. geoffroyi.* (Unfortunately, correlation coefficients are not significant for TAILL/TrL and FOOTL/TrL in *C. jacchus.*)

While TOTL and FOOTL also scale negatively to BW in most species, there are more exceptions and they are *not* always the same as for scaling of these features to TrL. Thus, while TOTL increases slowly relative to TrL in *Aotus,* it increases more quickly than BW during development, reflecting the strong positive allometry of TAILL/BW in this species. The same is true of *C. jacchus,* which exhibits relative isometry of TOTL/BW during growth (seen also in *Saimiri* and *A. palliata* females), due to the slight positive allometry of TAILL/BW. Most other species exhibit a fairly isometric relationship between TAILL/BW in ontogeny, except for the negative scaling in *A. palliata* total, female, and, especially, male samples, and *C. apella* total and male samples. FOOTL scales positively to TrL in *A. geoffroyi* (as noted above), but quite negatively to BW, with hind feet getting relatively smaller with increasing body bulk. FOOTL scales negatively to BW in most species, especially in *C. apella* combined- and single-sex samples, resulting in smaller feet in adults of this species compared to both body mass and body length. This relative decrease in FOOTL is significantly less marked in *C. capucinus* (discussed above). FOOTL maintains the same size relative to BW in *Aotus* and *A. palliata* total sample, but it exhibits significant, though slight, positive allometry to BW in *A. palliata* females and *C. jacchus.*

EARL correlates more poorly with measures of body size than do the other features, but in many cases it is significant enough to suggest some degree of a linear relationship during growth and development, to TrL more often than to BW. In many cases, the relationship is either isometric or slightly positive or negative but not significantly different from isometry. EARL does increase significantly more quickly than BW in *Callithrix jacchus* and less so in *Cebus apella* females, but scales significantly negatively in *C. apella* males and, especially, *A. geoffroyi* and *Aotus*. EARL also scales negatively to TrL in *A. geoffroyi* and slightly in *Aotus*, as well as in *A. palliata* combined- and single-sex samples. Positive allometry of EARL/TrL is exhibited by *Saimiri*, *C. apella* females, and especially *C. capucinus*.

Intraspecific: Adults Only

Separate Sex Samples

The separate male and female samples of *Alouatta palliata* adults did not have statistically significant correlations except for FOOTL/BW and TOTL/TrL, and so cannot be compared here (see Table 4). While *A. palliata* females do have a higher allometry coefficient than males for FOOTL/BW, as in the ontogenetic study, the difference is not significant at the 0.05 level.

Male and female samples can be compared for *Cebus apella*, and females consistently exhibit higher allometry coefficients than males—the same trend exhibited by the growth data. The differences in all cases except TrL/BW and FOOTL/BW are highly significant (0.01 level). Both TAILL and EARL show an extremely positive scaling relationship to BW, in contrast to the almost isometric or slightly positive scaling seen in the *C. apella* female ontogenetic sample. As in the ontogenetic study, *C. apella* females generally show lower correlations between features than do males, indicating a less linear relationship than that shown across adult males.

Species of Cebus

Very few traits examined show any significant correlation with body size in *Cebus capucinus* adults, including TAILL, FOOTL, and EARL to both BW and TrL. But TOTL scales very negatively to both BW and TrL in *C. capucinus*, significantly more so than in *C. apella* combined- and single-sex samples. This is in marked contrast to the ontogenetic study, where all *Cebus* samples are more or less the same for these relationships and most like the adult *C. apella* combined-sex and male samples. The TrL scales slightly more negatively to BW in *C. capucinus*, but the differences here are not significant at the 0.05 level.

Other

Adult samples for many species do not exhibit statistically significant correlations of the features used here with either BW or TrL, so comparisons

Table 4. Comparison of Body Weight (BW) and Trunk Length (TrL) to Other Parameters: Intraspecific, Adults[a]

Species	N, X = BW	N, X = TrL	Total length/BW r	Total length/BW LS	Total length/BW RMA	Trunk length/BW r	Trunk length/BW LS	Trunk length/BW RMA	Tail length/BW r	Tail length/BW LS	Tail length/BW RMA	Foot length/BW r	Foot length/BW LS	Foot length/BW RMA
Ateles geoffroyi	11	35	0.59^{NS}	0.590	—	0.30^{NS}	1.260	—	-0.02^{NS}	−0.048	—	0.07^{NS}	0.130	—
Alouatta palliata	35	83	0.49	0.354	—	0.43	0.710	1.651	0.10^{NS}	0.056	—	0.67	0.529	0.790
Alouatta palliata ♂	19	33	0.28^{NS}	0.274	—	0.35^{NS}	0.665	—	-0.08^{NS}	−0.047	—	0.49	0.425	0.867
Alouatta palliata ♀	16	50	0.43^{NS}	0.317	—	0.07^{NS}	0.140	—	0.41^{NS}	0.378	—	0.61	0.604	0.990
Cebus apella	185	194	0.59	0.442	—	0.66	0.561	0.850	0.34	0.344	1.012	0.63	0.423	0.671
Cebus apella ♂	95	102	0.54	0.389	0.720	0.58	0.569	0.981	0.24	0.230	0.958	0.47	0.347	0.738
Cebus apella ♀	90	92	0.35	0.392	1.120	0.37	0.379	1.024	0.24	0.424	1.767	0.33	0.254	0.770
Cebus capucinus	20	71	0.65	0.337	0.518	0.61	0.470	0.770	0.31^{NS}	0.227	—	0.18^{NS}	0.211	—
Saimiri sciureus	18	179	0.50	0.284	0.568	0.83	0.618	0.745	0.07^{NS}	0.045	—	0.41^{NS}	0.233	—
Aotus trivirgatus	25	111	-0.02^{NS}	−0.016	—	0.16^{NS}	0.115	—	-0.12^{NS}	−0.127	—	0.41	0.256	0.624
Saguinus midas	17	22	0.65	0.799	1.229	0.42^{NS}	0.988	—	0.65	0.600	0.923	0.81	0.642	0.793
Callithrix jacchus	25	49	0.74	0.671	0.907	0.26^{NS}	0.711	—	0.52	0.600	1.269	-0.02^{NS}	−0.051	—

Species	Ear length/BW r	Ear length/BW LS	Ear length/BW RMA	Total length/TrL r	Total length/TrL LS	Total length/TrL RMA	Tail length/TrL r	Tail length/TrL LS	Tail length/TrL RMA	Foot length/TrL r	Foot length/TrL LS	Foot length/TrL RMA	Ear length/TrL r	Ear length/TrL LS	Ear length/TrL RMA
Ateles geoffroyi	0.15^{NS}	0.462	—	0.78	0.395	0.506	0.04^{NS}	0.022	—	0.40	0.201	0.502	0.04^{NS}	0.032	—
Alouatta palliata	0.31^{NS}	0.570	—	0.88	0.471	0.535	0.12^{NS}	0.057	—	0.04^{NS}	0.055	—	0.17^{NS}	0.162	—
Alouatta palliata ♂	0.13^{NS}	0.305	—	0.94	0.558	0.594	0.46	0.189	0.411	0.33^{NS}	0.152	—	0.10^{NS}	0.106	—
Alouatta palliata ♀	0.26^{NS}	0.612	—	0.85	0.441	0.519	0.03^{NS}	0.013	—	-0.05^{NS}	−0.072	—	0.13^{NS}	0.112	—
Cebus apella	0.47	0.604	1.285	0.77	0.668	0.867	0.33	0.386	1.170	0.38	0.360	0.947	0.38	0.572	1.505
Cebus apella ♂	0.31	0.430	1.387	0.74	0.539	0.728	0.15^{NS}	0.138	—	0.12^{NS}	0.121	—	0.28	0.381	1.361
Cebus apella ♀	0.26	0.500	1.923	0.69	0.767	1.112	0.34	0.590	1.735	0.25	0.209	0.836	0.15^{NS}	0.319	—
Cebus capucinus	0.23^{NS}	0.457	—	0.54	0.283	0.524	0.09^{NS}	0.055	—	0.05^{NS}	0.035	—	0.03^{NS}	0.044	—
Saimiri sciureus	-0.36^{NS}	−1.125	—	0.73	0.621	0.851	0.32^{NS}	0.351	—	0.59	0.382	0.647	0.15^{NS}	0.231	—
Aotus trivirgatus	0.16^{NS}	0.190	—	0.80	0.520	0.650	0.00^{NS}	−0.001	—	0.11^{NS}	0.048	—	0.12^{NS}	0.139	—
Saguinus midas	0.46^{NS}	0.570	—	0.77	0.492	0.639	0.19^{NS}	0.130	—	0.46	0.174	0.378	0.24^{NS}	0.238	—
Callithrix jacchus	0.15^{NS}	0.542	—	0.49	0.183	0.373	−0.42	−0.225	—	-0.21^{NS}	−0.111	—	0.57	0.534	0.937

[a]LS, Least squares estimate of allometry coefficient. RMA, Reduced major axis estimate of allometry coefficient. NS, p not significant.

Table 5. Comparison of Body Weight (BW) and Trunk Length (TrL) to Other Parameters: Interspecific[a]

Group[b]	N	Total length/BW			Trunk length/BW			Tail length/BW			Foot length/BW		
		r	LS	RMA	r	LS	RMA	r	LS	RMA	r	LS	RMA
All platyrrhines	18	0.98	0.721	0.736	0.97	0.830	0.856	0.95	0.640	0.674	0.99	0.838	0.846
All platyrrhines, averaged by sex	18	0.98	0.720	0.735	0.97	0.828	0.854	0.96	0.646	0.673	0.99	0.834	0.842
All platyrrhines, male	18	0.97	0.719	0.741	0.98	0.842	0.859	0.94	0.626	0.666	0.98	0.842	0.859
All platyrrhines, female	18	0.98	0.730	0.745	0.96	0.808	0.842	0.96	0.670	0.698	0.98	0.827	0.844
Large-bodied	8	0.97	0.612	0.631	0.90	0.654	0.727	0.97	0.574	0.592	0.98	0.651	0.664
Large-bodied, averaged by sex	8	0.97	0.611	0.630	0.89	0.651	0.731	0.97	0.589	0.607	0.98	0.664	0.677
Small-bodied, plus *Aotus* and *Callicebus*	10	0.98	0.933	0.952	0.97	1.003	1.034	0.97	0.887	0.914	0.96	0.936	0.975
Small-bodied, plus *Aotus* and *Callicebus*, averaged by sex	10	0.98	0.922	0.941	0.98	1.000	1.020	0.97	0.870	0.897	0.97	0.924	0.953
Small-bodied	7	0.97	0.948	0.977	0.97	0.917	0.945	0.97	0.978	1.008	0.97	0.758	0.781
Small-bodied, averaged by sex	7	0.97	0.936	0.965	0.97	0.909	0.937	0.97	0.962	0.992	0.97	0.766	0.790

All platyrrhines	0.79	0.330	0.418	0.99	0.858	0.867	0.95	0.752	0.792	0.96	0.972	1.012	0.81	0.404	0.499
All platyrrhines, averaged by sex	0.79	0.327	0.414	0.99	0.857	0.866	0.95	0.756	0.796	0.96	0.965	1.005	0.82	0.398	0.485
All platyrrhines, male	0.77	0.362	0.470	0.99	0.856	0.865	0.97	0.747	0.770	0.97	0.966	0.996	0.83	0.449	0.541
All platyrrhines, female	0.79	0.296	0.375	0.97	0.858	0.884	0.91	0.754	0.829	0.94	0.948	1.008	0.76	0.342	0.450
Large-bodied	0.77	0.152	0.197	0.96	0.835	0.870	0.87	0.707	0.813	0.85	0.776	0.913	0.80	0.219	0.274
Large-bodied, averaged by sex	0.78	0.145	0.186	0.96	0.829	0.864	0.86	0.710	0.826	0.84	0.774	0.921	0.82	0.208	0.254
Small-bodied, plus *Aotus* and *Callicebus*	0.82	0.603	0.735	0.99	0.913	0.922	0.97	0.855	0.881	0.96	0.919	0.957	0.80	0.576	0.720
Small-bodied, plus *Aotus* and *Callicebus*, averaged by sex	0.82	0.590	0.719	0.99	0.906	0.915	0.96	0.841	0.876	0.97	0.908	0.936	0.79	0.559	0.708
Small-bodied	0.85	0.790	0.929	0.99	1.029	1.039	0.99	1.057	1.068	0.94	0.777	0.827	0.83	0.817	0.984
Small-bodied, averaged by sex	0.85	0.784	0.922	0.99	1.028	1.038	0.99	1.054	1.065	0.94	0.800	0.851	0.83	0.818	0.985

[a]LS, Least squares estimate of allometry coefficient. RMA, Reduced major axis estimate of allometry coefficient.
[b]See text.

of allometry coefficients would not be meaningful. Some comparisons can be made relative to the results of the ontogenetic study.

TOTL scales negatively to both BW and TrL in most species, as was the case for the developmental samples. The exceptions and extremes are not the same, however. In particular, *Saimiri* adults exhibit much stronger negative scaling of TOTL/BW than the developmental samples; *Cebus apella* adults (especially females) have much higher values of TOTL/TrL and *A. palliata* males much lower values than in the developmental samples. Scaling of TrL to BW is approximately the same for adult samples except *A. palliata,* which exhibits much stronger positive allometry than in the developmental sample.

Adult *Cebus apella,* especially females, have much higher allometry coefficients for TAILL to both BW and TrL. Unlike the developmental data, none of the species has significantly different allometry coefficients for FOOTL/BW. However, relationships in adults for FOOTL/TrL are just the reverse of those in the ontogenetic study. *Ateles* (and especially *Saguinus*) show extreme negative allometry, and *Cebus apella* is just the reverse, exhibiting a close to isometric relationship. Thus, while foot length grows quite rapidly relative to trunk length in *Ateles* in comparison to *C. apella,* it increases much more slowly across a static adult sample of *Ateles,* both absolutely and relative to *C. apella.* Only a few samples can be compared for EARL/ BW and EARL/TrL, and the relationships seen here are quite similar to those between the ontogenetic samples.

Thus, comparisons show that relationships between species are approximately the same for adult and ontogenetic samples for EARL relative to body size and TrL relative to BW, but there are some marked differences for the other features, especially FOOTL to body size. Results for EARL in particular are only tentative, as correlations with BW and TrL in adults are extremely low, usually not significant.

Interspecific

Determination of Species Mean Values

Differences in the allometry coefficients between samples with an overall mean versus those with a mean determined from the average of a species' male and female means are small for all variables and all groups (all platyrrhines, large-bodied platyrrhines, and small-bodied platyrrhines; see Table 5). In no cases are the differences significant at the 0.05 level. The remainder of the discussions will concentrate on the values for samples averaged by sex.

Differences Due to Sex

The differences between the allometry coefficient for male platyrrhines (using species means) and female platyrrhines are somewhat large in many cases. However, the difference is not significant at the 0.05 level for any variables. It was possible that the allometric relationships were best expressed by parallel lines, that is, ones with the same slope but different intercepts (values for *b* in the allometry equation). We examined this in two ways: (1) by making a direct comparison of the intercepts and (2) by comparing the correlation coefficients of the sample using the mean for all specimens, averaged by sex, to a sample using the mean for each sex of each species as a separate data point. In none of these cases do the differences appear to be significant. However, this may well have been affected by the fact that many of the species included are not strongly sexually dimorphic (Hershkovitz, 1977; Moynihan, 1976; Orlosky, 1973). In a study where the majority of species are sexually dimorphic, differences in the allometry coefficient and/or the intercept may be more marked and significantly affect interpretations of these values.

Clades of Increasing and Decreasing Size

In general, all variables are negatively allometric to body size (whether measured by BW or TrL) across all platyrrhines (the total sample), with the single exception of foot length to trunk length. Ear length is the most strongly negatively allometric. For most variables, the allometry coefficients for the clade of increasing size (the large-bodied atelines plus *Aotus* and *Callicebus*) are not significantly different from those for the entire platyrrhine sample. The exceptions are EARL to both BW and TrL, and FOOTL to BW, which are all significantly different from the total platyrrhine sample at the 0.01 level. In all three cases, the coefficients are much lower in the large-bodied (increasing size) sample. That is, ear and foot length increase even more slowly across adults in the large-bodied group relative to those indicators of body size. This is not true for FOOTL relative to TrL; the allometry coefficient of the large-bodied sample is not significantly different from the overall platyrrhine sample, and both values approximate isometry.

A comparison of the values for the dwarfed or small-bodied sample, however, yields many significant differences. Gould (1975, 1977) suggested that dwarfing lineages may not simply follow in reverse the same allometric trends as lineages increasing in size, but rather, may exhibit entirely different allometric relationships. Marshall and Corruccini (1978) showed this to be true in Australian marsupial lineages, as did Prothero and Sereno (1982) for limb proportions in Miocene dwarf rhinoceroses. In all cases except FOOTL to TrL, allometry coefficients for the small-bodied samples (both with and without *Aotus* and *Callicebus* spp.) are much larger than in either the total platyrrhine sample or the large-bodied platyrrhine sample, more closely approximating or exceeding isometry. In most cases, these differences are significant. In several cases, only the callitrichids alone or only the callitrichid-plus-*Callicebus* and *Aotus* sample is significantly different: for TrL/BW, the callitrichid-plus-*Callicebus* and *Aotus* sample is different from (larger than) both the total and large-bodied samples at the 0.05 level; while for TOTL,

TAILL, and EARL scaling to TrL as an indicator of body size, only the callitrichid-alone sample is significantly different from the total platyrrhine sample. The callitrichid-plus-*Callicebus* and *Aotus* sample is significantly different from the large-bodied sample for EARL/TrL and FOOTL/BW, in both cases reflecting the uniqueness of the large-bodied allometry coefficients as much as that of the small-bodied-plus-*Callicebus* and *Aotus* sample.

Therefore, total length, tail length, and ear length scaling to body size in platyrrhines support Gould's suggestion that dwarfing lineages will scale differently than lineages increasing in body size or static "mouse-to-elephant" across-species comparisons. Other than FOOTL/TrL, which scales approximately isometrically in all cross-species samples (it is lower in the callitrichid-only sample, but not significantly so), all traits scale negatively to body size across all platyrrhine species and in the large-bodied (increasing body size lineage) sample. Where the large-bodied sample differs significantly from the total platyrrhine sample, the relationship is even more strongly negatively allometric. In other words, the smaller platyrrhines have relatively larger total body length, longer tails, larger feet (relative to body weight), and larger ears. But within the dwarfed lineage (either including only callitrichids or callitrichids and species representing the possible ancestral condition), TOTL, TAILL, and EARL to either measure of body size, and TRUNKL to BW, have significantly larger coefficients of allometry, and exhibit close to an isometric relationship with body size. Therefore, the relative size of these body parts, which is large compared to other platyrrhines, remains fairly constant across all members of the dwarfed group, regardless of body size.

Discussion

Comparison of Ontogenetic and Adult Static Intraspecific Allometry

As reviewed above, many workers have suggested that static adult allometry might be similar to ontogenetic allometry, while others have strongly argued that no such relationship is to be expected as a norm. In this study, static adult allometry sample sizes are much smaller than in the cross-sectional developmental study, and therefore there are fewer cases of statistically significant correlation coefficients and corresponding RMA allometry coefficients. For those that are significant, some interesting comparisons can be made to the developmental data.

The intraspecific ontogenetic and adult static data sets are partially overlapping, and therefore statistical correlations cannot meaningfully be computed. However, simple inspection shows that the majority of coefficients for given species are quite different. In addition, values are split about equally between those where adult coefficients are higher and those where adult values are lower. This strongly suggests that static adult allometry cannot blindly be used to infer ontogenetic scaling and developmental processes, in agreement with Cheverud (1982), Cock (1966), Lande (1979), and Shea (1981).

However, grossly similar and different coefficients cluster. Static adult and ontogenetic allometry coefficients are essentially the same for all species for TrL/BW, and for most for TOTL/BW and FOOTL/BW. Large differences between static adult and developmental samples are much more common for all variables relative to TrL and for both EARL and TAILL to BW. All instances in which the adult allometry coefficients are much larger than those of the cross-sectional growth sample occur with *Cebus apella* combined- and single-sex samples. Although the number of comparisons is small, very real differences in the relationship of developmental and adult allometry may be revealed here. Cheverud (1982) has recently suggested that static adult allometric relationships due to environmental factors will be more like ontogenetic allometries than adult patterns due to genetic factors.

At this point we should digress to consider one methodological point in allometric analysis. Cheverud (1982) employs a log principal component approach in which allometry coefficients are estimated as proportional to the major axis eigenvector coefficients. Despite Jungers and German's (1981) critique of the principal component technique (see Hills, 1982; Corruccini, 1983), they failed to mention the main *caveat* involved in use of this technique: the necessity of collinearity (and level correlations) of all included variables with the major axis. Cheverud's major axis solutions for large environmental, genetic, phenotypic, and ontogenetic character sets varied from 64% included shared variance to only 16%. Such radically variable fit to the major axis signifies that the first principal components of the different analyses are literally different biological variables and no more comparable than, e.g., body weight allometry in one group with brain size allometry in another.

Pursuing this point in our own data, a principal component summary of platyrrhine allometry patterns is interesting. Table 6 shows the major component in intraspecific growth analyses of two of the larger samples. Both data sets share a salient characteristic: body weight, the base variable in so many allometry studies, is a unique outlier from the cluster of linear dimensions on the residual (second) axis. In other words, the error of measurement (or independent variation) in this variable away from the major axis is in a different direction or dimension from the relatively stable residual variation shown by the other measurements as a group. This is precisely what an independent variable to so many bivariate analyses should not do, and it renders principal component-based estimations of bivariate allometry to body weight quite untrustworthy in these data sets. It may be that the expression of body weight in different measurement units (mass versus linear dimension), its greater difficulty of measurement, and idiosyncratic measurement error, or actual biologically unique independent variation could explain its residual pattern. Whatever the cause, in a purely statistical sense body weight is not the ideal independent size variable in our data.

Comparing our results to Cheverud's (1982) suggestions, we find that the

Table 6. Principal Component Scalings (Logarithmic) of Cross-Sectional Intraspecific Data

	Eigenvectors			
	Cebus apella		*Aotus trivirgatus*	
Trait	PC1	PC2	PC1	PC2
BW[a]	0.908	−0.365	0.917	−0.396
TOTL	0.212	0.377	0.203	0.470
TAILL	0.170	0.521	0.216	0.569
FOOTL	0.129	0.258	0.172	0.370
EARL	0.130	0.588	0.076	0.201
TrL	0.260	0.203	0.188	0.349
Percent variance	85.9	6.5	83.7	14.7

[a]In raw state, not cube rooted.

relationship between various linear measures and trunk length may be based more on genetic factors than the relationship between those variables and body weight. And indeed the former show more major differences between ontogenetic and static adult intraspecific allometry coefficients. It seems likely that the relationship between both tail and ear length and body weight would be closely genetically programmed if these body parts function in part as heat dissipators and body temperature regulators, and they show the least absolute similarity to ontogenetic allometric patterns. If general patterns between species are examined, however (see above), similar relationships are seen in the two types of intraspecific samples for EARL to both BW and TrL, as well as for TrL to BW, but not for the remaining traits (especially FOOTL to body size). Thus, our data partially support Cheverud's contentions and suggest that there may be a strong environmental aspect to the static adult allometric relationship between trunk length and body weight.

Comparisons of Interspecific and Intraspecific Allometry

Obviously, if the pattern seen in lineages increasing in size is one of functional equivalence (Gould, 1975, 1977), then something quite different is happening in the dwarfed lineage. Marshall and Corruccini (1978) found that allometry of dwarfing lineages resembled, at least in some respects, static adult intraspecific allometry patterns. Gould (1977) has suggested that allometric patterns in phyletically dwarfed lineages may be more similar to ontogenetic intraspecific patterns than to static cross-species patterns, with dwarfing occurring as a reversal of ontogenetic development. Here, we can compare allometric scaling patterns in a dwarfed lineage not only to static cross-species and increasing-size samples (from which they clearly differ), but also to intraspecific static adult and ontogenetic scaling for a number of platyrrhine species. As intraspecific scaling varies greatly among different species, however, the search for similarities and general patterns is complex.

In general, the callitrichids (both with and without *Aotus* and *Callicebus*) exhibit scaling more similar to both intraspecific adult static and ontogenetic scaling than do the total and increasing-size-lineage cross-species samples for many features, but there are important exceptions. We will address each trait relative to body size (using both measures of size).

Trunk Length to Body Weight. In the small-bodied (decreasing-size) interspecific samples, TrL scales isometrically to BW, as it does in most intraspecific samples. The total platyrrhine and large-bodied platyrrhine samples have significantly lower coefficients. However, the small-bodied samples are significantly (0.01 level) different from the *Callithrix jacchus* growth sample, which exhibits strong positive allometry of TrL (as do *A. palliata* females and adult static *A. palliata*). Thus, in young *Callithrix,* body weight is fairly high relative to trunk length compared to most other platyrrhines. Increased body bulk and mass/area ratio in infant and juvenile callitrichids may assist in regulating body temperature, especially difficult in small mammals (Bourlière, 1975). It is not clear why infant and young *A. palliata* females should also exhibit larger body bulk relative to adult females, and smaller adult *A. palliata* individuals (presumably females) significantly larger body bulk (lower relative TrL) than larger adults (presumably males) (strong positive allometry of TrL to BW in the combined-sex static adult sample). There is presumably some key difference between the sexes of this species. Mean values, however, fall neatly near the regression line for all platyrrhine species.

Total Length to Body Size (BW or TrL). All interspecific samples are significantly different from most intraspecific samples, but especially the large-bodied sample for intraspecific static adult TOTL/BW allometries and the decreasing-size (small-bodied) samples for TOTL/TrL. For TOTL/BW, the callitrichids are significantly different from the growth and static adult species exhibiting negative allometry (many) and the positively allometric *Aotus* growth sample. The isometric scaling across callitrichids is similar only to that exhibited ontogenetically in *Callithrix jacchus, Saimiri,* and *Alouatta palliata* females, and across adult *C. jacchus* (also not significantly different from adult *Saguinus midas*). The dwarfed lineage sample is significantly different from almost all intraspecies samples for TOTL/TrL, as almost all intraspecies samples exhibit marked negative allometry (except *Cebus apella* females, which show positive scaling). Here, *C. jacchus* growth and static adult samples and adult *S. midas* all scale very negatively, quite different from the isometry of the dwarfed-species interspecific sample.

Tail Length to Body Size. Most intraspecies samples scale isometrically for TAILL/BW, as do the dwarfing-lineage cross-species samples. The latter are significantly different from those intraspecies ontogenetic and static adult samples exhibiting strong negative allometry (*A. palliata* total and male samples, *Cebus apella* total and male samples) or positive allometry (*Aotus, C. jacchus,* and adult *C. apella* females). However, TAILL scales negatively to TrL in most growth samples, like the total platyrrhine and large-bodied interspecific samples and significantly unlike the isometric relationship expressed across dwarfed species. The exceptions are *Saimiri* and *C. apella* females, which exhibit isometry as well, and adult *C. apella* (isometry of TAILL and TrL) and *C. apella* females (positive allometry of TAILL). (Correlations are not significant for *C. jacchus* ontogenetic data and most static adult samples.)

Foot Length to Body Size. As the interspecific samples all exhibit isometry or only slight negative allometry and most intraspecific samples exhibit marked negative allometry, none of the interspecific samples seem remarkably closer than others to ontogenetic or static adult intraspecific allometric patterns. For both FOOTL/BW and FOOTL/TrL, the callitrichid-only sample is slightly lower (more negative) than the others, which is in marked contrast to the highly positive scaling exhibited by *Callithrix* growth samples.

Ear Length to Body Size. For both EARL/BW and EARL/TrL, the dwarfed lineage samples scale significantly more like ontogenetic samples for most species (which are usually isometric or only slightly negatively allometric). Most adult static samples do not have significant correlations; the few that do (primarily *C. apella*) exhibit marked positive allometry. The allometry coefficients of the dwarfed cross-species samples are significantly different only from those few samples exhibiting strong positive scaling, which happens to include *C. jacchus* for EARL/BW, *Saimiri* and *C. capucinus* for EARL/TrL, and *C. apella* females for both.

Therefore, there are seven scaling comparisons in which the dwarfed lineage is significantly different from both the total sample and increasing-size lineage. The allometry coefficients and patterns of the dwarfed lineage are not noticeably similar to the patterns exhibited within most species—either across a static adult sample or ontogenetically—for scaling of total length to body size or scaling to trunk length (TOTL/BW, TOTL/TrL, and TAILL/TrL). For the other four regressions—the scaling of trunk length, tail length, and ear length to body weight and to trunk length—the patterns exhibited by the dwarfed lineage are more similar to the patterns exhibited by most intraspecific growth and static adult samples. This would appear to support, in part, the view of Huxley (1932), Gould (1966, 1975, 1977), and others (e.g., Alberch *et al.*, 1979; Shea, 1981) that evolutionary allometry and in particular the allometry within a dwarfing lineage is merely an extension of ontogenetic allometry and, alternatively, the findings of Marshall and Corruccini (1978) that allometry of dwarfing lineages resembles static adult allometry.

However, a closer look at the specific similarities and differences described above suggests otherwise. As noted previously, there is no good correlation between ontogenetic and static adult allometries for most traits, and even when there is (as for TOTL and FOOTL to BW), alternative species generally fall at the extremes in a comparison across platyrrhines. Thus, there is no necessary reason for interspecific allometries to be similar to *both* ontogenetic and static adult allometries; in fact, one might expect the opposite if, as suggested by Cheverud (1982) and Lande (1979), similar ontogenetic and static adult patterns are due to environmental control of variation and similar ontogenetic and evolutionary scaling are the result of underlying genetic controls being extended in their influence. In addition, if in fact dwarfing allometry is an extension of growth allometric patterns, it seems most likely that the intraspecific ontogenetic pattern being "extended" would be that of a species in the lineage (in this case, e.g., *Callithrix jacchus*) or of an ancestral species (or extant model, e.g., *Aotus*). But in this case, intraspecific patterns in *Aotus* and especially *C. jacchus* are least like the scaling seen across callitrichids.

Here, then, it is unclear why a few cases of relative scaling across species, especially within a dwarfed lineage, are superficially similar to scaling within the majority of platyrrhine species examined. Only the examination of ontogenetic and static adult allometry within additional platyrrhine species using more traits will answer whether this is a real phenomenon or the chance result of the sample of species included in the intraspecific studies.

Implications of Small Body Size

The advantages and disadvantages of small body size have been reviewed by Bourlière (1975). Primary among the problems posed is thermoregulation and the high metabolic cost of maintaining a constant body temperature. As expressed by Bourlière, it is generally assumed that large body size (and increased bulk and mass/area ratio) evolves to retard heat loss in colder environments, the basis of Bergmann's rule, and that small mammals are under much less stress in tropical environments (see also McNab, 1971; Riesenfeld, 1981; Schmidt-Nielsen, 1970; Schwartz and Rosenblum, 1981). However, there can be large temperature fluctuations within the neotropics. Hershkovitz (1977) has noted that temperature drops after sunset can be marked, that cold stress may be especially acute if these small monkeys are soaked by heavy night-time rain showers, and that *Callithrix* spp. and *Leontopithecus* at the southern end of their ranges are exposed to freezing temperatures at least a few days a year. Callitrichids must have several mechanisms for coping with cold stress, as Hershkovitz cites numerous examples of these small monkeys being exposed for prolonged periods to cold temperatures in captivity and showing no ill effects. Increased body bulk is clearly not one of these mechanisms, as body weight decreases relative to trunk length across progressively smaller platyrrhines. Callitrichids do exhibit at least two common approaches listed by Bourlière (1975), the ingesting of high-energy food, resins and saps (Coimbra-Filho and Mittermeier, 1977; Kinzey *et al.*, 1975), and marked daily fluctuation in body temperature, about twice that seen in other primates, reaching its maximum in late morning and dropping sharply after sunset (Morrison and Simões, 1962; Hershkovitz, 1977). In addition, they huddle and wrap themselves up in their tails when cold (Hershkovitz, 1977).

In fact, it appears that callitrichids are better adapted to withstand cold stress than excessive heat. Prostration and even death have been reported at temperatures of 35–39°C (Hampton *et al.*, 1966; Hershkovitz, 1977). Thus, although larger animals cannot dissipate body heat at the rate it is produced by metabolism (oxygen volume consumption scaling to mass to the 0.75 power), heat stress appears as severe in small monkeys as in the large platyrrhines, such as *Ateles* and *Alouatta*. Rosenblum and Schwartz (1982) have reported that *Saimiri* appears to have periodic problems of heat dissipation, and suggest that sneezing, urine washing, and tails may all serve thermoregulatory functions. Ripley (1977) found that langurs could control heat transfer by altering their positional behavior, and behavioral thermoregulation has been postulated for *Alouatta* (Young, 1982; see also Paterson, 1982). Hershkovitz (1977, p. 449) has described callitrichids adopting the "characteristic heat dissipating sprawl by lying along the length of a perch with limbs and tail dangling over the sides."

The trend of progressively longer total length, longer tails, and larger ears relative to body weight as body weight decreases across platyrrhines suggests strongly that tails and ears are serving a metabolic function, facilitating dissipation of excess body heat. Similar strategies are used by larger monkeys as well (e.g., Moynihan, 1976), but increasing relative size of these body parts in small-bodied forms may indicate their increased importance as thermoregulatory devices. This is contrary to the suggestion of Hershkovitz (1977) that diversity of ear size in primates is not related to thermoregulation but solely to social selection for display and recognition. However, Hershkovitz based this assumption on his observation of high variability in ear size and form only within callitrichids. As shown here, when the size range of the cross-species sample is extended, a clear relationship of negative scaling can be seen despite relatively poor correlation. This is not meant to imply that platyrrhines are the sole or a primary exception to Bergmann's rule, which has been defended recently by Riesenfeld (1981). Rather, temperature fluctuations in the neotropics may be sufficient to cause small primates (and other mammals) to develop numerous adaptations against periodic extremes of cold (against which larger primates are protected by their naturally higher body mass/area ratio), which in turn place them in jeopardy when temperatures soar during normal days, necessitating countermeasures. Larger primates may also have other means of cooling in addition to radiation through tails and ears.

Smaller animals increase oxygen consumption relatively rapidly when running. Smaller mammals have "less metabolic scope," as they can generally only increase maximum metabolic rate to 20 times normal, versus 50 times in medium-sized mammals (Taylor, 1973). Vertical running is relatively less costly in the small, while moving horizontally requires more energy. Since small mammals are at a heat load disadvantage in hot environments (Schmidt-Nielsen, 1964), why do they evolve? The answer, as well as the explanation for the 0.75 metabolic scaling principle, may be related to trunk elasticity in locomotion (Schmidt-Nielsen, 1975*a*). Thus the earlier cited locomotor requirements (Taylor, 1973) are more different than are metabolic requirements with respect to small mammals, and this may be particularly relevant to the adaptation of callitrichids.

The difference in allometric scaling between species in a dwarfed lineage (callitrichids) and the total and increasing-size lineage samples cannot totally be explained by simple functional necessities related to thermoregulation, however. Bourliére (1975) has shown that there is not a continuous gradient in body weight in mammals but a bimodal distribution, with "small" mammals weighing less than 3 kg and "large" mammals more than 5 kg. This division is clear in the platyrrhines examined here: mean body weight for *Ateles* and *Alouatta* spp. are equal to or greater than 5 kg, for all other species (including *Cebus*) equal to or less than 3 kg. If there is a break and change in scaling patterns due solely to alternate adaptive strategies for thermoregulation in small and large platyrrhines, it might be expected at the 3–5 kg gap. Others have suggested that 2 kg constitutes a critical threshold in body weight, with comparisons of scaling qualitatively different above and below this threshold (Schmidt-Nielsen, 1975*b;* Taylor, 1973). However, neither of these possible thresholds explains the differences seen here. While *Aotus* does fall closer to the callitrichid regression line for EARL/BW, other noncallitrichids weighing less than 2 or 3 kg do not. In fact, *Saimiri* falls below the total platyrrhine regressions for EARL and *Cebus* spp. for TAILL to BW and TrL. Thus, a different relationship exists between TOTL, TAILL, EARL, and TrL to BW across the dwarfed callitrichids. Apparently, once a critical increased relative length of these body parts was reached during dwarfing, that relative size was maintained during the radiation (and continued dwarfing) of this group. One possible hypothesis that could be generated from these data is that the critical relative length was reached in an animal approximately the size of *Callithrix jacchus,* which falls close to the total platyrrhine regression line. In this case, *Cebuella* would be the result of continued size reduction without the relative increase in tail, ear, and trunk length expected from extrapolation of the total platyrrhine regression. *Saguinus* could then be interpreted as the result of a slight reversal in the dwarfing trend, but with maintenance of the altered scaling of these body parts. It would be interesting to test this against the scaling of other body parts, and to acquire sufficient body weight data for *Leontopithecus* and *Callimico* to determine if they also fall on the callitrichid regression.

Summary and Conclusions

1. Although body weight is a prime determinant of metabolic requirements, for other studies of scaling other measures of "body size" or, especially, a combination of several size variables are more informative.

2. There are some differences due to sex in both ontogenetic and static adult allometric scaling, with females generally exhibiting larger allometry coefficients and lower correlation coefficients. No significant differences due to sex are found in the interspecific scaling patterns.

3. The ontogenetic sample of *Cebus capucinus* scales much the same as *Cebus apella* to trunk length but has much higher allometry coefficients for variables scaling to body weight. This results in adult *C. apella* that are heavier and stockier. Jungers and Fleagle (1981) found a similar difference between *C. apella* and *C. albifrons*. These differences may relate to different foraging strategies, and it is expected that all four *Cebus* species will scale similarly for features relative to trunk length, but *C. nigrivittatus* will be intermediate in scaling to body weight between *C. apella* and the other two species.

4. Comparison of interspecific samples shows that the combined platyrrhine sample and increasing-size (large-bodied) lineage scale similarly, and both exhibit negative allometry of all traits to both trunk length and body weight except for foot length to trunk length. The callitrichids, representing consecutive stages of a lineage that decreased in size through time, show very different patterns of scaling for all features except foot length. Across callitrichids, a fairly constant, isometric relationship exists for each variable to body size (measured by both body weight and trunk length).

5. Ontogenetic and static adult allometry coefficients do not show marked similarities for individual species; the latter cannot blindly be used to infer ontogentic scaling. There is some pattern in degree of similarity. Scaling of trunk length, total length, and foot length to body weight show the most similarities between ontogenetic and adult static samples; scaling of tail and ear length to body weight and of all features to trunk length are the most dissimilar. It appears that those relationships due to environmental factors (as opposed to genetic) show the most similarity between static adult and ontogenetic scaling, in partial support of Cheverud (1982).

6. For those scaling relationships where the decreasing-size lineage differs from the total and large-bodied samples, the dwarfed lineage is more like the majority of both ontogenetic and static adult intraspecies patterns for about half. However, there is no strong correlation between ontogenetic and static adult scaling for individual species for these relationships, and dwarfed lineage interspecific allometry is markedly unlike intraspecific allometry of individual callitrichids (*Callithrix jacchus*) and species potentially most like the ancestral form (*Aotus trivirgatus*). It is suggested that apparent similarities of interspecific allometry in dwarfed lineages and intraspecific allometric patterns may be stochastic, depending on fortuitous choice of species in the intraspecific sample.

7. Implications of small body size for thermoregulation have been reviewed, indicating that small-bodied platyrrhines are exposed to both cold and heat stress. Adaptations to heat stress include the functioning of ears and tails to dissipate heat, as reflected in the negative allometry of both features to body weight across all platyrrhines. However, proposed "natural" breaks between large- and small-bodied mammals in adaptive techniques in metabolic processes do not correspond to the break seen between callitrichids (a dwarfed lineage) and larger bodied platyrrhines and cannot provide the entire explanation for differences in scaling. Rather, dwarfing lineages may exhibit allometric patterns radically different from those exhibited by lineages

increasing in body size. In this case, a critical relative length of various features to body size was maintained during continued dwarfing in callitrichids, yielding a pattern of isometry of these features across the extant representatives of various stages of the dwarfing incident.

ACKNOWLEDGMENTS

We would like to thank Dr. R. W. Thorington, G. Blair, and G. Morgan for providing a copy of the catalogue data from the Smithsonian Institution and C. Asmann-Finch for transcribing the data. S. Spezia typed the manuscript and T. Thomas prepared the tables. In addition, we thank Drs. W. Jungers and W. Kinzey for discussion and comments.

References

Aiello, L. C. 1981. The allometry of primate body proportions. *Symp. Zool. Soc. Lond.* **48:**331–358.

Alberch, P., Gould, S. J., Oster, G. F., and Wake, D. B. 1979. Size and shape in ontogeny and phylogeny. *Paleobiology* **5:**296–317.

Albrecht, G. H. 1980. Latitudinal, taxonomic, sexual, and insular determinants of size variation in pigtail macaques, *Macaca nemestrina*. *Int. J. Primatol.* **1:**141–152.

Alexander, R. McN., Jayes, A. S., Maloiy, G. M. O., and Wathuta, E. M. 1979. Allometry of the limb bones of mammals from shrews (*Sorex*) to elephant (*Loxodonta*). *J. Zool. Lond.* **189:**305–314.

Anderson, D. L., Thompson, G. W., and Popovich, F. 1977. Tooth, chin, bone and body size correlations. *Am. J. Phys. Anthropol.* **46:**7–12.

Bauchot, R., and Stephan, H. 1969. Encéphalisation et niveau evolutif chez les simiens. *Mammalia* **33:**225–275.

Biegert, J., and Maurer, R. 1972. Rumpfskelettlange, Allometrien und Korperproportionen bei catarrhinen Primaten. *Folia Primatol.* **17:**142–156.

Bourlière, F. 1975. Mammals, small and large: The ecological implications of size, in: *Small Mammals: Their Productivity and Population Dynamics.* (F. B. Golley, K. Petrusewicz, and L. Ryszkowski, eds.), pp. 1–8, Cambridge University Press, Cambridge.

Bronson, R. T. 1981. Brain weight–body weight relationships in 12 species of non-human primates. *Am. J. Phys. Anthropol.* **56:**77–81.

Cheverud, J. M. 1982. Relationships among ontogenetic, static, and evolutionary allometry. *Am. J. Phys. Anthropol.* **59:**139–149.

Cock, A. G. 1966. Genetical aspects of metrical growth and form in animals. *Q. Rev. Biol.* **41:**131–190.

Coimbra-Filho, A. F., and Mittermeier, R. A. 1977. Tree-gouging, exudate-eating and the 'short-tusked' condition in *Callithrix* and *Cebuella,* in: *The Biology and Conservation of the Callitrichidae* (D. Kleiman, ed.), pp. 105–115, Smithsonian Institution, Washington, D.C.

Corruccini, R. S. 1978. Primate skeletal allometry and hominid evolution. *Evolution* **32:**752–758.

Corruccini, R. S. 1980. Size and positioning of the teeth and infratemporal fossa relative to taxonomic and dietary variation in primates. *Acta Anat.* **107:**231–235.

Corruccini, R. S. 1983. Principal components for allometric analysis. *Am. J. Phys. Anthropol.* **60:**451–453.

Corruccini, R. S., and Ciochon, R. L. 1979. Primate facial allometry and interpretations of australopithecine variation. *Nature* **28:**62–64.

Corruccini, R. S., and Henderson, A. M. 1978. Multivariate dental allometry in primates. *Am. J. Phys. Anthropol.* **48:**203–208.

Corruccini, R. S., and McHenry, H. M. 1979. Morphological affinities of *Pan paniscus. Science* **204:**1341–1343.

Dodson, P. 1975. Functional and ecological significance of relative growth in alligators. *J. Zool. Lond.* **175:**315–355.

Erikson, G. E. 1963. Brachiation in New World monkeys and in anthropoid apes. *Symp. Zool. Soc. Lond.* **10:**135–164.

Filippson, R., and Goldson, L. 1963. Correlation between tooth width, width of the head, length of the head, and stature. *Acta Odont. scand.* **21:**359–365.

Fooden, J. 1975. Taxonomy and evolution of liontail and pigtail macaques (Primates: Cercopithecidae). *Field Zool.* **67:**1–169.

Ford, S. M. 1980*a*. Callitrichids as phyletic dwarfs, and the place of the Callitrichidae in Platyrrhini. *Primates* **21:**31–43.

Ford, S. M. 1980*b*. A systematic revision of the Platyrrhini based on features of the postcranium, Ph.D. Diss., University of Pittsburgh, Pittsburgh.

Ford, S. M. 1981. Monophyly of the Platyrrhini and the place of *Aotus, Callicebus, Cebus* and *Saimiri* within the Platyrrhini. *Am. J. Phys. Anthropol.* **54:**220.

Garber, P. A. 1980. Locomotor behavior and feeding ecology of the Panamanian tamarin (*Saguinus oedipus geoffroyi,* Callitrichidae, Primates). *Int. J. Primatol.* **1:**185–201.

Garn, S. M., and Lewis, A. B. 1958. Tooth-size, body-size and 'giant' fossil man. *Am. Anthropol.* **60:**874–880.

Garn, S. M., Lewis, A. B., and Kerewsky, R. S. 1968. The magnitude and implications of the relationship between tooth size and body size. *Arch. Oral Biol.* **13:**129–131.

Gaulin, S. J. C., and Konner, M. 1977. On the natural diet of primates, including humans, in: *Nutrition and the Brain* (R. J. Wurtman and J. J. Wurtman, eds.), Vol. 1, pp. 1–86, Raven Press, New York.

Gingerich, P. D. 1977. Correlation of tooth size and body size in living hominoid primates, with a note on relative brain size in *Aegyptopithecus* and *Proconsul. Am. J. Phys. Anthropol.* **47:**395–398.

Goldstein, S., Post, D., and Melnick, D. 1978. An analysis of cercopithecoid odontometrics. I. The scaling of the maxillary dentition. *Am. J. Phys. Anthropol.* **49:**517–532.

Gould, S. J. 1966. Allometry and size in ontogeny and phylogeny. *Biol. Rev.* **41:**587–640.

Gould, S. J. 1975. On the scaling of tooth size in mammals. *Am. Zool.* **15:**351–362.

Gould, S. J. 1977. *Ontogeny and Phylogeny,* Harvard University Press, Cambridge.

Hampton, J. K., Jr., Hampton, S. H., and Landwehr, B. T. 1966. Observations on a successful breeding colony of the marmoset *Oedipomidas oedipus. Folia Primatol.* **4:**265–287.

Henderson, A. M., and Corruccini, R. S. 1976. Relationship between tooth size and body size in American Blacks. *J. Dent. Res.* **55:**94–96.

Hernandez-Camacho, J., and Cooper, R. W. 1976. The nonhuman primates of Colombia, in: *Neotropical Primates: Field Studies and Conservation* (R. W. Thorington and P. G. Heltne, eds.), pp. 35–69, National Academy of Sciences, Washington, D.C.

Hersh, A. H. 1934. Evolutionary relative growth in the titanotheres. *Am. Nat.* **168:**537–561.

Hershkovitz, P. 1977. *Living New World Monkeys (Platyrrhini) with an Introduction to Primates,* Vol. 1, University of Chicago Press, Chicago.

Hills, M. 1982. Bivariate versus multivariate allometry: A note on a paper by Jungers and German. *Am. J. Phys. Anthropol.* **59:**321–322.

Huxley, J. S. 1932. *Problems of Relative Growth,* Methuen, London.

Hylander, W. L. 1975. Incisor size and diet in anthropoids with special reference to Cercopithecidae. *Science* **189:**1095–1098.

Jerison, H. J. 1973. *Evolution of the Brain and Intelligence,* Academic Press, New York.

Jolicoeur, P. 1963. The multivariate generalization of the allometry equation. *Biometrics* **19:**497–499.

Jolicoeur, P. 1968. Interval estimation of the slope of the major axis of a bivariate normal distribution in the case of a small sample. *Biometrics* **24:**679–682.

Jolicoeur, P. 1973. Imaginary confidence limits of the slope of the major axis of a bivariate normal distribution: A sampling experiment. *J. Am. Stat. Assoc.* **68:**866–871.

Jolicoeur, P. 1975. Linear regressions in fishery research: Some comments. *J. Fish. Res. Board Can.* **32:**1491–1494.

Jungers, W. L. 1978. The functional significance of skeletal allometry in *Megaladapis* in comparison to living prosimians. *Am. J. Phys. Anthropol.* **49:**303–314.

Jungers, W. L. 1979. Locomotion, limb proportions, and skeletal allometry in lemurs and lorises. *Folia Primatol.* **32:**8–28.

Jungers, W. L., and Fleagle, J. G. 1980. Postnatal growth allometry of the extremities in *Cebus albifrons* and *Cebus apella:* A longitudinal and comparative study. *Am. J. Phys. Anthropol.* **53:**471–478.

Jungers, W. L., and German, R. Z. 1981. Ontogenetic and interspecific skeletal allometry in nonhuman primates: Bivariate versus multivariate analysis. *Am. J. Phys. Anthropol.* **55:**195–202.

Kay, R. F. 1975. The functional adaptations of primate molar teeth. *Am. J. Phys. Anthropol.* **43:**195–216.

Kermack, K. A., and Haldane, J. B. S. 1950. Organic correlation and allometry. *Biometrika* **37:**30–41.

Kinzey, W. G., Rosenberger, A. L., and Ramirez, M. 1975. Vertical clinging and leaping in a neotropical anthropoid. *Nature* **225:**327–328.

Klein, L. L., and Klein, D. J. 1976. Neotropical primates: Aspects of habitat usage, population density, and regional distribution in La Macarena, Colombia, in: *Neotropical Primates: Field Studies and Conservation* (R. W. Thorington and P. G. Heltne, eds.), pp. 70–78, National Academy of Sciences, Washington, D.C.

Kuhry, B., and Marcus, L. F. 1977. Bivariate linear models in biometry. *Syst. Zool.* **26:**201–209.

Kurtén, B. 1954. Observations on allometry in mammalian dentitions; its interpretation and evolutionary significance. *Acta Zool. Fenn.* **85:**2–13.

Lande, R. 1979. Quantitative genetic analysis of multivariate evolution applied to brain:body size allometry. *Evolution* **33:**402–416.

Lavelle, C. L. B. 1974. Relationship between tooth and skull size. *J. Dent. Res.* **53:**1301.

Lavelle, C. L. B. 1977. Relationship between tooth and long bone size. *Am. J. Phys. Anthropol.* **46:**423–426.

Leutenegger, W. 1973. Maternal–fetal weight relationships in Primates. *Folia Primatol.* **20:**280–293.

Leutenegger, W. 1980. Monogamy in callitrichids: A conseqence of phyletic dwarfism. *Int. J. Primatol.* **1:**95–99.

Lindstadt, S. L., and Calder, G. 1981. Body size, physiological time, and longevity of homeothermic animals. *Q. Rev. Biol.* **56:**1–16.

Lumer, H. 1939. Relative growth of the limb bones in the anthropoid apes. *Hum. Biol.* **11:**371–392.

Lumer, H., and Schultz, A. H. 1941. Relative growth of the limb segments and tail in macaques. *Hum. Biol.* **13:**283–305.

Lumer, H., and Schultz, A. H. 1947. Relative growth of the limb segments and tail in *Ateles geoffroyi* and *Cebus capucinus. Hum. Biol.* **19:**53–67.

McHenry, H. M., and Corruccini, R. S. 1981. *Pan paniscus* and human evolution. *Am. J. Phys. Anthropol.* **54:**355–367.

McNab, B. K. 1971. On the ecological significance of Bergmann's rule. *Ecology* **52:**845–854.

Manaster, B. J., and Manaster, S. 1975. Techniques for estimating allometric equations. *J. Morphol.* **147:**299–308.

Marshall, L. G., and Corruccini, R. S. 1978. Variability, evolutionary rates, and allometry in dwarfing lineages. *Paleobiology* **4:**101–119.

Morrison, P., and Simões, J., Jr. 1962. Body temperatures in two Brazilian primates. *Bol. Fac. Fil. Cien. Letr. Univ. São Paulo,* no. 261, *Zool.* no. 24, pp. 167–178.

Mosimann, J. E., and James, F. C. 1979. New statistical methods for allometry with applications to Florida red-winged blackbirds. *Evolution* **33:**444–459.

Moynihan, M. 1976. *The New World Primates,* Princeton University Press, Princeton.

Napier, J. R., and Napier, P. H. 1967. *Handbook of Living Primates,* Academic Press, New York.

Nie, N. H., Hull, C. H., Jenkins, J. G., Steinbrenner, K., and Bent, D. H. 1975. *SPSS: Statistical Package for the Social Sciences,* 2nd ed., McGraw-Hill, New York.

Orlosky, F. 1973. Comparative Dental Morphology of Extant and Extinct Cebidae, Ph.D. Diss., University of Washington, Seattle.

Paterson, J. D. 1982. Size, shape, behavior: Factors in thermoregulatory adaptation. *Int. J. Primatol.* **3:**322.

Pilbeam, D., and Gould, S. J. 1974. Size and scaling in human evolution. *Science* **186:**892–901.

Pirie, P. L. 1978. Allometric scaling in the postcanine dentition with reference to primate diets. *Primates* **19:**583–591.

Prothero, D. R., and Sereno, P. C. 1982. Allometry and paleoecology of medial Miocene dwarf rhinoceroses from the Texas Gulf Coastal Plain. *Paleobiology* **8:**16–30.

Radinsky, L. 1967. Relative brain size: A new measure. *Science* **155:**836–838.

Reeve, E. C. R., and Murray, P. D. F. 1942. Evolution in the horse's skull. *Nature* **150:**402–403.

Riesenfeld, A. 1981. The role of body mass in thermoregulation. *Am. J. Phys. Anthropol.* **55:**95–99.

Ripley, S. 1967. Intertroop encounters among Ceylon grey langurs (*Presbytis entellus*), in: *Social Communication among Primates* (S. A. Altmann, ed.), pp. 237–253, University of Chicago Press, Chicago.

Ripley, S. 1977. Gray zones and gray langurs: Is the "semi"-concept seminal? *Yearb. Phys. Anthropol.* **20:**376–394.

Rosenberger, A. L. 1977. *Xenothrix* and ceboid phylogeny. *J. Hum. Evol.* **6:**461–481.

Rosenberger, A. L. 1979. Phylogeny, evolution and classification of New World monkeys (Platyrrhini, Primates), Ph.D. Diss., City University of New York.

Rosenberger, A. L. 1981. Systematics: The higher taxa, in *Ecology and Behavior of Neotropical Primates* (A. F. Coimbra-Filho and R. A. Mittermeier, eds.), Vol. 1, pp. 9–27, Academia Brasileira de Ciencias, Rio de Janeiro.

Rosenblum, L. A., and Schwartz, G. G. 1982. Thermoregulation and its role as a determinant of behavior, Paper presented at the IXth Congress of the International Primatological Society, Atlanta, August 1982.

Roth, V. L., and Thorington, R. W., Jr. 1982. Relative brain size among African squirrels. *J. Mammal.* **63:**168–173.

Schmidt-Nielson, K. 1964. *Desert Animals. Physiological Problems of Heat and Water,* Clarendon, Oxford.

Schmidt-Nielsen, K. 1970. Energy metabolism, body size and problems of scaling. *Fed. Proc.* **29:**1524–1532.

Schmidt-Nielsen, K. 1975*a*. Scaling in biology: The consequences of size. *J. Exp. Zool.* **194:**287–308.

Schmidt-Nielsen, K. 1975*b*. *Animal Physiology. Adaptation and Environment,* Cambridge University Press, Cambridge.

Schultz, A. H. 1937. Proportions, variability and asymmetries of the long bones of the limbs and clavicles in man and apes. *Hum. Biol.* **9:**281–328.

Schultz, A. H. 1938. The relative length of the regions of the spinal column in Old World primates. *Am. J. Phys. Anthropol.* **24:**1–22.

Schultz, A. H. 1953. The relative thickness of the long bones and the vertebrae in primates. *Am. J. Phys. Anthropol.* **11:**277–312.

Schwartz, G. G., and Rosenblum, L. A. 1981. Allometry of primate hair density and the evolution of human hairlessness. *Am. J. Phys. Anthropol.* **55:**9–12.

Shea, B. T. 1981. Relative growth of the limbs and trunk in the African apes. *Am. J. Phys. Anthropol.* **56:**179–201.

Simpson, G. G. 1953. *The Major Features of Evolution,* Columbia University Press, New York.

Smith, R. J. 1981. On the definition of variables in studies of primate dental allometry. *Am. J. Phys. Anthropol.* **55:**323–329.

Sokal, R. R., and Rohlf, F. J. 1981. *Biometry,* 2nd ed., Freeman, San Francisco.

Steudel, K. 1981. Body size estimators in primate skeletal material. *Int. J. Primatol.* **2:**81–90.

Struhsaker, T. T., and Leland, L. 1977. Palm-nut smashing by *Cebus a. apella* in Colombia. *Biotropica* **9:**124–126.

Susman, R. L., and Creel, N. 1979. Functional and morphological affinities of the subadult hand (O.H. 7) from Olduvai Gorge. *Am. J. Phys. Anthropol.* **51:**311–332.

Taylor, C. R. 1973. Energy costs of animal locomotion, in: *Comparative Physiology. Locomotion, Respiration, Transport and Blood.* (L. Bolis, K. Schmidt-Nielsen, and S. H. P. Maddrell, eds.), pp. 23–42, North-Holland, Amsterdam.

Thorington, R. W., Jr. 1972. Proportions and allometry in the gray squirrel, *Sciurus carolinensis. Nemouria* **8:**1–17.

Von Bertalanffy, L., and Pirozynski, W. J. 1952. Ontogenetic and evolutionary allometry. *Evolution* **6:**387–392.

Wolpoff, M. 1973. Posterior tooth size, body size and diet in South African gracile australopithecines. *Am. J. Phys. Anthropol.* **39:**375–394.

Wood, B. A. 1979*a*. An analysis of tooth and body size relationships in five primate taxa. *Folia Primatol.* **31:**187–211.

Wood, B. A. 1979*b*. Models for assessing relative canine size in fossil hominids. *J. Hum. Evol.* **8:**493–502.

Wrazen, J. A., and Wrazen, L. A. 1982. Hoarding, body mass dynamics, and torpor as components of the survival strategy of the eastern chipmunk. *J. Mammal.* **63:**63–72.

Young, O. P. 1982. Aggressive interaction between howler monkeys and turkey vultures: The need to thermoregulate behaviorally. *Biotropica* **14:**228–231.

Zar, J. H. 1968. Calculation and miscalculation of the allometric equation as a model in biological data. *BioScience* **18:**1118–1120.

Zihlman, A. L., and Cramer, D. L. 1978. Skeletal differences between pygmy (*Pan paniscus*) and common (*Pan troglodytes*) chimpanzees. *Folia Primatol.* **29:**86–94.

The Present as a Key to the Past 19

Body Weight of Miocene Hominoids as a Test of Allometric Methods for Paleontological Inference

RICHARD J. SMITH

Introduction

Ramapithecus punjabicus, *Sivapithecus indicus*, and *Gigantopithecus bilaspurensis* are three medium- to large-sized apes from the middle Miocene (about 8–10 mya). In recent years, one or another of these species has been proposed to be the earliest known hominid (Simons, 1977; Kay, 1982), a common ancestor of all modern hominids and great apes (Greenfield, 1980), closely related to the ancestory of the orangutan (Andrews and Cronin, 1982; Landau *et al.*, 1982), or perhaps a side branch with no modern descendants (Zihlman and Lowenstein, 1979). These species are best known from deposits in Pakistan and India, and almost exclusively as teeth and facial bones. Their interrelationships are controversial, recent workers suggesting that *R. punjabicus* and *S. indicus* are simply males and females of a single species (de Bonis and Melentis, 1980), separate species from one genus (Greenfield, 1979; Andrews

RICHARD J. SMITH • Department of Orthodontics, School of Dental Medicine, Washington University, St. Louis, Missouri 63110.

and Cronin, 1982; Kay, 1982), or that they belong to different genera within a subfamily or family (Simons, 1981; Pickford, 1982; Landau *et al.*, 1982).

The problem addressed in this study is the estimation of their body weight, which is of interest for at least two reasons. First, body weight is an important ecological variable (Lindstedt and Calder, 1981; Gingerich *et al.*, 1982). Knowledge of body weight for these species would, for example, affect the plausibility of hypotheses concerning their arboreality or terrestriality, might indicate some broad limitations for diet, and would influence arguments concerning the relationship between *Ramapithecus* and *Sivapithecus*.

Second, questions of body weight can be used as an interesting test case for examining the limitations of some current methods for paleontological inference. Given the range of inferences that paleontologists attempt to make about the past, body weight should be, on *a priori* grounds, one of the more straightforward. The fossil record consists of morphological data and body weight is a morphological question. In addition, the known morphology is relevant: there is some kind of real biological relationship between tooth (and/or jaw) size and body weight. If there are problems with estimating body weight from tooth size, errors are likely to be even greater when morphological data are used to interpret social structure or diet of a fossil species.

In this chapter I will review a few previous attempts to estimate body weight of extinct hominoids, follow up on some problems with these analyses, present a new approach to the problem of fossil body weight estimation, and then briefly discuss the general implications of these findings for the use of allometric relationships in paleontological inference.

Specimens

Body weights will be estimated using an individual specimen for each species. *Ramapithecus punjabicus* is represented by GSP 4622/4857 (Pilbeam *et al.*, 1977). Measurements of GSP 6153 were used for tooth dimensions that could not be taken from the *Ramapithecus* mandible. *Sivapithecus indicus* is represented by mandible GSP 9564 (Pilbeam *et al.*, 1977). The *Gigantopithecus bilaspurensis* mandible (24261) was originally described by Simons and Chopra (1969). Measurements for *Ramapithecus* and *Sivapithecus* were taken from the original specimens, and for *Gigantopithecus* from a high-quality cast, all at the Harvard Peabody Museum and examined through the courtesy of David Pilbeam.

Evaluation of Some Previous Attempts to Estimate Body Weight of Fossil Hominoids from Tooth Size

Estimation of a feature such as body weight from incomplete fossil remains depends upon analogies with extant species. Although some types of morphological questions may be answered on paleontological material through direct mechanical interpretation (Rudwick, 1964), and are thus not limited by the range of variation in morphologies of extant animals, no such method exists at present for body weight.

Gingerich (1977) calculated the relationship between mandibular second molar length and body weight in living primates, in order to produce an equation for predicting body weight from tooth size of some fossil apes. Gingerich first attempted to develop an equation using only extant hominoids, but noted that because there are only a few hominoid species, the standard error of the resulting regression equation was large. He therefore also calculated an equation based on 38 primate species and used that equation for the principal estimates of fossil hominoid body weights (Table 1). However, the decision to use an equation based on hominoids or on a broader survey of the primates should not be based on the standard errors of the resulting equations. That the equation derived from 38 primate species has a smaller standard error than the equation for hominoids is a matter of the *precision* of the resulting estimates. The separate concern of *accuracy* depends upon the selection of appropriate species, which should be based on some assumptions about what living species are useful models for the fossil specimens in question (Radinsky, 1982). We could, for example, generate an equation relating tooth size to body weight using hundreds of species of rodents. Such an equation would have a small standard error, but probably would be of little use for predicting body weight of a hominoid.

Recently, Kay and Simons (1980) reported another attempt to infer body weight of a fossil hominoid. They also used M_2 length and body weight, taking data on 106 primate species, and obtained Eq. (3) listed in Table 1.

Kay and Simons note that their equation is "very close" to Gingerich's equation based on hominoids (Table 1). However, the two equations based on similar assumptions—that a broad survey of the primates can be used to infer body weight of a fossil hominoid—appear much less similar. Furthermore, when the two "similar" equations from Table 1 are used to estimate body weights for the *Ramapithecus, Sivapithecus,* and *Gigantopithecus* specimens, the results are quite different (Table 2), and the equations are thus not similar at all. The problem is that these equations have been calculated on log-transformed data, with which intuitive impressions of similarity and differences are likely to be incorrect (Smith, 1980).

Table 1. Previously Published Allometric Regressions Used to Estimate Fossil Hominoid Body Weights from M_2 Length

(1)	log body weight = 2.99 (log M_2 length) + 1.46	(Gingerich, 1977; hominoids)
(2)	log body weight = 3.29 (log M_2 length) + 1.10	(Gingerich, 1977; primates)
(3)	log body weight = 2.86 (log M_2 length) + 1.37	(Kay and Simons, 1980; primates)

Table 2. Estimates of Body Weight (in kg) Based on the Two "Similar" Equations from Table 1

	Table 1 Eq. (1)	Table 1 Eq. (3)
Gigantopithecus	175	97
Sivapithecus	72	42
Ramapithecus	50	29

A Closer Look at Old Methods with New Data

In order to explore further some of the difficulties with the use of tooth size–body weight allometry for inferring fossil body weights, I took tooth size data from Swindler's (1976) text and body weights from Harvey *et al.* (1978). Equations were calculated to examine the consequences of three decisions:

1. Include a wide variety of primates or only hominoids.
2. Use M_2 length or a more complete "functional" measure of tooth size (total postcanine area) as the independent variable.
3. Include males and females in the same data set [as in Gingerich (1977) and Kay and Simons (1980)] or generate separate tooth size–body weight equations for males and females.

Equations were calculated from log-transformed data by least squares regression and used to predict body weight for the three fossil hominoid specimens. The results are presented in Table 3. *Gigantopithecus* body weight can be estimated to be about 80 kg, based on the M_2 length of female primates, or 106 kg, based on the same measure of female hominoids, or 166 kg,

Table 3. Body Weight Predictions Based on Different Species and Different Measures of Tooth Size[a]

	Predicted body weight, kg							
	Based on M_2 length				Based on postcanine area			
	Primates		Hominoids		Primates		Hominoids	
	♀	♂	♀	♂	♀	♂	♀	♂
Gigantopithecus	80	115	106	166	109	155	114	167
Sivapithecus	37	50	49	66	43	59	49	64
Ramapithecus	27	36	35	45	29	39	34	42

[a] Tooth sizes from Swindler (1976) and body weights from Harvey *et al.* (1978). Equations for primates are based on 35 species, including prosimians, New and Old World monkeys, and hominoids. Hominoid equations are based on six species.

on the basis of male hominoids. *Sivapithecus* estimates vary from 37 to 66 kg and *Ramapithecus* from 27 to 45 kg. When males and females are combined in single equations, the estimated values are the averages of the separate estimates by sex. It should be noted that these predicted weights do not differ because of sampling differences in the species and specimens used to construct the equations. These additional sources of variation would increase the differences in results among typical studies.

Correlation Coefficients and the Method of Averaging Several Estimates

Two suggestions for improving fossil body weight estimates have been proposed recently. First, several workers have used the reasonable approach of averaging several body weight estimates from different features rather than depend upon the results of one equation (Martin, 1980; Kay and Simons, 1980; Aiello, 1981; Gingerich and Martin, 1981). Second, Martin (1982) has argued that the equations used need to have correlation coefficients of 0.98 or better in order to provide accurate estimates for an unknown species.

Martin (1980) has published data that allow the evaluation of these suggestions. He reported allometric regression equations between body weight and each of 14 skull measurements, using 36 extant primate species. From the resulting 14 equations, Martin selected four as best indicators of body weight. These measurements were maximum skull length, maximum width across the zygomatic arches, maximum length within the zygomatic arch on the right side, and width times length of the right occipital condyle. Three of these equations had correlations of 0.98 and one (zygomatic arch length) of 0.96. He then used each of these equations to estimate body weight of 12 extant primate species not used to generate the equations. These data are reproduced in Table 4, with the following changes:

1. *Hapalemur sinus* is excluded, because actual body weight is unknown.
2. Instead of reporting predicted body weights, the predicted values are compared to the actual (observed) body weights by computing

$$\frac{\text{Observed} - \text{predicted}}{\text{Predicted}} \times 100\%$$

   This results in a measurement of the accuracy of the predicted values.
3. A fifth estimate is computed for each species, based on the average body weight of the four individual estimates.

An inspection of Table 4 shows that for five out of 11 species, not a single estimate is within 20% of the actual value. For six of 11 species, the averaged estimate based on four equations is in error by about 30% or more. Also, the pattern of errors is not consistent. For some species, all equations overesti-

Table 4. Body Weight Predictions: Percent Deviations from Predicted Values[a]

	Maximum skull length (SL)	Maximum width across zygomatic arches (BZW)	Maximum length within a zygomatic arch (RZW)	Occipital condyle length × width (COW × COL)	Average of four estimates
Phaner furcifer	7.6	19.6	−11.3	−24.9	−5.3
Lemur mongoz	−32.1	16.0	−11.2	−4.6	−11.3
Euoticus elegantulus	8.2	−42.5	−3.0	7.4	−6.6
Galago alleni	−32.1	−36.9	−28.9	−43.7	−35.9
Cebuella pygmaea	101.4	75.0	79.3	250.0	109.3
Callimico goeldi	23.3	35.0	49.5	58.6	40.3
Papio anubis	−64.5	−50.9	28.5	−35.0	−44.2
Theropithecus gelada	−45.0	−45.4	24.2	−17.6	−29.4
Hylobates lar	4.5	−1.2	−1.2	−15.9	−4.2
Symphalangus syndactulus	−4.2	−1.6	−11.4	26.2	0.5
Pongo pygmaeus	−31.1	−45.3	20.6	−59.2	−39.0

[a]Modified from Martin (1980), Table 2.

mate or underestimate the actual value, while for other species, there are both positive and negative errors.

It is evident from these data that combining estimates from several equations does not ensure an accurate prediction. In addition, correlation coefficients are not a good indication of the value of an equation for prediction, since regressions with correlations of 0.98 can be inconsistent and inaccurate.

Another recent study demonstrates that these problems with body weight estimates are not unique to facial and dental measurements. Using 46 extant species, Aiello (1981) produced three equations relating postcranial measurements to body weight. The correlations for her three equations were 0.98, 0.98, and 0.99. Aiello used these equations to estimate body weight of several Miocene hominoids, and she also reported 95% confidence intervals for these estimates. All three equations could be used for only two of the species (Table 5). For *Dendropithecus,* the equation using femur midshaft diameter results in an estimate of 7121 g and the equation using humerus length 15,321 g. The 95% confidence intervals do not overlap. Clearly, one of these estimates is incorrect, in spite of the correlation between each of these variables and body weight among extant primates.

A New Approach: Narrow Allometry

Thus far, we have been working with the assumption that a 10-kg hominoid, or even a 1-kg monkey, provides some relevant information for the estimation of body weight in a species as large as *Gigantopithecus.* The other possibility is that size is such a fundamental feature of an animal's biology (e.g., Went, 1968; Haldane, 1965) that such a comparison is of little value; we are artificially imposing a statistical summary on a set of data with limited biological meaning.

From this point of view, instead of looking at features across a range of sizes, we should narrow in on the size range of concern. Since Gould (1966)

Table 5. Body Weight Estimates (with 95% Confidence Intervals) Based on Postcranial Measurements

	Body weight estimate, g		
	Based on femur midshaft diameter	Based on humerus midshaft circumference	Based on humerus length
Pliopithecus vindobonensis	8,023 (6,034–10,047)	7,823 (4,823–12,961)	11,428 (7,259–17,991)
Dendropithecus macinnesi	7,121 (5,642–8,884)	10,452 (6,446–16,947)	15,321 (9,728–24,131)

[a]Modified from Aiello (1981), Table 4.

Table 6. Body Weights of Species with Mandibular Postcanine Areas Similar to *Ramapithecus*

	Postcanine area, mm²	Body weight, kg
Melursus ursinus (sloth bear) ♂	397	100
Pan troglodytes (chimpanzee) ♂	415	44
Ramapithecus	424	?
Muntiacus muntjak (barking deer) ♀	450	15
♂	458	18
Helarctos malayanus (Malayan bear) ♀	459	45
Aepyceros melampus (impala) ♀	473	65
Antilocapra americana (proghorn antelope) ♀	472	40
♂	481	60

has defined allometry as "the study of size and its consequences," methods that depend upon a wide range of sizes can be identified as "broad allometry" and those that use intentionally restricted ranges as "narrow allometry" (Smith, 1980).

In order to estimate the body weights of the three fossil hominoids from the viewpoint of narrow allometry, the mammal collection of the United States National Museum (Smithsonian Institution) was searched for species with a mandibular posterior occlusal surface area about the same size as each of the fossil hominoids under consideration here. Tables 6–8 list the body weights of species whose posterior tooth sizes fall within narrow ranges (values in these tables are averages of two or three specimens for each sex of each species). These results confirm the preceding discussion, in that depending upon which species are selected, a wide range of body weights remain plausible for the fossil species. Mammals with postcanine dentitions about the size of *Ramapithecus* range from 15 to 100 kg in body weight; species found for *Sivapithecus* from 20 to 170 kg; and species for *Gigantopithecus* range from 30 to 250 kg.

While interspecific allometric regressions usually seem to confirm that there is a strong relationship between tooth size and body weight, the data in Tables 6–8 give a different impression. One species (*T. pecari*) uses a postcanine area of about 850 mm² to process food for a 30-kg body, while another species (*U. arctos*) uses the same postcanine area to process the food needed for a 250-kg body. There are clearly important factors other than body weight that affect tooth size. Analyses using the concept of narrow allometric comparisons are highly effective in revealing these differences.

Interpretation and Conclusions

It is always dispiriting to see Kunkel walk in. Sooner rather than later, you know, he is going to say, "Let's look at the facts." . . . Kunkel labors under the delusion that

Table 7. Body Weights of Species with Postcanine Areas Similar to *Sivapithecus*

	Postcanine area, mm²	Body weight, kg
Antilocapra americana (proghorn antelope) ♀	472	40
♂	481	60
Helarctos malayanus (Malayan bear) ♂	525	60
Ursus americanus (American black bear) ♀	544	100
Pongo pygmaeus (orangutan) ♀	546	37
Sivapithecus	563	?
Mazama americana (brocket, Cervidae) ♂	567	20
Odocoileus virginianus (white-tailed deer) ♂	584	110
Pongo pygmaeus (orangutan) ♂	650	69
Ursus arctos (European brown bear) ♀	696	170

facts pave the yellow brick road to the emerald city of Truth. . . . This is because when Kunkel looks at the facts he looks only at the facts he wants to look at in that particular glance.

(Baker, 1979)

Two general observations can be drawn from the preceding data. First, we can learn little about the body weight of *Ramapithecus, Sivapithecus,* or *Gigantopithecus* by comparing their skulls or dentitions to those of living animals, without making important additional assumptions. Second, we began with the argument that body weight should be a relatively simple inference. The present results may be considered to contradict that hypothesis, or they may (more likely) be considered to demonstrate that most statements made

Table 8. Body Weights of Species with Postcanine Areas Similar to *Gigantopithecus*

	Postcanine area, mm²	Body weight, kg
Tayassu pecari (white-lipped peccary) ♂	850	30
Ursus arctos (European brown bear) ♂	851	250
Blastocerus dichotomus (marsh deer) ♂	869	90
Phachochoerus aethiopicus (warthog) ♀	939	85
Gorilla gorilla (gorilla) ♂	975	160
Potamochoerus porcus (bushpig) ♀	990	75
Crocuta crocuta (spotted hyena) ♂	1014	80
Rangifer tarandus carabou (carabou) ♂	1054	225
Gigantopithecus	1071	?
Sus scrofa (European wild boar) ♀	1372	110

about the past—however straightforward they might appear—are at least equally, if not more so, capable of alternative interpretation.

An explanation for these problems with body weight inferences involves complex philosophical issues such as the uniqueness of historical events and the relationship between a statistical trend and biological law. However, in order to avoid an extensive (and possibly nonproductive) digression into these issues, for the purposes of this discussion I will accept the following two assumptions:

1. Some interesting and useful questions about the past can only be answered by analogy with the present.
2. One appropriate method for making analogies between the past and present is to seek empirical regularities in the present and directly apply these regularities to fossil specimens. It is often not possible to understand the nature of the biological process underlying the observed regularity.

Given these assumptions, the failure to accurately estimate fossil body weights from allometric regressions indicates simply that the equations used for these estimates are not good empirical regularities. In using the present to understand the past, we have been too lax in defining criteria by which we accept a bivariate allometric relationship as valid for paleontological inference.

As a first step in reevaluating these criteria, the following considerations are proposed.

1. Correlation coefficients are of little use in determining the value of an equation for paleontological inference. A low correlation will often increase the probability that estimates will be incorrect, but a high correlation does not ensure that they will be correct. This includes correlations of 0.98 and 0.99. Correlation coefficients are affected by the slope of the bivariate relationship and by the range of the variables (Smith, 1981). They are a measure of explained variance, not a measure of predictive error (Thorndike, 1978). Similarly, the statistical significance of the correlation or a confidence interval for a predicted value are not particularly relevant (Radinsky, 1982). We know that our estimate for a fossil specimen may be incorrect, but neither the correlation coefficient, its significance, or a confidence interval provides information on likely margins of error. Rather, the range of errors in prediction need to be documented directly.

2. The preceding point leads to the concept of testing an equation. This testing, by determining predicted values for species (or specimens) with known actual values, is best done with species not used to calculate the equation. If one has a data set with 30 species, it might be better to use perhaps 20–22 species to calculate the equation and the remainder to test the equation, rather than all 30 in the original calculations. An equation will always have a larger average prediction error for species not included in the calculation of the equation than for those in the equation, but the amount of this increased error will vary for different equations and different samples. The species

used to test the equation should be relevant. If interest is in predicting body weight of *Gigantopithecus*, the error in prediction of a *Galago* or *Saimiri* is irrelevant. Instead, gorillas, orangutans, and perhaps some large pigs or small bears would be useful. These data would provide a reference against which to judge potential errors in fossil inference.

3. There is little value in averaging the results of a few contradictory estimates in the hope that their average will be more accurate. When predicted values differ widely, one or more of them is wrong. One is thus including an incorrect value in the expectation of arriving at a correct average. This is only valid if errors are random about the mean. This might be true when 20–30 or more estimates are averaged from throughout the body, but not when seven or eight (or fewer) values are used from a restricted anatomical area, as will usually be the case for fossil specimens. For example, if *Gigantopithecus* has large teeth for its body weight, then it probably also has a large symphysis, a large mandibular ramus, a large zygomatic arch, and large mandibular condyles. Estimates from these features will not be random about the mean. It would be more useful to search for one good equation than to average the results of a few poor ones.

In conclusion, it is difficult to make certain kinds of paleontological inferences with any reasonable degree of confidence. This does not mean that we should stop trying. Rather, it means that we need to try harder.

Acknowledgments

This research was supported in part by the L. S. B. Leakey Foundation and NIH grant DE 05134. I am particularly grateful to David Pilbeam, Misia Landau, Rimas Vaisnys, Kathleen Gordon, and Alan Walker for many helpful discussions or comments on drafts of the manuscript. For courtesy and assistance at their respective institutions, I thank Peter Andrews and Prue Napier (British Museum, Natural History), Richard Thorington (United States National Museum), Sidney Anderson (American Museum of Natural History), and Maria Rutzmoser (Harvard Museum of Comparative Zoology). I am also grateful to Barbara Bass for typing several drafts of the manuscript.

References

Aiello, L. C. 1981. Locomotion in the Miocene Hominoidea, in: *Aspects of Human Evolution* (C. B. Stringer, ed.), pp. 63–97, Taylor & Francis, London.

Andrews, P., and Cronin, J. E. 1982. The relationships of *Sivapithecus* and *Ramapithecus* and the evolution of the orang-utan. *Nature* **297:**541–546.

Baker, R. 1979. In point of fact. *New York Times Sunday Magazine* **1979** (March 4).

De Bonis, L., and Melentis, J. 1980. Nouvelles remarques sur l'anatomie d'un Primate hominoïde du Miocène: *Ouranopithecus macedoniensis.* Implications sur la phylogénie des Hominidés. *C. R. Acad. Sci. Paris D* **290:**755–758.

Gingerich, P. D. 1977. Correlation of tooth size and body size in living hominoid primates, with a note on relative brain size in *Aegyptopithecus* and *Proconsul. Am. J. Phys. Anthropol.* **47:**395–398.

Gingerich, P. D., and Martin, R. D. 1981. Cranial morphology and adaptations in Eocene Adapidae. II. The Cambridge skull of *Adapis parisiensis. Am. J. Phys. Anthropol.* **56:**235–257.

Gingerich, P. D., Smith, B. H., and Rosenberg, K. 1982. Allometric scaling in the dentition of primates and prediction of body weight from tooth size in fossils. *Am. J. Phys. Anthropol.* **58:**81–100.

Gould, S. J. 1966. Allometry and size in ontogeny and phylogeny. *Biol. Rev.* **41:**587–640.

Greenfield, L. O. 1979. On the adaptive pattern of "*Ramapithecus.*" *Am. J. Phys. Anthropol.* **50:**527–548.

Greenfield, L. O. 1980. A late divergence hypothesis. *Am. J. Phys. Anthropol.* **52:**351–365.

Haldane, J. B. S. 1965. On being the right size, in: *The New Treasury of Science* (H. Shapley, S. Rapport, and H. Wright, eds.), pp. 474–478, Harper and Row, New York.

Harvey, P. H., Kavanaugh, M., and Clutton-Brock, T. H. 1978. Sexual dimorphism in primate teeth. *J. Zool. Lond.* **186:**475–485.

Kay, R. F. 1982. *Sivapithecus simonsi,* a new species of Miocene hominoid, with comments on the phylogenetic status of the Ramapithecinae. *Int. J. Primatol.* **3:**113–173.

Kay, R. F., and Simons, E. L. 1980. The ecology of Oligocene African Anthropoidea. *Int. J. Primatol.* **1:**21–37.

Landau, M., Pilbeam, D., and Richard, A. 1982. Human origins a century after Darwin.*BioScience* **32:**507–512.

Lindstedt, S. L., and Calder, W. A., III. 1981. Body size, physiological time, and longevity of homeothermic animals. *Q. Rev. Biol.* **56:**1–16.

Martin, R. D. 1980. Adaptation and body size in primates. *Z. Morphol. Anthropol.* **71:**115–124.

Martin, R. D. 1982. Allometric approaches to the evolution of the primate nervous system, in: *Primate Brain Evolution* (E. Armstrong and D. Falk, eds.), pp. 39–56, Plenum Press, New York.

Pickford, M. 1982. New higher primate fossils from the Middle Miocene deposits at Majiwa and Kaloma, Western Kenya. *Am. J. Phys. Anthropol.* **58:**1–19.

Pilbeam, D., Meyer, G. E., Badgley, C., Rose, M. D., Pickford, M. H. L., Behrensmeyer, A. K., and Ibrahim Shah, S. M. 1977. New hominoid primates from the Siwaliks of Pakistan and their bearing on hominoid evolution. *Nature* **270:**689–695.

Radinsky, L. 1982. Some cautionary notes on making inferences about relative brain size, in: *Primate Brain Evolution* (E. Armstrong and D. Falk, eds.), pp. 29–37, Plenum Press, New York.

Rudwick, M. J. S. 1964. The inference of function from structure in fossils. *Br. J. Philos. Sci.* **15:**27–40.

Simons, E. L. 1977. *Ramapithecus. Sci. Am.* **263:**28–35.

Simons, E. L. 1981. Man's immediate forerunners. *Philos. Trans. R. Soc. Lond. B* **292:**21–41.

Simons, E. L., and Chopra, S. R. K. 1969. *Gigantopithecus* (Pongidae, Hominoidea) a new species from North India. *Postilla* **138:**1–18.

Smith, R. J. 1980. Rethinking allometry. *J. Theor. Biol.* **87:**97–111.

Smith, R. J. 1981. Interpretation of correlations in intraspecific and interspecific allometry. *Growth* **45:**291–297.

Swindler, D. 1976. *Dentition of Living Primates,* Academic Press, London.

Thorndike, R. M. 1978. *Correlational Procedures for Research,* Gardner Press, New York.

Went, F. W. 1968. The size of man. *Am. Sci.* **56:**400–413.

Zihlman, A. L., and Lowenstein, J. M. 1979. False start of the human parade. *Nat. Hist.* **88**(7):86–91.

Allometric Perspectives on Fossil Catarrhine Morphology

20

KAREN STEUDEL

Introduction

Patterns of size and scaling are an important element in the analysis of fossil catarrhine morphology. Fossil representatives of this group vary substantially in size and in many cases related forms of different size are found in similarly dated strata so that size differences and their consequences become an important element in understanding the adaptive relationships among these forms as well as in establishing ancestor–descendant sequences. The removal or partitioning of the effects of size in morphological data becomes an important element in making correlations between structure and adaptation in fossil specimens.

The significance of this type of analysis is conspicuous in a number of fossil taxa. For example, early hominids are generally acknowledged to vary greatly in overall size (e.g., Robinson, 1972*a;* Pilbeam and Gould, 1974) and this has added to the difficulties of assessing the degree of similarity in the adaptations of the gracile and robust types (Brace, 1972; Wolpoff, 1971; Robinson and Steudel, 1973). Size differences are also readily apparent between the two fossil taxa considered as likely ancestors to the hominids, *Ramapithecus* and *Gigantopithecus;* and assessing the implications of these size differences for the adaptations and subsequent evolution of the two taxa is a

KAREN STEUDEL • Department of Zoology, University of Wisconsin, Madison, Wisconsin 53706.

substantial problem. There is also substantial size variation among the Miocene dryopithecines (e.g., Andrews, 1974), where it is often unclear how much of the morphological differences are due to size alone and how much to other adaptive shifts.

Given the small samples and usually fragmentary specimens characteristic of fossil data, one wonders how accurate allometric analysis can be. Any documentation of allometry requires a fairly reliable estimate of the body size of individual specimens if intrataxon allometry is to be studied. Species average body weights are of only limited use, since they can only be compared with averages of other variables, and with the small samples usually involved there is no reasonable assurance that the sample means accurately estimate the means of variables in the original population. Further, if one is able to obtain a reasonable estimate of intrataxon allometry for a fossil group, to what data on living species should this be compared for maximum insight? Since patterns of allometry for a single variable observed within various taxonomic units frequently differ from one another (e.g., Jungers, 1979; Steudel, 1982*a*), the conclusions one would draw about the fossils could differ depending on the modern sample chosen for comparison.

Consequently, before attempting an analysis of fossil primate allometry it seemed appropriate to consider the problems underlying this type of study to see how large an effect they have on the accuracy of results and if (or how) these problems can be circumvented.

What Range of Taxa Best Indicate the Effects of Size?

Since the adaptive implications of fossil morphology are usually interpreted based on structure–function correlations developed in modern relatives, description of the effects of size in modern primates is a necessary initial step in the study of allometry in fossils. Yet a major problem in accurately determining the morphological consequences of size variation in modern forms is the existence of substantial differences between patterns of allometry observed interspecifically (across a range of species showing substantial variation in size) as compared to static adult allometry within smaller taxonomic units—a single species or several closely related taxa. Comparisons of intraspecific versus interspecific allometry have focused primarily on brain size, although a few other characteristics have also been studied from this perspective (Meunier, 1959; Leutenegger, 1976; Echelle *et al.*, 1978). Brain weight and linear dimensions relating to brain size show the classic pattern of transpositional allometry (Meunier, 1959; Gould, 1971), where the intraspecific lines have lower slopes and higher y intercepts than does the interspecific line, so that, when plotted, the intraspecific lines appear as stepwise increments along the interspecific line. Perhaps because brain size allometry was so much better understood than that of other variables, the idea that transpositional allometry was a general phenomenon became common (e.g., Lande, 1979). This line of reasoning may also have been bolstered by casual perusal of results from least squares line fitting. Since correlations are frequently lower in intraspecific data [due partly to the smaller size range (Smith, 1980; Steudel, 1982*a*)] the least squares slopes also are frequently lower than their interspecific counterparts, giving a superficial appearance of transpositional allometry. Differences between intrataxon (within a species or group of closely related species) and intertaxon (broad sample) allometry make the role of size in producing structural adaptation very ambiguous. Furthermore, it is unclear which pattern is appropriate for use when one is attempting to partition the effects of other adaptive differences. Before adequate attempts can be made to document allometry in fossils, the ambiguities produced by apparently different consequences of size in samples differing in taxonomic inclusivity must be resolved.

Steudel (1982*a*) addressed this problem, using a sample of 249 Catarrhine primates representing the following taxa:

Hominoidea
- Pongidae
  - *Gorilla gorilla* ($N = 45$)
  - *Pan troglodytes* ($N = 39$)
  - *Pongo pygmaeus* ($N = 22$)
- Hylobatidea
  - *Hylobates lar* ($N = 32$)

Cercopithecoidea
- Cercopithecinae
  - *Papio* sp. ($N = 8$)
  - *Macaca mulatta* ($N = 27$)
  - *Cercopithecus mitis* ($N = 8$)
- Colobinae
  - *Nasalis larvatus* ($N = 15$)
  - *Presbytis cristatus* ($N = 32$)
  - *Colobus guereza* ($N = 21$)

Thirty-one specimens of *Homo sapiens* were also included for the present study. Intrataxon and intertaxon allometry were compared for a series of 22 skeletal variables from a variety of anatomical regions. (See Table 1 for a list of the variables.) In 60% of these comparisons, confidence testing indicated no significant difference between the pattern of allometry observed for a variable intraspecifically, over the entire sample, or within the major taxonomic divisions. Thus transpositional allometry appears not to be the rule in comparisons of allometry at different taxonomic levels. For cases of this type there is no theoretical difficulty in determining what pattern of allometry should be used for reference in studies of those variables in fossils.

In the remaining 40% of comparisons, however, a statistically significant difference ($p = 0.05$) between intrataxon and intertaxon allometry existed, so

Table 1. Least-Squares Slopes α, y Intercepts, and Correlation Coefficients r for the Relationship between Each Variable and Size (Skeletal Weight) for the Entire Sample and for Five Taxonomic Subsets, Together with Standard Errors (SE) for the Relationships Calculated over the Entire Sample

	Pongidae			Pongidae + *Hylobates*			Hominoidea			Cercopithecinae			Pongidae + *Homo*			Entire sample			
Variable	α	y	r	α	y	r	α	y	r	α	y	r	α	y	r	α	y	r	SE
1. Ramus height	0.41	−1.10	0.81	0.45	−1.41	0.97	0.42	−1.27	0.93	0.31	−0.34	0.91	0.40	−1.08	0.65	0.32	−0.41	0.91	0.19
2. Mandibular breadth	0.33	−2.02	0.84	0.41	−2.69	0.98	0.40	−2.61	0.96	0.33	2.03	0.79	0.32	−2.02	0.72	0.34	−2.19	0.95	0.14
3. External mandibular breadth	0.13	0.76	0.72	0.26	−0.25	0.97	0.26	−0.27	0.96	0.30	−0.55	0.96	0.13	0.78	0.64	0.29	−0.49	0.98	0.08
4. Mandibular depth	0.30	−1.10	0.71	0.40	−1.85	0.97	0.38	−1.79	0.96	0.35	−1.26	0.90	0.30	−1.13	0.70	0.28	−0.99	0.91	0.16
5. Palate length	0.26	−0.28	0.73	0.28	−0.47	0.96	0.25	−0.29	0.82	0.30	−0.35	0.78	0.24	−0.22	0.43	0.24	−0.26	0.85	0.19
6. Palate breadth	0.12	0.42	0.53	0.25	−0.66	0.95	0.25	−0.63	0.94	0.24	−0.54	0.88	0.12	0.40	0.48	0.25	−0.65	0.96	0.09
7. External maximum breadth	0.18	0.53	0.83	0.26	−0.17	0.98	0.26	−0.13	0.97	0.24	0.03	0.94	0.17	0.54	0.72	0.25	0.25	0.98	0.07
8. Bizygomatic breadth	0.28	0.51	0.88	0.28	0.50	0.98	0.27	0.55	0.97	0.29	0.52	0.96	0.27	0.54	0.82	0.26	0.64	0.98	0.07
9. Facial length	0.30	−0.06	0.84	0.40	−0.89	0.98	0.38	−0.76	0.94	0.54	−1.32	0.90	0.28	−0.01	0.61	0.34	−0.46	0.92	0.18
10. Orbit height	0.17	−0.01	0.63	0.19	−0.14	0.92	0.18	−0.11	0.90	0.08	0.35	0.54	0.16	0.03	0.57	0.20	−0.23	0.93	0.10
11. Orbit width	0.15	0.18	0.65	0.19	−0.12	0.94	0.19	−0.12	0.94	0.20	−0.20	0.91	0.15	0.21	0.62	0.20	−0.19	0.97	0.07
12. Skull height	0.16	1.05	0.79	0.22	0.57	0.97	0.24	0.46	0.90	0.22	0.49	0.93	0.16	1.08	0.44	0.28	0.16	0.95	0.12
13. Skull base	0.42	−1.17	0.88	0.33	−0.45	0.97	0.33	−0.46	0.96	0.30	−0.27	0.93	0.41	−1.06	0.83	0.34	−0.55	0.98	0.10
14. Skull length	0.26	0.65	0.80	0.24	0.81	0.96	0.25	0.75	0.94	0.21	1.02	0.96	0.26	0.72	0.68	0.26	0.68	0.97	0.09
15. Cranial breadth	0.11	1.53	0.68	0.19	0.88	0.96	0.21	0.80	0.91	0.22	0.63	0.94	0.11	1.56	0.41	0.25	0.51	0.96	0.10
16. Femoral length	0.22	1.73	0.81	0.16	2.16	0.94	0.19	2.00	0.78	0.31	1.11	0.93	0.23	1.76	0.46	0.24	1.64	0.90	0.15
17. Humeral length	0.31	1.11	0.89	0.17	2.23	0.91	0.17	2.26	0.90	0.33	0.91	0.96	0.31	1.12	0.87	0.29	1.25	0.91	0.17
18. Acetabular diameter	0.32	0.95	0.88	0.30	−0.83	0.98	0.31	−0.89	0.97	0.34	−1.24	0.95	0.32	−0.94	0.81	0.35	−1.19	0.98	0.10
19. Iliac breadth	0.46	−2.13	0.85	0.38	−1.52	0.96	0.41	−1.65	0.93	0.40	−1.57	0.94	0.46	−2.05	0.70	0.39	−1.53	0.96	0.15
20. Vertebral area	0.58	−1.03	0.87	0.63	−1.46	0.98	0.66	−1.61	0.96	0.69	−1.78	0.95	0.59	−1.04	0.75	0.64	−1.44	0.98	0.17
21. Femur width	0.36	−0.86	0.87	0.38	−1.05	0.98	0.39	−1.11	0.97	0.34	−0.89	0.97	0.36	−0.84	0.82	0.39	−1.11	0.99	0.09
22. Femur circumference	0.37	−0.69	0.81	0.37	−0.66	0.97	0.37	−0.70	0.96	0.32	−0.40	0.94	0.37	−0.66	0.78	0.36	−0.60	0.98	0.10

that some decision is necessary about which to use as the most fundamental indicator of the effects of size. Interestingly, the three variables that are heavily determined by brain (skull height, skull length, and cranial breadth) showed the classic pattern of transpositional allometry reported for the brain by previous workers. The other dimensions that differed within as compared to between species, however, did not show this pattern. For example, facial length (variable 9, measured from nasion to prosthion) showed positive allometry when measured over all nonhominids, but when tested within smaller taxonomic groupings there was a range from negative allometry to very strong positive allometry. There was no single consistent pattern of relationship between intrataxon as compared to intertaxon allometry in these data.

Of course in those cases where no difference between intrataxon and intertaxon allometry is apparent, no difficulties arise. But what conclusions can be drawn about the treatment of allometry in the cases where statistically demonstrable differences exist? Situations for which this was true fell into two categories. In one the slopes of some intraspecific regressions were very low and were associated with low correlation coefficients so that the 95% confidence intervals about the slopes overlapped zero. In the remaining cases the two allometric patterns differed in spite of reasonably high correlations in both intrataxon and intertaxon results. The biological causes underlying these two categories seem to be quite different.

When the 95% confidence interval about an allometric slope overlaps zero, it indicates that the variation of the dimension as a response to size is negligible. Size is not an important influence on the magnitude of the dimension in question. This situation was observed almost exclusively in intraspecific results and was especially common in those species (*P. troglodytes* and *H. lar*) with relatively little size variation due to low sexual dimorphism. As the range of size becomes smaller, individual error becomes a greater proportion of the total observed variation so that the amount of variation in the Y variate that can be explained by variation in size becomes smaller. Smith (1980) found that the coefficient of determination r^2 of correlation values by range of size was 0.31 for the 60 data sets that he examined. The same analysis for the present data yielded $r^2 = 0.34$. But this leaves about two-thirds of the variation in correlation coefficients unexplained. What elements in addition to a small size range might be producing the near independence between a variable and size? Perhaps the size of certain features need not vary in close response to body size in order to maintain optimal effectiveness intraspecifically. This argument is related to Gould's (1975) suggestion that intrapopulational variation in body weight may be primarily a function of nutrition and that there will be no correlation between brain size and nutrition, resulting in a lower brain size–body size correlation within a population than observed over a range of species. Such features might then have some optimal size within a species and variation about this target might not be closely tied to variation in body size. An interspecific line would result from differences in the target values characteristic of the series of species being studied. This seems especially plausible if intraspecific size variation is small so that features of "target" size would not show much proportional difference in the larger and smaller species members. Size of the eye orbit is the most common feature to show this pattern in my data and one can readily imagine that this might not be closely tied to body size in intraspecific adaption. Interestingly these variables, orbit height (#10) and orbit width (#11), are highly correlated with size interspecifically and within most of the taxonomic subsets.

Any allometic analysis based on intraspecific results of this type would be very unreliable since they reflect primarily an absence of allometric effects intraspecifically. On the other hand, orbital width is a variable showing little allometric variation within species but is highly correlated with size ($r = 0.97$) over the entire sample. This suggests that size estimates for fossils based on this variable might tend to give body size values closer to species means than would variables that showed similar allometry both within and between species.

In those cases where intrataxon and intertaxon results differ in spite of high correlations for both, this explanation does not apply. A number of other explanations have been suggested in the literature, generally focusing on the relationship of brain size to body size (Steudel, 1982*a*). These explanations have in common the conclusion that it is the intertaxon coefficients that most fundamentally indicate the adaptive response of a variable to a change in body size. The intrataxon coefficients are seen as caused by a variety of other factors. This viewpoint can be found in the literature as early as Huxley's (1932) classic book. While this view is probably correct for those cases in which the intraspecific correlations between a variable and size are low, none of these explanations satisfactorially deals with cases in which both intrataxon and intertaxon correlations between a variable and size are quite high but the coefficients of allometry differ.

An alternate possibility is that intertaxon allometric slopes differ from those measured over a wider taxonomic and size range because the optimal response to changing body size differs. Under this argument many dimensions do not vary in response to some extremely general influence of size, but rather as a result of a fairly specific influence of size produced by the adaptations relevant to the individual species. A series of related, functionally similar species showing a moderately wide range of body sizes would most accurately indicate the effects of size-required variation, while the intertaxon pattern, measured over a much greater variety of species, may be merely an artifact of changes in adaptation that accompany change in size in the particular sample.

An especially good illustration of this can be seen in Fig. 1, comparing intertaxon allometry calculated for facial length over the entire sample to intrataxon allometry in several subsets. It is apparent that patterns of allometry differ substantially between the various samples. Neither the slope for cercopithecines nor that for pongids overlaps that calculated over the entire sample at 95% confidence, although the hominoid lines does overlap at that confidence level. The intrataxon slope for cercopithecines is especially high, reflecting the fact that the larger species in this group have increasingly long

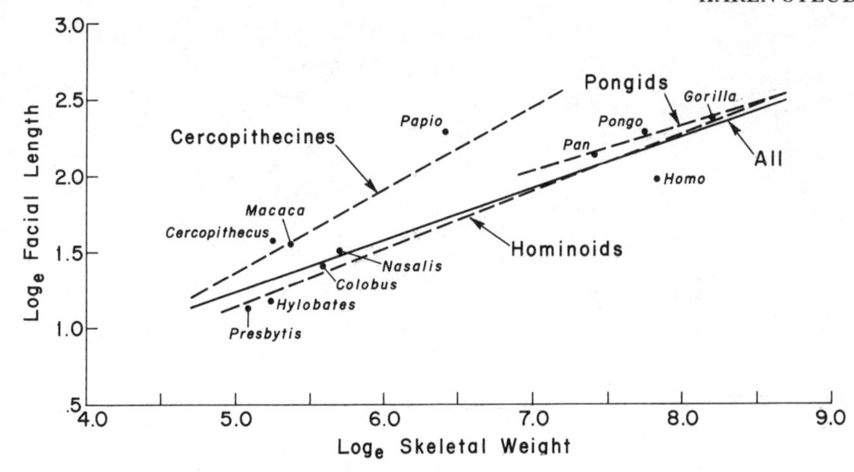

Fig. 1. Log–log plot of facial length relative to skeletal weight. The solid line represents the pattern of scaling found over the entire sample, while dashed lines indicate allometry within various subsets of the data. Species means are also included.

snouts, with *Papio* occupying the extreme position. No such tendency for elongation of the face is seen in the pongids. Among the Hominoidea there is slight positive allometry in facial length since gibbons are relatively shorter faced than are the great apes. Thus the three subsets have patterns of facial length allometry that differ quite a bit from one another. The intertaxon line takes its value of 0.34 because the largest species are among the shortest faced forms. Most of the smaller species represented are the relatively longer faced monkeys. The intertaxon line does not indicate fundamental effects of size but rather is heavily influenced by differing facial adaptations in taxonomic units representing different body sizes.

Another example of this can be seen in Fig. 2, showing femur length allometry. Here the apparent coefficient of allometry is again heavily determined by the adaptations of species included in the various subsets. The line determined by pongids + *Homo,* for example, is quite different from the line produced when pongids are combined with *Hylobates.* Adaptive differences between femur length allometry in colobine and cercopithecine monkeys have also been demonstrated (Steudel, 1982*b*).

If coefficients of allometry are determined over a series of species that differ in adaptations affecting the variables being studied, the allometric parameters will be determined partly by the necessary consequences of size and partly by the consequences of the different adaptations. The greater is the degree of adaptive shift in the species comprising the sample, the greater will be the effects of this on the allometric parameters. It may very well be impossible to totally eliminate non-size-required adaptive variation from the measurement of allometry, since there may be changes in adaptation correlated with a change in size even among conspecifics (Dodson, 1975). Within-species analysis also has the difficulty that the range of size will be relatively small so

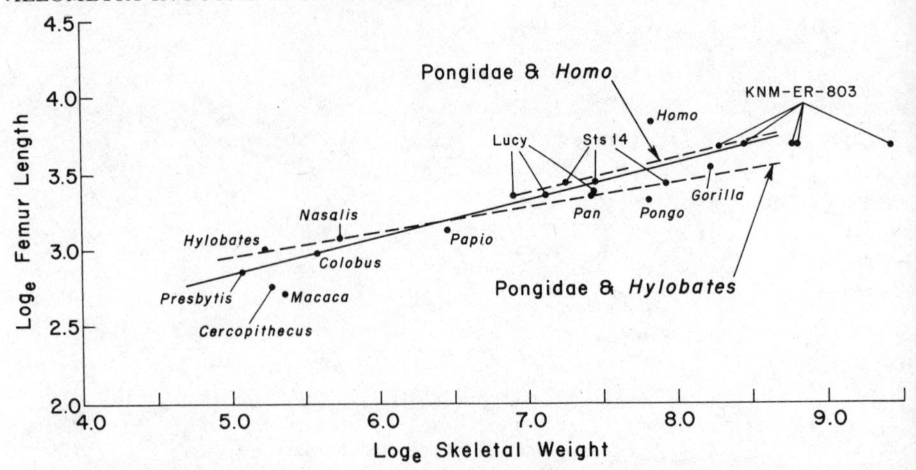

Fig. 2. Log–log plot of the scaling of femur length relative to skeletal weight. The solid line represents the pattern of scaling seen over the entire sample, while dashed lines indicate allometry within various subsets. In addition to species mean values, the values of skeletal weight for three fossil specimens estimated based on various reference samples are included.

that individual error will be relatively large, resulting in a tendency toward wider confidence intervals about any allometric parameter. The ideal of determining the effects of size increase alone on some structure without the added complication of adaptive shifts is probably unrealizable since most if not all data samples will include both types of variation. Nevertheless, one can approach the ideal situation by minimizing the amount of adaptive variation that is not directly related to size by including in a sample closely related species showing variation in body size while retaining similar function in the structures under study.

It is clear that increase in body size is accompanied by different adaptive responses in different groups for many variables, such as facial length, femur length, and humerus length. For such cariables, there is no single necessary response to size, but rather the effects of size are determined by the role of the structure in adaptation–information that is lacking in fossils. If a fossil specimen is to be seen in an allometric context, some decision must be made about which living forms are most likely to share a common adaptation with the fossils, and therefore be relevant for comparison.

The Choice of the Reference Population

For these reasons, a key decision that must be made in any study of fossil allometry is the makeup of the reference sample of extant taxa that will be used for comparison. In some cases single living species have been used for this purpose (e.g., Corruccini, 1976; McHenry, 1974, 1976). If the aim of the study is limited—simply to compare, say, size variation in the fossils to size variation in the particular modern taxon—this approach is sufficient. This is particularly true if a case can be made for the fossil in question being closely related, genetically and in adaptation, to the modern species used for comparison (e.g., *H. sapiens neanderthalensis* and *H. sapiens sapiens*). The difficulty with this method, however, is the uniqueness of fossil species. With the exception of very recent forms, each fossil species may be (and probably is) substantially different from any living species in at least certain aspects of its adaptation. Consequently a broader reference sample is usually necessary.

The composition of the broader sample can be a critical element in determining the outcome of the analysis. In general it seems appropriate to include living species that are assigned to the same taxon as the fossils at some level—say, the same superfamily. But this may be inadequate. For example, Miocene anthropoids from East Africa that are classified as hominoids appear to be more similar postcranially to cercopithecoids or ceboids (Corruccini *et al.*, 1976; Fleagle and Simons, 1978). Thus, representation from a larger taxonomic unit would be necessary in this case.

To determine to what extent the choice of a reference sample could influence size estimation in fossils, three skeletal variables were chosen from the 22-variable sample of modern catarrhines that could be measured on available fossil specimens and that have varying levels of correlation with size, measured as skeletal weight as described in Steudel (1980, 1981*b*). The lower the overall correlation values, the better is the chance that taxonomic subsets of the entire sample will show divergent allometric relationships. Femur length was chosen as a variable with a relatively low correlation with size over the entire sample ($r = 0.90$), but one that has previously been used in size estimation (McHenry, 1974; Mobb and Wood, 1977; Wood, 1976). Mandibular breadth has a relatively moderate correlation ($r = 0.95$) with size, and femoral circumference (below the lesser trochanter) has a very high correlation with size ($r = 0.98$). The least squares regression lines for each of these variables relative to size (skeletal weight) for the entire sample and several taxonomic subsets can be seen in Table 1. The position of each species mean was graphed and the regression lines based on the entire sample and on two subsets were drawn in. Since the three lines differed negligibly for femoral circumference, this graph is not reproduced here. The remaining two graphs can be seen in Figs. 2 and 3. Size was estimated using each of the variables for appropriate fossil specimens. The resulting size estimates for each fossil based on the several taxonomic subsets can be seen in Tables 2–4. Three of these specimens have been included in the graphs of femur length and mandibular breadth to give a visual indication of the magnitude of the difference resulting from the choice of the reference sample. The most cautionary results are those using femur length for size estimation. In spite of the correlation between this variable and size being 0.90 over the entire sample, the size estimates for fossils based on this variable deviated quite sharply from one another depending on which of the five reference samples was used. The greatest difference was between the results based on pongids and *Hylobates* in

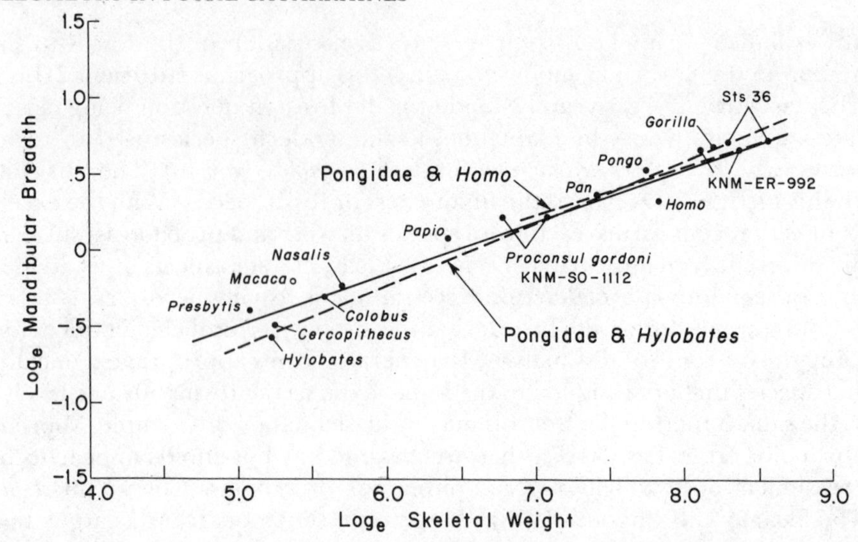

Fig. 3. Log–log plot of the scaling of mandibular breadth relative to skeletal weight. The solid line represents the pattern of scaling seen over the entire sample, while dashed lines indicate allometry within various subsets. In addition to species mean values, the values of skeletal weight for three fossil specimens estimated based on various reference samples are included.

comparison with pongids and *Homo.* This is expected since in the former, a relatively long-legged form is included in the smaller size range, while in the latter, a relatively long-legged form is included in the upper size range. Over a substantial size range such as *P. cristatus* to *Gorilla* a great deal of variation in a dependent variable can be accounted for by size (resulting in $r = 0.90$), while there are nevertheless substantial deviations from the regression line resulting from locomotor adaptations not directly related to size. Consequently, functional or taxonomic subsets of the sample can differ markedly from one another in their individual allometric ends and produce size estimates for a single fossil specimen that are at variance with one another. The size estimates for Sts 14 based on femur length suggest a size for this fossil anywhere from female *Pan* to male *Pongo.* The estimates for KNM-ER 803 range from an

Table 2. Femur Length: Skeletal Weight Estimates for Fossils Based on Regressions within Various Taxonomic Units[a]

	KNM-ER 993	KNM-ER 803	AL 288-1	Sts 14
Pongidae	8.55	8.82	7.36	7.73
Pongidae + *Hylobates*	9.06	9.44	7.44	7.94
Hominoidea	8.47	8.79	7.11	7.53
Pongidae + *Homo*	8.04	8.30	6.91	7.26
All	8.21	8.46	7.13	7.46

[a]Values are expressed as log skeletal weight (grams).

Table 3. Mandibular Breadth: Skeletal Weight Estimates for Fossils Based on Regressions within Various Taxonomic Units[a]

	SK 23	SK 34(av)	SK 12	SK 6	ER 729A	ER 1482A	ER 992	Sts 36	MLD 4
Pongidae	8.85	8.39	9.79	8.91	9.15	8.21	8.12	8.30	8.79
Pongidae + *Hylobates*	8.76	8.39	9.51	8.80	9.00	8.24	8.17	8.32	8.71
Hominoidea	8.78	8.40	9.55	8.83	9.03	8.25	8.18	8.33	8.73
Pongidae + *Homo*	9.13	8.66	10.09	9.19	9.44	8.47	8.38	8.56	9.06
All	9.09	8.65	10.00	9.15	9.38	8.47	8.38	8.56	9.03

[a]Values are expressed as log skeletal weight (grams).

approximately average-sized *Gorilla* to a value far in excess of the largest male *Gorilla*. Thus the choice of the reference sample has a very large influence on fossil size estimation even with a correlation of 0.90 over the entire sample. In the two other variables used as examples, the results are better. The overall correlations with size were higher, indicating less variation that is unrelated to size. For mandibular breadth (r = 0.95 over the entire sample), the smallest estimated value for a fossil specimen in percent of the largest estimated value for that specimen averaged 96.1% as compared to 90.2% for femur length. The effect this has on fossil size estimates is shown in Fig. 3. Femoral circumference, with its very high overall correlation of 0.98 with size, did not produce much variation in estimated fossil size even where different subsets were used in the calculation of regression parameters.

When mandibular breadth or femur length was used to estimate size, there was a marked tendency for the largest specimens to show the greatest percent variation among the five estimates of size. This results from the fact that the regression lines on which the estimates are based diverge in the upper size range for early hominids. Where the several lines cross, their respective size estimates will be much more similar. Body size estimates based on the various subsets will also tend to be more divergent in the small size range, below the point where the lines converge or cross. Consequently, the size of a fossil specimen will influence the extent to which size estimates based on different samples will diverge.

An especially conspicuous example of this problem has been attempted at size estimation in *Gigantopithecus* (see R. J. Smith, this volume). Since this form is known only from jaws and teeth, this is the only basis on which size can be estimated. Yet the very large size of the jaws and teeth together with wear patterns that suggest heavy grinding raises the possibility that this form may have had a diet unlike that of any living primate. If this is true, none of the taxonomic subsets would be especially suitable. The problem is exacerbated by the fact that the results from the various subsets diverge to increasingly greater extents to the larger sizes. The same problem would obtain in the estimation of body size in very small specimens, especially when using vari-

Table 4. Femoral Circumference: Skeletal Weight Estimates for Fossils Based on Regressions within Various Taxonomic Units[a]

	KNM-ER 999	KNM-ER 736	KNM-ER 815	KNM-ER 738	KNM-ER 737	OH 20	SK 82	SK 97
Pongidae	8.81	8.46	7.46	7.46	8.22	7.95	8.81	7.95
Pongidae + *Hylobates*	8.73	8.38	7.38	7.38	8.14	7.86	7.73	7.86
Hominoidea	8.84	8.49	7.49	7.49	8.24	7.97	7.84	7.97
Pongidae + *Homo*	8.73	8.38	7.38	7.38	8.14	7.86	7.73	7.86
All	8.81	8.44	7.42	7.42	8.19	7.92	7.78	7.92

[a] Values are expressed as log skeletal weight (grams).

ables for which divergence was greater in the small size range, e.g., mandibular breadth. Size estimations are substantially more reliable in fossils whose association with a modern taxonomic group is reasonably clear and/or which represent a size at which lines based on different modern groups overlap.

Choice of the Variable(s) to Estimate Size

The problem with estimating size in *Gigantopithecus* would be much less pronounced if it were possible to base the estimate on a variable or variables that showed very similar patterns of allometric variation in samples that differed in their taxonomic representation. If, for example, a *Gigantopithecus* femur was known so that femoral circumference could be measured, a much more accurate measure of size could be obtained, since all the living groups studied here had very similar allometric relationships for this variable. As can be seen from the fossil size estimates based on mandibular breadth and femur length, even variables with correlations of 0.90 or greater over the entire catarrhine sample can produce fairly divergent size estimates when different reference samples are used. If a particular fossil fits clearly into a particular modern group, it is only necessary that a variable be highly correlated with size within that group. But if, as is usually the case, the fossil in question is not systematically similar to a modern species or group, the choice of a reference sample becomes very problematic, so that it is preferable to estimate size using a variable or variables that show very similar patterns of allometry across a range of samples.

As is clear from the preceding discussion, variables showing moderately high correlations with size over the entire sample of catarrhine primates (r = 0.90 for femur length) can nevertheless produce substantially different values where size estimations are based on different subsets within the larger group. A number of variables in the present sample have very high correlations with size, $r \geq 0.97$. But even for some of these variables, a substantial difference between the several taxonomic divisions remains. For example, variable 3 (external mandibular breadth) has a correlation with size of 0.98 over the entire sample, yet the slopes vary from 0.13 to 0.26 between some of the subsets. Many previous studies have not been sufficiently concerned with establishing the utility of a variable as a size estimator before using it as such. For example, in their classic work, Biegert and Maurer (1972) relied on trunk length as an indicator of size. Trunk length was immediately suspect as a size estimator because of the variation in vertebral number (especially lumbar) in primates (e.g., Steudel, 1981*a*) and has recently been demonstrated to itself show negative allometry in primates as a result of a relatively short trunk in great apes (Jungers, 1984). Femur length was used to estimate size by Wood (1976), but, as shown above, this dimension shows different allometric patterns depending on the adaptations of the species comprising a subset. When such inappropriate variables are used, size estimates will be subject to undocumented error, and where allometry in other dimensions is calculated using these spurious size estimators, the results will be obfuscated by allometric variation in both variables. This unsuitability can be detected by comparing the regression equations for the several subsets or by using variables with extremely high correlation values and low standard errors only. The allometric regression equations together with the corresponding correlation coefficients for all 22 variables are listed in Table 1 for the entire sample of catarrhine primates and for five subsets. Steudel (1982*a*) contains the regression equations for the same series of variables for other subdivisions of the data.

Even when one has chosen a dimension that varies in a consistent manner with respect to size over a variety of living forms, however, the possibility remains that a fossil may be so different than any living species that the estimator produces dubious results. A surprising result apparent in Table 1 was the consistency of bizygomatic breadth (variable 8) as an indicator of size, since the degree of flaring and robusticity of the zygomatic processes would be expected to vary with diet. Not only was the correlation high and standard error low for this variable when measured over the entire sample, but a comparison of the regression lines calculated within the subsets reveals great similarity. Thus, within this series of catarrhines, bizygomatic breadth seems to vary largely as a result of size rather than other adaptive constraints.

Choice of a Line Fitting Technique

A substantial difference of opinion exists as to which of the various techniques for fitting lines describing the allometric relationship is preferable. The major contenders are major axis and reduced major axis (Jolicoeur, 1963, 1975; Kermack and Haldane, 1950; Ricker, 1973) and least squares regression (Huxley, 1932; Gould, 1975; Draper and Smith, 1981; Stahl and Gummerson, 1967), with a few workers preferring Bartlett's method (Bartlett, 1949). The latter method has not gained wide acceptance, presumably because of unreliability resulting from necessarily arbitrary decisions about how to divide the total sample into groups (Kuhry and Marcus, 1977). The major issue is whether it is preferable to regard one of the variables as independent and minimize error variance in the other, dependent, variable only, or rather to make no such assignment of dependence and independence and minimize error in both the x and y directions. The advantage of the latter approach is that body size—no matter what variable is used to represent this—is never measured without error, violating an assumption of the least squares method. In this situation many workers (e.g., Thorington, 1972; Creighton, 1980; Martin, 1980) feel that it is consequently preferable to minimize orthogonal deviations from the allometric equation. This argument is especially convincing when a variable is used to indicate "size" that has little or no theoretical justification, such as femur length (Wood, 1976), ischium length (Schultz, 1930, 1949; Leutenegger, 1974) or trunk length (Biegert and Maurer, 1972). Since these variables show substantial adaptive variation that is not directly related to size (Steudel, 1981*a,* 1982*b;* Jungers, 1984; Aiello, 1981), this should be minimized along with similar variation in the variable plotted along the *y* axis. On the other hand, inherent in the study of allometry is the idea that variation in size necessitates certain responses in the variation of other biological characteristics. Thus the dependence of these variates on size is implicit. Consequently, when some *x* variable is used that actually represents some fundamental element of body size, the treatment of these variables as dependent on the size variable is most consistent with the theory underlying the investigation of allometry. Use of this approach is further justified by the fact that major axis and reduced major axis techniques are not without their own statistical difficulties (Ricker, 1973; Kuhry and Marcus, 1977; Jolicoeur, 1975) and is consequently preferred by numerous workers (Goldstein *et al.,* 1978; Smith, 1980, 1981; Jungers, 1984; Steudel, 1982*a,b*).

Most authors regard body mass as the variable that best indicates the key element in body size (e.g., Jungers, 1984; Aiello, 1981; Smith, 1980). This position is justified by the fact that it is the total weight that must be supported and moved by the musculoskeletal system and the total metabolizing mass that must be supplied with nutrients through the digestive, respiratory, and circulatory systems. However, total body mass has been shown to vary substantially during the lifetime of an individual, even when the adult stage only is considered. Scollay (1980) found that adult male squirrel monkeys weighed an average of 14.4% more during their breeding season as compared to normal. Mori's (1979) data shows a similar fluctuation in Japanese macaque females. Not only does this produce a substantial error term, but it also weakens one of the theoretical justifications of this estimator as a measure of body size. Since most of the fluctuation in body weight will be in body fat, variation in this tissue, which makes such a negligible contribution to metabolism (Dobeln, 1956), will not significantly influence nutritional requirements. Another problem when females are considered is whether their condition when pregnant, lactating, or in neither state should be used. This is especially trenchant since female primates in the wild can spend most of their adult life in one or the other reproductive states. Consequently it is very possible that the body is adjusted to these increased weight and metabolic requirements. Given these factors, the use of body weight as an indicator of body size clearly has some drawbacks. An approach that avoids some of these difficulties is that of using mean body weight values for each species (e.g., Jungers, 1984). While this eliminates some of the error due to individual variation, it is completely valid only when the body weight data are from the same population from which data on the dependent variables is drawn, although perhaps variation in mean body weights between populations of the same species is not great, so that this condition need not be meet.

Because of these problems with body weight, the present data use skeletal weight as an indicator of body size (see also Trotter, 1954; Steudel, 1980, 1981*b*). While severe nutritional deprivation may eventually cause weight loss from the skeleton, this will occur with far lower frequency than loss of body weight in the form of fat. Thus the nutritional state of the animal population at the time size is measured will be less likely to have an effect when skeletal weight rather than body weight is used as an estimator. Furthermore, skeletal weight should be influenced by the reproductive demands on female primates in much the same way as will be other aspects of the musculoskeletal system. Therefore the use of skeletal weight to estimate size seems very appropriate. It has the additional advantage of being available on a much wider range of museum specimens than is body weight. The major disadvantage of skeletal weight is the effect of differences in skeletal preparation techniques. Bones vary in the extent to which they have been degreased and this will affect their weight. Yet data on *Pan troglodytes* that combined approximately equal representatives from a collection with very greasy bones and one with bones of a more typical level of greasiness showed a lower range of variation in skeletal weight than did a series of *H. sapiens* skeletons, all from the same collection and comparably prepared (Steudel, 1980). *Homo sapiens* was used for comparison because, like *Pan,* the degree of sexual dimorphism is not especially great compared to other hominoids of similar body size. Thus, while mode of preparation must influence skeletal weight, the magnitude of the effect does not seem especially large.

Since body size estimates for fossils are usually given in terms of body weight, skeletal weight values should be convertible to body weight so that comparisons are facilitated. The relationship between body weight BW (in kg) and skeletal weight SW (in kg) has been established for mammals as

$$SW = 0.061(BW)^{1.090}$$

(Prange *et al.,* 1979) or

$$SW = 0.093(BW)^{1.142}$$

(Kayser and Heusner, 1964), and for fish as

$$\log SW = 1.4794 + 1.0197 \log BW$$

(Reynolds and Karlotski, 1977). But it was unclear whether these values would apply to the series of primates studied here. To determine this relationship species mean values for body weights were taken from the literature and compared to the mean values for skeletal weights in the present data (see Table 5). Since these values represent different species populations, these results are only approximate. Values for both skeletal weight and body weight were expressed in kilograms and converted to natural logarithms (ln) before calculating the regression parameters. The resulting equation was

Table 5. Average Body Weights Taken from the Literature and Average Skeletal Weights from Present Data for Each Taxon in the Study

Taxon	Average body weight, kg	Average skeletal weight, kg
Gorilla gorilla	126.0[a]	3.641
Pan troglodytes	44.8[a,b]	1.669
Pongo pygmaeus	36.9[b]	2.322
Hylobates lar	5.5[a]	0.187
Cercopithecus mitis	6.3[a]	0.191
Papio sp.	21.0[a]	0.608
Colobus guereza	9.85[a]	0.268
Presbytis cristatus	7.0[a,c]	0.161
Nasalis larvatus	13.8[c]	0.299
Macaca mulatta	6.7[a]	0.215

[a]Post (1980).
[b]Sacher and Staffeldt (1974).
[c]Napier and Napier (1967).

$$\ln \text{SW} = -3.72 + 1.09 \ln \text{BW}, r = 0.971$$

This is extremely close to the results of Prange *et al.* (1979), especially when the *y*-intercept values are adjusted for my use in ln data as compared to their untransformed data. Their *y* intercept would be approximately −2.80 expressed in natural logs. This last equation can be used to express size estimates for skeletal weight in terms of body weight.

Equations for Predicting Body Size

For prediction of a value for body size based on some correlated variable, the general approach is to regress size (X) on the variable (Y), simply reversing the usual assignments of independence and dependence. The reason for this reversal of dependence lies in the calculation of confidence intervals about the predicted value. In least squares regression, all error is assigned to variation of the dependent variable, so that confidence intervals about a predicted Y include all observed error. The disadvantage of this is that two different lines must be calculated for every combination of X and Y, one indicating the size-dependent allometry of Y and one giving the relationship of X on Y. Furthermore, the regression of X on any Y assumes dependence of the variable that one in actuality regards as independent, a theoretically unsatisfying situation. Consequently it seemed useful to try to find a way of determining the confidence interval about an estimated value for X based on the standard Y on X regression. The estimated value for X from any Y would, of course, be determined by the point on the regression line corresponding to that value of Y. An upper confidence interval for the calculated X could be taken as the largest value of X for which the given value of Y falls within the 95% confidence limits of a predicted value for Y. The lower limit of X would be the smallest value of X at which the same were true. Figure 4 illustrates this diagrammatically. Unless the slope of the line is 1.0, the upper and lower limits for X will not be symmetrically arranged about the predicted value. This approach seems perfectly valid theoretically, so a test on actual data was carried out. Four variables that are useful for size estimation in fossils were chosen: bizygomatic breadth, mandibular breadth, palate breadth, and orbit width. The Y on X regression parameters for these variables have already been given in Table 1. Seven primate specimens (two macaques, two chimps, one baboon, and two gorillas) were found on which at least three of these variables and skeletal weight could be measured but which had not been included in the calculation of the regression parameters. Each of these specimens had one measurement that was not available. Each of the four variables was used to predict skeletal weight in the seven specimens by the two methods using results based on the (N = 280) sample: first the usual approach of calculating the regression of X on Y and second using the method just described. In 16 of the 25 cases the prediction for skeletal weight based on the regression of Y on X (X independent) was closer to the known value of skeletal weight than was the prediction from the regression of X on Y. In seven cases the regression of X on Y produced the better estimate and in two cases the predictions from the two methods were identical. In all 25 cases the confidence limits for X based on the regression of Y on X calculated as described above overlapped the known value for skeletal weight. This was true for all but two of the 95% confidence intervals about the estimate from the regression of X on Y. It is also interesting to note that the confidence intervals about predicted values calculated by these two quite different methods were in all cases very similar to one another in each of the 25 trials. The largest difference between confidence intervals calculated by the two different methods was ±0.049. While the suggested method for estimating values of X and corresponding confidence limits based on the regression of Y on X produced results quite similar to those obtained

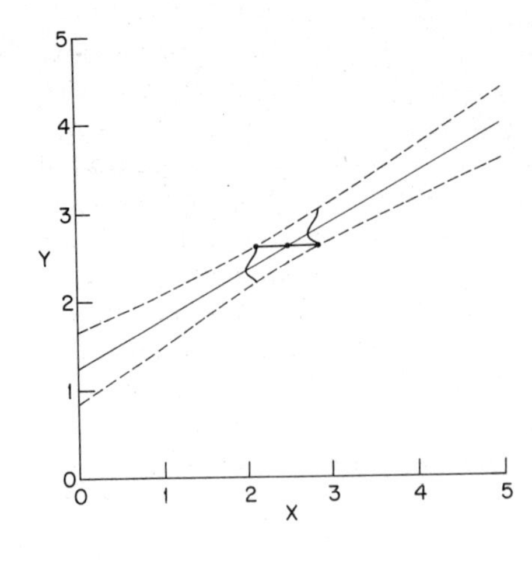

Fig. 4. The method of determining confidence intervals about an estimated value for body size based on regression equations in which size is the independent (x) variable. The dashed lines represent the 95% confidence limits about the allometric equation. See text for explanation.

from the regression of X on Y, they are in 16 of 25 cases actually better. Consequently this method of estimating X from the regression of Y on X has been used exclusively in the remaining analysis.

How Reliable Are Size Estimates on Fossils?

Given the potential pitfalls in fossil size estimation, one wonders if these estimates may be so inaccurate as to be virtually worthless. One way of testing this hypothesis is to estimate size for certain fossil specimens on which several estimating variables can be measured. A comparison of the similarity of the several estimates for each specimen would give some indication of their reliability. For this purpose three relatively complete early hominid crania were chosen. Two of these, SK48 and KNM-ER 406, are representative of the robust type, thought by many workers to show extreme dietary specializations (Robinson, 1956; Jolly, 1970; DuBrul, 1977). Thus difficulties with the choice of an appropriate reference sample are substantial. Furthermore, these specimens, especially KNM-ER 406, are large, in a size range where estimates based on different taxonomic groupings diverge. Thus these two specimens are especially good representations of the potential difficulties with fossils. The third specimen, Sts 5, represents gracile early hominids.

Three variables could be measured on these specimens that had high correlations and low standard errors when measured over the entire sample (Steudel, 1980): bizygomatic breadth ($r = 0.98$), palate breadth ($r = 0.96$), and orbital width ($r = 0.97$). The only taxonomic set, other than the entire sample, that seemed relevant for these specimens was pongids + *Homo*. Within this group, bizygomatic breadth retains a fairly good correlation with size ($r = 0.82$), but the other two correlations drop considerably. Orbital width has a correlation of 0.62 and palate breadth is even worse, with $r = 0.48$. The size estimates obtained for each specimen based on each variable can be seen in Tables 6 and 7 along with the corresponding confidence limits. These statistics are based on the regression of Y on X as discussed above. In all cases except the estimates for KNM-ER 406 based on the entire sample, the 95% confidence limits for all three estimates overlap one another. In the exceptional case, only the estimates for bizygomatic breadth and palate breadth fail to overlap. In addition each estimate for a specimen based on a given variable from the data over the entire sample falls within the confidence interval for the equivalent estimate based on pongids + *Homo*. Variation among the estimates for each specimen is expressed as the smallest estimate as a percentage of the largest estimate. These deviations are consistently smaller when based on size estimates based on the entire sample as compared to using pongids + *Homo*. The average size estimate for each fossil based on the three estimators is included in the tables. The palate breadth estimates for the pongids + *Homo* regression seemed suspicious since the correlations between this variable and

Table 6. Skeletal Weight Estimates with 95% Confidence Limits based on the Entire Sample for Three Fossil Skulls[a]

Estimator	Sts 5		SK 48		KNM-ER 406		r
		7.90		8.55		9.35	
Bizygomatic breadth	7.40		8.00		8.79		
		6.89		7.55		8.30	
		7.90		8.04		7.79	
Palate breadth	7.18		7.34		7.62		0.96
		6.49		6.64		6.90	
		7.69		8.00		8.98	
Orbit width	7.02		7.33		8.31		0.97
		6.39		6.70		7.66	
Average	7.20		7.56		8.24		
Smallest estimate as percent of largest	94.9%		91.6%		86.7%		

[a]Values are expressed in natural logarithms.

size is so low in this sample. Omitting estimates for that sample from palate breadth, averages are also computed for the two remaining variables in Table 7. The size estimates based on the several variables using two different reference samples are quite similar.

However, if one converts these estimates from skeletal weight to body weight using the equation given above, ln SW = −3.70 + 1.08 ln BW, the picture looks far less rosy. Using estimates based on the entire sample, one finds that the body weight estimates for Sts 5 vary from 33.1 to 47 kg, while

Table 7. Skeletal Weight Estimates with 95% Confidence Limits Based on Pongidae + *Homo* for Three Fossil Skulls[a]

Estimator	Sts 5		SK 48		KNM-ER 406		r
		8.03		8.60		9.38	
Bizygomatic breadth	7.45		8.03		8.78		0.82
		6.85		7.45		8.19	
		7.90		8.20		8.85	
Palate breadth	6.41		6.70		7.35		0.48
		4.64		5.00		5.71	
		7.74		8.18		9.54	
Orbit width	6.71		7.13		8.44		0.62
		5.58		6.05		7.42	
Average	6.86		7.29		8.19		
Average of bizygomatic breadth of orbit width	7.08		7.58		8.61		
Smallest estimate as percent of largest	86.0%		83.4%		83.7%		

[a]Values are expressed in natural logarithms.

estimates for SK 48 vary from 44 to 80 kg. Estimates for KNM-ER 406 vary from 57 to 161 kg! The amount of difference between the estimates for the gracile hominid is not unreasonable. But for SK 48 the maximum difference in nearly a factor of 2:1 and for KNM-ER 406 it is nearly 3:1. Here it seems that morphological divergence of robust early hominids has resulted in predictions based on more typical primates being quite inconsistent. Since the robust early hominids are well known for their widely flaring zygomatic processes (Robinson, 1972*b;* Walker and Leakey, 1978), one would expect estimates from bizygomatic breadth to be high (Steudel, 1980). Palate breadth is also closely tied to diet and shows a smaller increase through the three specimens than do the other two estimators. Thus the values for the robust specimens from this variable may be low. So for attempting estimates of body weight one is probably best off with the average for the three variables or the estimate based on orbit width. Nevertheless, the size estimates for the two robust hominids remain doubtful. This example should have a chastening influence on anyone engaged in fossil size estimation. But before one becomes hopelessly pessimistic, it should be remembered that these are just the kind of specimens one would expect to have trouble with because of their unique morphology, and that enough of their anatomy is known so that, for example, one would expect bizygomatic breadth to overestimate size in this form. One would expect to be better off using diameters of postcranial bones that seem to have a closer functional tie to body weight. The disadvantage here is that it is difficult to accurately assign long bones (especially fragmentary ones) to the correct taxon.

To see if postcranial bones indeed produced more consistent size estimates, I used data on three fossil hominid pelves on each of which two estimators of body size could be measured. The specimens were Sts 14, SK 3155, and SK 50. Acetabular diameter has a correlation of 0.98 with skeletal

Table 8. Skeletal Weight (SW) and Body Weight (BW) Estimates for Several Fossil Innominate Bones Based on the Entire Sample[a]

	Sts 14			SK 3155			SK 50		
Estimator	SW		BW	SW		BW	SW		BW
		7.85			7.75			8.07	
Acetabulum diameter	7.33		43.9	7.22		39.8	7.54		53.0
		6.80			6.68			7.00	
		8.20			8.27			8.77	
Iliac breadth	7.47		49.8	7.55		53.4	8.04		82.7
		6.75			6.83			7.43	
Average	7.40		46.8	7.38		45.9	7.79		66.2
Smaller estimate as percent of larger	98.1%			95.6%			93.8%		

[a]Skeletal weight in ln grams; body weight in kilograms. The 95% confidence limits are included for the skeletal weight estimates.

Table 9. Skeletal Weight (SW) and Body Weight (BW) Estimates for Several Fossil Innominate Bones Based on Pongidae + *Homo*

	Sts 14		SK 3155		SK 50	
Estimator	SW	BW	SW	BW	SW	BW
Acetabulum diameter	7.86 7.25 6.63	40.9	7.73 7.13 6.51	36.8	8.09 7.48 6.87	50.2
Iliac breadth	8.37 7.53 6.68	52.5	8.44 7.60 6.75	55.9	8.87 8.02 7.18	81.2
Average	7.39	46.4	7.36	45.1	7.75	63.9
Smaller estimate as percent of larger	96.3%		93.8%		93.3%	

[a]Skeletal weight in ln grams; body weight in kilograms. The 95% confidence limits are included for the skeletal weight estimates.

weight over the entire sample, and minimum breadth across the ilium (below the blade) has a correlation of 0.96. Again the two reference samples that seemed relevant were the entire sample and pongids + *Homo*. The resulting estimates can be seen in Tables 8 and 9. All four estimates for skeletal weight for each specimen fall within one anothers' 95% confidence intervals. A comparison of smallest to largest estimates also shows substantial consistency, the worst value being 93.3% (see Tables 8 and 9). But the variation in total body weight between the various estimates is nevertheless large. The estimates for Sts 14 are within 8.6 kg of one another, an acceptable level of uncertainty; but SK 3155, a pelvis of similar size, produced much greater variation, 19.1 kg. The estimates for the larger pelvis, SK 50, also shows great divergence, 32.5 kg. The divergence of these three early hominid specimens may be exacerbated by the relatively small acetabulae and sacral surfaces noted in these forms (Robinson, 1972*a*). Thus perhaps the pelvic proportions in these forms that had relatively recently made the transition from quadrupedal to bipedal progression were unlike those of any modern species. But the fact that this difference may be explicable does not mitigate the difficulties of size estimation in fossils.

The size estimates obtained here generally compare fairly well to previous estimates both in terms of values and confidence intervals. McHenry (1974, 1976) estimates body size of Sts 14 as 27.6 ± 10.5 kg based on vertebrae and concluded that this must be close to a minimum for the taxon. My estimates for Sts 14 from pelvic data produce much higher estimates, but this discrepancy may also be due to McHenry's estimate being based exclusively on *H. sapiens* as the reference population. In general the two values for gracile hominids, Sts 5 and Sts 14, are within the range expected for this taxon (e.g. Pilbeam and Gould, 1974), although they are a little higher than supposed by Robinson (1972*a*). The estimates for the SK 48 skull are within the range estimated for robust hominids by McHenry (1976). The average estimates for SK 50 are only a little higher. The only anomalous results among the robust specimens were for KNM-ER 406. Only the palate breadth estimates for this specimen fall within the range normally regarded as typical of this group (Pilbeam and Gould, 1974; Robinson, 1972*a*). It is clear that this skull represents a very large individual, but whether the estimates of 105–161 kg are accurate is difficult to say. The bizygomatic breadth estimates seem likely to be overestimates because of the unusual diet of this form. Yet no such argument can be made for the orbital width estimates. Perhaps these estimates of over 100 kg are accurate and this species was highly dimorphic. The estimates for the *H. erectus* pelvis from Swartkraus, SK 3155 (Brain, *et al.*, 1974) are somewhat smaller than the average for *H. erectus* suggested by Weidenreich (1941) of 53 kg but not unreasonably so.

Thus the size estimates for robust and gracile early hominids agree closely enough (with the possible exceptions of extremely large specimens like KNM-ER 406) with one another that reasonable estimates of average size for the taxa can be obtained. Furthermore, it seems fairly clear that there was a great range of variation, probably due to sexual dimorphism, in the robust group. But in spite of this, the error involved in estimating size for any given individual makes the analysis of allometry within these fossil taxa very dubious.

Because of the problems documented above, a rigorously quantitative analysis of allometry within fossil species does not appear to be possible. Given the relatively small samples available in fossil taxa, the amount of error inherent in size estimation for each specimen would make any calculated regression equation extremely suspect. This is particularly true if size were estimated from different variables on different specimens. It is also a greater problem for specimens in a size range, usually very large or very small, where equations based on different taxonomic units diverge.

One alternate application of estimates of fossil hominid body size is an analysis of trends showing increasing or decreasing size through time. If the same variable is used to indicate specimen size for the entire fossil sample, it seemed that this approach might be able to reveal some trend. Since there are many mandibles representing robust early hominids at many time zones, it seemed worth trying this using mandibular breadth to indicate size. Since this analysis makes no attempt to deal with rates of size change, absolute dates that are often controversial have been neglected. Instead a temporal sequence of specimens has been arranged (see Table 10). Mandibular breadth for each specimen is included. A visual trend in size is not apparent, and, in fact, the standard deviation of this sample of 14 specimens is 0.170. This is only slightly larger than the same value obtained for the 45 gorillas studied above, 0.12. One is left with the conclusion that the amount of variation through time is only slightly greater than that seen in a modern hominoid with substantial sexual dimorphism. Because of the substantial sexual dimorphism in robust early hominids (Steudel, 1980), any change in the average size calculated for specimens from one time period to the next will be heavily influenced by whether

Table 10. Robust Early Hominid Specimens in Approximate Temporal Sequence by Site

Site	Specimen	Mandibular breadth, mm
Swartkrans	SK 23	24.6
	SK 34	21.2
	SK 12	33.5
	SK 6	25.0
Below Chari Tuff—Ileret	KNM-ER 728	24.1
	KNM-ER 725	29.9
	KNM-ER 726	31.6
Below Middle Tuff—Ileret	KNM-ER 729	27.3
	KNM-ER 733	26.3
	KNM-ER 1468	36.1
Below Koobi Fora Tuff (Area 103)	KNM-ER 403	31.2
Below Lower Tuff—Ileret	KNM-ER 801	28.3
	KNM-ER 810	24.7
Below KBS Tuff—Koobi Fora	KNM-ER 1482A	20.0

larger or smaller members of the population happened to be preserved. Thus it appears that no conclusions can be drawn about the existence of change in body size through time in this taxon. Again the ambiguities involved in dealing with fossil populations seem too great for accurate conclusions on the evolutionary variation of body size.

Conclusions

Problems with the precision of size estimates for individual fossil specimens make studies of intrataxon allometry for fossils very dubious. The problems are especially great for fossil taxa that deviate markedly from adaptations characteristic of modern populations. Estimates for species in the extremely small or extremely large size range are also especially dubious, since it is at the extremes that allometric trends characteristic of various modern taxa tend to diverge. Size estimates for a single specimen based on different estimating variables can differ quite substantially. Consequently any size estimates that are given are at best only approximations. They are probably good enough to give a fairly clear idea of the general size and degree of sexual dimorphism for a reasonably well-documented fossil group, but are insufficient for accurate quantitative analysis of intrataxon allometry.

Acknowledgments

The author wishes to thank the curators of the museum collections utilized in this study for access to the specimens in their charge: P. H. Napier,

British Museum (Natural History); Prof. J. Biegert, Anthropologisches Institut, Universitat Zurich; Dr. Richard Thorington, Smithsonian Institute; Dr. Farish Jenkins, Museum of Comparative Zoology; Elizabeth Pillaert, University of Wisconsin Zoology Museum; Dr. R. E. F. Leakey, National Museums of Kenya, and Dr. C. K. Brain, Transvaal Museum.

Dr. H. J. Steudel suggested the technique for calculating a confidence interval about the estimated value for the *x* variable and provided the computer program that produced the results. His contribution is gratefully acknowledged. National Science Foundation grant GS-42706 provided the funds necessary for data collection. The University of Wisconsin Graduate School provided research time.

References

Aiello, L. C. 1981. The allometry of primate body proportions. *Symp. Zool. Soc. Lond.* **48:**331–358.

Andrews, P. 1974. New species of *Dryopithecus* from Kenya. *Nature* **249:**188–190.

Bartlett, M. S. 1949. Fitting a straight line when both variables are subject to error. *Biometrics* **5:**207–212.

Biegert, J., and Maurer, R. 1972. Rumpfskelettlange, Allometrien und Korperproportionen bei catarrhinen Primaten. *Folia Primatol.* **17:**142–156.

Brace, C. L. 1972. Sexual dimorphism in human evolution. *Yearb. Phys. Anthropol.* **16:**31–49.

Brain, C. K., Vrba, E. S., and Robinson, J. T. 1974. A new hominid innominate bone from Swartkrans. *Ann. Transvaal Mus.* **29:**55–66.

Corruccini, R. S. 1976. Multivariate allometry and australopithecine variation. *Evolution* **30:**558–563.

Corruccini, R. S., Ciochon, R. L., and McHenry, H. M. 1976. The postcranium of Miocene hominoids: Were dryopithecines merely "dental apes"? *Primates* **17:**205–223.

Creighton, G. K. 1980. Static allometry of mammalian teeth and the correlation of tooth size and body size in contemporary mammals. *J. Zool. Lond.* **191:**435–443.

Dobeln, W. V. 1956. Maximal oxygen uptake, body size, and total hemoglobin in normal man. *Acta Physiol. Scand.* **38:**193.

Dodson, P. 1975. Functional and ecological significance of relative growth in *Alligator*. *J. Zool. Lond.* **175:**315–355.

Draper, N. R., and Smith, H. 1981. *Applied Regression Analysis,* Wiley, New York.

DuBrul, E. L. 1977. Early hominid feeding machanisms. *Am. J. Phys. Anthropol.* **47:**305–320.

Echelle, A. F., Echelle, A. A., and Fitch, H. S. 1978. Inter- and intraspecific allometry in a display organ: The develop of *Anolis* (Iguanidae) species. *Copeia* **1978**(2)**:**245–250.

Fleagle, J. G., and Simons, E. L. 1978. Humeral morphology of the earliest apes. *Nature* **276:**705–707.

Goldstein, S., Post, D., and Melnick, D. 1978. An analysis of cercopithecoid odonotometric. I. The scaling of the maxillary dentition. *Am. J. Phys. Anthropol.* **49:**517–532.

Gould, S. J. 1971. Geometric similarity in allometric growth: A contribution to the problem of scaling in the evolution of size. *Am. Nat.* **105:**113–136.

Gould, S. J. 1975. Allometry in primates, with emphasis on scaling and evolution of the brain, in: *Approaches to Primate Paleobiology* (Contrib. Primatol., Vol. 5, F. Szalay, ed.), pp. 244–292, S. Karger, Basel.

Huxley, J. S. 1932. *Problems of Relative Growth,* Methuen, London.

Jolicoeur, P. 1963. The multivariate generalization of the allometry equation. *Biometrics* **19:**497–499.

Jolicoeur, P. 1975. Linear regression in fishery research: Some comments. *J. Fish. Res. Board Can.* **32:**1491–1494.

Jolly, C. E. 1970. The seed-eaters. *Man* **5:**5–24.

Jungers, W. L. 1979. Locomotion, limb proportions, and skeletal allometry in lemurs and lorises. *Folia Primatol.* **32:**8–28.

Jungers, W. L. 1984. Scaling of the hominoid locomotor skeleton with special reference to the lesser apes, in: *The Lesser Apes: Evolutionary and Behavioural Biology* (H. Preuschoft, D. Chivers, W. Brockelman, and N. Creel, eds.), Edinburgh University Press, Edinburgh (in press).

Kayser, C., and Heusner, A. 1964. Etude comparative du metabolisme energetique dans la serie animale. *J. Physiol. Paris* **56:**489–524.

Kermack, K. A., and Haldane, J. B. S. 1950. Organic correlation and allometry. *Biometrika* **37:**30–41.

Kuhry, B., and Marcus, L. F. 1977. Bivariate linear models in biometry. *Syst. Zool.* **26:**201–209.

Lande, R. 1979. Quantitative genetic analysis of multivariate evolution applied to brain:body size allometry. *Evolution* **33:**402–416.

Leutenegger, W. 1974. Functional aspects of pelvic morphology in simian primates. *J. Hum. Evol.* **3:**267–222.

Leutenegger, W. 1976. Allometry of neonatal size in eutherian mammals. *Nature* **263:**229–230.

McHenry, H. M. 1974. How large were the australopithecenes? *Am. J. Phys. Anthropol.* **40:**329–340.

McHenry, H. M. 1976. Early hominid body weight and cephalization. *Am. J. Phys. Anthropol.* **45:**77–84.

Martin, R. D. 1980. Adaptation and body size in primates. *Z. Morphol. Anthropol.* **71**(2):115–124.

Meunier, K. 1959. Die Allometrie des Vogelflügels. *Z. Wiss. Zool.* **161:**444–482.

Mobb, G. E., and Wood, B. A. 1977. Allometry and sexual dimorphism in the primate innominate bone. *Am. J. Anat.* **150:**531–538.

Mori, A. 1979. Analysis of population changes by measurement of body weight in the Koshima troop of Japanese macaques. *Primates* **20:**371–397.

Napier, J. R., and Napier, P. H. 1967. *A Handbook of Living Primates,* Academic Press, London.

Pilbeam, D., and Gould, S. J. 1974. Size and scaling in human evolution. *Science* **186:**892–901.

Post, D. G. 1980. Sexual dimorphism in the anthropoid primates: Some thoughts on causes, correlates, and the relationship to body size. (Unpublished manuscript).

Prange, H. D., Anderson, J. F., and Rahm, H. 1979. Scaling of skeletal mass to body mass in birds and mammals. *Am. Nat.* **113:**103–122.

Reynolds, W. W., and Karlotski, W. 1977. The allometric relationship of skeletal weight to body weight in teleost fishes: A preliminary comparison with birds and mammals. *Copeia* **1977:**160–163.

Ricker, W. E. 1973. Linear regressions in fishery research. *J. Fish. Res. Board Can.* **30:**409–434.

Robinson, J. T. 1956. The Dentition of the Australopithecinae, Transvaal Museum Mem., no. 9.

Robinson, J. T. 1972*a*. *Early Hominid Posture and Locomotion,* University of Chicago Press, Chicago.

Robinson, J. T. 1972*b*. The bearing of east Rudolf fossils on early hominid systematics. *Nature* **240:**239–240.

Robinson, J. T., and Steudel, K. 1973. Multivariate discriminant analysis of dental data bearing on early hominid affinities. *J. Hum. Evol.* **2:**509–527.

Sacher, G. A., and Staffeldt, E. F. 1974. Relation of gestation time to brain weight for placental mammals: Implications for the theory of vertebrate growth. *Am. Nat.* **108:**593–615.

Schultz, A. H. 1930. The skeleton of the trunk and limbs of higher primates. *Hum. Biol.* **11:**303–438.

Schultz, A. H. 1949. Sex differences in the pelves of primates. *Am. J. Phys. Anthropol.* **7:**401–424.

Scollay, P. 1980. Cross-sectional morphometric data on a population of semi-free ranging squirrel monkeys. *Am. J. Phys. Anthropol.* **53:**309–316.

Smith, R. J. 1980. Rethinking allometry. *J. Theor. Biol.* **87:**97–111.

Smith, R. J. 1981. On the definition of variables in studies of primate dental allometry. *Am. J. Phys. Anthropol.* **55:**323–329.

Stahl, W. R., and Gummerson, J. Y. 1967. Systematic allometry in five species of adult primates. *Growth* **31:**21–34.

Steudel, K. 1980. New estimates of early hominid body size. *Am. J. Phys. Anthropol.* **52:**63–70.

Steudel, K. 1981*a*. Functional aspects of primate pelvic structure: A multivariate approach. *Am. J. Phys. Anthropol.* **55:**399–410.

Steudel, K. 1981*b*. Body size estimators in primate skeletal material. *Int. J. Primatol.* **2:**81–90.

Steudel, K. 1982*a*. Patterns of intraspecific and interspecific allometry in Old World primates. *Am. J. Phys. Anthropol.* **59:**419–430.

Steudel, K. 1982*b*. Allometry and adaptation in the primate postcranial skeleton. *Am. J. Phys. Anthropol.* **59:**431–442.

Thorington, R. W. 1972. Proportions and allometry in the grey squirrel, *Sciurus carolinensis*. *Nemouria* **8:**1–17.

Trotter, M. 1954. A preliminary study of estimation of weight of the skeleton. *Am. J. Phys. Anthropol.* **12:**537–552.

Walker, A., and Leakey, R. E. F. 1978. The hominids of east Turkana. *Sci. Am.* **239:**54–66.

Weidenreich, F. 1941. The brain and its role in the phylogenetic transformation of the human skull. *Trans. Am. Philos. Soc. N.S.* **31**(5)**:**321–442.

Wolpoff, M. H. 1971. Metric Trends in Hominid Dental Evolution, Case Western Reserve University Studies in Anthropology, no. 2.

Wood, B. A. 1976. The nature and basis of sexual dimorphism in the primate skeleton. *J. Zool. Lond.* **180:**15–34.

Author Index

Subject Index

8096MF PA 286
10-30-06 32572 MS

DATE DUE

DEMCO INC 38-2971